Foundations for Microwave Circuits

Foundations for Microwave Circuits

Gilbert H. Owyang

Professor of Electrical Engineering
Worcester Polytechnic Institute

With 289 Figures

Springer-Verlag
New York Berlin Heidelberg
London Paris Tokyo Hong Kong

GILBERT H. OWYANG

Worcester Polytechnic Institute
Department of Electrical Engineering
Worcester, Massachusetts 01609
USA

Owyang, Gilbert H.
 Foundations for microwave circuits / Gilbert H. Owyang.
  p. cm.
 Includes index.
 ISBN 0-387-96989-6
 1. Microwave circuits. I. Title.
TK7876.O89 1989
621.381'32--dc20   89.6218

Printed on acid-free paper.

Text provided by the author and editors in camera-ready form.
Printed and bound by R.R. Donnelley & Sons, Harrisonburg, Virginia.
Printed in the United States of America.

9 8 7 6 5 4 3 2 1

ISBN 0-387-96989-6 Springer-Verlag New York Berlin Heidelberg
ISBN 3-540-96989-6 Springer-Verlag Berlin Heidelberg New York

In Memory of

My Parents

To

My wife, Lily
and our sons, Kevin and Colin

PREFACE

While many articles have been written on microwave devices, a great majority of them are prepared for specialists dealing in specific aspects of microwave engineering. At the same time, material at a fundamental level in tutorial form is extremely limited, especially for students who need to acquire basic knowledge in the field. Individuals seeking to gain a preliminary understanding of microwave circuits are usually relegated with little success to the endless search from one reference source to another. For non-experts, sequential derivations of basic relations are rarely available and extremely difficult to locate.

The purpose of this volume is to collect in one place the essential fundamental principles for a group of microwave devices. The chosen devices are those which form the basic modules found in practical microwave systems. Thus, these devices provide the crucial building blocks in common microwave systems, and their inherent characteristics are also the basis of some of the fundamental concepts in more complex devices. The material is presented in a continuous, self-contained manner. With the appropriate background, readers should be able to follow and understand the contents without the need for additional references.

The level of a textbook is a relative measure. In the area of electromagnetics, the treatment of electromagnetic field theory in the undergraduate curriculum varies greatly from university to university. This book is intended to serve students who have had a one-year introductory course in electromagnetic field theory in addition to the freshman-level general physics course in electricity, magnetism and waves.

The reader of this book will require a basic knowledge of propagation in unbounded regions and of guided waves and some background in electromagnetic radiation. In the field of electrical engineering education, these topics are usually covered in a one-year undergraduate course in electromagnetic theory.

All topics covered in this volume exists in the literature. Many of these valuable pioneering works play an important part of the author's education, hence, their inspiration and influence upon the writing of this book cannot be discounted. The author acknowledges his sincere appreciation. With limited resources, it is almost impossible to acknowledge all original contributors. Some of the author's most frequently consulted references are listed in the bibliography at the end of this book.

The writer is grateful to Worcester Polytechnic Institute for a sabbatical leave of absence which permitted him to complete seven chapters of the first draft of the manuscript. The author also acknowledges the encouragement and support of Dr. Kevin A. Clements, Department Head, Department of Electrical Engineering.

Finally, I express my sincere appreciation to my wife Lily and our children for their patience and understanding during the preparation of this book.

Gilbert H. Owyang

Table of Contents

xvi

CHAPTER I Review of Transmission Line Theory

1. Transmission Line Equations

This chapter will briefly review transmission line theory. This is not intended to be a comprehensive treatment on this topic. However, it will provide the necessary background needed for the understanding of the material in the book. More detailed information on this subject can be found in textbooks on transmission line theory, and some of these are listed in the bibliography [B].

Consider an infinitely long, two-conductor transmission line. Let a short section Δz of the line be represented by any of the equivalent circuits shown in Fig. 1-1. When the length Δz is small, all three equivalent circuits produce identical results. The transmission line parameters are:

$\quad\quad$ L = distributed inductance in henrys per unit length of line
$\quad\quad$ C = distributed capacitance in farads per unit length of line
$\quad\quad$ R = distributed resistance in ohms per unit length of line
$\quad\quad$ G = distributed conductance in mhos per unit length of line

The equivalent L-network, Fig. 1-1B, will be used in the following analysis. The output voltage across terminals 2–2′ is given by

$$\underline{v}(z + \Delta z, t) \approx \underline{v}(z, t) - \Delta z \left[R + L\frac{\partial}{\partial t} \right] \underline{i}(z + \Delta z, t)$$

or

$$\frac{1}{\Delta z}[\, \underline{v}(z + \Delta z, t) - \underline{v}(z, t)\,] \approx - \left[R + L\frac{\partial}{\partial t} \right] \underline{i}(z + \Delta z, t) \tag{1}$$

Equation (1) is an approximation since a finite section of line with distributed parameters is represented by a circuit with lumped elements. This approximation improves as the length Δz is decreased. Equation (1) becomes an exact description in the limit as $\Delta z \to 0$, i.e.,

$$\frac{\partial}{\partial z}\underline{v}(z, t) = \lim_{\Delta z \to 0} \left\{ - \left[R + L\frac{\partial}{\partial t} \right] \left[\underline{i}(z, t) + \Delta z\frac{\partial}{\partial t}\underline{i}(z, t) + \cdots \right] \right\}$$

$$= - \left[R + L\frac{\partial}{\partial t} \right] \underline{i}(z, t) \tag{2a}$$

$$\underline{i}(z + \Delta z, t) = \underline{i}(z, t) + \Delta z \frac{\partial}{\partial t} \underline{i}(z, t) + \cdots \tag{2b}$$

where Taylor's expansion is employed to obtain (2a).

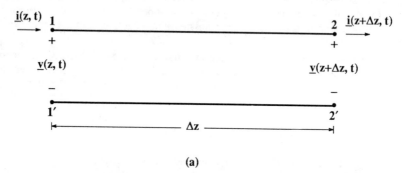

(a)

Figure 1-1(a): A short section of transmission line

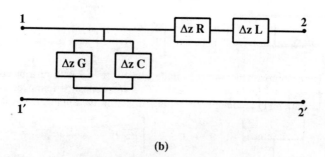

(b)

Figure 1-1(b): Equivalent L-network

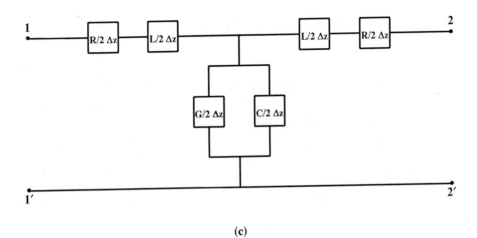

(c)

Figure 1-1(c): Equivalent T-network

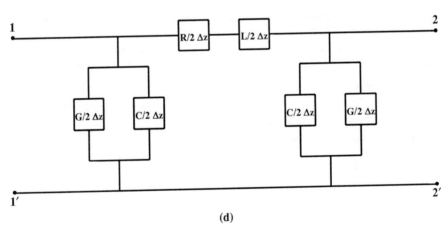

(d)

Figure 1-1(d): Equivalent π-network

The relation between the input and the output current is obtained similarly.

$$\underline{i}(z + \Delta z, t) \approx \underline{i}(z, t) - \Delta z \left[G + C\frac{\partial}{\partial t} \right] \underline{v}(z, t)$$

or

$$\frac{\partial}{\partial z} \underline{i}(z, t) = - \left[G + C\frac{\partial}{\partial t} \right] \underline{v}(z, t) \tag{3}$$

The following shorthand notation will be used:

$$\underline{v} \equiv \underline{v}(z, t) \tag{4a}$$

$$\underline{i} \equiv \underline{i}(z, t) \tag{4b}$$

The symbol $\equiv$ means by definition. The transmission line equations, (2) and (3), then become

$$\frac{\partial}{\partial z} \underline{v} = -R\underline{i} - L\frac{\partial}{\partial t}\underline{i} \tag{5a}$$

$$\frac{\partial}{\partial z}\underline{i} = -G\underline{v} - C\frac{\partial}{\partial t}\underline{v} \tag{5b}$$

For exponential excitation with the form $e^{j\omega t}$, the response is known to have the same time dependency, $e^{j\omega t}$. Let the trial solution for (5) be

$$\underline{v}(z, t) = ve^{j\omega t} \tag{6a}$$

$$\underline{i}(z, t) = ie^{j\omega t} \tag{6b}$$

$$v \equiv v(z), \qquad i \equiv i(z) \tag{6c}$$

The substitution of (6) into (5) yields

$$\frac{dv}{dz} = -Zi \tag{7a}$$

$$\frac{di}{dz} = -Yv \tag{7b}$$

$$Z \equiv R + j\omega L \tag{7c}$$

$$Y \equiv G + j\omega C \tag{7d}$$

The set of partial differential equations, (5), is thus reduced to a set of ordinary differential equations.

To determine the voltage v, the current i is solved from (7a), and then substituted into (7b).

$$\frac{di}{dz} = \frac{d}{dz} \frac{-1}{Z} \frac{dv}{dz} = -Yv$$

or

$$\frac{d^2v}{dz^2} = \gamma^2 v \tag{8a}$$

$$\gamma^2 \equiv YZ \tag{8b}$$

This is a second-order homogeneous differential equation with constant coefficients. Let the trial solution be

$$v(z) = V_o e^{pz} \tag{9}$$

where V_o and p are some constants to be determined. The substitution of (9) into (8a) yields

$$p_{1,2} = \pm\gamma \tag{10}$$

This indicates that there are two possible solutions which have the form of (9). A general solution for (8) is given by the sum of two independent solutions,

$$v = V^+e^{-\gamma z} + V^-e^{\gamma z} \tag{11}$$

where $V^{\pm}$ are some constants to be determined by the boundary conditions.

The corresponding current can be obtained from (7a).

$$i = \frac{-1}{Z} \frac{dv}{dz}$$

$$= \frac{\gamma}{Z} [V^+e^{-\gamma z} - V^-e^{\gamma z}]$$

$$i = \frac{1}{Z_o} [V^+e^{-\gamma z} - V^-e^{\gamma z}] \tag{12a}$$

$$Z_o \equiv \frac{\gamma}{Y} = \sqrt{\frac{Z}{Y}} \tag{12b}$$

where Z_o is known as the characteristic impedance and γ is the propagation constant of the line.

The complete expressions for the voltage and the current are obtained by substituting (11) and (12) into (6).

$$\underline{v}(z, t) = V^+e^{j\omega t - \gamma z} + V^-e^{j\omega t + \gamma z} \tag{13a}$$

$$\underline{i}(z, t) = \frac{1}{Z_o} [V^+e^{j\omega t - \gamma z} - V^-e^{j\omega t + \gamma z}] \tag{13b}$$

$$Z_o = \sqrt{\frac{R + j\omega L}{G + j\omega C}} \qquad (13c)$$

$$\gamma = \sqrt{(R + j\omega L)(G + j\omega C)} \equiv \alpha + j\beta \qquad (13d)$$

2. Wave Parameters and Characteristic Impedance

The steady-state solutions of the transmission line equations have been obtained in the previous section, Eq. (1-13). The propagation constant γ governs the manner in which $\underline{v}$ and $\underline{i}$ vary with z. In general, γ is a complex quantity, $\gamma = \alpha + j\beta$. The real part of γ determines the manner in which the waves diminish, or attenuate, as a function of the spatial variable z. Hence, α is known as the attenuation constant and its unit is nepers per unit length. The imaginary part of γ determines the variation in phase of the function along the axis of the spatial variable z. Thus, β is known as the phase constant and is measured in radians per unit wavelength.

The characteristic impedance, Z_0, is a characteristic of the line and has the dimension of an impedance. Characteristic impedance does not involve the length of the line, nor does it involve the properties of the termination.

The expressions for propagation constant and characteristic impedance will be obtained for the two most practical cases: the lossless line and the low-loss line.

(a) **Lossless Line**

A lossless line is defined by the conditions $R \equiv 0$ and $G \equiv 0$.

For the conditions of a lossless line, (1-13c) and (1-13d) become

$$\gamma = j\omega\sqrt{LC} = j\beta \tag{1}$$

$$Z_0 = \sqrt{\frac{L}{C}} \tag{2}$$

(b) **Low-Loss Line**

The specifications for a low-loss line are $R \ll \omega L$ and $G \ll \omega C$.

In this case, the propagation constant can be arranged as

$$\gamma = \sqrt{j\omega L \left[1 + \frac{R}{j\omega L} \right] j\omega C \left[1 + \frac{G}{j\omega C} \right]} \tag{3}$$

This expression can be expanded in a binomial series.

$$(1 + x)^n = 1 + nx + \frac{n(n-1)}{2!} x^2 + \cdots + \frac{n!}{(n-r)!\, r!} x^r + \cdots \tag{4}$$

As a first-order approximation, only the first two terms in the expansion will be retained. Equation (3) then becomes

$$\gamma \approx j\omega\sqrt{LC}\left[1 + \frac{1}{2}\left[\frac{R}{j\omega L} + \frac{G}{j\omega C}\right]\right]$$

$$= \frac{1}{2}\left[R\sqrt{\frac{C}{L}} + G\sqrt{\frac{L}{C}}\right] + j\omega\sqrt{LC} = \alpha + j\beta \tag{5}$$

The chartacteristic impedance can be expanded similarly.

$$Z_0 \approx \sqrt{\frac{L}{C}} + j\frac{1}{2\omega}\sqrt{\frac{L}{C}}\left[\frac{G}{C} - \frac{R}{L}\right] \tag{6}$$

2.1. Example: Determination of Propagation Constant - General Case

Determine α and β in terms of line parameters of an arbitrary line.

Solution:

The propagation constant of a transmission line is given by (1-13d). α and β can be obtained by squaring both sides of this expression and equating the corresponding real and imaginary parts, which yields

$$\alpha^2 - \beta^2 + j2\alpha\beta \equiv M + jN$$

$$\alpha^2 - \beta^2 = M \tag{1a}$$

$$2\alpha\beta = N \quad \text{or} \quad \alpha = \frac{N}{2\beta} \tag{1b}$$

$$M \equiv RG - \omega^2 LC \tag{1c}$$

$$N \equiv \omega(LG + RC) \tag{1d}$$

α can be eliminated from (1a) by means of (1b).

$$\beta^4 + M\beta^2 - (N/2)^2 = 0$$

$$\beta^2 = \frac{M}{2}\left[\pm\sqrt{1 + (N/M)^2} - 1\right] \tag{2}$$

The sign in front of the square root is chosen to be positive so that β will be real by assumption. Then

$$\beta = \sqrt{\frac{M}{2}\left[\sqrt{1 + (N/M)^2} - 1\right]} \tag{3a}$$

$$\alpha = \frac{\dfrac{N}{2}}{\sqrt{\dfrac{M}{2}\left[\sqrt{1 + (N/M)^2} - 1\right]}} \times \left[\frac{M/2\left[\sqrt{1 + (N/M)^2} + 1\right]}{M/2\left[\sqrt{1 + (N/M)^2} + 1\right]}\right]^{1/2}$$

$$= \sqrt{\frac{M}{2}\left[\sqrt{1 + (N/M)^2} + 1\right]} \tag{3b}$$

3. Interpretation of the Solution

The general solution of the transmission line equation with an exponential excitation, $e^{j\omega t}$, is found to be [Eq. (1-13)]

$$\underline{v}(z, t) = V^+ e^{j\omega t - (\alpha + j\beta)z} + V^- e^{j\omega t + (\alpha + j\beta)z}$$

$$= V^+ e^{-\alpha z} [\cos(\omega t - \beta z) + j\sin(\omega t - \beta z)]$$

$$+ V^- e^{\alpha z} [\cos(\omega t + \beta z) + j\sin(\omega t + \beta z)] \tag{1}$$

The exponential factor in the above expression will cause the magnitude to vary as a function of z. To obtain a physical meaning for the terms within the brackets, it is necessary to study each of the following functions separately.

$$g^+ \equiv \cos(\omega t - \beta z) \tag{2a}$$

$$g^- \equiv \cos(\omega t + \beta z) \tag{2b}$$

It is unnecessary to investigate the corresponding sine functions since they are simply shifted by 90 degrees from the corresponding cosine functions.

When $z = 0$, the function g^+ becomes

$$g^+ = \cos \omega t \tag{3}$$

and this function is plotted in Fig. 3-1. The period T of the plot is defined by

$$\omega t \big|_{t=T} \equiv 2\pi \tag{4}$$

or

$$T = \frac{2\pi}{\omega} = \frac{1}{f} \tag{5}$$

where f is the frequency in hertz.

When $t = 0$, the function g^+ takes the form

$$g^+ = \cos \beta z \tag{6}$$

This function is plotted in Fig. 3-2. The period of this plot is known as the wavelength, λ, and it is defined by

$$\beta z \big|_{z=\lambda} \equiv 2\pi$$

or

$$\lambda = \frac{2\pi}{\beta} \tag{7}$$

and a more frequently used expression is

$$\beta = \frac{2\pi}{\lambda} \tag{8}$$

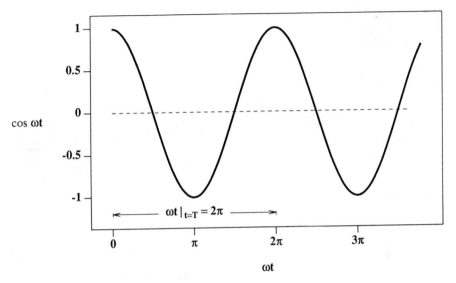

Figure 3-1: cos ωt vs ωt

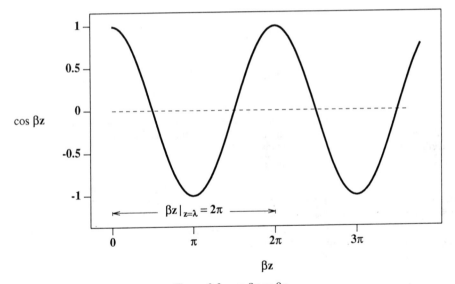

Figure 3-2: cos βz vs βz

Equation (2a) can be rearranged as

$$g^+ = \cos\left[-\beta(z - \frac{\omega}{\beta}t)\right] = \cos\beta(z - \lambda ft) = \cos\beta(z - v_p t)$$

$$\equiv f^+(z - v_p t) \tag{9}$$

where $v_p \equiv f\lambda$ is the velocity of propagation and its meaning will be clarified promptly.

A function of two variables can be easily investigated by keeping one of the variables fixed at some value and studying its variation with respect to the other variable. At $t = 0$, one has

$$f^+(z - v_p t)|_{t=0} = f^+(z) \tag{10}$$

At a later time, $t = t_1$, then

$$f^+(z - v_p t)|_{t=t_1} = f^+(z - v_p t_1) \equiv f^+(z_1) \tag{11a}$$

$$z_1 \equiv z - v_p t_1 \tag{11b}$$

Since z and z_1 are different by a constant value, $v_p t_1$, therefore $f^+(z)$ and $f^+(z_1)$ should have the same shape. In Fig. 3-3, $f^+(z_1)$ is the same curve as $f^+(z)$, displaced by a distance of $v_p t_1$ to the right of $f^+(z)$, along the z-axis. This can be verified by locating the point $z_1 = 0$ with respect to the point $z = 0$. From (11b), one has

$$z_1 = 0 = z - v_p t_1$$

or

$$z_1 = 0 \quad \text{when} \quad z = v_p t_1 \quad \text{or} \quad v_p = \frac{z}{t_1} \tag{12}$$

If time is allowed to increase continuously, then the function $f^+(z)$ will appear at an increasing distance $v_p t$ from the origin, $z = 0$, as if the function $f^+(z)$ is traveling continuously in the positive z-direction. It is for this reason that the function $f^+(z - v_p t)$ is known as a traveling wave function in the positive z-direction. The speed of traveling is given by $\frac{z}{t} = v_p$, or v_p is the velocity of propagation of the wave function f^+.

The function $g^-(\omega t + \beta z) \equiv f^-(z + v_p t)$ can be similarly shown to be a traveling wave but it is traveling in the negative z-direction. This is because the displacement is given by $z = -v_p t$, by virtue of the positive sign in the function $f^-(z + v_p t)$.

The traveling wave concept may be obtained from another point of view. Attention is focused on some point of the function $e^{j(\omega t - \beta z)}$ where the phase is fixed at some constant value. Let this "constant phase" be chosen at an initial time t_0 and at an initial position z_0, then

$$\omega t_0 - \beta z_0 = K \tag{13}$$

where K is an arbitrary constant. The problem, now, is to locate this same constant phase point at some later time $t_1 \equiv t_0 + \Delta t$ and its new position $z_1 \equiv z_0 + \Delta z$. That is,

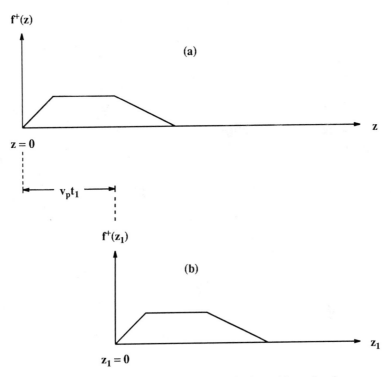

Figure 3-3: Traveling-wave function in the positive z-direction

$$\omega t_1 - \beta z_1 = K \tag{14}$$

In other words, it is desired to find the location, both in the time domain and in the spatial domain, of the same constant phase point as defined by (13). The substitution of the definitions for t_1 and z_1 into (14) results in

$$(\omega t_0 - \beta z_0) + \omega \Delta t - \beta \Delta z = K \tag{15}$$

By virtue of (13), (15) becomes

$$\omega \Delta t - \beta \Delta z = 0$$

or

$$\frac{\Delta z}{\Delta t} = \frac{\omega}{\beta} \tag{16}$$

The value of Δt and Δz may be chosen to be as small or as large as one pleases. In order to obtain a continuous variation of this constant phase point, one would choose Δt, and consequently Δz, as small as possible. Therefore,

$$\lim_{\Delta t \to 0} \frac{\Delta z}{\Delta t} = \frac{dz}{dt} = \frac{\omega}{\beta} \equiv v_p \tag{17}$$

where v_p is known as the phase velocity, since it is the velocity of the constant phase point of the wave function.

One can similarly show that the phase velocity of the wave function $e^{j(\omega t + \beta z)}$ is

$$v_p = -\frac{\omega}{\beta} = \frac{dz}{dt} \tag{18}$$

which indicates that its value is directed toward the negative z-direction.

4. Terminated Line

A transmission line of infinite length has been investigated in previous sections. In practical situations, lines are terminated by a source at one end and a load at the other. In analyzing a transmission line terminated by a passive load, it is usually more convenient to use the load end as the zero reference.

In previous sections, z is used as the position variable and its zero reference is located at the source end. In the analysis to follow, s will be used as the position variable and its zero reference is placed at the load end (Fig. 4-1).

The voltage and current distributions as functions of the variable z are found to be [Eq. (1-13)]

$$v(z) = V^+ e^{-\gamma z} + V^- e^{\gamma z} \tag{1a}$$

$$i(z) = \frac{1}{Z_0} [V^+ e^{-\gamma z} - V^- e^{\gamma z}] \tag{1b}$$

To convert v and i into functions of s, make the following change of variable (Fig. 4-1).

$$z = L_1 - s \tag{2}$$

where L_1 is the length of the line. Thus

$$v(s) = V_- e^{\gamma s} + V_+ e^{-\gamma s} \tag{3a}$$

$$V_- \equiv V^+ e^{-\gamma L_1} \tag{3b}$$

$$V_+ \equiv V^- e^{\gamma L_1} \tag{3c}$$

The current distribution as a function of s is obtained similarly.

$$i(s) = \frac{1}{Z_0} [V_- e^{\gamma s} - V_+ e^{-\gamma s}] \tag{4}$$

The values of V_- and V_+ are to be determined from the boundary conditions. At the load, s = 0, one has

$$v(s = 0) = i_L Z_L \tag{5}$$

where $i_L \equiv i(s = 0)$ is the current through load impedance Z_L and it can be obtained from (4) by setting s = 0.

$$v(0) = \frac{Z_L}{Z_0} (V_- - V_+) \tag{6}$$

v(0) can also be obtained from (3) by setting s = 0 .

$$v(0) = V_- + V_+ \tag{7}$$

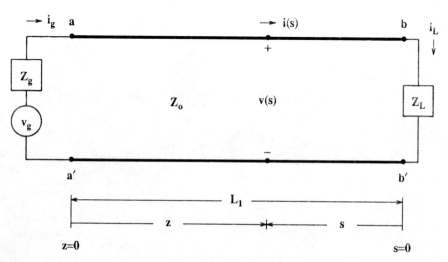

Figure 4-1: A terminated transmission line

Equating (6) and (7) yields

$$\frac{V_+}{V_-} = \frac{Z_L - Z_0}{Z_L + Z_0} \tag{8}$$

The traveling wave which impinges upon the termination is known as the incident wave at the termination. The wave which bounces back from the termination is known as the reflected wave. The ratio of the reflected wave to the incident wave at the junction is known as the reflection coefficient.

$$\text{Reflection coefficient } \Gamma = \frac{\text{Reflected wave}}{\text{Incident wave}} \tag{9}$$

For the present case, one has, at the load end, the following relations:

$$\text{Incident wave } v_i(s = 0) = V_- e^{\gamma s}|_{s=0} = V_- \tag{10a}$$

$$\text{Reflected wave } v_r(s = 0) = V_+ e^{-\gamma s}|_{s=0} = V_+ \tag{10b}$$

Then

$$\text{Reflection coefficient } \Gamma(s = 0) \equiv \Gamma_L = \frac{V_+}{V_-} \tag{11}$$

The combination of (8) and (11) produces

$$\Gamma_L = \frac{Z_L - Z_0}{Z_L + Z_0} = \frac{V_+}{V_-} \tag{12}$$

where Γ_L is the reflection coefficient at the load Z_L.

Equation (8) relates the two unknowns V_+ and V_-. Another equation is needed to determine these unknowns. This is obtained by applying the boundary condition at the source end. At $s = L_1$ one has

$$v(s = L_1) = v_g - i_g Z_g \tag{13}$$

where v_g is the voltage of the generator, i_g is the current of the source, and Z_g is the internal impedance of the generator.

$$i_g = i(s = L_1) = \frac{1}{Z_0} [V_- e^{\gamma L_1} - V_+ e^{-\gamma L_1}] \tag{14}$$

Equation (14) is obtained from (4) with $s = L_1$. The substitution of (14) into (13) gives

$$v(L_1) = v_g - \frac{Z_g}{Z_0} [V_- e^{\gamma L_1} - V_+ e^{-\gamma L_1}] \tag{15}$$

The voltage at the source end can also be obtained from (3) by setting $s = L_1$.

$$v(L_1) = V_- e^{\gamma L_1} + V_+ e^{-\gamma L_1} \tag{16}$$

Identifying (16) with (15) and collecting terms, one gets

$$V_- e^{\gamma L_1} \left[1 + \frac{Z_g}{Z_o} \right] = v_g + V_+ e^{-\gamma L_1} \left[\frac{Z_g}{Z_o} - 1 \right]$$

$$V_- e^{\gamma L_1} = \frac{Z_o}{Z_g + Z_o} v_g + V_+ e^{-\gamma L_1} \frac{Z_g - Z_o}{Z_g + Z_o} \tag{17}$$

For a passive termination at $s = L_1$, i.e., $v_g = 0$, then (17) becomes

$$\frac{V_- e^{\gamma L_1}}{V_+ e^{-\gamma L_1}} = \frac{Z_g - Z_o}{Z_g + Z_o} \tag{18}$$

It is to be noted that at the source end, one has

$$v_i(s = L_1) = V_+ e^{-\gamma L_1} \tag{19a}$$

$$v_r(s = L_1) = V_- e^{\gamma L_1} \tag{19b}$$

By definition of (9) and (19), Eq. (18) implies

$$\Gamma_g \equiv \Gamma(s = L_1) = \frac{Z_g - Z_o}{Z_g + Z_o} \tag{20}$$

where Γ_g is the reflection coefficient at the generator end. With (20), Eq. (17) becomes

$$V_- e^{\gamma L_1} = \frac{Z_o}{Z_g + Z_o} v_g + \Gamma_g V_+ e^{-\gamma L_1} \tag{21}$$

With the source, the load, and the line specified, the constants V_+ and V_- can be determined from (8) and (21). The substitution of (11) into (21) yields

$$V_- e^{\gamma L_1} = \frac{Z_o}{Z_g + Z_o} v_g + \Gamma_L \Gamma_g V_- e^{-\gamma L_1} \tag{22}$$

V_- can be solved from (22).

$$V_- = \frac{Z_o v_g}{Z_g + Z_o} \times \frac{1}{e^{\gamma L_1} - \Gamma_L \Gamma_g e^{-\gamma L_1}} \tag{23}$$

and

$$V_+ = \Gamma_L V_- = \frac{Z_o v_g}{Z_g + Z_o} \frac{\Gamma_L}{e^{\gamma L_1} - \Gamma_L \Gamma_g e^{-\gamma L_1}} \tag{24}$$

With V_+ and V_- determined, (3a) and (4) become

$$v(s) = \frac{Z_o}{Z_g + Z_o} \, v_g \, \frac{e^{\gamma s} + \Gamma_L e^{-\gamma s}}{e^{\gamma L_1} - \Gamma_g \Gamma_L e^{-\gamma L_1}} \tag{25a}$$

$$i(s) = \frac{v_g}{Z_g + Z_o} \times \frac{e^{\gamma s} - \Gamma_L e^{-\gamma s}}{e^{\gamma L_1} - \Gamma_g \Gamma_L e^{-\gamma L_1}} \tag{25b}$$

5. The Crank Diagram

The voltage and current distributions are given by (4-3) and (4-4), and can be rearranged as follows.

$$v(s) = V_-e^{\gamma s} [1 + \Gamma_L e^{-2\gamma s}] \tag{1a}$$

$$i(s) = \frac{V_-e^{\gamma s}}{Z_0} [1 - \Gamma_L e^{-2\gamma s}] \tag{1b}$$

$$\Gamma_L = \frac{V_+}{V_-} \equiv |\Gamma_L| e^{j\Theta_r} = \frac{Z_L - Z_0}{Z_L + Z_0} \tag{1c}$$

With the polar form of Γ_L and $\gamma = \alpha + j\beta$, one has

$$v(s) = V_-e^{(\alpha+j\beta)s} [1 + |\Gamma_L| e^{-2\alpha s - j\Psi}] \tag{2a}$$

$$i(s) = \frac{V_-}{Z_0} e^{(\alpha+j\beta)s} [1 - |\Gamma_L| e^{-2\alpha s - j\Psi}] \tag{2b}$$

$$\Psi \equiv 2\beta s - \Theta_r \tag{2c}$$

The analysis of a transmission line with losses is rather complicated. For many practical lines, the overall effect of the factor $e^{\pm\alpha s}$ is negligible when s is of the order of laboratory dimensions. It is therefore helpful to introduce the lossless model, which provides a simplified picture.

For the lossless case, $\alpha = 0$, (2) becomes

$$v(s) = V_-e^{j\beta s} [1 + |\Gamma_L| e^{-j\Psi}] \tag{3a}$$

$$i(s) = \frac{V_-}{Z_0} e^{j\beta s} [1 - |\Gamma_L| e^{-j\Psi}] \tag{3b}$$

The terms within brackets describe the variation of $|v(s)|$ and $|i(s)|$ as a function of Ψ, since the factor outside the brackets has constant amplitude.

Inside the brackets is the sum of a real phasor of unity amplitude and a complex phasor $|\Gamma_L| e^{-j\Psi}$ (Fig. 5-1).

Equation (3a) implies that the voltage v(s) assumes a maximum value when the two phasors are in phase.

$$\Psi = 2\beta s_M - \Theta_r = \pm 2n\pi, \qquad n = 0, 1, 2, \ldots \tag{4a}$$

or

$$s_M = \frac{2n\pi + \Theta_r}{2\beta} = n\frac{\lambda}{2} + \frac{\Theta_r}{2\beta} \tag{4b}$$

where s_M is the location of voltage maxima. Positive n is chosen to avoid the negative value of s which lies exterior of the line. The voltage maxima appear $\lambda/2$ apart and the amplitude is given by

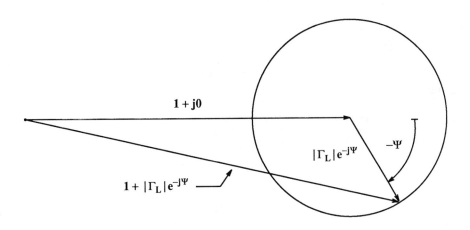

$$\Psi = 2\beta s - \Theta_L$$

Figure 5-1: The crank diagram

$$|v_{MAX}| = |V_-| (1 + |\Gamma_L|) \tag{5}$$

The voltage v(s) assumes a minimum value when the two phasors within brackets of (3a) are in phase opposition. That is, the voltage is minimum when

$$\Psi = 2\beta s_m - \Theta_\Gamma = (2n + 1)\pi, \qquad n = 0, 1, 2, \ldots \tag{6a}$$

or

$$s_m = (2n + 1) \frac{\lambda}{4} + \frac{\Theta_\Gamma}{2\beta} \tag{6b}$$

where s_m is the location of voltage minima. Voltage minina are also $\lambda/2$ apart and the magnitude is given by

$$|v_{min}| = |V_-| (1 - |\Gamma_L|) \tag{7}$$

Equations (4) and (6) indicate that adjacent voltage maxima and minima are separated by a distance of $\lambda/4$.

The current distribution also possesses maxima and minima. An inspection of (3b) shows that the current is maximum when phasors within brackets are in phase. The required condition is given by (6), and the amplitude of maximum current is

$$|i_{MAX}| = \frac{|V_-|}{|Z_0|} (1 + |\Gamma_L|) \tag{8}$$

A minimum current occurs when (4) is satisfied and its magnitude is

$$|i_{min}| = \frac{|V_-|}{|Z_0|} (1 - |\Gamma_L|) \tag{9}$$

It has just been shown that both voltage and current vary periodically from a maximum value to a minimum value along the line. The location of voltage maximum coincides with that of current minimum, while the location of voltage minimum coincides with that of current maximum. The locations of maxima and minima, in fact, any particular value, are fixed along the line and thus produce a standing wave pattern.

The ratio of the maximum amplitude of the voltage standing wave to the magnitude of its minimum is known as the standing-wave ratio and is denoted by S. From (5) and (7), the voltage standing-wave ratio, VSWR, is given by

$$S \equiv VSWR = \frac{|V_{MAX}|}{|V_{min}|} = \frac{1 + |\Gamma_L|}{1 - |\Gamma_L|} \tag{10}$$

Another useful parameter is the ratio of v(s) to i(s); this is designated by Z(s). From Eq. (1), one has

$$Z(s) \equiv \frac{v(s)}{i(s)} = Z_o \times \frac{1 + \Gamma_L e^{-2\gamma s}}{1 - \Gamma_L e^{-2\gamma s}} \qquad \text{general case} \qquad (11a)$$

$$= Z_o \times \frac{1 + |\Gamma_L| e^{-j\Psi}}{1 - |\Gamma_L| e^{-j\Psi}} \qquad \text{lossless case} \qquad (11b)$$

$Z(s)$ is the input impedance of the line at the location s, looking toward the load. $Z(s)$ is maximum at the location where $v(s)$ is maximum, i.e., at $\Psi = 2n\pi$, $n = 0, 1, 2, \ldots$

$$Z_{MAX} = \frac{v_{MAX}}{i_{min}} = Z_o \times \frac{1 + |\Gamma_L|}{1 - |\Gamma_L|} = S\,Z_o \qquad \text{lossless line} \qquad (12a)$$

$Z(s)$ is minimum when $\Psi = (2n + 1)\pi$, $n = 0, 1, 2, \ldots$

$$Z_{min} = \frac{v_{min}}{i_{MAX}} = \frac{Z_o}{S} \qquad \text{lossless line} \qquad (12b)$$

For a lossless line, both Z_{MAX} and Z_{min} are purely resistive regardless of the termination.

When the line is lossy, terms within brackets of (2) can still be interpreted as a sum of two phasors. The amplitude of the second phasor decreases as the distance s increases. The effect of losses in the line will be clarified in the next section when the terminations are specified.

6. The Short-Circuited Line

One of the simplest terminated lines is the short-circuited line. Not only is this simpler to analyze than a line with general termination, but it also has many practical applications. Short-circuited lines are extensively used as circuit elements at high frequencies.

The general expression for input impedance of a terminated line is expressed in Eq. (5-11). For a short-circuited line, $Z_L = 0$, the reflection coefficient, (5-1c), is

$$\Gamma_L = -1e^{j0^\circ} \tag{1}$$

The expressions for the voltage and the current, (5-1), become

$$v(s) = 2V_- \sinh \gamma s \tag{2a}$$

$$i(s) = \frac{2V_-}{Z_0} \cosh \gamma s \tag{2b}$$

$$Z(s) = Z_0 \tanh \gamma s \tag{2c}$$

$$\sinh x = (1/2)\,(\,e^x - e^{-x}\,) \tag{2d}$$

$$\cosh x = (1/2)\,(\,e^x + e^{-x}\,) \tag{2e}$$

To obtain some physical properties of the above results, the case of the lossless line will be investigated first.

(a) Lossless Line

The ideal lossless line is defined by $\alpha = 0$; then

$$\gamma = j\beta \tag{3}$$

and (2) is simplified to

$$v(s) = V_-\,(\,e^{j\beta s} - e^{-j\beta s}\,) = 2jV_- \sin \beta s \tag{4a}$$

$$i(s) = \frac{2V_-}{Z_0} \cos \beta s \tag{4b}$$

$$Z(s) = jZ_0 \tan \beta s \tag{4c}$$

Plots of magnitude and phase angle of voltage and current are shown in Fig. 6-1. Equations (4a) and (4b) indicate that voltage and current are 90 degrees out of phase in the spatial domain (the s-coordinate). They are also in phase quadrature in the time domain owing to the presence of the factor j in (4a). Both v(s) and i(s) exhibit a standing-wave pattern in the s-domain.

Equation (4c) implies that the input impedance of the short-circuited line is purely reactive and its value varies with the electrical length βs of the line. The range of the input impedance is from $-\infty$ to $+\infty$ (Fig. 6-2).

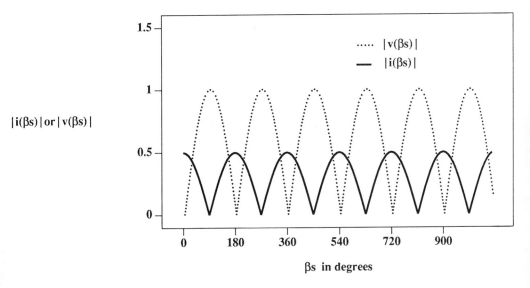

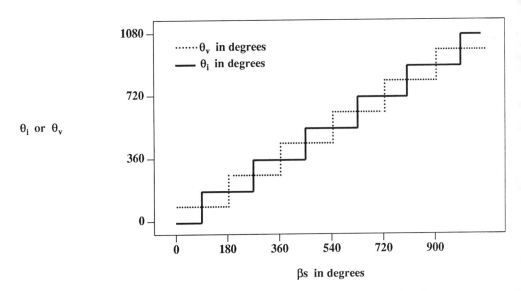

Figure 6-1: Voltage and current distributions along a lossless line terminated
by short-circuit

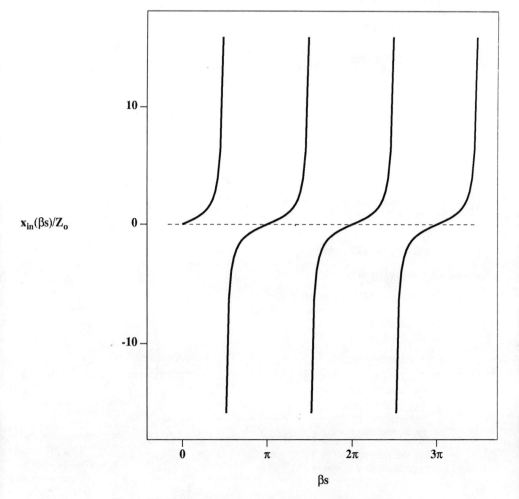

Figure 6-2: The input impedance of a short-circuited lossless line

(b) Line with Finite Losses

For a line with finite losses, the voltage and current, (2), can be rearranged as follows.

$$v(s) = V_- (e^{\alpha s}e^{j\beta s} - e^{-\alpha s}e^{-j\beta s})$$

$$= V_- [e^{\alpha s} (\cos \beta s + j \sin \beta s) - e^{-\alpha s} (\cos \beta s - j \sin \beta s)]$$

$$= 2V_- [\sinh \alpha s \cos \beta s + j \cosh \alpha s \sin \beta s] \tag{5a}$$

$$i(s) = \frac{V_-}{Z_0} [(e^{\alpha s} + e^{-\alpha s}) \cos \beta s + j(e^{\alpha s} - e^{-\alpha s}) \sin \beta s]$$

$$= \frac{2V_-}{Z_0} [\cosh \alpha s \cos \beta s + j \sinh \alpha s \sin \beta] \tag{5b}$$

The voltage and current distributions for the line with losses are rather complicated to plot. However, it is not difficult to obtain an approximation of such distributions for the case of low loss. The inspection of (5a) indicates that the coefficient of the sin βs term is always greater than that of the cos βs term. Therefore, the upper bound of $|v(s)|$ is determined by the sin βs term.

$$|v(s)|_{MAX} = |2V_- \cosh \alpha s \sin \beta s|_{MAX} = 2V_- \cosh \alpha s \tag{6a}$$

Similarly, the lower bound of $|v(s)|$ is determined by the cos βs term.

$$|v(s)|_{min} = |2V_- \sinh \alpha s \cos \beta s|_{min} = 2V_- \sinh \alpha s \tag{6b}$$

Plots of v(s) and i(s) of a short-circuited lossy line are quite different from those for a lossless line (Fig. 6-3 and Fig. 6-4).

(i) The standing-wave patterns are not recurrent with distance s along the line.

(ii) The curves of the magnitude are no longer touching the zero axis at minima and they are rounded at those locations rather than forming sharp cusps.

(iii) The plots of phase angles are neither discontinuous at half-wavelength intervals nor uniform in between; rather they change continuously at varying but always finite, nonzero rates.

By taking the ratio of v(s) to i(s) in the distribution plot, it is possible to plot the magnitude of the impedance as a function of position on the line. The upper bound of Z(s) is given by

$$|Z_{MAX}| = \frac{|v_{MAX}|}{|i_{min}|} = |Z_0 \coth \alpha s| \tag{7a}$$

The lower bound of Z_0 is given by

$$|Z_{min}| = \frac{|v_{min}|}{|i_{MAX}|} = |Z_0 \tanh \alpha s| \tag{7b}$$

The fluctuations in impedance are considerably greater than the voltage and current fluctuations, because they are the ratios of the maximum voltage to the minimum current (Fig. 6-4).

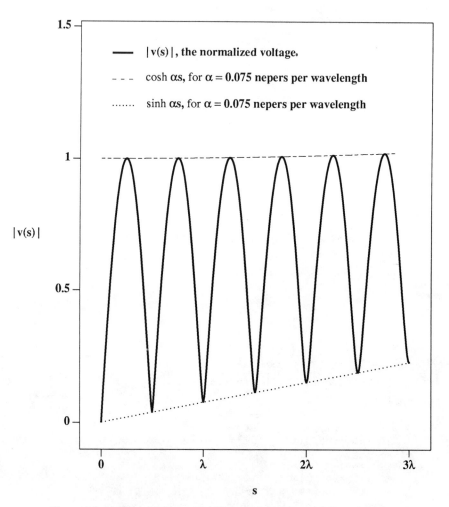

Figure 6-3: Voltage and current distributions along a low-loss line terminated by short-circuit

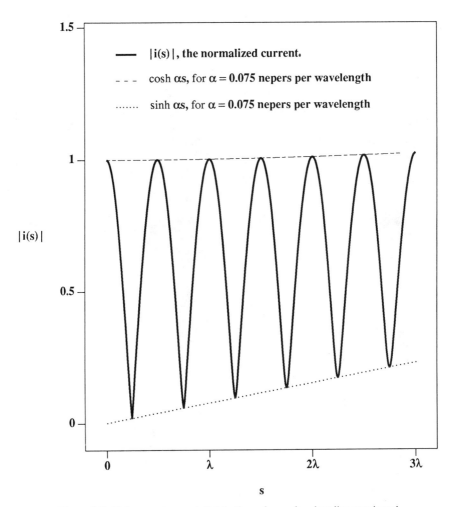

Figure 6-3: Voltage and current distributions along a low-loss line terminated by short-circuit

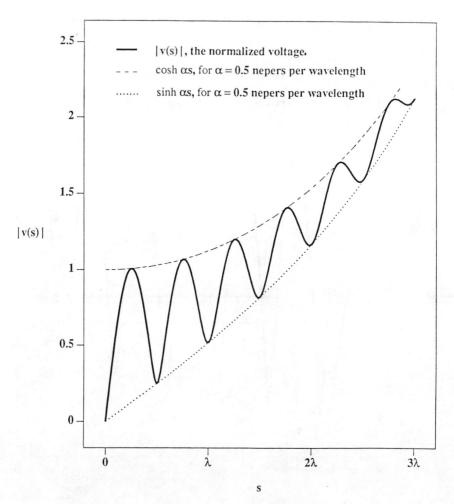

Figure 6-4: Voltage and current distributions along a line with moderate loss.

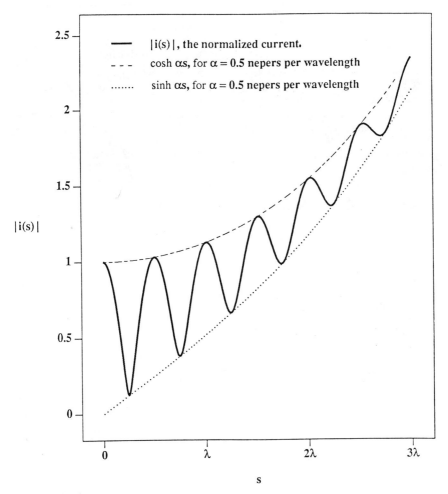

Figure 6-4: Voltage and current distributions along a line with moderate loss.

Close to the load, $\alpha s \ll 1$, the minimum impedance can be approximated by

$$Z_{min} \approx jZ_o\alpha s \tag{8a}$$

since tanh $\alpha s \approx \alpha s$. Likewise, the maximum impedance is approximated by

$$Z_{MAX} \approx j \frac{Z_o}{\alpha s} \tag{8b}$$

The small value of αs can be achieved by either having very small α or very small s or both.

It is to be noted that both the maximum $Z(s)$ and the minimum $Z(s)$ tend toward Z_o for large values of αs.

The voltage standing-wave ratio was defined for a lossless line as the ratio of the maximum to the minimum voltage on the line. Since this ratio is a function of distance in the case of a lossy line, the VSWR is, strictly speaking, undefined for a lossy line. However, in the practical low-loss line, an average standing-wave ratio is defined as the ratio of the average of two successive maxima to the minimum in between or the ratio of the maximum to the average of the two minima on either side of it.

7. Quarter-Wave Transformer

Both single-stub and double-stub tuners provide matching for various load impedances. For a fixed load, it is possible to obtain the matching by inserting a short section of line with proper characteristic impedance.

The input impedance of a lossless line terminated by a load impedance Z_L, Fig. 7-1, is given by

$$Z(s) = Z_o \frac{e^{j\beta s} + \Gamma_L e^{-j\beta s}}{e^{j\beta s} - \Gamma_L e^{-j\beta s}} \tag{1a}$$

$$\Gamma_L = \frac{Z_L - Z_o}{Z_L + Z_o} \tag{1b}$$

Equation (1a) can be arranged to a more convenient form by eliminating Γ_L.

$$Z(s) = Z_o \frac{(Z_L + Z_o) e^{j\beta s} + (Z_L - Z_o) e^{-j\beta s}}{(Z_L + Z_o) e^{j\beta s} - (Z_L - Z_o) e^{-j\beta s}}$$

$$= Z_o \frac{Z_L (e^{j\beta s} + e^{-j\beta s}) + Z_o (e^{j\beta s} - e^{-j\beta s})}{Z_o (e^{j\beta s} + e^{-j\beta s}) + Z_L (e^{j\beta s} - e^{-j\beta s})}$$

$$= Z_o \frac{Z_L \cos \beta s + j Z_o \sin \beta s}{Z_o \cos \beta s + j Z_L \sin \beta s} \tag{2}$$

Consider the arrangement shown in Fig. 7-2. The input impedance across terminals a–a′ is given by (2), with an appropriate change of notation.

$$Z_2 \equiv Z(s = L_1) = Z_{o2} \frac{Z_L \cos \beta L_1 + j Z_{o2} \sin \beta L_1}{Z_{o2} \cos \beta L_1 + j Z_L \sin \beta L_1} \tag{3}$$

When the line length is chosen to be a quarter-wavelength, $L_1 \equiv \lambda/4$, then

$$\beta L_1 = \frac{2\pi}{\lambda} \frac{\lambda}{4} = \frac{\pi}{2} \tag{4}$$

Consequently,

$$Z_2 = Z(s = \lambda/4) = \frac{Z_{o2} Z_{o2}}{Z_L}$$

or

$$Z_{o2} = \sqrt{Z_L Z_2} \tag{5}$$

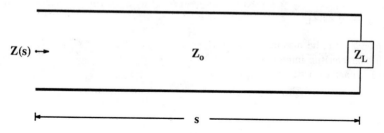

Figure 7-1: A terminated line

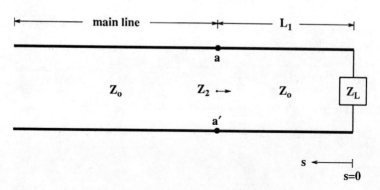

Figure 7-2: A line in cascade with a terminated line

If the line is to be matched at terminals a–a', then one must have

$$Z_2 = Z_{o1} \tag{6}$$

Equation (5) then becomes

$$Z_{o2} = \sqrt{Z_{o1} Z_L} \tag{7}$$

The load Z_L can be matched to the main line, Z_{o1}, if a quarter-wavelength-long lossless line with a characteristic impedance Z_{o2}, defined by (7), is inserted between the load and the main line. The characteristic impedance Z_{o2} is the geometric mean of Z_{o1} and Z_L.

8. Power Calculation - Complex Notation

Complex functions have been used in the solution of problems involving steady-state sinusoids. The resulting simplification of linear problems in steady state is well recognized. However, care should be exercised for nonlinear expressions. One of the most frequently used nonlinear expressions appears in the calculation of power - which is a product of two complex terms.

Let the sinusoidal voltage across and current through a circuit element be (Fig. 8-1)

$$v(t) = V_0 \cos \omega t \tag{1a}$$

$$i(t) = I_0 \cos (\omega t + \theta) \tag{1b}$$

where θ is the phase angle between the voltage and the current. The instantaneous power supplied to the circuit element is

$$P(t) = v(t)i(t) = V_0 I_0 \cos \omega t \cos (\omega t + \theta)$$

$$= \frac{1}{2} V_0 I_0 [\cos (2\omega t + \theta) + \cos \theta] \tag{2}$$

where the identity

$$\cos A \cos B = \frac{1}{2} [\cos (A + B) + \cos (A - B)] \tag{3}$$

is used to obtain (2).

The average power over a period of time is obtained by integrating (2) over the period T and the result is then divided by T.

$$<P> \equiv P_{avg} = \frac{1}{T} \int_0^T P(t) \, dt = \frac{V_0 I_0}{2T} \int_0^T [\cos (2\omega t + \theta) + \cos \theta] \, dt$$

$$= \frac{V_0 I_0}{2T} \left[\frac{1}{2} [\sin (2\omega T + \theta) - \sin \theta] + T \cos \theta \right]$$

$$= \frac{1}{2} V_0 I_0 \cos \theta \tag{4}$$

since $\omega T = 2\pi$.

When the voltage and the current are expressed by complex exponential functions,

$$v_e = V_0 e^{j\omega t} \tag{5a}$$

$$i_e = I_0 e^{j(\omega t + \theta)} \tag{5b}$$

then the instantaneous power, (2), becomes

$$P(t) = v(t)i(t) = Re [v_e(t)] Re [i_e(t)] \tag{6}$$

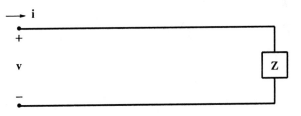

Figure 8-1: Current and voltage in a circuit element

The complex functions, $v_e(t)$ and $i_e(t)$, can be decomposed into the real and the imaginary components.

$$v_e(t) \equiv v_r + jv_i \quad \text{and} \quad i_e(t) \equiv i_r + ji_i \tag{7a}$$

where the subscripts r and i signify the real and imaginary components, respectively. The corresponding complex-conjugate functions are

$$v_e^*(t) = v_r - jv_i \quad \text{and} \quad i_e^*(t) = i_r - ji_i \tag{7b}$$

The real parts of $v_e(t)$ and $i_e(t)$ are then given by

$$\text{Re}\,[v_e(t)] = \frac{1}{2}\,[\,v_e(t) + v_e^*(t)\,] \tag{8a}$$

$$\text{Re}\,[i_e(t)] = \frac{1}{2}\,[\,i_e(t) + i_e^*(t)\,] \tag{8b}$$

The substitution of (8) into (6) yields

$$P(t) = \frac{1}{4}\,[\,v_e(t) + v_e^*(t)\,]\,[\,i_e(t) + i_e^*(t)\,]$$

$$= \frac{1}{4}\left[\,[\,v_e i_e + (\,v_e i_e\,)^*\,] + [\,v_e i_e^* + (\,v_e i_e^*\,)^*\,]\,\right]$$

$$= \frac{1}{2}\left[\,\text{Re}\,[\,v_e i_e\,] + \text{Re}\,[\,v_e i_e^*\,]\,\right]$$

$$= \frac{1}{2}\,V_o I_o\,[\,\cos(2\omega t + \theta) + \cos\theta\,] \tag{9}$$

Equation (9) is identical to (2), and the time-average power can be expressed as

$$<P> = \frac{1}{2}\,V_o I_o \cos\theta = \frac{1}{2}\,\text{Re}\,[\,v_e(t)i_e^*(t)\,] \tag{10}$$

This is the inportant relation of the time-average power when voltage and current are expressed in terms of exponential functions. The time-average power is equal to one-half the real part of the product of the voltage and the complex conjugate of the associated current.

9. Problems

1. A 600-ohm lossless open-circuited line has a length of 3.35 wavelengths. If the voltage across the input is 150 volts, find the input current and the output voltage.

2. A 300-ohm lossless line is terminated by a load of 250 + j350 ohms.

    (a) Determine the voltage distribution along the line.

    (b) If the voltage at the input is 10 volts, find the current and the voltage at the load for a line of length 3.51 wavelengths.

    (c) Determine the input impedance of the line.

3. A 50-ohm lossless line has a voltage standing-wave ratio of 3.2. The first voltage minimum is located at 1.1 m from the load and the next occurs at 2.6 m from the load. Determine the load impedance and the operating wavelength.

4. Two lossless lines are connected in cascade. The first line has a length of 3.6 wavelengths and its characteristic impedance is 100 ohms. The second one is a 50-ohm line and its length is 1.46 wavelengths. The second line is terminated by a resistive load of 150 ohms. If the input voltage is 1 volt, determine the voltage across the load.

5. A 25-ohm resistive load is connected to a 100-ohm lossless line through a quarter-wavelength transformer. Determine and sketch the variation of the reactance in the frequency range of ±20% of the designed frequency.

CHAPTER II   Review on Waveguides

1. Maxwell's Equations

This chapter will review some of the basic principles in waveguide theory. Additional details are available in many textbooks and some of these are listed in the bibliography [C].

James Clerk Maxwell, a Scottish physicist and mathematician, collected and postulated that the following four relations will prescribe all electromagnetic phenomena.

$$\nabla \times \underline{\vec{E}} = - \frac{\partial}{\partial t} \underline{\vec{B}} \tag{1a}$$

$$\nabla \times \underline{\vec{H}} = \underline{\vec{J}}_s + \sigma \underline{\vec{E}} + \frac{\partial}{\partial t} \underline{\vec{D}} \tag{1b}$$

$$\nabla \cdot \underline{\vec{D}} = \rho \tag{1c}$$

$$\nabla \cdot \underline{\vec{B}} = 0 \tag{1d}$$

where $\underline{\vec{J}}_s$ is the current density of the source. This set of relations is known as Maxwell's equations. The constitutive relations are

$$\underline{\vec{B}} = \mu \underline{\vec{H}} \tag{1e}$$

$$\underline{\vec{D}} = \varepsilon \underline{\vec{E}} \tag{1f}$$

where μ is the permeability and ε is the permittivity of the medium.

The notation $\vec{G} \equiv \vec{G}(\vec{r},t)$ represents a function of position and time, where $\vec{r}$ is a vector pointing from the origin to the field point.

For time-harmonic fields at a single frequency, it is convenient to use an exponential representation, i.e.,

$$\underline{\vec{G}}(\vec{r}, t) = \vec{G}(\vec{r})e^{j\omega t} \equiv \vec{G}e^{j\omega t} \qquad \vec{G} \equiv \vec{G}(\vec{r}) \tag{2a}$$

Scalar functions are defined similarly.

$$\underline{f}(\vec{r}, t) = fe^{j\omega t} \qquad f \equiv f(\vec{r}) \tag{2b}$$

Although practical fields are either sinusoidal or cosinusoidal time functions, exponential functions are simpler to manipulate mathematically. The actual field is obtained by taking the real part of the exponential field for the sinusoidal source.

Maxwell's equations for fields varying as $e^{j\omega t}$ are as follows.

$$\nabla \times \vec{E} = -j\omega\mu\vec{H} \tag{3a}$$

$$\nabla \times \vec{H} = \vec{J}_s + j\omega\varepsilon\vec{E} \tag{3b}$$

$$\nabla \cdot \vec{D} = \rho \tag{3c}$$

$$\nabla \cdot \vec{B} = 0 \tag{3d}$$

$$\underline{\varepsilon} = \varepsilon \left(1 + \frac{\sigma}{j\omega\varepsilon} \right) \tag{3e}$$

2. Guided Waves

Maxwell's equations for monochromatic fields in a medium without any sources are

$$\nabla \times \vec{E} = -j\omega\mu\vec{H} \tag{1a}$$

$$\nabla \times \vec{H} = j\omega\underline{\varepsilon}\vec{E} \qquad \underline{\varepsilon} = \varepsilon \left[1 + \frac{\sigma}{j\omega\varepsilon} \right] \tag{1b}$$

$$\nabla \cdot \vec{D} = 0 \tag{1c}$$

$$\nabla \cdot \vec{B} = 0 \tag{1d}$$

To obtain an equation involving only one of the fields, calculate the curl of (1a) and then use (1b) to eliminate $\vec{H}$.

$$\nabla \times \nabla \times \vec{E} = -j\omega\mu \, \nabla \times \vec{H} = -j\omega\mu \, (\, j\omega\underline{\varepsilon} \, \vec{E} \,)$$

$$= \gamma_0^2 \vec{E}$$

$$\nabla \, \nabla \cdot \vec{E} - \nabla^2\vec{E} - \gamma_0^2\vec{E} = 0$$

$$\nabla^2\vec{E} + \gamma_0^2 \, \vec{E} = 0 \tag{2a}$$

$$\gamma_0^2 \equiv \omega^2\mu\underline{\varepsilon} \tag{2b}$$

$$\nabla \times \nabla \times \vec{G} = \nabla\nabla \cdot \vec{G} - \nabla^2\vec{G} \tag{2c}$$

$$\vec{A} \times (\vec{B} \times \vec{C}) = \vec{B}(\vec{A} \cdot \vec{C}) - \vec{C}(\vec{A} \cdot \vec{B}) \tag{2d}$$

since $\nabla \cdot \vec{D} = \varepsilon \, \nabla \cdot \vec{E} = 0$. Equation (2c) can be obtained from the "BAC-CAB" (pronounced as back-cab) rule, (2d), provided the operators are always placed on the left-hand side of the function to be operated on.

Similarly, one can take the curl of (1b) and use (1a) to eliminate $\vec{E}$; one then gets

$$\nabla^2\vec{H} + \gamma_0^2\vec{H} = 0 \tag{2e}$$

The analysis can be greatly simplified if all fields are decomposed into transverse and axial components. Let

$$\vec{E}(x, y, z) = \vec{e}(x, y) \, e^{\mp j\gamma_z z} = (\, \vec{e}_t + \vec{e}_z \,) \, e^{\mp j\gamma_z z} \tag{3a}$$

$$\vec{H}(x, y, z) = \vec{h}(x, t) \, e^{\mp j\gamma_z z} = (\, \vec{h}_t + \vec{h}_z \,) \, e^{\mp j\gamma_z z} \tag{3b}$$

$$\vec{e}_t \equiv \hat{x}e_x + \hat{y}e_y, \qquad \vec{e}_z \equiv \hat{z}e_z \tag{3c}$$

$$\vec{h}_t \equiv \hat{x}h_x + \hat{y}h_y, \qquad \vec{h}_z = \hat{z}h_z \tag{3d}$$

$$\nabla = \nabla_t + \hat{z}\frac{\partial}{\partial z} = \nabla_t \mp \hat{z}j\gamma_z \tag{3e}$$

$$\nabla_t \equiv \hat{x}\frac{\partial}{\partial x} + \hat{y}\frac{\partial}{\partial y} \tag{3f}$$

where γ_z is the propagation constant of the guide along the longitudinal z-direction of the guide. It is to be noted that the decomposing of fields into transverse and axial components is a general technique. The use of Cartesian coordinates is to clarify any uncertainty. Equations (3a) and (3b) imply that fields within the guide are traveling waves along the axial direction of the guide. Then (1a) becomes

$$\nabla \times \vec{E} = [\nabla_t \mp j\gamma_z\hat{z}] \times (\vec{e}_t + \vec{e}_z) = -j\omega\mu(\vec{h}_t + \vec{h}_z)$$

$$\nabla_t \times \vec{e}_t + \nabla_t \times \vec{e}_z \mp j\gamma_z\hat{z} \times \vec{e}_t = -j\omega\mu(\vec{h}_t + \vec{h}_z)$$

Axial component: $\quad \nabla_t \times \vec{e}_t = -j\omega\mu\vec{h}_z \tag{4a}$

Transverse components: $\quad \nabla_t \times \vec{e}_z \mp j\gamma_z\hat{z} \times \vec{e}_t = -j\omega\mu\vec{h}_t \tag{4b}$

One has from (1b)

$$\nabla \times \vec{H} = [\nabla_t \mp j\hat{z}\gamma_z] \times (\vec{h}_t + \vec{h}_z) = j\omega\varepsilon(\vec{e}_t + \vec{e}_z)$$

$$\nabla_t \times \vec{h}_t + \nabla_t \times \vec{h}_z \mp j\gamma_z\hat{z} \times \vec{h}_t = j\omega\varepsilon(\vec{e}_t + \vec{e}_z)$$

Axial component: $\quad \nabla_t \times \vec{h}_t = j\omega\varepsilon\vec{e}_z \tag{5a}$

Transverse components: $\quad \nabla_t \times \vec{h}_z \mp j\gamma_z\hat{z} \times \vec{h}_t = j\omega\varepsilon\vec{e}_t \tag{5b}$

Equation (1c) yields

$$\nabla \cdot \vec{D} = [\nabla_t \mp j\gamma_z\hat{z}] \cdot \varepsilon(\vec{e}_t + \vec{e}_z) = 0$$

$$\nabla_t \cdot \vec{e}_t = \pm j\gamma_z e_z \tag{6}$$

and (1d) gives

$$\nabla \cdot \vec{B} = [\nabla_t \mp \gamma_z\hat{z}] \cdot \mu(\vec{h}_t + \vec{h}_z) = 0$$

$$\nabla_t \cdot \vec{h}_t = \pm j\gamma_z h_z \tag{7}$$

For all practical interests, it is convenient to consider the following three types of problems:

(a) Transverse electromagnetic waves - TEM waves are characterized by the absence of field components in the direction of propagation. In this analysis the direction of propagation is chosen to be along the z-axis, and in such a case, $E_z = 0 = H_z$.

(b) Transverse electric waves - TE waves are characterized by having no electric field along the direction of propagation, i.e., $E_z = 0$.

(c)   Transverse magnetic waves - TM waves are defined by having $H_z = 0$.

These three types of waves will be treated separately.

3. Transverse Electromagnetic (TEM) Waves

Transverse electromagnetic waves are characterized by $E_z = 0 = H_z$. For vanishing axial field components, Maxwell's equations, (2-4) to (2-7), become

$$\nabla_t \times \vec{e}_t = 0 \tag{1a}$$

$$\pm \gamma_z \hat{z} \times \vec{e}_t = \omega \mu \vec{h}_t \tag{1b}$$

$$\nabla_t \times \vec{h}_t = 0 \tag{1c}$$

$$\mp \gamma_z \hat{z} \times \vec{h}_t = \omega \underline{\varepsilon} \vec{e}_t \tag{1d}$$

$$\nabla_t \cdot \vec{e}_t = 0 \tag{1e}$$

$$\nabla_t \cdot \vec{h}_t = 0 \tag{1f}$$

By virtue of (1a), one can define a scalar potential function U_e,

$$\vec{e}_t \equiv -\nabla_t U_e \tag{2}$$

since $\nabla \times \nabla f = 0$ for any scalar function f. Then (2-2a) gives

$$\nabla^2 (-\nabla_t U_e) + \gamma_o^2 (-\nabla_t U_e) = 0 \quad \text{or} \quad (-\nabla_t) [\nabla^2 U_e + \gamma_o^2 U_e] = 0$$

$$\nabla_t^2 U_e + (\gamma_o^2 - \gamma_z^2) U_e = \text{constant} \tag{3a}$$

$$\nabla^2 \equiv \nabla \cdot \nabla = \nabla_t^2 - \gamma_z^2 \tag{3b}$$

The substitution of (2) into (1e) yields

$$\nabla_t^2 U_e = 0 \tag{4}$$

With (4), (3a) reduces to

$$(\gamma_o^2 - \gamma_z^2) U_e = \text{constant} \tag{5}$$

If

$$\gamma_o^2 - \gamma_z^2 \neq 0$$

then

$$U_e = \frac{\text{constant}}{(\gamma_o^2 - \gamma_z^2)} = \text{another constant} \tag{6a}$$

since both γ_o^2 and γ_z^2 are constant at a given frequency. From (2) and (6a), one has

$$\vec{e}_t = -\nabla_t U_e = 0 \tag{6b}$$

which is a trivial solution.

For a nontrivial solution, one must have

$$\gamma_o^2 - \gamma_z^2 = 0$$

or

$$\gamma_z = \pm\gamma_o \tag{7}$$

Equation (3a) then becomes

$$\nabla_t^2 U_e = \text{constant} \tag{8}$$

Since the scalar potential U_e must also satisfy (4), one concludes that the constant in (8) should be zero. In other words, the scalar potential for TEM waves satisfies the Laplace equation

$$\nabla_t^2 U_e = 0 \tag{9}$$

In summary, the procedure to obtain the solution for TEM waves is:

(1) Find the scalar potential which satisfies (9) and the associated boundary conditions.
(2) The fields are then given by (1b) and (2).

$$\vec{e}_t = -\nabla_t U_e \tag{10a}$$

$$\vec{h}_t = -Y_{TEM}\hat{z} \times \vec{e}_t \tag{10b}$$

$$Y_{TEM} \equiv \pm\frac{\gamma_z}{\omega\mu} \tag{10c}$$

4. Transverse Electric (TE) Waves

Transverse electric waves are characterized by the vanishing axial component of the electric field, $E_z = 0$. The axial component of the magnetic field, h_z, acts as a potential function for TE waves. The Helmholtz equation for the $\vec{H}$-field is

$$\nabla^2\vec{H} + \gamma_o^2\vec{H} = 0 \tag{2-2e}$$

$$(\nabla_t^2 - \gamma_z^2)\vec{H} + \gamma_o^2\vec{H} = 0$$

$$\nabla_t^2\vec{H} + \gamma^2\vec{H} = 0 \tag{1a}$$

$$\gamma^2 \equiv \gamma_o^2 - \gamma_z^2 \tag{1b}$$

since

$$\nabla^2 = \nabla_t^2 - \gamma_z^2 \tag{3-3b}$$

and

$$\vec{H} = (\vec{h_t} + \vec{h_z})\, e^{\mp j\gamma_z z} \tag{2-3b}$$

Equation (1a) can be decomposed into two equations according to the spatial direction of each term.

$$\nabla_t^2 h_z + \gamma^2 h_z = 0 \tag{2a}$$

$$\nabla_t^2\vec{h_t} + \gamma^2\vec{h_t} = 0 \tag{2b}$$

The general procedure is to find the solution of (2a) which satisfies the associated boundary conditions and then express transverse components of fields in terms of the axial component h_z.

Maxwell's equations for TE waves are (2-4) to (2-7) subject to the condition $e_z = 0$.

$$\nabla_t \times \vec{e_t} = -j\omega\mu\vec{h_z} \tag{3a}$$

$$\pm\gamma_z\hat{z} \times \vec{e_t} = \omega\mu\vec{h_t} \tag{3b}$$

$$\nabla_t \times \vec{h_t} = 0 \tag{3c}$$

$$\nabla_t \times \vec{h_z} \mp j\gamma_z\hat{z} \times \vec{h_t} = j\omega\varepsilon\vec{e_t} \tag{3d}$$

$$\nabla_t \cdot \vec{e_t} = 0 \tag{3e}$$

$$\nabla_t \cdot \vec{h_t} = \pm j\gamma_z h_z \tag{3f}$$

The transverse components of $\vec{h_t}$ are, from (3b) and (3d),

$$\omega\mu\vec{h_t} = \pm\gamma_z\hat{z}\times\vec{e_t}$$

$$= \pm\gamma_z\hat{z}\times\frac{1}{j\omega\varepsilon}\left[\nabla_t\times(\hat{z}h_z)\mp j\gamma_z\hat{z}\times\vec{h_t}\right]$$

$$j\gamma_0^2\vec{h_t} = \pm\gamma_z\hat{z}\times(\nabla_t h_z\times\hat{z}) - j\gamma_z^2\hat{z}\times(\hat{z}\times\vec{h_t})$$

$$= \pm\gamma_z\nabla_t h_z + j\gamma_z^2\vec{h_t}$$

$$\vec{h_t} = \frac{\mp j\gamma_z}{\gamma^2}\nabla_t h_z \tag{4}$$

where γ^2 is defined by (1b).

The transverse components of the E-field are obtained from (3b).

$$\hat{z}\times\vec{e_t} = Z_{TE}\vec{h_t} \qquad \text{and} \qquad Z_{TE} \equiv \frac{\mp\omega\mu}{\gamma_z}$$

or

$$\hat{z}\times(\hat{z}\times\vec{e_t}) = Z_{TE}\hat{z}\times\vec{h_t}$$

$$\vec{e_t} = Z_{TE}\hat{z}\times\vec{h_t} \tag{5}$$

The procedure for obtaining the TE-wave solution is as follows.

(1) Find the solution of h_z which satisfies (2a) and the associated boundary conditions.

(2) The transverse components of TE-fields are then given by (4) and (5).

The complete expressions for the fields are

$$\vec{H} = \vec{H_t} + \vec{H_z} = \pm\vec{h_t}e^{\mp j\gamma_z z} + \vec{h_z}e^{\mp j\gamma_z z} \tag{6a}$$

$$\vec{E} = \vec{E_t} = Z_{TE}\hat{z}\times\vec{h_t}e^{\mp j\gamma_z z} \tag{6b}$$

$$Z_{TE} \equiv \frac{\mp\omega\mu}{\gamma_z} \tag{6c}$$

The sign in front of $\vec{h_t}$ indicates a positive (upper sign) or a negative (lower sign) traveling wave. The sign for the $\vec{E_t}$ field remains unchanged since it involves the factor γ_z twice, once in the expression for $\vec{h_t}$ and again in Z_{TE}.

Only the sign of either $\vec{e_t}$ or $\vec{h_t}$ can change when a reversal in the direction of energy flow is expected to occur. Thus, the solution for a TE wave propagating in the negative z-direction can be chosen either as

$$\vec{E}_t = -\vec{e}_t e^{j\gamma_z z} \tag{7a}$$

$$\vec{H} = (\vec{h}_t - \vec{h}_z)\, e^{j\gamma_z z} \tag{7b}$$

or

$$\vec{E}_t = \vec{e}_t e^{j\gamma_z z} \tag{8a}$$

$$\vec{H} = (-\vec{h}_t + \vec{h}_z)\, e^{j\gamma_z z} \tag{8b}$$

One choice is the negative of the other. The latter choice, (8), is arbitrarily chosen as the convention for this analysis.

5. Transverse Magnetic (TM) Waves

The transverse magnetic waves are characterized by the vanishing of the axial component of the magnetic field, $H_z = 0$. The axial component e_z plays the role of the potential function for TM waves. The Helmholtz equation for the $\vec{E}$-field is

$$\nabla^2 \vec{E} + \gamma_0^2 \vec{E} = 0 \tag{2-2a}$$

or

$$\nabla_t^2 e_z + \gamma^2 e_z = 0 \tag{1a}$$

$$\nabla_t^2 \vec{e}_t + \gamma^2 \vec{e}_t = 0 \tag{1b}$$

$$\vec{E} = (\vec{e}_t + \vec{e}_z)\, e^{\mp j\gamma_z z} \tag{1c}$$

$$\gamma^2 \equiv \gamma_0^2 - \gamma_z^2 \tag{1d}$$

The axial component of the electric field, e_z, is determined by (1a) and the associated boundary conditions.

Maxwell's equations for TM waves are (2-4) to (2-7) subject to the condition $h_z = 0$.

$$\nabla_t \times \vec{e}_t = 0 \tag{2a}$$

$$\nabla_t \times \vec{e}_z \mp j\gamma_z \hat{z} \times \vec{e}_t = -j\omega\mu \vec{h}_t \tag{2b}$$

$$\nabla_t \times \vec{h}_t = j\omega\underline{\varepsilon} \vec{e}_z \tag{2c}$$

$$\mp \gamma_z \hat{z} \times \vec{h}_t = \omega\underline{\varepsilon} \vec{e}_t \tag{2d}$$

$$\nabla_t \cdot \vec{e}_t = \pm j\gamma_z e_z \tag{2e}$$

$$\nabla_t \cdot \vec{h}_t = 0 \tag{2f}$$

The transverse components of $\vec{e}_t$ can be obtained from (2d) and (2b).

$$\omega\underline{\varepsilon} \vec{e}_t = \mp \gamma_z \hat{z} \times \vec{h}_t$$

$$= \mp \gamma_z \hat{z} \times \frac{1}{-j\omega\mu}\, [\, \nabla_t \times (\hat{z} e_z) \mp j\gamma_z \hat{z} \times \vec{e}_t \,]$$

$$-j\gamma_0^2\vec{e}_t \;=\; \mp\gamma_z\hat{z}\times(\,\nabla_t e_z\times\hat{z}\,)\;+\;j\gamma_z^2\hat{z}\times(\,\hat{z}\times\vec{e}_t\,)$$

$$\qquad\;=\; \mp\gamma_z\,\nabla_t e_z \;-\; j\gamma_z^2\vec{e}_t$$

$$\vec{e}_t \;=\; \mp\,\frac{j\gamma_z}{\gamma^2}\,\nabla_t e_z \tag{3}$$

The transverse components of the H-field are, from (2d),

$$\hat{z}\times\vec{h}_t \;=\; Y_{TM}\,\vec{e}_t$$

$$\hat{z}\times(\,\hat{z}\times\vec{h}_t\,)\;=\; Y_{TM}\,\hat{z}\times\vec{e}_t$$

$$\vec{h}_t \;=\; -Y_{TM}\hat{z}\times\vec{e}_t \tag{4a}$$

$$Y_{TM} \;\equiv\; \frac{\mp\omega\varepsilon}{\gamma_z} \;=\; Z_{TM}^{-1} \tag{4b}$$

To obtain the TM solution:

    (1)   Solve for e_z from (1a) subject to the associated boundary conditions.

    (2)   The transverse components of the fields are given by (3) and (4).

The complete expressions for the fields are:

$$\vec{E} \;=\; \vec{E}_t + \vec{E}_z \;=\; \left[\,\frac{\mp j\gamma_z}{\gamma^2}\,\nabla_t e_z \;\pm\; \vec{e}_z\,\right] e^{\mp j\gamma_z z} \tag{5a}$$

$$\vec{H} \;=\; \vec{H}_t \;=\; -Z_{TM}^{-1}\,\hat{z}\times\vec{e}_t e^{\mp j\gamma_z z} \tag{5b}$$

It is convenient to keep the sign of $\vec{e}_t$ the same for propagation in both the positive and the negative z-direction. Then, since

$$\nabla\cdot\vec{E} \;=\; 0 \;=\; \nabla_t\cdot\vec{E}_t \;+\; \frac{\partial E_z}{\partial z} \tag{6}$$

this requires that the z-component of the electric field be

$$-\vec{e}_z e^{j\gamma_z z}$$

for a wave propagating in the negative z-direction. This is because $\nabla_t\cdot\vec{E}_t$ does not change sign whereas $\partial E_z/\partial z$ does, in view of the change in the sign in front of γ_z in the exponential factor $e^{\pm j\gamma_z z}$.

The transverse magnetic field must also change sign upon reversal of the direction of propagation in order to obtain a change in the direction of energy flow.

This sign convention can be summarized as follows. The transverse variations of the fields are represented by $\vec{e}_t$, $\vec{h}_t$, $\vec{e}_z$, and $\vec{h}_z$; and these are independent of the direction of propagation.

Waves propagating in the positive z-direction are given by

$$\vec{E}^+ = (\vec{e}_t + \vec{e}_z)\, e^{-j\gamma_z z} \tag{7a}$$

$$\vec{H}^+ = (\vec{h}_t + \vec{h}_z)\, e^{-j\gamma_z z} \tag{7b}$$

For the wave propagating in the negative z-direction,

$$\vec{E}^- = (\vec{e}_t - \vec{e}_z)\, e^{j\gamma_z z} \tag{8a}$$

$$\vec{H}^- = (-\vec{h}_t + \vec{h}_z)\, e^{j\gamma_z z} \tag{8b}$$

6. General Case

Solutions for guided waves have been obtained for each of the following cases: TEM, TE, and TM waves. Such classification is entirely artificial and is done to simplify the procedure of solution. In the general case where axial components of both electric and magnetic fields exist, the solution is obtained by the summation of TE and TM solutions.

The total transverse components of the electric field are given by the sum of (4-6b) and (5-3).

$$\vec{e}_t = \frac{\omega\mu}{\mp\gamma_z} \hat{z} \times \vec{h}_t + \frac{\mp j\gamma_z}{\gamma^2} \nabla_t e_z$$

$$= \frac{\omega\mu}{\mp\gamma_z} \hat{z} \times \frac{\mp j\gamma_z}{\gamma^2} \nabla_t h_z + \frac{\mp j\gamma_z}{\gamma^2} \nabla_t e_z$$

$$= \frac{\mp j\gamma_z}{\gamma^2} \nabla_t e_z + \frac{j\gamma_0}{\gamma^2} \hat{z} \times \nabla_t(\eta h_z) \tag{1}$$

where

$$\eta \equiv \sqrt{\frac{\mu}{\varepsilon}} \tag{2a}$$

$$\gamma_0 = \omega \sqrt{\mu\varepsilon} \tag{2b}$$

$$\gamma^2 = \gamma_0^2 - \gamma_z^2 \tag{2c}$$

Equation (4-4) is used to eliminate $\vec{h}_t$.

The total transverse components of the magnetic fields are given by the sum of (4-4) and (5-4a).

$$\vec{h}_t = \frac{\mp j\gamma_z}{\gamma^2} \nabla_t h_z - \frac{\omega}{\mp\gamma_z} \varepsilon \hat{z} \times \vec{e}_t$$

$$\sqrt{\frac{\mu}{\varepsilon}} \vec{h}_t = \frac{\mp j\gamma_z}{\gamma^2} \nabla_t(\eta h_z) - \frac{\omega}{\mp\gamma_z} \varepsilon \sqrt{\frac{\mu}{\varepsilon}} \hat{z} \times \frac{\mp j\gamma_z}{\gamma^2} \nabla_t e_z$$

$$\eta\vec{h}_t = \frac{\mp j\gamma_z}{\gamma^2} \nabla_t(\eta h_z) - \frac{j\gamma_0}{\gamma^2} \hat{z} \times \nabla_t e_z \tag{3}$$

Equation (5-3) is used to eliminate $\vec{e}_t$.

7. Group Velocity

The phase velocity of an equiphase front of a single-frequency uniform plane wave is given by

$$v_p = \frac{\omega}{\beta} \quad \text{m/s} \tag{1}$$

In a lossless medium, the phase constant, $\beta = \omega\sqrt{\mu\varepsilon}$, is a linear function of the frequency ω. Consequently, the phase velocity, $v_p = 1/\sqrt{\mu}\,\varepsilon = $ constant, is independent of the frequency.

However, in lossy media, with ε replaced by $\underline{\varepsilon} = \varepsilon\,[1 - (j\sigma/\omega\varepsilon)]$, it is clearly indicated by (1) that the phase constants are not linear functions of ω. In such cases, waves at different frequencies will propagate with different phase velocities. Since all information-carrying signals are made of a band of frequencies, each frequency component of the signal wave travels with a different phase velocity and produces distortions in the signal waveform at the destination. This distortion of signal is known as dispersion.

Consider a wave packet made up of two traveling waves of equal magnitude but operating at different frequencies, $\omega_1 = \omega + \Delta\omega$ and $\omega_2 = \omega - \Delta\omega$ with $\Delta\omega \ll \omega$. Then

$$E = E_0 \{\cos [(\omega + \Delta\omega)t - (\beta + \Delta\beta)s] + \cos [(\omega - \Delta\omega)t - (\beta - \Delta\beta)s]\}$$

$$= E_0 \{\cos [(\omega t - \beta s) + (\Delta\omega t - \Delta\beta s)] + \cos [(\omega t - \beta s) - (\Delta\omega t - \Delta\beta s)]\} \tag{2}$$

where s is the distance of propagation. For a dispersive medium, β is a function of ω and $\Delta\beta$ is associated with the change of frequency $\Delta\omega$. By the use of the trigonometric identity

$$\cos (A \pm B) = \cos A \cos B \mp \sin A \sin B \tag{3}$$

Equation (2) becomes

$$E = 2E_0 \cos (\omega t - \beta s) \cos (\Delta\omega t - \Delta\beta s)$$

$$= 2E_0 \cos \omega(t - \frac{\beta}{\omega}s) \cos \Delta\omega(t - \frac{\Delta\beta}{\Delta\omega}s) \equiv 2E_0 F(\omega)G(\Delta\omega) \tag{4}$$

This is a wave of high angular frequency ω modulated by a wave of low angular frequency $\Delta\omega$. The high-frequency wave is represented by the factor

$$F(\omega) \equiv \cos \omega(t - \frac{\beta}{\omega}s) \tag{5}$$

and its phase velocity at any constant phase point, $\omega t - \beta s = $ constant, is given by

$$v_p = \frac{ds}{dt} = \frac{\omega}{\beta} \tag{6}$$

The slow-varying envelope is given by the factor

$$G(\Delta\omega) \equiv \cos (\Delta\omega t - \Delta\beta s) \tag{7}$$

and the corresponding phase velocity of the constant phase point, $\Delta\omega t - \Delta\beta s = $ constant, is

$$v_g = \frac{ds}{dt} = \frac{\Delta\omega}{\Delta\beta} = \frac{1}{\Delta\beta/\Delta\omega} \tag{8}$$

This is the velocity of a point on the envelope of the wave pocket and is known as the group velocity of the wave pocket. In the limit that $\Delta\omega \to 0$, (8) becomes

$$v_g = \frac{1}{d\beta/d\omega} \tag{9}$$

The relation between v_p and v_g can be obtained as follows.

$$\frac{d\beta}{d\omega} = \frac{d}{d\omega}\frac{\omega}{v_p} = \frac{1}{v_p} - \frac{\omega}{v_p^2}\frac{dv_p}{d\omega}$$

$$v_g = \frac{1}{d\beta/d\omega} = \left[\frac{1}{v_p} - \frac{\omega}{v_p^2}\frac{dv_p}{d\omega}\right]^{-1} = v_p\left[1 - \frac{\omega}{v_p}\frac{dv_p}{d\omega}\right]^{-1} \tag{10}$$

There are three possible cases:

(a)   Non-dispersive case - The phase velocity is independent of ω, or

$$\frac{dv_p}{d\omega} = 0 \tag{11a}$$

and (10) yields

$$v_g = v_p \tag{11b}$$

(b)   Normal dispersion - This is defined by the condition

$$\frac{dv_p}{d\omega} < 0 \tag{12a}$$

$$v_g < v_p \tag{12b}$$

The phase velocity decreases with the angular frequency ω.

(c)   Anomalous dispersion - This case is defined by

$$\frac{dv_p}{d\omega} > 0 \tag{13a}$$

and

$$v_g > v_p \tag{13b}$$

The phase velocity increases with the angular frequency ω.

8. Propagation Constant

The propagation constant along the axis of the guide is [Eq. (4-1b)]

$$\gamma_z = \sqrt{\gamma_o^2 - \gamma^2} = \sqrt{\omega^2 \mu \varepsilon - \gamma^2} \tag{1}$$

where γ is a parameter depending upon the geometry of the cross section of the guide. For a given guide, γ remains constant for a specific mode. When the angular frequency ω is allowed to vary, three possible cases arise.

(a) Cutoff - The cutoff is defined by the condition $(\omega^2 \mu \varepsilon - \gamma^2)|_{\omega = \omega_c} = 0$. Then at cutoff frequency, $\omega = \omega_c$ or $f = f_c$, the propagation constant, (1), vanishes. Consequently, there will be no propagation at this frequency.

$$\omega_c = \gamma c = 2\pi f_c \tag{2a}$$

$$c \equiv \frac{1}{\sqrt{\mu \varepsilon}} \tag{2b}$$

(b) Propagation mode - When the frequency is high enough such that $\omega^2 \mu \varepsilon > \gamma^2$, the propagation constant is real and propagation without attenuation exists.

$$\gamma_z = \omega \sqrt{\mu \varepsilon} \sqrt{1 - (f_c/f)^2} \equiv \beta_z \tag{3a}$$

$$\gamma^2 = \omega_c^2 \mu \varepsilon \tag{3b}$$

where f_c is the cutoff frequency which is defined by (2a).

(c) Evanescent mode - At frequencies below cutoff, $\omega^2 \mu \varepsilon - \gamma^2 < 0$, the propagation constant becomes purely imaginary.

$$\gamma_z = j\omega_c \sqrt{\mu \varepsilon} \sqrt{1 - (f/f_c)^2} \tag{4}$$

For the propagation mode, $f > f_c$, the phase velocity is [Eq. (3a)]

$$v_p = \frac{\omega}{\beta_z} = \frac{c}{\sqrt{1 - (f_c/f)^2}} \tag{5}$$

The group velocity can be obtained from (3a).

$$\beta_z = \sqrt{\mu \varepsilon} \sqrt{\omega^2 - \omega_c^2}$$

$$d\beta_z = \frac{1}{c} \frac{\omega d\omega}{\sqrt{\omega^2 - \omega_c^2}}$$

$$v_g = \frac{d\omega}{d\beta_z} = c \sqrt{1 - (f_c/f)^2} \tag{6}$$

9. Electromagnetic Energy

The work done, $\Delta \underline{w}_f$, by the electromagnetic field to move a charged particle $\Delta \underline{q}$ a distance $\Delta \vec{\underline{L}}_q$ is

$$\Delta \underline{w}_f = \Delta \vec{\underline{F}} \cdot \Delta \vec{\underline{L}}_q = \Delta \underline{q} \, (\vec{\underline{E}} + \vec{\underline{v}}_q \times \vec{\underline{B}}) \cdot \Delta \vec{\underline{L}}_q \tag{1}$$

where the terms within parentheses represent the Lorentz force, and $\underline{v}_q$ is the velocity of the charged particle. Let the work be accomplished within the time interval Δt; then the power $\Delta \underline{P}_f$ supplied by the field is

$$\Delta \underline{P}_f = \frac{\Delta \underline{w}_f}{\Delta t} = \Delta \underline{q} \, (\vec{\underline{E}} + \vec{\underline{v}}_q \times \vec{\underline{B}}) \cdot \frac{\Delta \vec{\underline{L}}_q}{\Delta t} \tag{2}$$

In the limit as Δt approaches zero,

$$\delta \underline{P}_f = \frac{d \underline{w}_f}{dt} = \Delta \underline{q} \, (\vec{\underline{E}} + \vec{\underline{v}}_q \times \vec{\underline{B}}) \cdot \vec{\underline{v}}_q$$

$$= \Delta \underline{q} \, \vec{\underline{E}} \cdot \vec{\underline{v}}_q \tag{3a}$$

$$\vec{\underline{v}}_q \equiv \lim_{\Delta t \to 0} \frac{\Delta \vec{\underline{L}}_q}{\Delta t} \tag{3b}$$

For a continuous distribution of charges, $\Delta \underline{q}$ is the total charge within the volume $\Delta \tau$. The power density $\underline{P}_f$ delivered by the field to charges $\Delta \underline{q}$ is

$$\underline{P}_f = \lim_{\Delta \tau \to 0} \frac{\delta \underline{P}_f}{\Delta \tau} = \lim_{\Delta \tau \to 0} \frac{\Delta \underline{q}}{\Delta \tau} \, \vec{\underline{v}}_q \cdot \vec{\underline{E}}$$

$$= \underline{P} \vec{\underline{v}}_q \cdot \vec{\underline{E}} = \vec{\underline{J}} \cdot \vec{\underline{E}} \tag{4}$$

where $\vec{\underline{J}} = \underline{P} \vec{\underline{v}}_q$ is the current density and $\underline{P}$ is the volume charge density.

The total power delivered by the field to the entire volume Λ is

$$\underline{P} = \int_\Lambda \underline{P}_f \, d\tau = \int_\Lambda \vec{\underline{J}} \cdot \vec{\underline{E}} \, d\tau \tag{5}$$

where $d\tau$ is the differential element of volume.

By the principle of energy conservation, any power supplied by the field should be either equal to the increase in the net energy in the system or equal to the flow of power carried by the system from its sources.

10. Poynting Theorem

(a) Instantaneous Poynting Vector

The power supplied by the field is

$$\underline{P} = \int_\Lambda \vec{\underline{J}} \cdot \vec{\underline{E}} \, d\tau$$

$$= \int_\Lambda (\nabla \times \vec{\underline{H}} - \frac{\partial}{\partial t}\vec{\underline{D}}) \cdot \vec{\underline{E}} \, d\tau$$

$$= \int_\Lambda (\nabla \times \vec{\underline{H}} \cdot \vec{\underline{E}} - \frac{\partial \underline{w}_e}{\partial t}) \, d\tau \tag{1a}$$

$$\frac{\partial \underline{w}_e}{\partial t} \equiv \frac{\partial}{\partial t}(\frac{1}{2} \vec{\underline{D}} \cdot \vec{\underline{E}}) = \vec{\underline{E}} \cdot \frac{\partial}{\partial t}\vec{\underline{D}} \tag{1b}$$

where $\underline{w}_e$ is the instantaneous electric energy density. The first term in (1a) is one term in the following identity.

$$\nabla \cdot (\vec{\underline{E}} \times \vec{\underline{H}}) = \vec{\underline{H}} \cdot \nabla \times \vec{\underline{E}} - \vec{\underline{E}} \cdot \nabla \times \vec{\underline{H}} \tag{2}$$

Then (1a) becomes

$$\underline{P} = \int_\Lambda \left[(\vec{\underline{H}} \cdot \nabla \times \vec{\underline{E}} - \nabla \cdot \vec{\underline{E}} \times \vec{\underline{H}}) - \frac{\partial \underline{w}_e}{\partial t} \right] d\tau$$

$$= \int_\Lambda \left[-\vec{\underline{H}} \cdot \frac{\partial}{\partial t}\vec{\underline{B}} - \frac{\partial \underline{w}_e}{\partial t} \right] d\tau - \oint_S \vec{\underline{E}} \times \vec{\underline{H}} \cdot \hat{n} \, da$$

$$= -\int_\Lambda \frac{\partial \underline{w}_s}{\partial t} \, d\tau - \oint_S \vec{\underline{S}} \cdot \hat{n} \, da \tag{3a}$$

where $\oint_S$ means the integration over a closed surface S.

$$\frac{\partial \underline{w}_m}{\partial t} \equiv \frac{1}{2} \frac{\partial}{\partial t} (\vec{\underline{B}} \cdot \vec{\underline{H}}) = \vec{\underline{H}} \cdot \frac{\partial}{\partial t} \vec{\underline{B}} \tag{3b}$$

$$\underline{w}_s \equiv \underline{w}_m + \underline{w}_e \tag{3c}$$

$$\vec{\underline{S}} \equiv \vec{\underline{E}} \times \vec{\underline{H}} = \text{Poynting vector} \tag{3d}$$

Equation (3) can be rearranged as [Eq. (9-5)]

$$\underline{P} = \int_\Lambda \frac{\partial \underline{w}_f}{\partial t} \, d\tau \tag{4a}$$

$$= \int_\Lambda \underline{\vec{J}} \cdot \underline{\vec{E}} \, d\tau \tag{4b}$$

$$= -\int_\Lambda \frac{\partial \underline{w}_s}{\partial t} \, d\tau - \oint_S \underline{\vec{S}} \cdot \hat{n} \, da \tag{4c}$$

Equating (4a) to (4c) yields

$$-\oint_S \underline{\vec{S}} \cdot \hat{n} \, da = \frac{\partial}{\partial t} \int_\Lambda (\underline{w}_f + \underline{w}_s) \, d\tau \tag{5}$$

The Poynting vector may be interpreted as the power density of the electromagnetic field. The total inflow of electromagnetic power through the closed surface S is equal to the time rate of increase of the total energy in the volume Λ enclosed by the closed surface S.

Equating (4b) to (4c) gives

$$-\oint_S \underline{\vec{S}} \cdot \hat{n} \, da = \int_\Lambda \underline{\vec{J}} \cdot \underline{\vec{E}} \, d\tau + \frac{\partial}{\partial t} \int_\Lambda \underline{w}_s \, d\tau \tag{6}$$

Equation (6) is equivalent to (5) expressed in terms of field quantities.

The current density is composed of the current density from the sources, $\underline{\vec{J}}_{so}$, and the conduction current density, $\underline{\vec{J}}_c = \sigma \underline{\vec{E}}$. Hence,

$$-\oint_S \underline{\vec{S}} \cdot \hat{n} \, da = \int_\Lambda \underline{\vec{J}}_{so} \cdot \underline{\vec{E}} \, d\tau + \sigma \int_\Lambda \underline{\vec{E}} \cdot \underline{\vec{E}} \, d\tau + \frac{\partial}{\partial t} \int_\Lambda \underline{w}_s \, d\tau \tag{7a}$$

or

$$-\int_\Lambda \underline{\vec{J}}_{so} \cdot \underline{\vec{E}} \, d\tau = \oint_S \underline{\vec{S}} \cdot \hat{n} \, da + \sigma \int_\Lambda \underline{\vec{E}} \cdot \underline{\vec{E}} \, d\tau + \frac{\partial}{\partial t} \int_\Lambda \underline{w}_s \, d\tau \tag{7b}$$

The term on the left-hand side of (7b) represents the power supplied by the source and this is a positive quantity since $\underline{\vec{J}}_{so}$ and $\underline{\vec{E}}$ are oppositely directed in the source region. The first term on the right is the total outflow of the Poynting flux through the surface S. The second term on the right is the joule losses in the volume Λ. The last term is the rate of increase of stored energy within the volume. In other words, (7) is the energy balance relation for the system.

(b) Time Average Poynting Vector

The harmonic time dependence $e^{j\omega t}$ has been used for all field quantities. That is, fields have the general form

$$\underline{\vec{G}}(\vec{r}, t) \equiv \vec{G}(\vec{r}) e^{j\omega t} \tag{8}$$

The real physical field, $\underline{\vec{G}}^P(\vec{r}, t)$, is the real (or imaginary) part of a complex vector parameter.

$$\underline{\vec{G}}^P(\vec{r}, t) \equiv \vec{G}(\vec{r}) \cos \omega t = \text{Re} \, [\vec{G}(\vec{r}) e^{j\omega t}] \tag{9}$$

The instantaneous Poynting vector is then

$$\underline{\vec{S}} = \text{Re} \, [\, \underline{\vec{E}} \,] \times \text{Re} \, [\, \underline{\vec{H}} \,]$$

$$= \frac{1}{2} \, (\underline{\vec{E}} + \underline{\vec{E}}^*) \times \frac{1}{2} \, (\underline{\vec{H}} + \underline{\vec{H}}^*)$$

$$= \frac{1}{4} \, \{ \, [\, \underline{\vec{E}} \times \underline{\vec{H}} + (\underline{\vec{E}} \times \underline{\vec{H}})^* \,] + [\, \underline{\vec{E}} \times \underline{\vec{H}}^* + (\underline{\vec{E}} \times \underline{\vec{H}}^*)^* \,] \, \}$$

$$= \frac{1}{2} \, \text{Re} \, [\, \underline{\vec{E}} \times \underline{\vec{H}} + \underline{\vec{E}} \times \underline{\vec{H}}^* \,] \tag{10}$$

The time-average Poynting vector is the average value of the integral of $\underline{\vec{S}}$ over a period T.

$$\langle \underline{\vec{S}} \rangle \equiv \frac{1}{T} \int_0^T \underline{\vec{S}} \, dt = \frac{1}{T} \int_0^T \frac{1}{2} \, \text{Re} \, [\, \underline{\vec{E}} \times \underline{\vec{H}}^* + \underline{\vec{E}} \times \underline{\vec{H}} \,] \, dt \tag{11}$$

With field vectors expressed as

$$\underline{\vec{E}}(\vec{r}, t) = \vec{E}(\vec{r}) e^{j\omega t} \tag{12a}$$

$$\underline{\vec{H}}(\vec{r}, t) = \vec{H}(\vec{r}) e^{j(\omega t + \theta)} \tag{12b}$$

then (11) becomes

$$\langle \underline{\vec{S}} \rangle = \frac{1}{T} \int_0^T \frac{1}{2} \, \text{Re} \, [\, \vec{E}(\vec{r}) \times \vec{H}^*(\vec{r}) e^{-j\theta} + \vec{E}(\vec{r}) \times \vec{H} e^{j(2\omega t + \theta)}] \, dt$$

$$= \frac{1}{2} \, \text{Re} \, [\, \underline{\vec{E}} \times \underline{\vec{H}}^*] \tag{13}$$

(c) Complex Poynting Vector

The complex Poynting vector $\underline{\vec{E}} \times \underline{\vec{H}}^*$ can be obtained from the following expression.

$$\nabla \cdot \underline{\vec{E}} \times \underline{\vec{H}}^* = \underline{\vec{H}}^* \cdot \nabla \times \underline{\vec{E}} - \underline{\vec{E}} \cdot \nabla \times \underline{\vec{H}}^*$$

$$= \underline{\vec{H}}^* \cdot (-\partial \vec{B}/\partial t) - \underline{\vec{E}} \cdot (\vec{J}^* + \partial \vec{D}^*/\partial t)$$

$$= -j\omega \vec{B} \cdot \vec{H}^* - \underline{\vec{E}} \cdot \vec{J}^* + j\omega \underline{\vec{E}} \cdot \vec{D}^* \tag{14}$$

The above expression is for fields which have the form given in (8); and Maxwell's equations, (1-1), are used to obtain the final form in (14).

Let the medium in the volume Λ be characterized by permittivity $\varepsilon = \varepsilon_r - j\varepsilon_i$, permeability $\mu = \mu_r - j\mu_i$ and conductivity σ. Then (14) becomes

$$\nabla \cdot \vec{\underline{E}} \times \vec{\underline{H}}^* = \omega[-j(\mu_r - j\mu_i)\,\vec{\underline{H}} \cdot \vec{\underline{H}}^* + j(\varepsilon_r - j\varepsilon_i)^*\,\vec{\underline{E}} \cdot \vec{\underline{E}}^*] - \vec{\underline{E}} \cdot (\vec{\underline{J}}_{so}^* + \vec{\underline{J}}_c^*)$$

$$= \omega[-(j\mu_r + \mu_i)\,\vec{\underline{H}} \cdot \vec{\underline{H}}^* + (j\varepsilon_r - \varepsilon_i)\vec{\underline{E}} \cdot \vec{\underline{E}}^*]$$

$$- \vec{\underline{E}} \cdot \vec{\underline{J}}_{so}^* - \sigma \vec{\underline{E}} \cdot \vec{\underline{E}}^*$$

$$= j\omega(\varepsilon_r \vec{\underline{E}} \cdot \vec{\underline{E}}^* - \mu_r \vec{\underline{H}} \cdot \vec{\underline{H}}^*) - \omega(\varepsilon_i \vec{\underline{E}} \cdot \vec{\underline{E}}^* + \mu_i \vec{\underline{H}} \cdot \vec{\underline{H}}^*)$$

$$- \vec{\underline{E}} \cdot \vec{\underline{J}}_{so}^* - \sigma \vec{\underline{E}} \cdot \vec{\underline{E}}^* \tag{15}$$

The integration of (15) over the volume Λ gives

$$-\oint_S \vec{\underline{E}} \times \vec{\underline{H}}^* \cdot \hat{n}\,da = \int_\Lambda [\vec{\underline{E}} \cdot \vec{\underline{J}}_{so}^* + \sigma \vec{\underline{E}} \cdot \vec{\underline{E}}^* + \omega(\varepsilon_i \vec{\underline{E}} \cdot \vec{\underline{E}}^* + \mu_i \vec{\underline{H}} \cdot \vec{\underline{H}}^*)]\,d\tau$$

$$- j\int_\Lambda \omega(\varepsilon_r \vec{\underline{E}} \cdot \vec{\underline{E}}^* - \mu_r \vec{\underline{H}} \cdot \vec{\underline{H}}^*)\,d\tau \tag{16}$$

The above result may be decomposed into its real and imaginary parts.

$$- \operatorname{Re}\oint_S \vec{\underline{E}} \times \vec{\underline{H}}^* \cdot \hat{n}\,da = \int_\Lambda \operatorname{Re}[\vec{\underline{E}} \cdot \vec{\underline{J}}_{so}^*]\,d\tau + \sigma\int_\Lambda \vec{\underline{E}} \cdot \vec{\underline{E}}^*\,d\tau$$

$$+ \omega\int_\Lambda [\varepsilon_i \vec{\underline{E}} \cdot \vec{\underline{E}}^* + \mu_i \vec{\underline{H}} \cdot \vec{\underline{H}}^*]\,d\tau \tag{17a}$$

$$- \operatorname{Im}\oint_S \vec{\underline{E}} \times \vec{\underline{H}}^* \cdot \hat{n}\,da = \omega\int_\Lambda [\mu_r \vec{\underline{H}} \cdot \vec{\underline{H}}^* - \varepsilon_r \vec{\underline{E}} \cdot \vec{\underline{E}}^*]\,d\tau + \int_\Lambda \operatorname{Im}[\vec{\underline{E}} \cdot \vec{\underline{J}}_{so}^*]\,d\tau \tag{17b}$$

Equation (17a) implies that the real electromagnetic power transmitted through the closed surface plus the power supplied by the sources is equal to the losses produced by the conduction current density plus the losses from the damping forces of polarizations. Because μ_i and ε_i must be positive to represent energy loss, the imaginary part of μ and ε should be negative as defined.

Equation (17b) shows that the imaginary part of the inflow of the Poynting flux through the closed surface is proportional to the net reactive energy stored in the magnetic and electric field within the volume.

Note that $(1/2)\varepsilon\vec{\underline{E}} \cdot \vec{\underline{E}}^*$ is the time-average electric energy density for fields having the form given in (8). For a positive traveling wave,

$$\vec{\underline{E}}(\vec{r}, t) = \vec{e}(x, y)\cos(\omega t - \beta z) \tag{18}$$

The energy density average over a wavelength along the z-direction is

$$\frac{1}{\lambda}\int_0^\lambda \frac{1}{2}\varepsilon \vec{\underline{E}} \cdot \vec{\underline{E}}^*\,dz = \frac{1}{2\lambda}\vec{e} \cdot \vec{e}^*\int_0^\lambda \cos^2\beta z\,dz = \frac{1}{4}\varepsilon\,R[\vec{\underline{E}} \cdot \vec{\underline{E}}^*] \tag{19}$$

11. Method of Separation of Variables

(a) Rectangular Coordinates

The solution of the scalar partial differential equation

$$\frac{\partial^2 U}{\partial x^2} + \frac{\partial^2 U}{\partial y^2} + \frac{\partial^2 U}{\partial z^2} + \gamma_o^2 U = 0 \tag{1a}$$

$$\gamma_o^2 \equiv \omega^2 \mu \varepsilon \tag{1b}$$

may be obtained by assuming a trial solution which is the product of three functions, each of which is a function of only one variable.

$$U(x, y, z) = X(x)Y(y)Z(z) \tag{2}$$

The substitution of the trial solution into (1) gives

$$YZ\frac{d^2 X}{dx^2} + XZ\frac{d^2 Y}{dy^2} + XY\frac{d^2 Z}{dz^2} + \gamma_o^2 XYZ = 0 \tag{3}$$

Dividing (3) by U yields

$$\frac{1}{X}\frac{d^2 X}{dx^2} + \frac{1}{Y}\frac{d^2 Y}{dy^2} + \frac{1}{Z}\frac{d^2 Z}{dz^2} + \gamma_o^2 = 0 \tag{4}$$

Each of the first three terms in (4) can at most be a function of only one variable. Since the sum of all these three terms must be a constant for all values of x, y, and z, it is reasoned that each term in (4) must be a constant. This may also be verified by differentiating (4) with respect to any one of the variables, say x, and yields

$$\frac{d}{dx}\left[\frac{1}{X}\frac{d^2 X}{dx^2}\right] = 0 \tag{5}$$

since only the first term in (4) contains a function of x. Equation (5) implies that

$$\frac{1}{X}\frac{d^2 X}{dx^2} = \text{constant} \equiv -\gamma_x^2 \tag{6a}$$

By a similar procedure, one can show

$$\frac{1}{Y}\frac{d^2 Y}{dy^2} = \text{constant} \equiv -\gamma_y^2 \tag{6b}$$

and

$$\frac{1}{Z}\frac{d^2 Z}{dz^2} = \text{constant} \equiv -\gamma_z^2 \tag{6c}$$

The substitution of (6) into (4) yields the condition on the constants of separation.

$$\gamma_o^2 = \gamma_x^2 + \gamma_y^2 + \gamma_z^2 \tag{7}$$

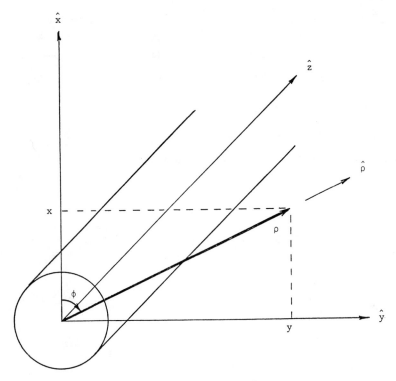

Figure 11-1: Cylindrical coordinates

Equation (6) can be written in a more familiar form.

$$\frac{d^2X}{dx^2} + \gamma_x^2 X = 0 \tag{8a}$$

$$\frac{d^2Y}{dy^2} + \gamma_y^2 Y = 0 \tag{8b}$$

$$\frac{d^2Z}{dz^2} + \gamma_z^2 Z = 0 \tag{8c}$$

These are homogeneous ordinary differential equations and the solutions are known to be

$$X(x) = X^+ e^{-j\gamma_x x} + X^- e^{j\gamma_x x} \tag{9a}$$

$$Y(y) = Y^+ e^{-j\gamma_y y} + Y^- e^{j\gamma_y y} \tag{9b}$$

$$Z(z) = Z^+ e^{-j\gamma_z z} + Z^- e^{j\gamma_z z} \tag{9c}$$

The solution of (1) is obatined by the substitution of (9) into (2).

$$U(x, y, z) = (X^+ e^{-j\gamma_x x} + X^- e^{j\gamma_x x})(Y^+ e^{-j\gamma_y y} + Y^- e^{j\gamma_y y})(Z^+ e^{-j\gamma_z z} + Z^- e^{j\gamma_z z}) \tag{10a}$$

$$\gamma_o^2 = \gamma_x^2 + \gamma_y^2 + \gamma_z^2 \tag{10b}$$

This is the general form of field solutions which are separable in rectangular coordinates. The constants of integration, $X^\pm$, $Y^\pm$, and $Z^\pm$, are to be determined from the boundary conditions.

(b) Cylindrical Coordinates

The Helmholtz equation in cylindrical coordinates, (ρ, ϕ, z), can be obtained from (1a) by a change of variables (Fig. 11-1).

$$\rho = \sqrt{x^2 + y^2} \tag{11a}$$

$$\phi = \tan^{-1}\frac{y}{x} = \cot^{-1}\frac{x}{y} \tag{11b}$$

It has the following form (see Section 11.1):

$$\frac{\partial^2 U}{\partial \rho^2} + \frac{1}{\rho}\frac{\partial U}{\partial \rho} + \frac{1}{\rho^2}\frac{\partial^2 U}{\partial \phi^2} + \frac{\partial^2 U}{\partial z^2} + \gamma_o^2 U = 0 \tag{12}$$

Let the trial solution for (12) be

$$U(\rho, \phi, z) \equiv R(\rho)\Phi(\phi)Z(z) \tag{13}$$

The substitution of (13) into (12) gives

$$\rho^2\left[\frac{1}{R}\left(\frac{d^2R}{d\rho^2} + \frac{1}{\rho}\frac{dR}{d\rho}\right) + \frac{1}{Z}\frac{d^2Z}{dz^2} + \gamma_0^2\right] = -\frac{1}{\Phi}\frac{d^2\Phi}{d\phi^2} \equiv \nu^2 \tag{14}$$

The last equality is arrived at by the same reasoning as used to obtain (6). Equation (14) can now be separated into two.

$$\frac{d^2\Phi}{d\phi^2} + \nu^2\Phi = 0 \tag{15a}$$

$$\frac{1}{R}\left[\frac{d^2R}{d\rho^2} + \frac{1}{\rho}\frac{dR}{d\rho}\right] + \gamma_0^2 - \frac{\nu^2}{\rho^2} = -\frac{1}{Z}\frac{d^2Z}{dz^2} \equiv \gamma_z^2 \tag{15b}$$

Equation (15b) can be further separated as follows.

$$\frac{d^2R}{d\rho^2} + \frac{1}{\rho}\frac{dR}{d\rho} + \left[\gamma^2 - \frac{\nu^2}{\rho^2}\right]R = 0 \tag{16a}$$

$$\frac{d^2Z}{dz^2} + \gamma_z^2 Z = 0 \tag{16b}$$

$$\gamma^2 \equiv \gamma_0^2 - \gamma_z^2 \tag{16c}$$

The solutions for (15a) and (16b) are well known.

$$\Phi(\phi) = \Phi_{1\nu}\sin\nu\phi + \Phi_{2\nu}\cos\nu\phi \tag{17a}$$

$$Z(z) = Z^+e^{-j\gamma_z z} + Z^-e^{j\gamma_z z} \tag{17b}$$

Equation (16a) can be converted into standard form by a change of variable.

$$\zeta \equiv \gamma\rho \tag{18}$$

Then

$$\frac{d^2R}{d\zeta^2} + \frac{1}{\zeta}\frac{dR}{d\zeta} + \left(1 - \frac{\nu^2}{\zeta^2}\right)R = 0 \tag{19}$$

This is known as the Bessel differential equation. Since this equation will not be used extensively in this book, only a very brief outline of the solution is presented here.

One solution of this equation can be obtained by the method of power series. The solution is assumed to have the form of a series.

$$R(\zeta) = a_0 + a_1\zeta + a_2\zeta^2 + \cdots \tag{20}$$

Substitute (20) into (19) and rearrange the resulting expression by collecting together terms of ζ with identical power. In order that the resultant series is to vanish, the sum of coefficients of each term ζ^k, $k = 1, 2, \ldots$, must also vanish. Imposing the condition that the coefficient of each ζ^k is zero, one obtains the special function known as the Bessel function of the first kind of order ν, $J_\nu(\zeta)$, where ν is any integer. For $\nu = 0$, this is

$$J_o(\zeta) = \sum_{k=o}^{\infty} (-1)^k \frac{(\zeta/2)^{2k}}{k!^2} \tag{21}$$

The second linear independent solution of (19) can be obtained by techniques of variation of parameters. In this method, the trial solution is assumed as

$$R \equiv fg \qquad f \equiv J_\nu(\zeta) \tag{22}$$

where g is an unknown function to be determined. The substitution of (22) into (19) yields

$$g\left[\frac{d^2f}{d\zeta^2} + \frac{1}{\zeta}\frac{df}{d\zeta} + \left(1 - \frac{\nu^2}{\zeta^2}\right)f\right] + f\frac{d^2g}{d\zeta^2} + \frac{dg}{d\zeta}\left(2\frac{df}{d\zeta} + \frac{f}{\zeta}\right) = 0$$

or

$$-\frac{d^2g/d\zeta^2}{dg/d\zeta} = \frac{2}{f}\frac{df}{d\zeta} + \frac{1}{\zeta} \tag{23}$$

by virtue of (19). The integration of (23) with respect to ζ gives

$$-\ln\frac{dg}{d\zeta} = 2\ln f + \ln \zeta$$

or

$$\frac{dg}{d\zeta} = \frac{1}{\zeta f^2} \tag{24}$$

and

$$g = \int \frac{d\zeta}{\zeta f^2} \tag{25}$$

Thus the second solution for (19) is

$$R = fg = J_\nu(\zeta) \int \frac{d\zeta}{\zeta f^2} \equiv N_\nu(\zeta) \tag{26}$$

This is known as the Bessel function of the second kind of order ν.

The complete solution of (19) is therefore

$$R(\zeta) = R_1 J_\nu(\zeta) + R_2 N_\nu(\zeta)$$

or

$$R(\rho) = R_1 J_\nu(\gamma\rho) + R_2 N_\nu(\gamma\rho) \tag{27}$$

where R_1 and R_2 are constants of integration.

Values of Bessel functions are tabulated in tables and textbooks. These functions have the shape of attenuated sine or cosine curves. The periodicities are approximately 2π for an argument greater than ten. Some of the lower-order Bessel functions are sketched in Fig. 11-2.

The solution of (12) is obtained by substituting (17) and (27) into (13).

$$U(\rho, \phi, z) = [R_1 J_v(\gamma\rho) + R_2 N_v(\gamma\rho)] [\Phi_1 \sin v\phi + \Phi_2 \cos v\phi] [Z^+ e^{-j\gamma_z z} + Z^- e^{j\gamma_z z}] \quad (28)$$

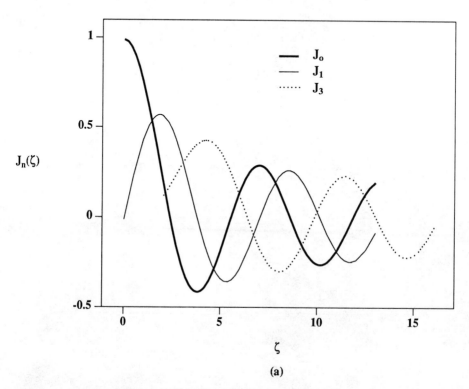

(a)

Figure 11-2(a): Bessel function of the first kind

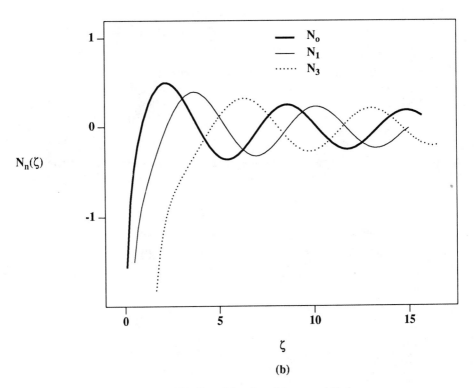

Figure 11-2(b): Bessel function of the second kind

11.1 Example - Laplace Equation in Cylindrical Coordinates

Show that the Laplace equation in cylindrical coordinates is given by

$$\nabla^2 U = \frac{1}{\rho} \frac{\partial}{\partial \rho} \left[\rho \frac{\partial}{\partial \rho} \right] U + \frac{1}{\rho^2} \frac{\partial^2 U}{\partial \phi^2} + \frac{\partial^2 U}{\partial z^2} \tag{1}$$

Solution:

The Laplace equation in rectangular coordinates is

$$\nabla^2 U = \frac{\partial^2 U}{\partial x^2} + \frac{\partial^2 U}{\partial y^2} + \frac{\partial^2 U}{\partial z^2} \tag{2}$$

Parameters in rectangular and cylindrical coordinates are related as follows (Fig. 11-1).

$$\rho = \sqrt{x^2 + y^2} \tag{2a}$$

$$\phi = \tan^{-1} \frac{y}{x} \tag{2b}$$

The partial derivatives in rectangular coordinates are transformed into derivatives in cylindrical coordinates by the following relations.

$$\frac{\partial}{\partial x} = \frac{\partial}{\partial \rho} \frac{d\rho}{dx} + \frac{\partial}{\partial \phi} \frac{d\phi}{dx} = \cos \phi \frac{\partial}{\partial \rho} - \frac{\sin \phi}{\rho} \frac{\partial}{\partial \phi} \tag{3a}$$

$$\frac{\partial}{\partial y} = \frac{\partial}{\partial \rho} \frac{d\rho}{dy} + \frac{\partial}{\partial \phi} \frac{d\phi}{dy} = \sin \phi \frac{\partial}{\partial \rho} + \frac{\cos \phi}{\rho} \frac{\partial}{\partial \phi} \tag{3b}$$

$$\frac{d\rho}{dx} = \frac{d}{dx} \sqrt{x^2 + y^2} = \frac{x}{\rho} = \cos \phi \tag{3c}$$

$$\frac{d\rho}{dy} = \frac{y}{\rho} = \sin \phi \tag{3d}$$

$$\frac{d\phi}{dx} = \frac{d}{dx} \left[\cot^{-1} \frac{x}{y} \right] = \frac{-y}{x^2 + y^2} = \frac{-\sin \phi}{\rho} \tag{3e}$$

$$\frac{d\phi}{dy} = \frac{x}{x^2 + y^2} = \frac{\cos \phi}{\rho} \tag{3f}$$

The second partial derivatives are therefore

$$\frac{\partial^2}{\partial x^2} = \left[\frac{\partial}{\partial x} \right] \left[\frac{\partial}{\partial x} \right] = \left[\cos \phi \frac{\partial}{\partial \rho} - \frac{\sin \phi}{\rho} \frac{\partial}{\partial \phi} \right]^2$$

$$\frac{\partial^2}{\partial x^2} = \cos^2 \phi \, \frac{\partial^2}{\partial \rho^2} + \frac{2 \cos \phi \sin \phi}{\rho} \left[\frac{1}{\rho} \frac{\partial}{\partial \phi} - \frac{\partial^2}{\partial \rho \, \partial \phi} \right]$$

$$+ \frac{\sin^2 \phi}{\rho} \left[\frac{\partial}{\partial \rho} + \frac{1}{\rho} \frac{\partial^2}{\partial \phi^2} \right] \tag{4a}$$

$$\frac{\partial^2}{\partial y^2} = \left[\sin \phi \, \frac{\partial}{\partial \rho} + \frac{\cos \phi}{\rho} \frac{\partial}{\partial \phi} \right]^2$$

$$= \sin^2 \phi \, \frac{\partial^2}{\partial \rho^2} + \frac{2 \sin \phi \cos \phi}{\rho} \left[\frac{\partial^2}{\partial \rho \, \partial \phi} - \frac{1}{\rho} \frac{\partial}{\partial \phi} \right] \tag{4b}$$

$$+ \frac{\cos^2 \phi}{\rho} \left[\frac{\partial}{\partial \rho} + \frac{1}{\rho} \frac{\partial^2}{\partial \phi^2} \right] \tag{4b}$$

The sum of (4a) and (4b) yields

$$\frac{\partial^2}{\partial x^2} + \frac{\partial^2}{\partial y^2} = (\cos^2 \phi + \sin^2 \phi) \frac{\partial^2}{\partial \rho^2} + \frac{1}{\rho} (\sin^2 \phi + \cos^2 \phi) \frac{\partial}{\partial \rho}$$

$$+ \frac{1}{\rho^2} (\sin^2 \phi + \cos^2 \phi) \frac{\partial^2}{\partial \phi^2}$$

$$= \frac{\partial^2}{\partial \rho^2} + \frac{1}{\rho} \frac{\partial}{\partial \rho} + \frac{1}{\rho^2} \frac{\partial^2}{\partial \phi^2}$$

$$= \frac{1}{\rho} \frac{\partial}{\partial \rho} \left[\rho \frac{\partial}{\partial \rho} \right] + \frac{1}{\rho^2} \frac{\partial^2}{\partial \phi^2} \tag{5}$$

The substitution of (5) into (2) yields (1).

12. Rectangular Waveguide

Hollow conducting pipes of rectangular cross section, Fig. 12-1, are the most commonly used waveguides. Fields within the guide are obtained by solving the appropriate wave equation.

(a) TM Wave

The wave equation to be solved is (5-1a).

$$\frac{\partial^2 e_z}{\partial x^2} + \frac{\partial^2 e_z}{\partial y^2} + \gamma^2 e_z = 0 \tag{1a}$$

$$\gamma^2 = \gamma_0^2 - \gamma_z^2, \qquad \gamma_0^2 = \omega^2 \mu \varepsilon \tag{1b}$$

The solution of (1a) is obtained by the method of separation of variables. For a positive traveling wave in the axial direction, Fig. 12-1, the solution is [Eq. (11-10)]

$$E_z(x, y, z) = e_z(x, y)e^{-j\gamma_z z} = (X^+ e^{-j\gamma_x x} + X^- e^{j\gamma_x x})(Y^+ e^{-j\gamma_y y} + Y^- e^{j\gamma_y y})\, e^{-j\gamma_z z} \tag{2}$$

The constants $X^{\pm}$ and $Y^{\pm}$ will be determined by imposing the boundary conditions. These conditions are that tangential components of the electric fields must vanish at the surface of the conducting walls at $x = 0$, $x = a$, $y = 0$, and $y = b$. By applying these conditions one at a time, (2) becomes

$$E_z(x, y, z) = E_{mn}^+ \sin \gamma_x x \sin \gamma_y y\, e^{-j\gamma_z z} \tag{3a}$$

$$\gamma_x = \frac{m\pi}{a} \qquad m = 1, 2, 3, \ldots, \tag{3b}$$

$$\gamma_y = \frac{n\pi}{b} \qquad n = 1, 2, 3, \ldots, \tag{3c}$$

$$\gamma_{zmn}^2 = \gamma_0^2 - \gamma_x^2 - \gamma_y^2 \tag{3d}$$

The corresponding transverse components of the fields are given by (5-3) and (5-4a).

(b) TE Wave

The Helmholtz equation for the TE wave is [Eq. (4-2a)]

$$\frac{\partial^2 h_z}{\partial x^2} + \frac{\partial^2 h_z}{\partial y^2} + \gamma^2 h_z = 0 \tag{4}$$

The solution which satisfies the appropriate boundary conditions at the conducting wall of the guide is

$$H_z(x, y, z) = h_z(x,y)\, e^{-j\gamma_z z} = H_{mn}^+ \cos \gamma_x x \cos \gamma_y y\, e^{-j\gamma_z z} \tag{5a}$$

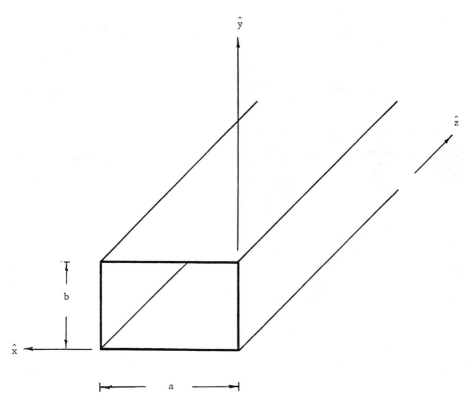

Figure 12-1: Rectangular waveguide

$$\gamma_x = \frac{m\pi}{a} \qquad m = 0, 1, 2, \ldots, \tag{5b}$$

$$\gamma_y = \frac{n\pi}{b} \qquad n = 0, 1, 2, \ldots, \tag{5c}$$

$$\gamma_{zmn}^2 = \gamma_o^2 - \gamma_x^2 - \gamma_y^2 \tag{5d}$$

Either m or n can be zero but not simultaneously. The corresponding transverse components of fields are given by (4-4) and (4-5).

(c) Propagation Constant

The propagation constant along the z-direction is given by (3d) or (5d) for TM or TE waves, respectively.

$$\gamma_{zmn} = \sqrt{\omega^2 \mu \underline{\varepsilon} - \gamma_x^2 - \gamma_y^2} \tag{6}$$

An inspection of the right-hand side of (6) reveals the important characteristic of guided waves. For a guide with specific dimensions, a and b, γ_x and γ_y are real numbers for a given mode, m and n. As the frequency is allowed to vary, terms under the square root can change from less than to equal to, to greater than zero. The cutoff condition is defined by the situation when the terms under the root add up to zero.

$$[\, \omega^2 \mu \underline{\varepsilon} - \gamma_x^2 - \gamma_y^2 \,]|_{\omega=\omega_c} = 0 \tag{7a}$$

where ω_c is the cutoff angular frequency. At cutoff, there is neither phase shift nor attenuation. There is also no propagation at cutoff.

At frequencies below cutoff frequency, $\omega < \omega_c$, the resultant quantity under the root is negative, and hence, γ_{zmn} is a purely imaginary number.

$$\gamma_{zmn} = \sqrt{\omega^2 \mu \underline{\varepsilon} - \gamma_x^2 - \gamma_y^2} \equiv -j\alpha_{zmn} \tag{7b}$$

and the factor $e^{-j\gamma_{zmn}z} = e^{-\alpha_{zmn}z}$ represents attenuation.

At frequencies above cutoff, $\omega > \omega_c$, the quantity under the root is positive, and γ_{zmn} is purely real.

$$\gamma_{zmn} = \sqrt{\omega^2 \mu \underline{\varepsilon} - \gamma_x^2 - \gamma_y^2} \equiv \beta_{zmn} \tag{7c}$$

The factor $e^{j\gamma_{zmn}z} = e^{j\beta_{zmn}z}$ represents propagation without attenuation.

The cutoff frequency is given by [Eq. (7a)]

$$\beta_{cmn} \equiv \omega_{cmn} \sqrt{\mu \underline{\varepsilon}} = \sqrt{\gamma_x^2 + \gamma_y^2} \tag{8a}$$

$$\lambda_{cmn} = \frac{2\pi}{\beta_{cmn}} = \frac{2}{\sqrt{(m/a)^2 + (n/b)^2}} \tag{8b}$$

$$f_{cmn} = \frac{\beta_{cmn}}{2\pi\sqrt{\mu\varepsilon}} = \frac{c}{\lambda_{cmn}} = \frac{c}{2}\sqrt{(m/a)^2 + (n/b)^2} \tag{8c}$$

$$c \equiv \frac{1}{\sqrt{\mu\varepsilon}} \tag{8d}$$

The propagation constant can be expressed in terms of the cutoff frequency.

$$\beta_{zmn} = \sqrt{\omega^2\mu\varepsilon - \omega_{cmn}^2\mu\varepsilon} = \omega\sqrt{\mu\varepsilon}\sqrt{1 - (f_{cmn}/f)^2} \tag{9a}$$

$$\lambda_{zmn} = \frac{2\pi}{\beta_{zmn}} = \frac{2\pi}{2\pi f\sqrt{\mu\varepsilon}\sqrt{1 - (f_{cmn}/f)^2}} = \frac{\lambda}{\sqrt{1 - (f_{cmn}/f)^2}} \tag{9b}$$

$$\lambda = \frac{c}{f} \tag{9c}$$

$$v_{zmn} = \frac{\omega}{\beta_{zmn}} = \frac{c}{\sqrt{1 - (f_{cmn}/f)^2}} \tag{9d}$$

The wave impedances are given by (4-6c) and (5-4b).

$$Z_{TE_{mn}} = \mp\frac{\omega\mu}{\gamma_{zmn}} = \frac{\omega\mu}{\omega\sqrt{\mu\varepsilon}\sqrt{1 - (f_{cmn}/f)^2}} = \frac{\sqrt{\mu/\varepsilon}}{\sqrt{1 - (f_{cmn}/f)^2}} \tag{10a}$$

$$Z_{TM_{mn}} = \mp\frac{\gamma_{zmn}}{\omega\varepsilon} = \sqrt{\frac{\mu}{\varepsilon}}\sqrt{1 - (f_{cmn}/f)^2} \tag{10b}$$

These impedances are real quantities for frequencies above cutoff, and are purely imaginary for frequencies below cutoff. Both of them approach the value of intrinsic impedance, $\eta = \sqrt{\mu/\varepsilon}$, as the operating frequencies become very large.

13. Circular Waveguide

A circular waveguide is a cylindrical guide with a circular cross section of radius b (Fig. 13-1). In view of the geometry of the guide, it will be most convenient to use cylindrical coordinates, (ρ, ϕ, z), for such problem.

(a) TM Waves

As in the case of rectangular waveguides, TM waves are obtained by finding the solution of the Helmholtz equation for e_z.

$$\nabla_t^2 e_z + \gamma^2 e_z = 0 \tag{1a}$$

$$\gamma^2 \equiv \gamma_0^2 - \gamma_z^2 \qquad \gamma_0^2 = \omega^2 \mu \varepsilon \tag{1b}$$

In cylindrical coordinates, (1a) has the following form [Eq. (11-12)]

$$\frac{\partial^2 e_z}{\partial \rho^2} + \frac{1}{\rho} \frac{\partial e_z}{\partial \rho} + \frac{1}{\rho^2} \frac{\partial^2 e_z}{\partial \phi^2} + \gamma^2 e_z = 0 \tag{2}$$

The solution of (2) is given by the product of Eqs. (11-17a) and (11-27).

$$e_z(\rho, \phi) = [\, R_1 J_\nu(\gamma \rho) + R_2 N_\nu(\gamma \rho) \,][\, \Phi_1 \sin \nu\phi + \Phi_2 \cos \nu\phi \,] \tag{3}$$

Since $N_\nu(\gamma \rho)$ is infinite at $\rho = 0$, it is not an acceptable solution for the present problem; therefore, (3) becomes

$$e_z(\rho, \phi) = [\, K_{1\nu} \sin \nu\phi + K_{2\nu} \cos \nu\phi \,] J_\nu(\gamma \rho) \tag{4}$$

The relative amplitudes of $K_{1\nu}$ and $K_{2\nu}$ determine the orientation of the fields in the guide. In a circular guide, for any value of ν, the reference $\phi = 0$ can be chosen to make either $K_1\nu$ or $K_2\nu$ disappear. To simplify the analysis, the reference is arbitrarily chosen to make $K_{1\nu} = 0$

$$e_z(\rho, \phi) = K_\nu \cos \nu\phi \, J_\nu(\gamma \rho) \tag{5}$$

The Bessel function of the first kind demonstrates a variation which approximates a damped sinusoid and equals zero at the roots $q_{\nu m}$ of the expression $J_\nu(\gamma \rho) = 0$. The subscript ν refers to the order of $J_\nu(\gamma \rho)$, and m refers to the roots in the order of their magnitudes. These roots are tabulated in many books and articles, such as Beatti (Bell Tech. J. Vol. 37, p. 689).

The boundary condition requires that e_z be zero at $\rho = b$.

$$J_\nu(\gamma b) = 0 = J_\nu(q_{\nu m})$$

That is,

$$\gamma_{\nu m}^2 = \frac{q_{\nu m}^2}{b^2} = \gamma_0^2 - \gamma_{z\nu m}^2 \tag{6a}$$

or

$$\gamma_{zvm} = \sqrt{\gamma_o^2 - (q_{vm}/b)^2} \tag{6b}$$

Additional subscripts are needed to identify various solutions.

At cutoff, $\omega = \omega_{cvm}$,

$$[\omega^2 \mu \underline{\varepsilon} - (q_{vm}/b)^2]|_{\omega=\omega_{cvm}} = 0$$

$$\beta_{cvm}^2 \equiv \omega_{cvm}^2 \mu \underline{\varepsilon} = \frac{(2\pi)^2}{\lambda_{cvm}^2} = \frac{q_{vm}^2}{b^2}$$

Therefore,

$$\lambda_{cvm} = \frac{2\pi b}{q_{vm}} \tag{7}$$

The integer v describes the number of circumferential variations of the fields. The largest value of the cutoff wavelength, λ_{cvm}, occurs at the lowest root. The lowest value of q_{vm} is the first root of the zero-order Bessel function, $q_{01} = 2.14048$. The TM_{01} wave is the dominant TM mode and the cutoff wavelength $\lambda_{c01} = 2.6129b$.

The wave impedance is given by (5-4b), with appropriate subscripts,

$$Z_{TM_{vm}} = \frac{\mp \gamma_{zvm}}{\omega \varepsilon} \tag{8}$$

With (6a), e_z can be expressed as [Eq. (5)]

$$e_{zvm}(\rho, \phi) = K_v \cos v\phi \, J_v(\gamma_{vm}\rho) \tag{9a}$$

Transverse components of fields are given by [Eqs. (5-3) and (5-4a)]

$$\vec{e}_{tvm} = \frac{\mp j\gamma_{zvm}}{\gamma_{vm}^2} [\, \hat{\rho}\frac{\partial}{\partial\rho} + \hat{\phi}\frac{1}{\rho}\frac{\partial}{\partial\phi} \,] e_z(\rho, \phi)$$

$$= \mp j \frac{\gamma_{zvm}}{\gamma_{vm}^2} K_v [\, \hat{\rho}\gamma_{vm}J_v'(\gamma_{vm}\rho) \cos v\phi - \hat{\phi}\frac{v}{\rho} \sin v\phi \, J_v(\gamma_{vm}\rho) \,] \tag{9b}$$

$$\gamma_{vm} = \frac{q_{vm}}{b} \tag{9c}$$

$$\vec{h}_{tvm} = -Y_{TM_{vm}}\hat{z} \times \vec{e}_t \tag{9d}$$

$$J_v'(k\zeta) \equiv \frac{d}{d(k\zeta)} J_v(k\zeta) = \frac{1}{k}\frac{d}{d\zeta} J_v(k\zeta) \tag{9e}$$

The complete expressions for TM_{vm} traveling waves are as follows.

$$\underline{E}_{zvm}(\vec{r}, t) = e_{zvm}(\rho, \phi) \, e^{j(\omega t \, \mp \, \gamma_{zvm}z)} \tag{10a}$$

$$\underline{\vec{E}}_{tvm}(\vec{r}, t) = \vec{e}_{tvm}(\rho, \phi) \, e^{j(\omega t \, \mp \, \gamma_{zvm}z)} \tag{10b}$$

$$\underline{\vec{H}}_{tvm}(\vec{r}, t) = - Y_{TM_{vm}} \hat{z} \times \underline{\vec{E}}_{tvm}(\vec{r}, t). \tag{10c}$$

(b) TE Waves

For transverse electric waves, the problem is to solve (2) with e_z replaced by h_z. The appropriate expression for h_z is

$$h_z(\rho, \phi) = K_v \cos v\phi \, J_v(\gamma\rho) \tag{11}$$

The transverse components of the fields are given by (4-4) and (4-5).

$$\vec{h}_t(\rho, \phi) = \frac{\mp j\gamma_z}{\gamma^2} \left[\hat{\rho}\frac{\partial}{\partial\rho} + \hat{\phi}\frac{1}{\rho}\frac{\partial}{\partial\phi} \right] h_z(\rho, \phi)$$

$$= \mp j\frac{\gamma_z}{\gamma^2} K_v [\hat{\rho}\gamma J_v'(\gamma\rho) \cos v\phi - \hat{\phi} \frac{v}{\rho} J_v(\gamma\rho) \sin v\phi] \tag{12a}$$

$$\vec{e}_t(\rho, \phi) = Z_{TE} \, \hat{z} \times \vec{h}_t$$

$$= Z_{TE}\frac{\mp j\gamma_z}{\gamma^2} K_v [\hat{\phi}\gamma J_v'(\gamma\rho) \cos v\phi + \hat{\rho} \frac{v}{\rho} J_v(\gamma\rho) \sin v\phi] \tag{12b}$$

The boundary condition requires that the tangential components of the electric field be zero at the conducting wall, $\rho = b$, i.e., $e_\phi(b, \phi) = 0$.

$$J_v'(\gamma b) = 0 \tag{13}$$

With the relations,

$$\zeta J_v'(\zeta) = v J_v(\zeta) - \zeta J_{v+1}(\zeta) \tag{14a}$$

$$J_v'(\zeta) = \frac{v}{\zeta} J_v(\zeta) - J_{v+1}(\zeta) \tag{14b}$$

$$J_0'(\zeta) = - J_1(\zeta) \tag{14c}$$

one can evaluate the roots q_{vm}' for expression (13). By virtue of (14c), TE_{0m} and TM_{1m} modes are degenerate.

At the wall of the guide, $\rho = b$,

$$J_v'(\gamma b) = 0 = J_v'(q_{vm}')$$

$$\gamma_{vm} = \frac{q'_{vm}}{b} = \sqrt{\gamma_o^2 - \gamma_{zvm}^2} \tag{15a}$$

and

$$\gamma_{zvm} = \sqrt{\gamma_o^2 - (q'_{vm}/b)^2} \tag{15b}$$

The cutoff wavelength is

$$\lambda_{cvm} = \frac{2\pi b}{q'_{vm}} \tag{16}$$

The lowest value of q'_{vm} is $q'_{11} = 1.8412$ and the corresponding cutoff wavelength is $\lambda_{c11} = 3.4125b$.

The wave impedance is [Eq. (4-6c)]

$$Z_{TE_{vm}} = \frac{\mp\omega\mu}{\gamma_{zvm}} \tag{17}$$

In view of the multiple solutions of the problem, expressions (12) should be modified as follows.

$$\vec{h}_{tvm}(\rho, \phi) = \frac{\mp j\gamma_{zvm}}{\gamma_{vm}^2} K_v [\hat{\rho}\gamma_{vm}J'_v(\gamma_{vm}\rho) \cos v\phi - \hat{\phi} \frac{v}{\rho} J_v(\gamma_{vm}\rho) \sin v\phi] \tag{18a}$$

$$\vec{e}_{tvm}(\rho, \phi) = Z_{TE_{vm}}\hat{z} \times \vec{h}_{tvm}(\rho, \phi) \tag{18b}$$

$$\gamma_{vm} = \frac{q'_{vm}}{b} \tag{18c}$$

The complete expressions for TE waves are:

$$\underline{H}_{zvm}(\vec{r}, t) = K_v \cos v\phi \, J_v(\gamma_{vm}\rho)e^{j(\omega t \mp \gamma_{zvm}z)} \tag{19a}$$

$$\underline{\vec{H}}_{tvm}(\vec{r}, t) = \frac{\mp j\gamma_{zvm}}{\gamma_{vm}^2} K_v [\hat{\rho}\gamma_{vm}J'_v(\gamma_{vm}\rho) \cos v\phi - \hat{\phi} \frac{v}{\rho} J_v(\gamma_{vm}\rho) \sin v\phi] e^{j(\omega t \mp \gamma_{zvm}z)} \tag{19b}$$

$$\underline{\vec{E}}_{tvm}(\vec{r}, t) = Z_{TE_{vm}}\hat{z} \times \underline{\vec{H}}_t(\vec{r}, t) \tag{19c}$$

13.1. Field Distribution - TM_{01} Mode in Circular Guide

Determine and sketch the field distribution of the TM_{01} mode within a circular guide.

Solution:

The field distribution for an individual mode can be obtained from the real part of the general expression, (13-10). For the TM_{01} mode, $q_{01} = 2.405$, the fields are

$$E_{z01} = K_0 J_o(\gamma_{01}\rho) \cos(\omega t \mp \gamma_{z01}z) \qquad \gamma_{01} = \frac{q_{01}}{b} \tag{1a}$$

$$E_{\rho01} = \pm E_{\rho0} J'_o(\gamma_{01}\rho) \sin(\omega t \mp \gamma_{z01}z)$$

$$= \mp E_{\rho0} J_1(\gamma_{01}\rho) \sin(\omega t \mp \gamma_{z01}z) \qquad E_{\rho0} \equiv \frac{\mp j\gamma_{z01} K_0}{\gamma_{01}} \tag{1b}$$

$$E_{\phi01} = 0 \tag{1c}$$

$$H_{\rho01} = 0 \tag{1d}$$

$$H_{\phi01} = -Y_{TM_{01}} E_{\rho01} \tag{1e}$$

where Eq. (13-9) yields

$$e_{z01} = K_0 J_o(\gamma_{01}\rho) \tag{2a}$$

$$\vec{e}_{t01} = \mp \frac{\gamma_{z01}}{\gamma_{01}} K_0 \rho J'_o(\gamma_{01}\rho) \equiv \hat{\rho} e_{\rho01} \tag{2b}$$

and

$$\gamma_{01} = \frac{q_{01}}{b} \tag{2c}$$

The field distribution can now be sketched as shown in Fig. 13.1-1.

81

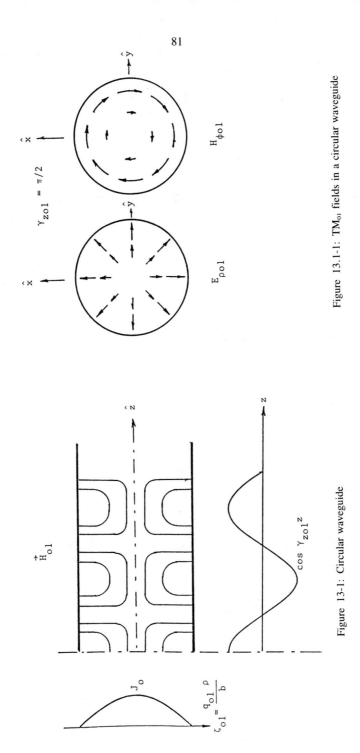

Figure 13.1-1: TM_{01} fields in a circular waveguide

Figure 13-1: Circular waveguide

13.2. TE Field Distribution - Circular Guide

Determine the field distribution of the TE_{11} mode in a circular waveguide.

Solution:

The actual fields of the TE_{11} mode are given by the real part of (13-19).

$$\underline{H}_{z11} = K_1 \cos \phi \, J_1(\gamma_{11}\rho) g(z, t) \tag{1a}$$

$$\underline{H}_{\rho11} = H_{\rho o} J_1'(\gamma_{11}\rho) \cos \phi \, f(z, t), \qquad H_{\rho o} \equiv K_1 \frac{\gamma_{z11}}{\gamma_{11}} \tag{1b}$$

$$\underline{H}_{\phi11} = H_{\phi o} J_1(\gamma_{11}\rho) \sin \phi \, f(z, t), \qquad H_{\phi o} \equiv \frac{H_{\rho o}}{\rho} \tag{1c}$$

$$\vec{\underline{E}}_{t11} = Z_{TE_{11}} \hat{z} \times (\rho \underline{H}_{\rho11} + \hat{\phi} \underline{H}_{\phi11}) = \rho \underline{E}_{\rho11} + \hat{\phi} \underline{E}_{\phi11} \tag{1d}$$

$$\underline{E}_{\rho11} = -Z_{TE_{11}} H_{\phi11} = E_{\rho o} J_1(\gamma_{11}\rho) \sin \phi \, f(z, t), \qquad E_{\rho o} = Z_{TE_{11}} H_{\phi o} \tag{1e}$$

$$\underline{E}_{\phi11} = Z_{TE_{11}} H_{\rho11} = E_{\phi o} J_1'(\gamma_{11}\rho) \cos \phi \, f(z, t), \qquad E_{\phi o} = Z_{TE_{11}} H_{\rho o} \tag{1f}$$

$$g(z, t) = \cos (\omega t \mp \gamma_{z11}z) \qquad \text{and} \qquad f(z, t) = \sin (\omega t \mp \gamma_{z11}z) \tag{1g}$$

The field distribution of the electric field in the transverse plane is determined by the first two factors in (1e) and (1f).

$$\underline{E}_{\rho11} = e_\rho f(z, t)$$

$$e_\rho = E_{\rho o} J_1(\gamma_{11}\rho) \sin \phi \tag{2a}$$

$$\underline{E}_{\phi11} = e_\phi f(z, t)$$

$$e_\phi = E_{\phi o} J_1'(\gamma_{11}\rho) \cos \phi \tag{2b}$$

The distribution of the total transverse field

$$\vec{e}_t = \rho e_\rho + \hat{\phi} e_\phi \tag{3}$$

in the cross-sectional plane is sketched in Fig. 13.2-1. The factor $f(z, t)$ represents the traveling wave phenomena. That is, the field

$$\vec{\underline{E}}_t = \rho \underline{E}_{\rho11} + \hat{\phi} \underline{E}_{\phi11} \tag{4}$$

is the same as $\vec{e}_t$ modulated by the traveling wave function $f(z, t)$.

The magnetic field can be expressed as

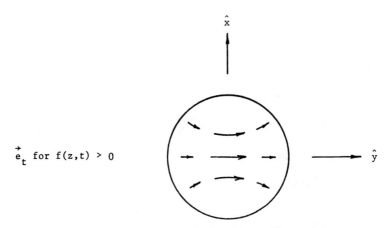

Figure 13.2-1: Transverse $\vec{e}_{11}$-field in a circular waveguide

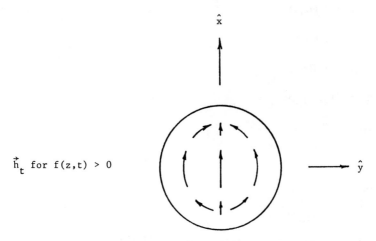

Figure 13.2-2: Tranverse $\vec{h}_{11}$-field in circular waveguide

$$\underline{H}_{z11} = h_z g(z, t)$$

$$h_z = K_1 \cos \phi \, J_1(\gamma_{11}\rho) \tag{5a}$$

$$\underline{H}_{\rho 11} = h_\rho f(z, t)$$

$$h_\rho = H_{\rho o} J_1'(\gamma_{11}\rho) \cos \phi \tag{5b}$$

$$\underline{H}_{\phi 11} = h_\phi f(z, t)$$

$$h_\phi = H_{\phi o} J_1(\gamma_{11}\rho) \sin \phi \tag{5c}$$

The field distribution of

$$\vec{h}_t = \hat{\rho} h_\rho + \hat{\phi} h_\phi \tag{6}$$

in the transverse plane is sketched in Fig. 13.2-2.

The field distribution of

$$\vec{\underline{H}}_{11} = \hat{\rho} \underline{H}_{\rho 11} + \hat{\phi} \underline{H}_{\phi 11} + \hat{z} \underline{H}_{z11} \tag{7}$$

at typical values of z and t can be obtained for a specified value of ϕ. At $\phi = 0$, (2) and (5) become

$$\underline{H}_{z11} = h_z g(z, t) \tag{8a}$$

$$h_z = K_1 J_1(\gamma_{11}\rho) \tag{8b}$$

$$\underline{H}_{\rho 11} = h_\rho f(z, t) \tag{8c}$$

$$h_\rho = H_{\rho o} J_1'(\gamma_{11}\rho) \tag{8d}$$

$$\underline{H}_{\phi 11} = 0 \tag{8e}$$

$$\underline{E}_{\rho 11} = 0 \tag{8f}$$

$$E_{\phi 11} = e_\phi f(z, t) \tag{8g}$$

$$e_\phi = E_{\phi o} J_1'(\gamma_{11}\rho) \tag{8h}$$

At $t = 0$, one has

$$f(z, t)|_{t=0} = \mp \sin \gamma_{z11} z, \qquad g(z, t)|_{t=0} = \cos \gamma_{z11} z \tag{9}$$

The distribution of $h_{\rho 11}$ and $e_{\phi 11}$ at $t = 0$ and $\phi = 0°$ or $180°$ is sketched in Fig. 13.2-3a for a typical location along the z-axis. The distribution of

$$\vec{\underline{H}}_{11} = \hat{\rho}\underline{H}_{\rho 11} + \hat{z}\underline{H}_{z11} \tag{10}$$

is sketched in Fig. 13.2-3b.

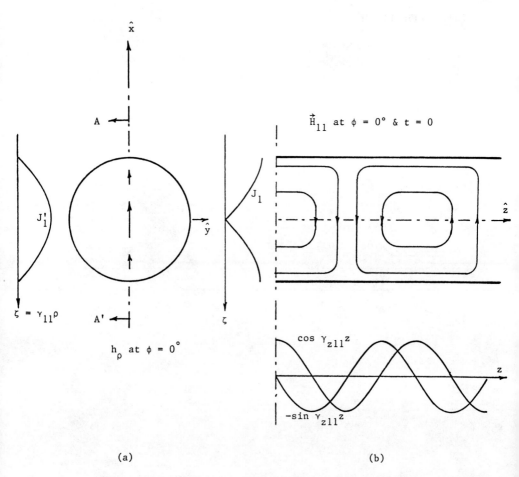

(a)   (b)

Figure 13.2-3: Distribution of TEsub11-field along circular waveguide

14. Problems

1. The axial component of the electric field in a rectangular waveguide is given as

$$E_z(r, t) = E_o \sin \frac{\pi}{a}x \sin \frac{\pi}{b}y \cos (\omega t - \gamma_z z)$$

The waveguide has a width of a m in the x-direction and b m in the y-direction. Determine the propagation coefficient γ_z and all field components of the guided wave.

2. A rectangular waveguide has the dimensions a = 0.625 in. and b = 1.25 in. Determine all possible propagating TM modes for each of the following operating frequences:
   (a) 10^{10} Hz.
   (b) 11×10^9 Hz.

3. It is desired to transmit a 1-kw TE wave by an air-filled rectangular guide at an energy transport velocity of 60% of the velocity of light. The maximum instantaneous peak of the axial magnetic field is 100 a/m. Determine the minimum required cross section of the waveguide.

4. Determine the time-average power transmitted by the TE_{10} mode in a rectangular air-filled waveguide with a width of a m. in the x-direction and a height of b m. in the y-direction.

5. Design an air-filled rectangular waveguide meeting the following specifications:
   (a) Operating frequency = 11×10^9 Hz for the dominant TE_{10} mode.
   (b) The operating frequency is 20% above the cutoff value.
   (c) The mode with the next higher cutoff should be operating at 20% below its cutoff.

6. The standard x-band rectangular waveguide has the dimensions 2.29 cm x 1.20 cm.
   (a) Find the cutoff frequencies for the four lowest propagating modes.
   (b) Calculate the guide wavelength for the TE_{10} mode operating at 8.5×10^9 Hz.
   (c) Determine the wave impedance of the TE_{10} mode at 8.5×10^9 Hz.

CHAPTER III   The Scattering Matrix

1. Introduction

The scattering matrix can be used to describe a large class of passive microwave components. In many cases, it can provide to a complete understanding of microwave devices. Thus, one may avoid the complex analysis of the structure via the technique of boundary value solution.

The scattering matrix of an m-port network is defined to relate the incident and the reflected waves at each individual port. The performance of a network is described by the scattering matrix under any specified terminating conditions. Scattering matrices exist for all linear, passive, and time-invariant networks.

For a network which is symmetric, reciprocal, and conservative, it is possible to deduce the scattering coefficients purely from the general properties of the scattering matrix and from symmetry considerations of the fields.

2. The Scattering Matrix

The scattering matrix [S] of a two-port network is defined to relate the incident waves and the reflected waves at each port (Fig. 2-1).

$$[b] \ = \ [S] \ [a] \tag{1a}$$

where

$$[S] \ = \ \begin{bmatrix} s_{11} & s_{12} \\ s_{21} & s_{22} \end{bmatrix} \tag{1b}$$

$$[a] \ = \ [\ a_1 \quad a_2\ ]^t \ = \ \text{column matrix of incident waves} \tag{1c}$$

$$[b] \ = \ [\ b_1 \quad b_2\ ]^t \ = \ \text{column matrix of reflected waves} \tag{1d}$$

For a two-port device, [S] is a 2 by 2 square matrix. Elements along the main diagonal of the scattering matrix are reflection coefficients, and the off-diagonal elements are transmission coefficients.

The expansion of (1a) yields the following relations.

$$b_1 \ = \ s_{11}a_1 \ + \ s_{12}a_2 \tag{2a}$$
$$b_2 \ = \ s_{21}a_1 \ + \ s_{22}a_2 \tag{2b}$$

The scattering coefficients can be defined as:

$$s_{11} \ = \ \left. \frac{b_1}{a_1} \right]_{a_2=0} \ = \ \text{reflection coefficient at port 1} \tag{3a}$$

$$s_{21} \ = \ \left. \frac{b_2}{a_1} \right]_{a_2=0} \ = \ \text{forward transmission coefficient from port 1 to port 2} \tag{3b}$$

$$s_{12} \ = \ \left. \frac{b_1}{a_2} \right]_{a_1=0} \ = \ \text{reverse transmission coefficient from port 2 to port 1} \tag{3c}$$

$$s_{22} \ = \ \left. \frac{b_2}{a_2} \right]_{a_1=0} \ = \ \text{reflection coefficient at port 2} \tag{3d}$$

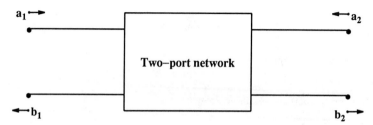

Figure 2-1: Incident and reflected waves of a two-port network

2.1. Scattering Matrix of a Series Impedance

A reciprocal and symmetrical two-port device is characterized by a series impedance Z (Fig. 2.1-1). Determine the scattering matrix of this device.

Solution:

To facilitate the formulation, let the two-port device be connected to a source and a load as shown in Fig. 2.1-2. Then one has

$$\begin{bmatrix} b_1 \\ b_2 \end{bmatrix} = \begin{bmatrix} s_{11} & s_{12} \\ s_{21} & s_{22} \end{bmatrix} \begin{bmatrix} a_1 \\ a_2 \end{bmatrix} \tag{1}$$

with

$$s_{11} = s_{22} \qquad \text{symmetrical}$$

$$s_{12} = s_{21} \qquad \text{reciprocal}$$

and the expansion of (1) yields

$$b_1 = s_{11}a_1 + s_{21}a_2 \tag{1a}$$

$$b_2 = s_{21}a_1 + s_{11}a_2 \tag{1b}$$

$$s_{11} = \left. \frac{b_1}{a_1} \right]_{a_2=0} \tag{2}$$

The voltage across terminals 1–1′ is

$$v_{11'} = a_1 + b_1 \qquad \text{in general} \tag{3a}$$

$$= a_1 \qquad \text{if } b_1 = 0 \tag{3b}$$

and

$$b_1 = 0 \qquad \text{when} \qquad Z_{11'} = Z_0 \tag{3c}$$

Therefore,

$$a_1 = v_{11'}\big|_{b_1=0} = iZ_{11'}\big|_{Z_{11'}=Z_0} = \left. \frac{v_0}{Z_0 + Z_{11'}} Z_{11'} \right|_{Z_{11'}=Z_0} = \frac{v_0}{2} \tag{4}$$

In general, from Fig. 2.1-2,

$$v_{11'} = v_0 - iZ_0 = v_0 - \frac{v_0}{Z_0 + Z_{11'}} Z_0 = v_0 \frac{Z_{11'}}{Z_0 + Z_{11'}} \tag{5}$$

From (3a),

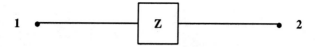

Figure 2.1-1: A series element as a two-port network

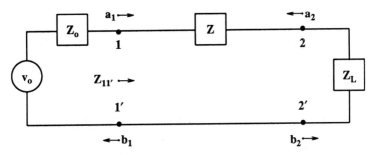

Figure 2.1-2: A two-port network excited by a source and terminated by a load

$$b_1 = v_{11'} - a_1 = v_o \frac{Z_{11'}}{Z_o + Z_{11'}} - \frac{v_o}{2} = \frac{v_o}{2} \frac{Z_{11'} - Z_o}{Z_o + Z_{11'}}$$

With $Z_{11'} = Z + Z_L$,

$$b_1 = \frac{v_o}{2} \frac{Z + Z_L - Z_o}{Z_o + Z + Z_L} \quad \text{in general} \tag{6}$$

For $a_2 = 0$, this implies no reflection from the load, or $Z_L = Z_o$. Then

$$b_1 |_{a_2=0} = b_1 |_{Z_L=Z_o} = \frac{v_o}{2} \frac{Z}{Z + 2Z_o} \tag{7}$$

Therefore

$$s_{11} = \frac{\text{Eq. (7)}}{\text{Eq. (4)}} = \frac{Z}{Z + 2Z_o} \tag{8}$$

The voltage across terminals 2–2′ is

$$v_{22'} = a_2 + b_2 = iZ_L$$

and

$$v_{22'} = iZ_L = \frac{v_o}{Z_o + Z + Z_L} Z_L$$

$$b_2 = v_{22'} |_{a_2=0} = v_{22'} |_{Z_L=Z_o} = \frac{v_o Z_o}{Z + 2Z_o} \tag{9}$$

and

$$s_{21} = \frac{\text{Eq. (9)}}{\text{Eq. (4)}} = \frac{2Z_o}{Z + 2Z_o} \tag{10}$$

3. Definition of Scattering Coefficients

In network analysis at low frequencies, the conventional variables are total currents and total voltages at terminals of the network. In the scattering matrix formulation, the variables are incident voltages, reflected voltages, incident currents, and reflected currents. At any port, there exist linear relations between the total voltage or current and the corresponding scattered voltages and currents. That is,

$$a_k = \alpha_{11}v_k + \alpha_{12}i_k \tag{1a}$$

$$b_k = \alpha_{21}v_k + \alpha_{22}i_k \tag{1b}$$

where the α_{jk}'s are constants of proportionality which define the transformation. These constants are chosen for the convenience in the analysis. One such choice is to make $(a_k a_k^*/2)$ equal to the incident power at port k and $(b_k b_k^*/2)$ equal to the emergent power at the same port.

The above transformation, (1), can be conveniently obtained from the transmission line relations. Consider the lossless transmission line as shown in Fig. 3-1. The voltage and current distribution are known to be

$$v(z) = V^+e^{-j\beta z} + V^-e^{j\beta z} \tag{2a}$$

$$i(z) = \frac{1}{R_0} [V^+e^{-j\beta z} - V^-e^{j\beta z}] \tag{2b}$$

where $j\beta$ is the propagation coefficient and R_0 is the characteristic resistance of the lossless line; $V^\pm$ are arbitrary constants of integration which are to be determined by the boundary conditions.

Equation (2a) relates the total voltage $v(z)$ at the point of observation to the incident and reflected waves at the same point. Equation (2b) relates the corresponding current variables to the incident and the reflected waves. The quantities in (2a) [or (2b)] are measured in volts [or amperes].

Equation (2) can be rearranged as follows:

$$v(z) = \sqrt{R_0} \left[\frac{V^+}{\sqrt{R_0}} e^{-j\beta z} + \frac{V^-}{\sqrt{R_0}} e^{j\beta z} \right]$$

$$= \sqrt{R_0} \left[V_n^+e^{-j\beta z} + V_n^-e^{j\beta z} \right] \tag{3a}$$

$$i(z) = \frac{1}{\sqrt{R_0}} \left[V_n^+e^{-j\beta z} - V_n^-e^{j\beta z} \right] \tag{3b}$$

where the normalized coefficients $V_n^\pm$ are defined by

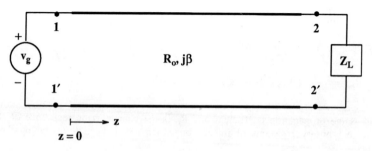

Figure 3-1: A lossless transmission line

$$V_n^{\pm} \equiv \frac{V^{\pm}}{\sqrt{R_o}} \tag{4}$$

These normalized coefficients are defined such that

$$V_n^+ V_n^{+*} = \frac{(V^+)^2}{R_o} = \text{power of incident wave} \tag{5a}$$

$$V_n^- V_n^{-*} = \frac{(V^-)^2}{R_o} = \text{power of reflected wave} \tag{5b}$$

Equation (3) can be expressed in terms of incident and reflected parameters, a and b,

$$\frac{v(z)}{\sqrt{R_o}} = a + b \tag{6a}$$

$$\sqrt{R_o}\, i(z) = a - b \tag{6b}$$

where

$$a \equiv a(z) = V_n^+ e^{-j\beta z} = \text{normalized incident voltage wave} \tag{6c}$$

$$b \equiv b(z) = V_n^- e^{j\beta z} = \text{normalized reflected voltage wave} \tag{6d}$$

The parameters a and b can be expressed in terms of $v(z)$ and $i(z)$.

$$a = \frac{1}{2} \left[\frac{v(z)}{\sqrt{R_o}} + i(z)\sqrt{R_o} \right] \tag{7a}$$

$$b = \frac{1}{2} \left[\frac{v(z)}{\sqrt{R_o}} - i(z)\sqrt{R_o} \right] \tag{7b}$$

For a two-port network arranged as shown in Fig. 3-2, the scattering parameters can be defined similarly. The terminating resistances, R_{o1} and R_{o2}, are normally used as the normalizing resistances for port 1 and port 2, respectively.

$$a_1 = \frac{1}{2} \left[\frac{v_1}{\sqrt{R_{o1}}} + i_1\sqrt{R_{o1}} \right] \tag{8a}$$

$$b_1 = \frac{1}{2} \left[\frac{v_1}{\sqrt{R_{o1}}} - i_1\sqrt{R_{o1}} \right] \tag{8b}$$

$$a_2 = \frac{1}{2} \left[\frac{v_2}{\sqrt{R_{o2}}} + i_2\sqrt{R_{o2}} \right] \tag{8c}$$

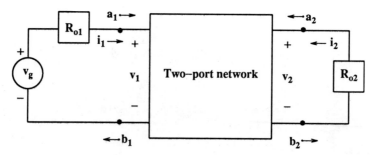

Figure 3-2: A two-port network

$$b_2 = \frac{1}{2} \left[\frac{v_2}{\sqrt{R_{o2}}} - i_2\sqrt{R_{o2}} \right] \tag{8d}$$

For practical microwave circuits, each port is connected to a waveguide with identical characteristic resistance R_o. Then (8) becomes

$$a_1 = \frac{1}{2} \left[\frac{v_1}{\sqrt{R_o}} + i_1\sqrt{R_o} \right] \tag{9a}$$

$$b_1 = \frac{1}{2} \left[\frac{v_1}{\sqrt{R_o}} - i_1\sqrt{R_o} \right] \tag{9b}$$

$$a_2 = \frac{1}{2} \left[\frac{v_2}{\sqrt{R_o}} + i_2\sqrt{R_o} \right] \tag{9c}$$

$$b_2 = \frac{1}{2} \left[\frac{v_2}{\sqrt{R_o}} - i_2\sqrt{R_o} \right] \tag{9d}$$

3.1. Appendix

(a) In the following, it will be shown that

$$\frac{1}{2} a_1 a_1^* = \text{available power at port 1}$$

The terminal voltage v_1 is given by (Fig. 3.2)

$$v_1 = v_g - i_1 R_{o1} \tag{1}$$

The substitution of (1) into (3-8a) yields

$$a_1 = \frac{1}{2} \left[\frac{v_g - i_1 R_{o1}}{\sqrt{R_{o1}}} + i_1\sqrt{R_{o1}} \right]$$

$$= \frac{1}{2} \left[\frac{v_g}{\sqrt{R_{o1}}} - i_1\sqrt{R_{o1}} + i_1\sqrt{R_{o1}} \right]$$

$$= \frac{v_g}{2\sqrt{R_{o1}}} \tag{2}$$

and

$$\frac{1}{2} a_1 a_1^* = \frac{1}{2} \left[\frac{v_g}{2\sqrt{R_{o1}}} \right]^2$$

$$= \frac{v_g^2}{8R_{o1}} = \frac{(1/2)\,(v_g/2)^2}{R_{o1}} = \frac{1}{2} \frac{v_1^2}{R_{o1}} \tag{3}$$

which is the maximum available power for the source when it is match-terminated, Fig. 3.1-1.

(b) In the following, it will be shown that

$$\frac{1}{2} b_2 b_2^* = \text{emergent power at port 2}$$

With port 2 match-terminated, i.e., $a_2 = 0$, (3-8c) gives

$$a_2 = 0 = \frac{1}{2} \left[\frac{v_2}{\sqrt{R_{o2}}} + i_2\sqrt{R_{o2}} \right]$$

or

$$i_2\sqrt{R_{o2}} = - \frac{v_2}{\sqrt{R_{o2}}} \tag{4}$$

The substitution of (4) into (3-8d) yields

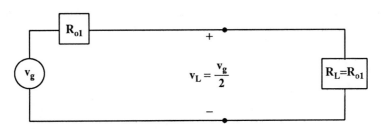

Figure 3.1-1: A matched-terminated source

$$b_2 = \frac{1}{2}\left[\frac{v_2}{\sqrt{R_{o2}}} - \left(-\frac{v_2}{\sqrt{R_{o2}}}\right)\right]$$

$$= \frac{v_2}{\sqrt{R_{o2}}} \tag{5}$$

Therefore,

$$\frac{1}{2}b_2b_2^* = \frac{1}{2}\frac{v_2^2}{R_{o2}}$$

$$= \text{maximum output power at port 2} \tag{6}$$

3.2. Example: Scattering Matrix for a Two-Port Device

Determine the scattering matrix for the two-port device in Section 2.1 by the concept of section 3.

Solution:

The arrangement is shown in Fig. 3.2-1.

$$s_{11} = \left. \frac{b_1}{a_1} \right]_{a_2=0} \tag{1}$$

For $a_2 = 0$, terminals 2–2' must be terminated by Z_0. The parameter a_1 and b_1 are related by

$$b_1 = \rho_1 a_1 \tag{2}$$

and the reflection coefficient ρ_1 is given by

$$\rho_1 = \frac{Z_{in} - Z_0}{Z_{in} + Z_0} = \frac{Z}{Z + 2Z_0} \tag{3a}$$

where

$$Z_{in} = Z + Z_0 \tag{3b}$$

Therefore, from (2),

$$s_{11} = \left. \frac{b_1}{a_1} \right]_{a_2=0} = \rho_1 = \frac{Z}{Z + 2Z_0} \tag{4}$$

$$s_{21} = \left. \frac{b_2}{a_1} \right]_{a_2=0} \tag{5}$$

$$b_2|_{a_2=0} = iZ_0 = \frac{v_0 Z_0}{Z + 2Z_0} \tag{6}$$

But

$$-v_0 - iZ_0 = a_1 + b_1 = \text{total voltage across } 1\text{–}1'$$
$$= a_1 (1 + \rho_1) = a_1 \frac{2(Z + Z_0)}{Z + 2Z_0} \tag{7}$$

With

$$v_0 - iZ_0 = v_0 - \frac{v_0 Z_0}{Z + 2Z_0} = v_0 \frac{Z + Z_0}{Z + 2Z_0} \tag{8}$$

Therefore, by equating (7) and (8), one has

$$v_0 \frac{Z + Z_0}{Z + 2Z_0} = a_1 \frac{2(Z + Z_0)}{Z + 2Z_0} \quad \text{or} \quad a_1 = \frac{v_0}{2} \tag{9}$$

$$s_{21} = \left. \frac{b_2}{a_1} \right|_{a_2=0} = \frac{\text{Eq. (6)}}{\text{Eq. (9)}} = \frac{2Z_0}{Z + 2Z_0} \tag{10}$$

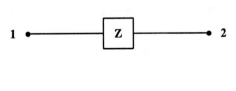

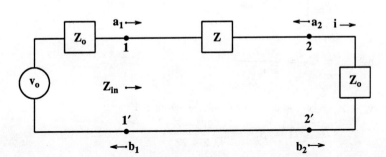

Figure 3.2-1: A series element as a two-port network

3.3. Example: Scattering Matrix for a Two-Port Network - Alternate Approach

Determine the scattering matrix for the two-port network arranged as shown in Fig. 3.3-1.

Solution:

The forward transmission coefficient s_{21} can be evaluated by considering that the input port A–A' is excited by a voltage source v_g with an internal impedance Z_g and the output port B–B' is terminated by a load Z_L (Fig. 3.3-1).

The voltage and current at the input port are

$$v_1 = v_g - i_1 Z_g \tag{1a}$$

$$i_1 = \frac{v_g}{Z_g + Z_1} \tag{1b}$$

The forward transmission coefficient is given by (2-3b). The condition $a_2 = 0$ implies that (3-8c) must vanish.

$$a_2 = \frac{1}{2} \left[\frac{v_2}{\sqrt{R_{o2}}} + \sqrt{R_{o2}}\, i_2 \right] = 0 \quad \text{or} \quad \frac{v_2}{\sqrt{R_{o2}}} = -\sqrt{R_{o2}}\, i_2 \tag{2}$$

The incident parameter a_1 is obtained from (3-8a) by imposing (1).

$$
\begin{aligned}
a_1 &= \frac{1}{2\sqrt{R_{o1}}} \left[\, (v_g - i_1 Z_g) + i_1 R_{o1} \,\right] \\[2mm]
&= \frac{1}{2\sqrt{R_{o1}}} \left[v_g + \frac{v_g}{Z_g + Z_1} (R_{o1} - Z_g) \right] \\[2mm]
&= \frac{v_g}{2\sqrt{R_{o1}}} \left[1 + \frac{R_{o1} - Z_g}{Z_1 + Z_g} \right] = \frac{v_g}{2\sqrt{R_{o1}}} \frac{Z_1 + R_{o1}}{Z_1 + Z_g} \tag{3a} \\[2mm]
&= \frac{v_g}{2\sqrt{R_{o1}}} \quad \text{for } Z_g = R_{o1} \tag{3b}
\end{aligned}
$$

The reflection parameter b_2 is obtained from (3-8d) subject to (2).

$$b_2 = \frac{v_2}{\sqrt{R_{o2}}} \tag{4}$$

The coefficient s_{21} is given by (2-3b) with the use of (3) and (4).

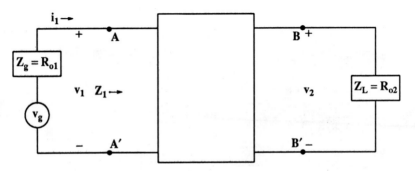

Figure 3.3-1: A two-port network

$$s_{21} = \left.\frac{b_2}{a_1}\right]_{a_2=0} = \frac{2v_2}{v_g} \sqrt{\frac{R_{o1}}{R_{o2}}} \frac{Z_1 + Z_g}{Z_1 + R_{o1}} \tag{5a}$$

$$= \frac{2v_2}{v_g} \sqrt{\frac{R_{o1}}{R_{o2}}} \qquad \text{for } Z_g = R_{o1} \tag{5b}$$

The coefficient s_{12} can be obtained similarly by driving port B–B′ by a voltage source v_{g2} with an internal impedance Z_{g2} with port A–A′ terminated by a load Z_L (Fig. 3.3-2). The current and voltage at port B–B′ are

$$v_2 = v_{g2} - i_2 Z_{g2} \tag{6a}$$

$$i_2 = \frac{v_{g2}}{Z_2 + Z_{g2}} \tag{6b}$$

The incident and reflected wave parameters are

$$a_2 = \frac{1}{2\sqrt{R_{o2}}} \left[v_{g2} + i_2 (R_{o2} - Z_{g2}) \right]$$

$$= \frac{v_{g2}}{2\sqrt{R_{o2}}} \left[1 + \frac{R_{o2} - Z_{g2}}{Z_2 + Z_{g2}} \right] = \frac{v_{g2}}{2\sqrt{R_{o2}}} \frac{Z_2 + Z_{o2}}{Z_2 + Z_{g2}} \tag{7a}$$

$$b_1 = \frac{v_1}{\sqrt{R_{o1}}} \qquad \text{for } Z_{g2} = R_{o2} \tag{7b}$$

The coefficient s_{12} is

$$s_{12} = \left.\frac{b_1}{a_2}\right]_{a_1=0} = \frac{\dfrac{v_1}{\sqrt{R_{o1}}}}{\dfrac{v_{g2}}{2\sqrt{R_{o2}}} \left[1 + \dfrac{R_{o2} - Z_{g2}}{Z_2 + Z_{g2}} \right]}$$

$$= \frac{2v_1}{v_{g2}} \sqrt{\frac{R_{o2}}{R_{o1}}} \frac{1}{1 + \dfrac{R_{o2} - Z_{g2}}{Z_2 + Z_{g2}}} \tag{8a}$$

$$= \frac{2v_1}{v_{g2}} \sqrt{\frac{R_{o2}}{R_{o1}}} \qquad \text{for } Z_{g2} = R_{o2} \tag{8b}$$

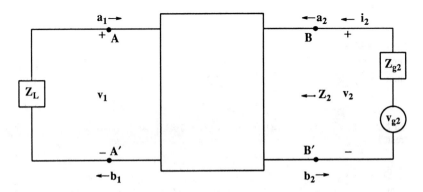

Figure 3.3-2: A two-port network with load and source interchanged

4. Characteristic Equation of the Scattering Matrix

Given a square matrix [F] of order m and a column matrix $[X_m]$ of order m, then the set of homogeneous equations

$$[F] [X_m] = k_m [X_m] \qquad (1)$$

has a nontrivial solution only if

$$\left| [F] - k_m [u] \right| = 0 \qquad (2)$$

where [u] is a unitary matrix. Then (1) is known as the eigenvalue equation of the matrix [F], k_m is the nontrivial solution known as the eigenvalue, and $[X_m]$ is the corresponding eigenfunction.

The scattering matrix of an m-port network is given by (2-1)

$$[S] [a] = [b] \qquad (3)$$

A direct comparison of (1) and (3) reveals that the eigenvector represents a possible excitation (the incident parameter matrix) and the eigenvalue λ_m represents the corresponding reflection coefficient (since $b_k = s_{jk}a_k$).

Equation (3) can be rearranged as

$$([S] - \lambda_m [u]) [X_m] = 0 \qquad (4)$$

This is a homogeneous equation; besides having the trivial solution $[X_m] = 0$, which is of little interest, it will have a nontrivial solution, non-vanishing $[X_m]$, only if

$$\det \left| [S] - \lambda_m [u] \right| = 0 \qquad (5)$$

This equation is known as the characteristic equation of the scattering matrix [S] and it will determine the values of λ_m for which nonvanishing $[X_m]$ exists.

5. Eigenvalues and Eigenvectors

For an m-port network, the scattering matrix [S] is an m x m square matrix.

$$[S] = \begin{bmatrix} s_{11} & s_{12} & \cdots & s_{1m} \\ s_{21} & s_{22} & \cdots & s_{2m} \\ \cdots & \cdots & \cdots & \cdots \\ s_{m1} & s_{m2} & \cdots & s_{mm} \end{bmatrix} \tag{1}$$

The eigenvalue equation of [S] is

$$([S] - \lambda [u]) [X] = 0 \tag{2}$$

and it has a nontrivial solution only if

$$\det | [S] - \lambda [u] | = 0 \tag{3}$$

For an m-port network, (3) has the form

$$\det \left[\begin{bmatrix} s_{11} & s_{12} & \cdots & s_{1m} \\ s_{21} & s_{22} & \cdots & s_{2m} \\ \cdots & \cdots & \cdots & \cdots \\ s_{m1} & s_{m2} & \cdots & s_{mm} \end{bmatrix} - \begin{bmatrix} \lambda & 0 & \cdots & 0 \\ 0 & \lambda & \cdots & 0 \\ \cdots & \cdots & \cdots & \cdots \\ 0 & 0 & \cdots & \lambda \end{bmatrix} \right] = 0$$

or

$$\det \begin{bmatrix} s_{11} - \lambda & s_{12} & \cdots & s_{1m} \\ s_{12} & s_{22} - \lambda & \cdots & s_{2m} \\ \cdots & \cdots & \cdots & \cdots \\ s_{m1} & s_{m2} & \cdots & s_{mm} - \lambda \end{bmatrix} = 0 \tag{4}$$

The expansion of (4) gives rise to an mth-degree polynomial in λ, $f(\lambda)$.

$$f(\lambda) = [(-\lambda)^m + d_1 (-\lambda)^{m-1} + d_2 (-\lambda)^{m-2} + \cdots + d_m] \tag{5}$$

This is known as the characteristic polynomial. It can be expressed in terms of the roots, λ_1, $\lambda_2, \ldots, \lambda_m$, of the characteristic equation.

$$f(\lambda) = \det | [S] - \lambda [u] | = 0$$
$$= (\lambda_1 - \lambda) (\lambda_2 - \lambda) \cdots (\lambda_m - \lambda) \tag{6}$$

These roots λ_i, $i = 1, 2, \ldots, m$, are known as the eigenvalues of the matrix [S]. For each root λ_i, there exists a nontrivial solution $[X_i]$ known as the eigenvector. For any value of λ other than λ_i, (2) has only a trivial solution, $[X] = 0$.

It is important to note that, since (3) is homogeneous, if $[X_i]$ is an eigenvector corresponding to the eigenvalue λ_i, then $C_i [X_i]$ is also an eigenvector, where C_i is an arbitrary nonzero constant. This implies that the magnitude of the eigenvector is not specified by the eigenvalue equation (3). The magnitude of the eigenvector is given by

$$\left| C [X_i] \right| = \sqrt{(C [X_i])^t (C [X_i])^*} = C\sqrt{[X_i]^t [X_i]^*} \tag{7}$$

It is a common practice to choose the constant C such that the eigenvector has unit magnitude. In other words, eigenvectors are usually normalized to unit magnitude.

5.1. Example: Eigenvalues and Eigenvectors

Determine the eigenvalues and eigenvectors of the matrix [S].

$$[S] = \begin{bmatrix} 3 & 4 \\ 2 & 1 \end{bmatrix} \tag{1}$$

Solution:

The characteristic equation is

$$f(\lambda) = \left| [S] - \lambda [u] \right| = \det \begin{bmatrix} 3 - \lambda & 4 \\ 2 & 1 - \lambda \end{bmatrix} = (\lambda - 5)(\lambda + 1) \tag{2}$$

The eigenvalues are

$$\lambda_1 = 5 \quad \text{and} \quad \lambda_2 = -1$$

The eigenvector equation for $\lambda = \lambda_1 = 5$ is

$$[S][X_1] = \lambda_1 [X_1]$$

$$\begin{bmatrix} 3 & 4 \\ 2 & 1 \end{bmatrix} \begin{bmatrix} x_{11} \\ x_{12} \end{bmatrix} = 5 \begin{bmatrix} x_{11} \\ x_{12} \end{bmatrix}$$

The expansion of the above relation gives

$$3x_{11} + 4x_{12} = 5x_{11} \quad \text{or} \quad x_{12} = \frac{x_{11}}{2}$$

$$2x_{11} + x_{12} = 5x_{12} \quad \text{or} \quad x_{11} = 2x_{12}$$

The eigenvector for $\lambda_1 = 5$ is

$$[X_1] = x_{11} \begin{bmatrix} 1 & 1/2 \end{bmatrix}^t \quad \text{or} \quad [X_1] = x_{11} \begin{bmatrix} 2 & 1 \end{bmatrix}^t$$

$$\left| [X_1] \right| = \left| \sqrt{x_{11}^2 + x_{12}^2} \right| = \sqrt{5}\, x_{11} = 1$$

Therefore,

$$x_{11} = \frac{1}{\sqrt{5}}$$

The normalized eigenvector is

$$[X_1] = \frac{1}{\sqrt{5}} \begin{bmatrix} 2 & 1 \end{bmatrix}^t$$

The eigenvector equation for $\lambda_2 = -1$ is

$$\begin{bmatrix} 3 & 4 \\ 2 & 1 \end{bmatrix} \begin{bmatrix} x_{21} \\ x_{22} \end{bmatrix} = -1 \begin{bmatrix} x_{21} \\ x_{22} \end{bmatrix}$$

and

$$3x_{21} + 4x_{22} = -x_{21} \quad \text{or} \quad x_{21} = -x_{22}$$

$$2x_{21} + x_{22} = -x_{22} \quad \text{or} \quad x_{21} = -x_{22}$$

The eigenvector for $\lambda_2 = -1$ is

$$[X_2] = x_{21} [1 \quad -1]^t$$

and

$$|[X_2]| = 1 = x_{21}\sqrt{2} \quad \text{or} \quad x_{21} = \frac{1}{\sqrt{2}}$$

The normalized eigenvector is

$$[X_2] = \frac{1}{\sqrt{2}} [1 \quad -1]^t$$

6. Some Properties of Eigenvalues

The characteristic equation of the matrix [S] for an m-port device is

$$\det \left| [S] - \lambda [u] \right| = f(\lambda) \equiv (-\lambda)^m + d_1 (-\lambda)^{m-1} + d_2 (-\lambda)^{m-2} + \cdots + d_m = 0 \quad (1)$$

The characteristic polynomial can also be expressed in terms of its eigenvalues, λ_i.

$$f(\lambda) = (\lambda_1 - \lambda)(\lambda_2 - \lambda) \cdots (\lambda_m - \lambda) \quad (2)$$

By direct expansion of (2), one can show that

$$f(\lambda) = (-\lambda)^m + (\lambda_1 + \lambda_2 + \cdots + \lambda_m)(-\lambda)^{m-1}$$
$$+ [\lambda_1\lambda_2 + \lambda_1\lambda_3 + \cdots + \lambda_1\lambda_m + \lambda_2\lambda_3 + \lambda_2\lambda_4 + \cdots + \lambda_2\lambda_m$$
$$+ \cdots + \lambda_{m-1}\lambda_m] (-\lambda)^{m-2} + \cdots + \lambda_1\lambda_2\lambda_3 \cdots \lambda_m \quad (3)$$

A comparison of (1) and (3) results in

$$d_1 = \lambda_1 + \lambda_2 + \lambda_3 + \cdots + \lambda_m \quad (4a)$$

$$d_2 = \lambda_1\lambda_2 + \lambda_1\lambda_3 + \cdots + \lambda_1\lambda_m + \lambda_2\lambda_3 + \lambda_2\lambda_4 + \cdots + \lambda_2\lambda_m$$
$$+ \cdots + \lambda_{m-1}\lambda_m \quad (4b)$$

$$\cdots \cdots \cdots$$

$$d_m = \lambda_1\lambda_2\lambda_3 \cdots \lambda_m \quad (4m)$$

(a) By putting $\lambda = 0$ in (1), one has

$$\det \left| [S] - \lambda [u] \right|_{\lambda=0} = \det \left| [S] \right| = f(\lambda = 0) = d_m = \lambda_1 \lambda_2 \cdots \lambda_m \quad (5)$$

That is, the determinant of [S] is equal to the product of the m eigenvalues of the matrix [S]. Therefore, a matrix is singular if it has a zero eigenvalue, and is nonsingular if all its eigenvalues are nonzero.

(b) It can be shown by direct expansion that

$$d_1 = s_{11} + s_{22} + \cdots + s_{mm} \quad (6)$$

This will be obvious by inspecting (5-4), which is repeated here.

$$\det \left| \ [S] - \lambda \ [u] \ \right| \ = \det \begin{bmatrix} s_{11} - \lambda & s_{12} & \cdots & s_{1m} \\ s_{21} & s_{22} - \lambda & \cdots & s_{2m} \\ \cdots & \cdots & \cdots & \cdots \\ s_{m1} & s_{m2} & \cdots & s_{mm} - \lambda \end{bmatrix}$$

$$= (s_{11} - \lambda) \begin{bmatrix} s_{22} - \lambda & s_{23} & \cdots & s_{2m} \\ s_{32} & s_{33} - \lambda & \cdots & s_{3m} \\ \cdots & \cdots & \cdots & \cdots \\ s_{m2} & s_{m3} & \cdots & s_{mm} - \lambda \end{bmatrix}$$

$$- \ s_{21} \begin{bmatrix} s_{21} & \cdots & s_{2m} \\ s_{31} & \cdots & s_{3m} \\ \cdots & \cdots & \cdots \\ s_{m1} & \cdots & s_{mm} - \lambda \end{bmatrix} + \ \cdots \tag{7}$$

d_1 is the coefficient of the term λ^{m-1}, and therefore focus will be on this term. The λ^{m-1} term exists only in the first term of (7); all other terms have at most the λ^{m-2} term, which is of little interest at this moment. Further expansion of the first term gives

$$\text{1st term of (7)} \ = \ (s_{11} - \lambda)(s_{22} - \lambda) \begin{bmatrix} s_{33} - \lambda & \cdots & s_{3m} \\ \cdots & \cdots & \cdots \\ s_{m3} & \cdots & s_{mm} - \lambda \end{bmatrix} + \ \cdots \tag{8}$$

By the same argument, the λ^{m-1} term is contained in the first term of (8). This process can be continued and eventually one obtains

$$\text{1st term of } \det \left| \ [S] - \lambda \ [u] \ \right| \ = \ (s_{11} - \lambda) \ (s_{22} - \lambda) \ \cdots \ (s_{mm} - \lambda) \tag{9}$$

Since (9) has the same form as (2), consequently the coefficient of the λ^{m-1} term in (9) is given by

$$d_1 \ = \ s_{11} \ + \ s_{22} \ + \ s_{33} \ + \ \cdots \ + \ s_{mm} \tag{10}$$

Since d_1 is also given by (4a), therefore,

$$d_1 \ = \ s_{11} \ + \ s_{22} \ + \ \cdots \ + \ s_{mm} \ = \ \lambda_1 \ + \ \lambda_2 \ + \ \cdots \ + \ \lambda_m \tag{11}$$

Hence,

$$\text{trace of } [S] \ \equiv \ \text{Tr } [S] \ = \ \sum_{i=1}^{m} s_{ii} \ = \ \sum_{i=1}^{m} \lambda_i \tag{12}$$

The trace of matrix [S], Tr [S], is the sum of the elements s_{ii} along the main diagonal of the matrix. In other words, the sum of the eigenvalues of a matrix is equal to its trace.

(c) For an mth order matrix [S], one has

$$\det \left| \ [S] - \lambda \ [u] \ \right| \ = \det \left| \ [S]^t - \lambda \ [u] \ \right| \tag{13}$$

This can be shown by direct expansion; see Section 6.1. Therefore, $[S]$ and its transpose $[S]^t$ have the same eigenvalues. $[S]$ and $[S]^t$ will have different eigenvectors except when $[S]$ is a symmetric matrix; then $[S] = [S]^t$.

(d) If $\lambda_1, \lambda_2, \ldots, \lambda_m$ are eigenvalues of $[S]$, then the matrix $c[S]$, where c is an arbitrary scalar constant, has $c\lambda_1, c\lambda_2, \ldots, c\lambda_m$ as its eigenvalues, since

$$\left| \ c[S] - c\lambda \ [u] \ \right| \ = \ \left| \ c \ (\ [S] - \lambda \ [u] \ ) \ \right| \ = \ c^m \left| \ [S] - \lambda \ [u] \ \right| \tag{14}$$

(e) It can also be shown that the eigenvalues of $[S]^{-1}$ are the inverse of eigenvalues of $[S]$, provided none of the eigenvalues of $[S]$ are zero.

$$\left| \ [S] - \lambda \ [u] \ \right| \ = \ \left| -\lambda \ [S] \ (\ [S]^{-1} - \frac{[u]}{\lambda} \ ) \right| \ = \ \left| -\lambda \ [S] \right| \ \left| [S]^{-1} - \frac{[u]}{\lambda} \right| \tag{15}$$

This shows that if $[S]$ has eigenvalue λ, then $[S]^{-1}$ has $1/\lambda$ as its eigenvalue.

(f) The eigenvectors with distinct (nonrepeated) eigenvalues are linearly independent. Mathematically, this means that no linear relationship of the following type can exist.

$$c_1 \ [X_1] + c_1 \ [X_2] + \cdots + c_m \ [X_m] = 0 \tag{16}$$

where $c_1, c_2, \ldots, c_m$ are constants, except when $c_1 = c_2 = \cdots = c_m = 0$.

Since

$$[S] \ [X_i] \ = \ \lambda_i \ [X_i] \qquad i = 1, 2, \ldots, m \tag{17}$$

then

$$(\ [S] - \lambda_j \ [u] \ ) \ [X_i] \ = \ [S] \ [X_i] \ - \ \lambda_j \ [X_i] \ = \ (\lambda_i - \lambda_j) \ [X_i] \tag{18}$$

Suppose the linear relation (16) does exist for some nonzero c_i. Consider

$$(\ [S] - \lambda_2 \ [u] \ ) \ (c_1 \ [X_1] + c_2 \ [X_2] + \cdots + c_m \ [X_m] \ ) = 0 \tag{19}$$

The use of (18) in (19) yields

$$c_1 \ (\lambda_1 - \lambda_2) \ [X_1] + c_3 \ (\lambda_3 - \lambda_2) \ [X_3] + \cdots + c_m \ (\lambda_m - \lambda_2) \ [X_m] = 0 \tag{20}$$

where $[X_2]$ has been removed by this operation. Proceeding with operations $([S] - \lambda_3 \ [u])$, $([S] - \lambda_4 \ [u])$, $\ldots$, $([S] - \lambda_m \ [u])$, successively, then $[X_3], [X_4], \ldots, [X_m]$ will be eliminated accordingly and one arrives at

$$c_1 \ (\lambda_1 - \lambda_2) \ (\lambda_1 - \lambda_3) \ \cdots \ (\lambda_1 - \lambda_m) \ [X_1] = 0 \tag{21}$$

Since the λ_i's are different by assumption and $[X_1]$ is nonzero, (21) implies $c_1 = 0$.

In a similar procedure, one can eliminate $[X_1], [X_3], [X_4], \ldots$,

$$c_2 \ (\lambda_2 - \lambda_1) \ (\lambda_2 - \lambda_3) \ \cdots \ (\lambda_2 - \lambda_m) \ [X_2] = 0 \tag{22}$$

which implies $c_2 = 0$.

In this manner, one can show that

$$c_1 = c_2 = c_3 = \cdots = c_m = 0 \tag{23}$$

provided all λ's are different.

If, however, two or more eigenvalues are equal, then not all c_i's need to be zero. Consequently, the eigenvectors may be either linearly dependent or linearly independent.

(g) One of the most important characteristics of eigenvectors is the orthogonality. It was shown in Section (c) that $[S]$ and $[S]^t$ have identical eigenvalues; and they have, in general, different eigenvectors unless $[S]$ is symmetric.

For a nonsymmetric square $[S]$ matrix, let

$$[S] [X_i] = \lambda_i [X_i] \tag{24a}$$

$$[S]^t [Y_k] = \lambda_k [Y_k] \tag{24b}$$

where

$$[X_i] = [x_{i1} \quad x_{i2} \quad \cdots \quad x_{1m}]^t = \text{column eigenvector of } [S] \text{ corresponding to } \lambda_i \tag{24c}$$

$$[Y_k] = [y_{k1} \quad y_{k2} \quad \cdots \quad y_{km}]^t = \text{column eigenvector of } [S]^t \text{ corresponding to } \lambda_k \tag{24d}$$

Taking the transpose of (24b), one has

$$[Y_k]^t [S] = \lambda_k [Y_k]^t \tag{25}$$

Pre-multiply (24a) by $[Y_k]^t$ to get

$$[Y_k]^t [S] [X_i] = \lambda_i [Y_k]^t [X_i] \tag{26}$$

Post-multiply (25) by $[X_i]$ to obtain

$$[Y_k]^t [S] [X_i] = \lambda_k [Y_k]^t [X_i] \tag{27}$$

The difference between (26) and (27) is

$$(\lambda_i - \lambda_k) [Y_k]^t [X_i] = 0 \tag{28}$$

Hence, for λ_i different from λ_k, one has

$$[Y_k]^t [X_i] = 0 \tag{29}$$

That is, the row eigenvector $[Y_k]^t$ corresponding to any eigenvalue λ_k of a general square matrix is orthogonal to the column eigenvector $[X_i]$ corresponding to any different eigenvalue λ_i.

If $[S]$ is a symmetric matrix, $[S]^t = [S]$, then (24b) implies that $[Y_k]$ is an eigenvector of $[S]$ and consequently (29) yields

$$[X_k]^t [X_i] = 0 \tag{30}$$

The eigenvectors of a symmetric matrix are orthogonal for distinct eigenvalues.

(h) A real symmetric square matrix has real eigenvalues and real eigenvectors. For a real symmetric matrix $[S]$ of order m, one has

$$[S] [X_i] = \lambda_i [X_i] \tag{31}$$

and the complex conjugate of (31) is

$$[S] [X_i]^* = \lambda_i^* [X_i^*] \tag{32}$$

since $[S] = [S^*] = $ real. Pre-multiplying (31) by $[X_i^*]^t$ yields

$$[X_i^*]^t [S] [X_i] = \lambda_i [X_i^*]^t [X_i] \tag{33}$$

Pre-multiplying (32) by $[X_i]^t$ gives

$$[X_i]^t [S] [X_i^*] = \lambda_i^* [X_i]^t [X_i^*] \tag{34}$$

Takng the difference of (33) and (34), one obtains

$$[X_i^*]^t [S] [X_i] - [X_i]^t [S] [X_i^*] = (\lambda_i - \lambda_i^*) [X_i]^t [X_i^*] \tag{35}$$

because

$$[X_i^*]^t [X_i] = [X_i]^t [X_i^*]$$

Since $[X_i]^t [S] [X_i^*]$ is a number and the transpose of a number is the number itself, therefore,

$$[X_i]^t [S] [X_i^*] = [X_i]^t [S] [X_i^*]^t = [X_i^*]^t [S]^t [X_i] = [X_i^*]^t [S] [X_i] \tag{36}$$

because $[S]$ is symmetric. The substitution of (36) into (35) yields

$$(\lambda_i - \lambda_i^*) [X_i]^t [X_i^*] = 0 \tag{37}$$

Since $[X_i]^t [X_i^*] = \left| [X_i] \right|^2$ is not zero, in general, therefore

$$\lambda_i - \lambda_i^* = 0$$

or

$$\lambda_i = \lambda_i^* \tag{38}$$

This implies that λ_i is real since only a real number is equal to its complex conjugate.

6.1 Example

To show the validity of Eq.(6.13) for a 3 x 3 matrix.

Solution:

Consider a general 3 x 3 matrix [S]

$$[S] = \begin{bmatrix} s_{11} & s_{12} & s_{13} \\ s_{21} & s_{22} & s_{23} \\ s_{31} & s_{32} & s_{33} \end{bmatrix} \tag{1}$$

$$\det | [S] - \lambda [u] | = \det \begin{bmatrix} s_{11}-\lambda & s_{21} & s_{13} \\ s_{21} & s_{22}-\lambda & s_{23} \\ s_{31} & s_{32} & s_{33}-\lambda \end{bmatrix}$$

$$= (s_{11} - \lambda) [(s_{22} - \lambda) (s_{33} - \lambda) - s_{23} s_{32}]$$

$$- s_{12} [s_{21} (s_{33} - \lambda) - s_{23} s_{31}]$$

$$+ s_{13} [s_{21} s_{32} - s_{31} (s_{22} - \lambda)] \tag{2}$$

The above expression is obtained by expanding along the first row.

$$\det |[S]^t - \lambda [u] | = \det \begin{bmatrix} s_{11}-\lambda & s_{21} & s_{31} \\ s_{12} & s_{22}-\lambda & s_{32} \\ s_{13} & s_{23} & s_{33}-\lambda \end{bmatrix}$$

$$= (s_{11} - \lambda) [(s_{22} - \lambda) (s_{33} - \lambda) - s_{33} s_{32}]$$

$$- s_{12} [s_{21} (s_{33} - \lambda) - s_{23} s_{31}]$$

$$+ s_{13} [s_{21} s_{32} - s_{31} (s_{22} - \lambda)] \tag{3}$$

Equation (3) is obtained by expanding along the first column. Direct comparison of (2) and (3), yields

$$\det | [S] - \lambda [u] | = \det | [S]^t - \lambda [u] | = \det | [S] - \lambda [u] |^t \qquad \text{Q.E.D.}$$

7. Multiple Eigenvalues

(a) Consider a matrix [S] which has an eigenvalue with multiplicity m. One has only one equation,

$$[S] [X_i] = \lambda_i [X_i] \tag{1}$$

to solve for all m eigenvectors. A new technique will be developed to find these m eigenvectors.

The case of distinct eigenvalues will be considered first and the results will be extended to the case of repeated eigenvalues.

(b) Let the characteristic equation be

$$f(\lambda) = (-1)^m [\lambda^m - d_1\lambda^{m-1} + d_2\lambda^{m-2} - \cdots + (-1)^m d_m] \tag{2}$$

Define

$$g(\lambda, \mu) \equiv \frac{f(\lambda) - f(\mu)}{\lambda - \mu} \tag{3}$$

$$= (-1)^m \left[\frac{\lambda^m - \mu^m}{\lambda - \mu} - d_1 \frac{\lambda^{m-1} - \mu^{m-1}}{\lambda - \mu} - \cdots + (-1)^m d_m \right] \tag{4}$$

It is to be noted that $[f(\lambda) - f(\mu)]$ is exactly divisible by $(\lambda - \mu)$.

Next, make the following substitutions in (4):

(i) replace λ by $\lambda [u]$

(ii) replace μ by [S]

Then one obtains

$$g(\lambda [u], [S]) = \frac{f(\lambda[u]) - f([S])}{\lambda[u] - [S]} \equiv [F(\lambda)] \tag{5}$$

or

$$(\lambda [u] - [S]) [F(\lambda)] = f(\lambda[u]) - f([S]) = f(\lambda[u])$$

because $f([S]) = 0$ [Section 8],

$$f(\lambda[u]) = (\lambda[u] - [S]) [F(\lambda)] \tag{6}$$

Since, by definition [Eq. (5-6)]

$$f(\lambda) = \det \left| [S] - \lambda [u] \right|$$

and

$$f([S]) = \det \left| [S] - [S] [u] \right| = \det \left| [S] - [S] \right| = 0 \tag{7}$$

this implies that every square matrix satisfies its own characteristic equation. This is called the Cayley-Hamilton theorem (more details of this theorem are given in Section 8).

(c) It is to be noted that

$$f(\lambda[u]) = f(\lambda) [u] \tag{8}$$

Then (6) becomes

$$f(\lambda) [u] = (\lambda [u] - [S]) [F(\lambda)] \tag{9}$$

For each eigenvalue λ_i, one has

$$f(\lambda_i) [u] = (\lambda_i [u] - [S]) [F(\lambda_i)] = 0 \tag{10}$$

since $f(\lambda_i) = 0$. Equation (10) has the identical form to the eigenvalue equation, (4-4), which defines the eigenvector. Therefore, the columns of $[F(\lambda_i)]$ are proportional to the eigenvectors. For the case of distinct roots, that is, only one independent eigenvector corresponds to each eigenvalue, the columns of $[F(\lambda_i)]$ are expected to be linearly dependent or zero.

(d) For the case that the eigenvalue λ_i has multiplicity m, the $f(\lambda_i)$ should have a factor $(\lambda - \lambda_i)^m$. Rewriting (9) to be

$$(\lambda [u] - [S]) [F(\lambda)] = f(\lambda) [u] \tag{11a}$$

for $\lambda = \lambda_i$

$$(\lambda_i [u] - [S]) [F(\lambda_i)] = 0 \tag{11b}$$

Differentiating (11a) with respect to λ, one has

$$(\lambda [u] - [S]) [F(\lambda)]' + [F(\lambda)] = f'(\lambda) [u] \tag{12a}$$

$$(\lambda [u] - [S]) [F(\lambda)]'' + 2[F(\lambda)]' = f''(\lambda) [u] \tag{12b}$$

$$(\lambda [u] - [S]) [F]''' + 3[F]'' = f''' [u] \tag{12c}$$

$$\cdots\cdots\cdots\cdots$$

$$(\lambda [u] - [S]) [F]^{(m-1)} + (m - 1) [F]^{(m-2)} = f^{(m-1)} [u] \tag{12m}$$

Since $f(\lambda)$ has a factor $(\lambda - \lambda_i)^m$, therefore

$$f^{(k-1)}(\lambda_i) = 0 \qquad \text{for } k = 1, 2, \ldots, m \tag{13}$$

Equations (13) and (12a) yield

$$(\lambda_i [u] - [S]) [F(\lambda_i)]' + [F(\lambda_i)] = 0 \tag{14}$$

or

$$([S] - \lambda_i [u]) [F(\lambda_i)]' = [F(\lambda_i)] \tag{15a}$$

Similarly, the remaining relations in (12) can be expressed as follows.

$$([S] - \lambda_i [u]) \; [F(\lambda_i)]'' = 2 \; [F(\lambda_i)]' \tag{15b}$$

$$([S] - \lambda_i [u]) \; [F(\lambda_i)]''' = 3 \; [F(\lambda_i)]'' \tag{15c}$$

$$\cdots \cdots \cdots$$

$$([S] - \lambda_i [u]) \; [F(\lambda_i)]^{(m-1)} = (m - 1) \; [F(\lambda_i)]^{(m-2)} \tag{15m}$$

Multiply (15a) by ($[S] - \lambda_i [u]$) to get

$$([S] - \lambda_i [u])^2 \; [F(\lambda_i)]' = ([S] - \lambda_i [u]) \; [F(\lambda_i)] = 0 \tag{16}$$

where (11b) is used to obtain the final result. The remaining relations in (15) can be simplified similarly. Summarizing, the final expressions are

$$([S] - \lambda_i [u]) \; [F(\lambda_i)] = 0 \tag{17a}$$

$$([S] - \lambda_i [u])^2 \; [F(\lambda_i)]' = 0 \tag{17b}$$

$$\cdots \cdots \cdots$$

$$([S] - \lambda_i [u])^{m-1} \; [F(\lambda_i)]^{(m-2)} = 0 \tag{17m}$$

Hence, for the eigenvalue λ_i of multiplicity m, the m equations in (17) should be satisfied. These m equations generate m vectors which are known as the generalized eigenvectors.

(e) Sometimes it is possible to obtain all m vectors without the use of all m relations in (17). This can be demonstrated by considering a matrix which has an eigenvalue λ_k with a multiplicity of 3. These eigenvalues can "break" as

(i) $[(\lambda_k), (\lambda_k), (\lambda_k)]$
(ii) $[(\lambda_k, \lambda_k), (\lambda_k)]$
(iii) $[(\lambda_k, \lambda_k, \lambda_k)]$

The meaning of these cases will be discussed in the following subsections.

(f) The eigenvalues "break" as

$$[(\lambda_k), (\lambda_k), (\lambda_k)]$$

means that the eigenvalues, though repeated, behave as if each were distinct. This is the case when one of the matrices, $[F(\lambda_k)]$ or $[F(\lambda_k)]'$ or $[F(\lambda_k)]''$, has three linearly independent columns, and three linearly independent eigenvectors can be obtained from only one of these matrices.

(g) The eigenvalues "break" as

$$[(\lambda_k, \lambda_k), \lambda_k]$$

means that two of the eigenvalues splits together and the remaining one splits separately. In this case, one of the matrices, $[F(\lambda_k)]$ or $[F(\lambda_k)]'$ or $[F(\lambda_k)]''$, has two linearly independent columns and these can be used as eigenvectors. The remaining vector must be determined from one of the other two matrices.

(h) The eigenvvalues "break" as

$$[\, (\lambda_k, \lambda_k, \lambda_k) \,]$$

means that none of the three matrices, $[F(\lambda_k)]$, $[F(\lambda_k)]'$, and $[F(\lambda_k)]''$, possesses linearly independent columns. That is, each matrix can supply only one eigenvector. This is the case when all three relations are needed to determine all three eigenvectors, or in the case of multiplicity m, all m relations are needed.

(i) The procedure to find the eigenvectors for an eigenvalue λ_i with a multiplicity m is:

(1) Construct the function $[F(\lambda_i)]$, Eq. (5).

(2) Find the derivatives $[F(\lambda_i)]^{(k)}$, for $k = 1, 2, \ldots, (m - 1)$.

(3) The linearly independent columns of $[F(\lambda_i)]^{(k)}$ provide the generalized eigenvectors.

7.1. Example: Eigenvectors for Repeated Roots

Determine the eigenvectors of a matrix [A] with repeated roots.

$$[A] = \begin{bmatrix} 2 & 2 & 1 \\ 1 & 3 & 1 \\ 1 & 2 & 2 \end{bmatrix} \tag{1}$$

Solution:

The characteristic equation is

$$f(\lambda) = \det \left| [A] - \lambda [u] \right| = \det \begin{bmatrix} 2-\lambda & 2 & 1 \\ 1 & 3-\lambda & 1 \\ 1 & 2 & 2-\lambda \end{bmatrix}$$

$$= 5 - 11\lambda + 7\lambda^2 - \lambda^3 \tag{2}$$

$$= (1-\lambda)(5-\lambda)(1-\lambda)$$

From (2), one has

$$f(\lambda) = \lambda^3 - 7\lambda^2 + 11\lambda - 5 \tag{3}$$

$$f(\lambda[u]) = \lambda^3 [u] - 7\lambda^2 [u] + 11\lambda [u] - 5 [u] \tag{4}$$

$$f([A]) = [A]^3 - 7 [A]^2 + 11 [A] - 5 [u] \tag{5}$$

The subtraction of (5) from (4) yields

$$f(\lambda[u]) - f([A]) = (\lambda^3 [u] - [A]^3) - 7 (\lambda^2 [u] - [A]^2) + 11 (\lambda [u] - [A]) \tag{6}$$

The g function is then given by [Eq. (7.4)]

$$g = \frac{\lambda^3 [u] - [A]^3}{\lambda [u] - [A]} - 7 \frac{\lambda^2 [u] - [A]^2}{\lambda [U] - [A]} + 11 \frac{\lambda [u] - [A]}{\lambda [u] - [A]}$$

$$= (\lambda^2 [u] + \lambda [A] + [A]^2) - 7 (\lambda [u] + [A]) + 11 [u]$$

$$= (\lambda^2 - 7\lambda + 11) [u] + (\lambda - 7) [A] + [A]^2 \tag{7}$$

From (1), one has

$$[A]^2 = [A][A] = \begin{bmatrix} 2 & 2 & 1 \\ 1 & 3 & 1 \\ 1 & 2 & 2 \end{bmatrix} \begin{bmatrix} 2 & 2 & 1 \\ 1 & 3 & 1 \\ 1 & 2 & 2 \end{bmatrix}$$

$$= \begin{bmatrix} 7 & 12 & 6 \\ 6 & 13 & 6 \\ 6 & 12 & 7 \end{bmatrix}$$

Then

$$g = \begin{bmatrix} \lambda^2 - 7\lambda + 11 & 0 & 0 \\ 0 & \lambda^2 - 7\lambda + 11 & 0 \\ 0 & 0 & \lambda^2 - 7\lambda + 11 \end{bmatrix}$$

$$+ (\lambda - 7) \begin{bmatrix} 2 & 2 & 1 \\ 1 & 3 & 1 \\ 1 & 2 & 2 \end{bmatrix} + \begin{bmatrix} 7 & 12 & 6 \\ 6 & 13 & 6 \\ 6 & 12 & 7 \end{bmatrix}$$

$$= \begin{bmatrix} \lambda^2 - 5\lambda + 4 & 2\lambda - 2 & \lambda - 1 \\ \lambda - 1 & \lambda^2 - 4\lambda + 3 & \lambda - 1 \\ \lambda - 1 & 2\lambda - 2 & \lambda^2 - 5\lambda + 4 \end{bmatrix} \qquad (8)$$

For $\lambda_1 = 5$, (8) yields

$$g(\lambda_1) = \begin{bmatrix} 4 & 8 & 4 \\ 4 & 8 & 4 \\ 4 & 8 & 4 \end{bmatrix} \qquad (9)$$

Hence

$$[X_1] = [1 \quad 1 \quad 1]^t \qquad \text{or} \qquad [X_1] = \frac{1}{\sqrt{3}}[1 \quad 1 \quad 1]^t \qquad (10)$$

For $\lambda_2 = 1$, then

$$g(\lambda_2) = \begin{bmatrix} 0 & 0 & 0 \\ 0 & 0 & 0 \\ 0 & 0 & 0 \end{bmatrix} \quad \text{trivial case}$$

$$\frac{dg}{d\lambda} = \begin{bmatrix} 2\lambda - 5 & 2 & 1 \\ 1 & 2\lambda - 4 & 1 \\ 1 & 2 & 2\lambda - 5 \end{bmatrix}$$

$$\left. \frac{dg}{d\lambda} \right]_{\lambda=1} = \begin{bmatrix} -3 & 2 & 1 \\ 1 & -2 & 1 \\ 1 & 2 & -3 \end{bmatrix} \tag{11}$$

The sum of all three columns in (11) is zero. Therefore, there can be at most two independent columns. Choose the first two columns as eigenvectors.

$$[X_2] = [\, -3 \quad 1 \quad 1 \,]^t \quad \text{or} \quad [X_2] = \frac{1}{\sqrt{11}} [\, -3 \quad 1 \quad 1 \,]^t$$

$$[X_3] = [\, 2 \quad -2 \quad 2 \,]^t \quad \text{or} \quad [X_2] = \frac{1}{\sqrt{3}} [\, 1 \quad -1 \quad 1 \,]^t$$

8. Cayley-Hamilton Theorem

This theorem states that every square matrix satisfies its own characteristic equation. If the characteristic polynomial of an mth-order matrix [A] is given by

$$f(\lambda) = \det \left| [A] - \lambda [u] \right|$$

then

$$f([A]) = \det \left| [A] - [A] [u] \right| = 0$$

Consider a polynomial in [A] of degree p having the general form

$$F([A]) = d_p [A]^p + d_{p-1} [A]^{p-1} + \cdots + d_0 [u] \tag{1}$$

where

$$[A]^2 = [A] [A]$$
$$[A]^3 = [A] [A] [A]$$

$$\cdots \cdots$$

Post-multiplying (1) by any eigenvector $[X_i]$ of matrix [A] produces

$$F([A]) [X_i] = (d_p [A]^p + d_{p-1} [A]^{p-1} + \cdots + d_0 [u]) [X_i] \tag{2}$$

The eigenvalue equation of [A] is

$$[A] [X_i] = \lambda_i [X_i] \tag{3a}$$

Then

$$[A]^2 [X_i] = [A] [A] [X_i] = [A] (\lambda_i [X_i]) = \lambda_i^2 [X_i] \tag{3b}$$

$$\cdots \cdots \cdots \cdots$$

$$[A]^m [X_i] = \lambda_i^m [X_i] \tag{3m}$$

Thus

$$F([A]) [X_i] = (d_p\lambda_i^p + d_{p-1}\lambda_i^{p-1} + \cdots + d_0) [X_i] = F(\lambda_i) [X_i] \tag{4a}$$

and

$$F(\lambda_i) \equiv d_p \lambda_i^p + d_{p-1} \lambda_i^{p-1} + \cdots + d_0 \tag{4b}$$

or

$$\{F([A]) - F(\lambda_i) [u]\} [X_i] = 0 \tag{5}$$

The last equation can be rearranged more explicitly as

$$\{F([A]) - \gamma_i [u]\} [X_i] = 0 \tag{6a}$$
$$\gamma_i \equiv F(\lambda_i) \tag{6b}$$

This has the form of an eigenvalue equation. It implies that $[X_i]$ is an eigenvector of the matrix function $F([A])$ and the corresponding eigenvalue is γ_i. The function of a matrix $[A]$, $F([A])$, has the same eigenvector as the matrix $[A]$; its eigenvalue is related to the eigenvalue of $[A]$ by the same function, $\gamma_i = F(\lambda_i)$. For a nontrivial solution of the general equation

$$\{F([A]) - \gamma [u] \} [X] = 0 \tag{7}$$

it is necessary that

$$\det \left| \; F([A]) - \gamma [u] \; \right| = 0 \tag{8}$$

since $[X]$ does not equal zero in general.

When $F(\lambda_i)$ is chosen to be the characteristic polynomial of matrix $[A]$, i.e.,

$$F(\lambda) \equiv f(\lambda) = \det \left| \; [A] - \lambda [u] \; \right| \tag{9}$$

then

$$F(\lambda_i) = f(\lambda_i) = 0 \tag{10}$$

Consequently, (6a) becomes

$$\{ F([A]) - F(\lambda_i) [u] \} [X_i] = \{ f([A]) - f(\lambda_i) [u] \} [X_i] = f([A]) [X_i] = 0 \tag{11}$$

Therefore,

$$f([A]) = 0 \tag{12}$$

This implies that the matrix $[A]$ satisfies its own characteristic equation.

9. Eigenvectors and Eigenvalues of a Two-Port Device

(a) The scattering matrix of a two-port device is a 2 x 2 matrix and the eigenvalue equation is

$$[S] [X] = \lambda [X] \tag{1a}$$

where

$$[S] = \begin{bmatrix} s_{11} & s_{12} \\ s_{21} & s_{22} \end{bmatrix} \tag{1b}$$

$$[X] = [x_1 \quad x_2]^t \tag{1c}$$

(b) The characteristic equation is

$$\det \left| [S] - \lambda [u] \right| = \det \begin{bmatrix} s_{11} - \lambda & s_{12} \\ s_{21} & s_{22} - \lambda \end{bmatrix} \tag{2a}$$

A symmetrical two-port device is a device which has identical behavior when its ports are interchanged. Mathematically, this property is specified by $s_{12} = s_{21}$ and $s_{11} = s_{22}$.

A device is reciprocal if the ratio of the excitation to its response is independent of the location of the source, $s_{21} = s_{12}$. Note that the structure does not have to be symmetrical in order for its matrices to be symmetrical.

For symmetrical and reciprocal devices, then

$$\det \left| [S] - \lambda[u] \right| = \det \begin{bmatrix} s_{11} - \lambda & s_{21} \\ s_{21} & s_{11} - \lambda \end{bmatrix} = 0 \tag{2b}$$

or

$$(s_{11} - \lambda)^2 - s_{21}^2 = 0 \tag{2c}$$

That is,

$$s_{11} - \lambda = \pm s_{21} \tag{2d}$$

The solutions are

$$\lambda_1 = s_{11} + s_{21} \tag{3a}$$
$$\lambda_2 = s_{11} - s_{21} \tag{3b}$$

The eigenvalues are linear combinations of the scattering coefficients for symmetrical and reciprocal devices.

(c) The scattering coefficients for symmetrical and reciprocal devices can also be expressed in terms of the eigenvalues by taking the sum and the difference of (3a) and (3b).

$$s_{11} = (1/2) (\lambda_1 + \lambda_2) \tag{4a}$$
$$s_{21} = (1/2) (\lambda_1 - \lambda_2) \tag{4b}$$

Thus, if either set of variables is known, the other set is also specified. Consequently, the boundary conditions of junctions may be established in terms of either set of variables.

(d) For $\lambda = \lambda_1 = (s_{11} + s_{21})$, the eigenvalue equation, (1a), becomes

$$[S] [X_1] = \lambda_1 [X_1] \tag{5a}$$

where

$$[X_1] = [x_{11} \quad x_{12}]^t \tag{5b}$$

Equation (5a) can be expanded as follows.

$$\begin{bmatrix} s_{11} & s_{21} \\ s_{21} & s_{11} \end{bmatrix} \begin{bmatrix} x_{11} \\ x_{12} \end{bmatrix} = (s_{11} + s_{21}) \begin{bmatrix} x_{11} \\ x_{12} \end{bmatrix} \tag{6a}$$

or

$$s_{11}x_{11} + s_{21}x_{12} = (s_{11} + s_{21}) x_{11} \tag{6b}$$

$$s_{21}x_{11} + s_{11}x_{12} = (s_{11} + s_{21}) x_{12} \tag{6c}$$

Both (6b) and (6c) reduce to

$$x_{12} = x_{11} \tag{6d}$$

Therefore,

$$[X_1] = x_{11} [1 \quad 1]^t \tag{6e}$$

where x_{11} has an arbitrary magnitude. For the normalized eigenvector, one has

$$|[X_1]| = 1 = \sqrt{x_{11}^2 + x_{12}^2} = \sqrt{2}x_{11}$$

or

$$x_{11} = \frac{1}{\sqrt{2}} \tag{6f}$$

Therefore,

$$[X_1] = \frac{1}{\sqrt{2}} [1 \quad 1]^t \tag{7}$$

(e) It was shown in Section 4 that an eigenvector represents the excitation of a device. The eigenvector $[X_1]$, (7), represents the excitation of the two-port device by two in-phase incident waves (Fig. 9-1). The in-phase excitation produces an open circuit at the plane of symmetry of the device. This will be obvious by considering the low-frequency network excited by in-phase voltage sources (Fig. 9-2). The currents in each of the wires at the mid-point will be equal because of the symmetry but will be oppositely directed so that the net current at the midpoint is zero. With no current in any of the wires crossing the midpoint plane, each wire may be cut and no change will result at either of the driving-point terminals.

131

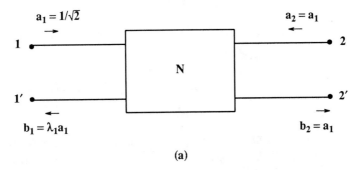

(a)

Figure 9-1(a): Two-port network excited by in-phase sources

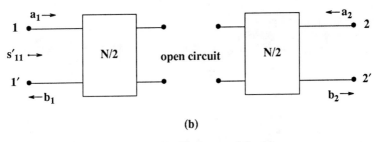

(b)

Figure 9-1(b): Eigen-network for (a)

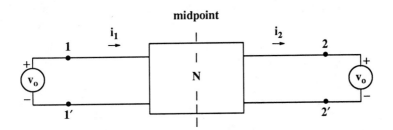

Figure 9-2: Two-port network excited by in-phase voltage sources

The eigenvector $[X_1]$ is shown in Fig. 9-1a, i.e., the two-port network is excited by in-phase incident waves with magnitude of $2^{-1/2}$. The one-port network obtained by bisecting the two-port network along the plane of symmetry with the bisected terminals left open is called an eigen-network. The reflected wave parameter b_1 of the eigen-network is equal to the corresponding eigenvalue λ_1 times a_1. Figure 9-1b is the eigen-network of the in-phase eigen-solution. Therefore, for the one-port eigen-network Fig. 9-1b, one has

$$b_1 = s_{11}'a_1 = \lambda_1 a_1 \qquad \text{or} \qquad s_{11}' = \lambda_1$$

where s_{11}' is the reflection coefficient at the input of the short-circuited eigen-network.

At high frequencies, the electric open circuit is interpreted as a magnetic wall (H-field = 0).

(f) For $\lambda = \lambda_2 = s_{11} - s_{21}$, one has

$$\begin{bmatrix} s_{11} & s_{21} \\ s_{21} & s_{11} \end{bmatrix} \begin{bmatrix} x_{21} \\ x_{22} \end{bmatrix} = (s_{11} - s_{21}) \begin{bmatrix} x_{21} \\ x_{22} \end{bmatrix} \tag{8a}$$

$$s_{11}x_{21} + s_{21}x_{22} = (s_{11} - s_{21})x_{21} \tag{8b}$$

$$s_{21}x_{21} + s_{11}x_{22} = (s_{11} - s_{21})x_{22} \tag{8c}$$

or

$$x_{21} = -x_{22} \tag{8d}$$

That is,

$$[X_2] = x_{22}[\,1 \qquad -1\,]^t \tag{8e}$$

The normalized eigenvector is

$$[X_2] = \frac{1}{\sqrt{2}}[\,1 \qquad -1\,]^t \tag{9}$$

(g) The excitation by $[X_2]$ represents the excitation of the device by two oppositely phased incident waves, Fig. 9-3. The equivalent-low frequency case is shown in Fig. 9-4.

The currents in each of the wires at the midpoint are again equal because of the symmetry of the network but are co-directional in each wire in view of the oppositely phased driving sources. Short-circuiting all wires at the midpoint plane will not cause any disturbance at the driving terminals. In other words, the relative potential of any wire at midpoint caused by the driving source at port 1 is equal to (because of symmetry) but oppositely directed from that caused by the driving source at port 2 (because of the relative phase of the two driving sources). The net potential of any wire at the midpoint is therefore zero. The short-circuiting of points at zero potential will produce no disturbance.

The eigen-network of the oppositely phased eigen-solution is shown in Fig. 9-3b. For this one-port network, one has

$$b_1 = s_{11}''a_1 = \lambda_2 a_1 \qquad \text{or} \qquad s_{11}'' = \lambda_2$$

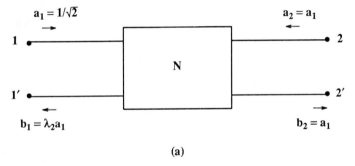

(a)

Figure 9-3(a): Two-port network excited by oppositely phased sources

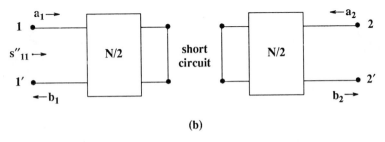

(b)

Figure 9-3(b): Eigen-network for (a)

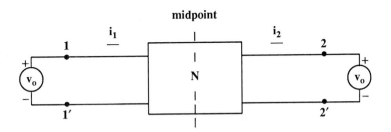

Figure 9-4: Two port network excited by oppositely phased voltage sources

where s_{11}'' is the reflection coefficient at the input of the short-circuited eigen-network. At high frequencies, the short circuit is equivalent to an electric wall (E-field = 0).

9.1. Example: Scattering Matrix of a Shunt Admittance

Determine the scattering matrix using the eigen-network concept. Consider a two-port device which is characterized as a shunt admittance Y, Fig. 9.1-1.

Solution:

The eigen-network is obtained by expressing the original network by two symmetrical parts, Fig. 9.1-2, and then bisecting it into two equal portions.

λ_1 is obtained from the open-circuited eigen-network, Fig. 9.1-3 [see Section 9(e)].

$\lambda_1 = s_{11}' = $ reflection coefficient of the open-circuited eigen-network

$$= \frac{Y_o - Y/2}{Y_o + Y/2} \tag{1}$$

λ_2 is obtained from the short-circuited eigen-network, Fig. 9.1-4 [see Section 9(g)].

$\lambda_2 = s_{11}'' = $ reflection coefficient of the short-circuited eigen-network

$$= \frac{Y_o - \infty}{Y_o + \infty} = -1 \tag{2}$$

The scattering coefficients for the original network are:

$$s_{11} = \frac{\lambda_1 + \lambda_2}{2} = \frac{1}{2}\left[\frac{Y_o - Y/2}{Y_o + Y/2} - 1 \right]$$

$$= \frac{-Y}{Y + 2Y_o} \tag{3a}$$

$$s_{21} = \frac{\lambda_1 - \lambda_2}{2} = \frac{1}{2}\left[\frac{Y_o - Y/2}{Y_o + Y/2} - (-1) \right]$$

$$= \frac{2Y_o}{Y + 2Y_o} \tag{3b}$$

And the scattering matrix is

$$[S] = \begin{bmatrix} s_{11} & s_{21} \\ s_{21} & s_{11} \end{bmatrix} \tag{4}$$

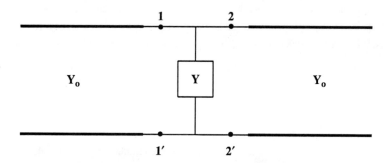

Figure 9.1-1: A shunt admittance as a two-port network

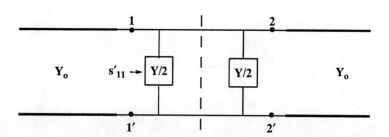

Figure 9.1-2: Equivalent circuit for a shunt element

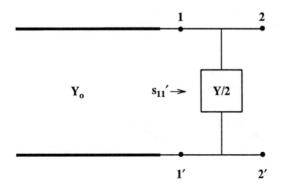

Figure 9.1-3: Open-circuited eigen-network

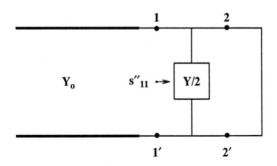

Figure 9.1-4: Short-circuited eigen-network

10. Diagonalization of Scattering Matrix - Distinct Eigenvalues

A diagonal matrix is highly desirable because of the ease in manipulation. For example, the power of a diagonal matrix is obtained by taking the power of the diagonal elements, i.e., if

$$[D_1] = \begin{bmatrix} \alpha & 0 \\ 0 & \beta \end{bmatrix}$$

then

$$[D_1]^k = \begin{bmatrix} \alpha^k & 0 \\ 0 & \beta^k \end{bmatrix}$$

For a general mth-order matrix [S] with distinct eigenvalues, there are m linearly independent eigenvectors, $[X_1], [X_2], \ldots, [X_m]$.

$$[S][X_i] = \lambda_i [X_i] \qquad i = 1, 2, \ldots, m \tag{1a}$$

$$[X_i] = [x_{i1} \quad x_{i2} \quad \cdots \quad x_{im}]^t \tag{1b}$$

Construct a square matrix [P] whose columns are the eigenvectors of [S].

$$[P] = [[X_1] \quad [X_2] \quad \cdots \quad [X_m]]$$

$$= \begin{bmatrix} x_{11} & x_{21} & \cdots & x_{m1} \\ x_{12} & x_{22} & \cdots & x_{m2} \\ \cdots & \cdots & \cdots & \cdots \\ x_{1m} & x_{2m} & \cdots & x_{mm} \end{bmatrix} \tag{2}$$

Let the diagonal matrix [D] be defined as

$$[D] \equiv \begin{bmatrix} \lambda_1 & 0 & \cdots & 0 \\ 0 & \lambda_2 & \cdots & 0 \\ \cdots & \cdots & \cdots & \cdots \\ 0 & 0 & \cdots & \lambda_m \end{bmatrix} \tag{3}$$

Consequently,

$$[P][D] = \begin{bmatrix} \lambda_1 x_{11} & \lambda_2 x_{21} & \cdots & \lambda_m x_{m1} \\ \lambda_1 x_{12} & \lambda_2 x_{22} & \cdots & \lambda_m x_{m2} \\ \cdots & \cdots & \cdots & \cdots \\ \lambda_1 x_{1m} & \lambda_2 x_{2m} & \cdots & \lambda_m x_{mm} \end{bmatrix}$$

$$= [\lambda_1 [X_1] \quad \lambda_2 [X_2] \quad \cdots \quad \lambda_m [X_m]] \tag{4}$$

Equation (1) can be written for all eigenvectors,

$$[S][P] = [[S][X_1] \quad [S][X_2] \quad \cdots \quad [S][X_m]]$$
$$= [\lambda_1 [X_1] \quad \lambda_2 [X_2] \quad \ldots \quad \lambda_m [X_m]] \tag{5}$$

by virtue of (1a). The substitution of (4) into (5) yields

$$[S][P] = [P][D] \tag{6}$$

Pre-multiplying (6) by $[P]^{-1}$ gives

$$[D] = [P]^{-1} [S][P] \tag{7}$$

where [D] is a diagonal matrix by definition. $[P]^{-1}$ exists since its columns are linearly independent.

Thus, a square matrix [S] with distinct eigenvalues can be diagonalized by the transformation of (7), where [P] is a square matrix whose columns are the eigenvectors of the matrix [S].

11. Diagonalization of a Symmetric Matrix

It was shown in Section 6(g) [Eq. 6-30] that the eigenvectors of a real symmetric matrix are orthogonal. For normalized eigenvectors, this is expressed as

$$[X_i]^t [X_k] = \delta_{ki} \tag{1}$$

where δ_{ki} is the Kronecker delta function, i.e.,

$$
\begin{aligned}
\delta_{ki} &= 1 && \text{for } k = i \\
&= 0 && \text{for } k \neq i
\end{aligned}
\tag{2}
$$

For this case, the [P] matrix [Eq. 10-2]

$$[P] = [[X_1] \ [X_2] \ \cdots \ [X_m]] \tag{3}$$

satisfies the relation

$$[P]^t [P] = [u] = [P]^{-1} [P] \tag{4}$$

by virtue of (1); therefore

$$[P]^t = [P]^{-1} \tag{5}$$

for normalized eigenvectors of a real symmetric matrix.

The diagonalization relation, (10-7), becomes

$$[D] = [P]^{-1} [S] [P] = [P]^t [S] [P] \tag{6}$$

A real symmetric matrix [S] [with distinct or repeated eigenvalues and with eigenvectors satisfying (1)] may be diagonalized by the orthogonal transformation (6). The orthogonal matrix [P] is a matrix whose columns are the normalized eigenvectors of the [S] matrix. The diagonal elements of the matrix [D] equal the eigenvalues of the [S] matrix.

11.1 Example: Diagonalization

Diagonalize the matrix

$$[S] = \begin{bmatrix} 3 & 4 \\ 2 & 1 \end{bmatrix} \tag{1}$$

Solution:

The eigenvectors of this matrix are found to be (see Section 5.1)

$$[X_1] = \frac{1}{\sqrt{5}} \begin{bmatrix} 2 & 1 \end{bmatrix}^t \quad \text{and} \quad [X_2] = \frac{1}{\sqrt{2}} \begin{bmatrix} 1 & -1 \end{bmatrix}^t \tag{2}$$

The [P] matrix is

$$[P] = [\ [X_1] \quad [X_2]\ ] = \begin{bmatrix} \dfrac{2}{\sqrt{5}} & \dfrac{1}{\sqrt{2}} \\ \dfrac{1}{\sqrt{5}} & \dfrac{-1}{\sqrt{2}} \end{bmatrix} \tag{3}$$

Since [S] is not symmetric, therefore, $[P]^{-1} \neq [P]^t$. The inverse of [P] can be determined according to Section 11.2.

$$[P]_{cofactor} = \begin{bmatrix} \dfrac{-1}{\sqrt{2}} & \dfrac{-1}{\sqrt{5}} \\ \dfrac{-1}{\sqrt{2}} & \dfrac{2}{\sqrt{5}} \end{bmatrix}$$

$$\text{Adj } [P] = [P]_{cofactor}^t = \begin{bmatrix} \dfrac{-1}{\sqrt{2}} & \dfrac{-1}{\sqrt{2}} \\ \dfrac{-1}{\sqrt{5}} & \dfrac{2}{\sqrt{5}} \end{bmatrix}$$

$$\Delta_P = \frac{-2}{\sqrt{10}} - \frac{1}{\sqrt{10}} = \frac{-3}{\sqrt{10}}$$

The inverse of [P] is therefore

$$[P]^{-1} = \frac{\text{Adj } [P]}{\Delta_P} = \frac{1}{\Delta_P} \begin{bmatrix} \dfrac{-1}{\sqrt{2}} & \dfrac{-1}{\sqrt{2}} \\ \dfrac{-1}{\sqrt{5}} & \dfrac{2}{\sqrt{5}} \end{bmatrix}$$

Then the diagonal matrix is

$$[D] = [P]^{-1} [S] [P] = \frac{1}{\Delta_P} \begin{bmatrix} 3 & 4 \\ 2 & 1 \end{bmatrix} \begin{bmatrix} \dfrac{2}{\sqrt{5}} & \dfrac{1}{\sqrt{2}} \\ \dfrac{1}{\sqrt{5}} & \dfrac{-1}{\sqrt{2}} \end{bmatrix}$$

$$= \frac{-\sqrt{10}}{3} \begin{bmatrix} \dfrac{-15}{\sqrt{10}} & 0 \\ 0 & \dfrac{3}{\sqrt{10}} \end{bmatrix} = \begin{bmatrix} 5 & 0 \\ 0 & -1 \end{bmatrix}$$

11.2. Inverse of a Matrix

The formula for the inverse of a matrix [A] can be found in most textbooks on matrices, and it is repeated here for convenience. Consider the matrix [A]

$$[A] = \begin{bmatrix} a_{11} & a_{12} & \cdots & a_{1m} \\ a_{21} & a_{22} & \cdots & a_{2m} \\ \cdots & \cdots & \cdots & \cdots \\ a_{m1} & a_{m2} & \cdots & a_{mm} \end{bmatrix} \tag{1}$$

Its inverse is given by the relation

$$[A]^{-1} = \frac{\text{Adj } [A]}{\Delta_A} \tag{2}$$

where Δ_A is the determinant of [A]. The adjoint of [A] is defined by

$$\text{Adj } [A] \equiv [A]^t_{\text{cofactor}} \tag{3a}$$

where

$$[A]_{\text{cofactor}} \equiv \begin{bmatrix} A_{11} & A_{12} & \cdots & A_{1m} \\ A_{21} & A_{22} & \cdots & A_{2m} \\ \cdots & \cdots & \cdots & \cdots \\ A_{m1} & A_{m2} & \cdots & A_{mm} \end{bmatrix} \tag{3b}$$

and

$$A_{jk} = \text{cofactor of element } a_{jk} = (-1)^{j+k} \left| [A_{jk}] \right|$$

$$[A_{jk}] = [A] \text{ matrix with jth row and kth column deleted}$$

12. Diagonalization - Multiple Eigenvalues

(a) A nonsymmetric matrix with repeated eigenvalues is not diagonalizable, but it may be reduced to Jordan normal form.

An m by m matrix with repeated eigenvalues possesses generalized eigenvectors which can be determined by the following set of equations, (7-15), for $\lambda = \lambda_i$.

$$([S] - \lambda_i [u]) [F(\lambda_i)] = 0 \tag{1a}$$

$$([S] - \lambda_i [u]) [F^{(1)}(\lambda_i)] = [F(\lambda_i)] \tag{1b}$$

$$([S] - \lambda_i [u]) [F^{(2)}(\lambda_i)] = 2 [F^{(1)}(\lambda_i)] \tag{1c}$$

$$([S] - \lambda_i [u]) [F^{(3)}(\lambda_i)] = 3 [F^{(2)}(\lambda_i)] \tag{1d}$$

$$. . . . .$$

$$([S] - \lambda_i [u]) [F^{(m-1)}(\lambda_i)] = (m - 1) [F^{(m-2)}(\lambda_i)] \tag{1m}$$

For the case where the repeated eigenvalues all "break" together,

$$(\lambda_i, \lambda_i, \ldots, \lambda_i),$$

then one generalized eigenvector is provided by each of these functions,

$$[F(\lambda_i)], \ [F^{(1)}(\lambda_i)], \ldots, [F^{(m-1)}(\lambda_i)]$$

Rearrange (1) into the following form.

$$([S] - \lambda_i [u]) (1/0!) [F(\lambda_i)] = 0 \tag{2a}$$

$$([S] - \lambda_i [u]) (1/1!) [F^{(1)}(\lambda_i)] = (1/0!) [F(\lambda_i)] \tag{2b}$$

$$([S] - \lambda_i [u]) (1/2!) [F^{(2)}(\lambda_i)] = (1/1!) [F^{(1)}(\lambda_i)] \tag{2c}$$

$$([S] - \lambda_i [u]) (1/3!) [F^{(3)}(\lambda_i)] = (1/2!) [F^{(2)}(\lambda_i)] \tag{2d}$$

$$. . . . .$$

$$([S] - \lambda_i [u]) 1/(m-1)! [F^{(m-1)}(\lambda_i)] = 1/(m-2)! [F^{(m-2)}(\lambda_i)] \tag{2m}$$

Let the generalized eigenvectors be defined as follows.

$$[a_1] = (1/0!) [F(\lambda_i)] \tag{3a}$$

$$[a_2] = (1/1!) [F^{(1)}(\lambda_i)] \tag{3b}$$

$$[a_3] = (1/2!) [F^{(2)}(\lambda_i)] \tag{3c}$$

$$. . . . .$$

$$[a_m] = 1/(m-1)! [F^{(m-1)}(\lambda_i)] \tag{3m}$$

Then (2) becomes

$$([S] - \lambda_i [u]) [a_1] = 0 \qquad \text{or} \qquad [S] [a_1] = \lambda_i [a_1] \tag{4a}$$

$$([S] - \lambda_i [u]) [a_2] = [a_1] \quad \text{or} \quad [S] [a_2] = \lambda_i [a_2] + [a_1] \tag{4b}$$

$$([S] - \lambda_1 [u]) [a_3] = [a_2] \quad \text{or} \quad [S] [a_3] = \lambda_i [a_3] + [a_2] \tag{4c}$$

. . . .

$$([S] - \lambda_i [u]) [a_m] = [a_{m-1}] \quad \text{or} \quad [S] [a_m] = \lambda_i [a_m] + [a_{m-1}] \tag{4m}$$

Define the [P] matrix as

$$[P] = [\, [a_1] \quad [a_2] \quad \ldots \quad [a_m] \,] \tag{5}$$

Then (4) can be expressed in matrix form.

$$[S] [P] = [P] [J] \tag{6a}$$

where

$$[J] = \begin{bmatrix} \lambda_i & 1 & 0 & \cdots & 0 & 0 \\ 0 & \lambda_i & 1 & \cdots & 0 & 0 \\ 0 & 0 & \lambda_i & \cdots & 0 & 0 \\ \cdots & \cdots & \cdots & \cdots & \cdots & \cdots \\ 0 & 0 & 0 & \cdots & \lambda_i & 1 \\ 0 & 0 & 0 & \cdots & 0 & \lambda_i \end{bmatrix} \tag{6b}$$

[J] is known as the Jordan normal form. It is a square matrix with λ_i as its leading diagonal elements; elements immediately above the leading diagonal are unity and all other elements are zero. The unity element will be replaced by zero if the eigenvalues break separately, i.e., the Jordan form will then reduce to a diagonal matrix.

Equation (6) can be written as

$$[J] = [P]^{-1} [S] [P]$$

(b) The Jordan form of the general case when a matrix has some distinct and some repeated eigenvalues can be illustrated by considering an example. Suppose the matrix has the following eigenvalues:

$$\lambda_1, \lambda_2, (\lambda_3), (\lambda_3), (\lambda_4, \lambda_4), (\lambda_4), (\lambda_5, \lambda_5, \lambda_5).$$

Then the Jordan form is

$$[J] = \begin{bmatrix} \lambda_1 & 0 & 0 & 0 & 0 & 0 & 0 & 0 & 0 & 0 \\ 0 & \lambda_2 & 0 & 0 & 0 & 0 & 0 & 0 & 0 & 0 \\ 0 & 0 & \lambda_3 & 0 & 0 & 0 & 0 & 0 & 0 & 0 \\ 0 & 0 & 0 & \lambda_3 & 0 & 0 & 0 & 0 & 0 & 0 \\ 0 & 0 & 0 & 0 & \lambda_4 & 1 & 0 & 0 & 0 & 0 \\ 0 & 0 & 0 & 0 & 0 & \lambda_4 & 0 & 0 & 0 & 0 \\ 0 & 0 & 0 & 0 & 0 & 0 & \lambda_4 & 0 & 0 & 0 \\ 0 & 0 & 0 & 0 & 0 & 0 & 0 & \lambda_5 & 1 & 0 \\ 0 & 0 & 0 & 0 & 0 & 0 & 0 & 0 & \lambda_5 & 1 \\ 0 & 0 & 0 & 0 & 0 & 0 & 0 & 0 & 0 & \lambda_5 \end{bmatrix} \qquad (7)$$

13. Unitary Property

By virtue of conservation of energy, the total power incident at all ports of a passive structure, P_{in}, should equal the sum of power absorbed by the structure, P_{diss}, and power emergent from all ports, P_{out}.

$$P_{in} = P_{diss} + P_{out} \tag{1a}$$

or

$$P_{diss} = P_{in} - P_{out}$$

$$= \frac{1}{2} \sum_{k}^{m} (a_k a_k^* - b_k b_k^*) = \frac{1}{2} ([a^*]^t [a] - [b^*]^t [b])$$

$$= \frac{1}{2} ([a^*]^t \{ [u] - [S^*]^t [S] \} [a]) \equiv \frac{1}{2} [a^*]^t [Q] [a] \tag{1b}$$

$$[a] = [a_1 \quad a_2 \quad \cdots \quad a_m]^t \tag{1c}$$

$$[b] = [b_1 \quad b_2 \quad \cdots \quad b_m]^t \tag{1d}$$

$$[b] = [S] [a] \tag{1e}$$

$$[Q] \equiv [u] - [S^*]^t [S] = \text{dissipation matrix} \tag{1f}$$

For a lossless structure,

$$P_{diss} = 0 = [a^*]^t \{ [u] - [S^*]^t [S] \} [a]$$

or

$$[S^*]^t [S] = [u] \tag{2}$$

The scattering matrix of a lossless structure is unitary. The unitary property plays an important role in establishing relations between the entries of the scattering matrix $[S]$.

For a two-port structure, the unitary condition

$$[S^*]^t [S] = \begin{bmatrix} s_{11}^* & s_{21}^* \\ s_{12}^* & s_{22}^* \end{bmatrix} \begin{bmatrix} s_{11} & s_{12} \\ s_{21} & s_{22} \end{bmatrix} = [u]$$

yields the following relations.

$$|s_{11}|^2 + |s_{21}|^2 = 1 \tag{3a}$$

$$|s_{22}|^2 + |s_{12}|^2 = 1 \tag{3b}$$

$$s_{11}^* s_{12} + s_{21}^* s_{22} = 0 \tag{3c}$$

$$s_{12}^* s_{11} + s_{22}^* s_{21} = 0 \tag{3d}$$

Equation (3a) [or (3b)] states that the unit incident power at port 1 [or port 2] is equal to the reflected power at port 1 [or port 2] plus the transmitted power at port 2 [or port 1]. This will be more obvious when both sides of (3a) [or (3b)] are multiplied by $|a_1|^2$ [or $|a_2|^2$].

$$|s_{11}a_1|^2 + |s_{21}a_1|^2 = |a_1|^2 \tag{4a}$$

or

$$P_{1ref} + P_{2trans} = P_{1in} \tag{4b}$$

Similarly,

$$|s_{22}a_2|^2 + |s_{12}a_2|^2 = |a_2|^2 \tag{5a}$$

$$P_{2ref} + P_{1trans} = P_{2in} \tag{5b}$$

Equations (3c) and (3d) represent the situation when both ports are excited by their eigenvectors.

$$[b] = [S][a] \tag{6a}$$

$$[a] = [\, 1 \quad 1\,]^t \tag{6b}$$

and

$$\begin{bmatrix} b_1 \\ b_2 \end{bmatrix} = \begin{bmatrix} s_{11} & s_{12} \\ s_{21} & s_{22} \end{bmatrix} \begin{bmatrix} 1 \\ 1 \end{bmatrix}$$

This gives

$$b_1 = s_{11} + s_{12}, \qquad b_2 = s_{21} + s_{22} \tag{6c}$$

The expansion of

$$[b^*]^t [b] = [a^*]^t [a]$$

yields

$$|s_{11}|^2 + |s_{12}|^2 + [\, s_{11}^* s_{12} + s_{11} s_{12}^* \,]$$
$$+ |s_{21}|^2 + |s_{22}|^2 + [\, s_{21}^* s_{22} + s_{21} s_{22}^* \,] = 2$$

or

$$[\, s_{11}^* s_{12} + s_{22} s_{21}^* \,] + [\, s_{12}^* s_{11} + s_{21} s_{22}^* \,] = 0 \tag{7}$$

where (3a) and (3b) are used to obtain (7). Equation (7) will be satisfied provided the quantity within each set of brackets vanishes individually. These are identical to conditions (3c) and (3d). In other words, (3c) and (3d) are the energy balance relations for the case when both ports are excited by corresponding engenvectors.

Identical relations will be obtained when the oppositely phased eigenvector is used, i.e.,

$$[a] = [1 \quad -1]$$

14. Dissipation Matrix

(a) The dissipation matrix is defined by Eq. (13-1f).

$$[Q] = [u] - [S*]^t [S] \tag{1}$$

The dissipation matrix is a function of the scattering matrix [S]. The eigenvectors of a matrix function [F([A])] are identical to those of the matrix [A] and the eigenvalues of [F([A])], q_i, are related to those of [A] by the Cayley-Hamilton theorem, (8-6b);

$$q_i = F(\lambda_i^A) \tag{2}$$

where λ_i^A is the eigenvalue of [A].

Let the eigenvalue equation of [S] be

$$[S] [X_i] = s_i [X_i] \tag{3}$$

where $[X_i]$ is the eigenvector corresponding to the eigenvalue x_i. (λ_i was previously used to represent eigenvalue. In the present analysis, eigenvalues for more than one function are involved. To avoid complicated notation such as λ_i^S for the eigenvalue of [S], the simpler notation s_i is adopted.) Then

$$[Q] [X_i] = q_i [X_i] \tag{4a}$$

$$q_i = 1 - s_i^* s_i \tag{4b}$$

where q_i is the eigenvalue of [Q].

(b) For a two-port symmetrical device, the dissipation matrix is

$$
[Q] = [u] - \begin{bmatrix} s_{11}^* & s_{21}^* \\ s_{21}^* & s_{11}^* \end{bmatrix} \begin{bmatrix} s_{11} & s_{21} \\ s_{21} & s_{11} \end{bmatrix}
$$

$$
= \begin{bmatrix} 1 - (|s_{11}|^2 + |s_{21}|^2) & -(s_{11}^* s_{21} + s_{21}^* s_{11}) \\ -(s_{21}^* s_{11} + s_{11}^* s_{21}) & 1 - (|s_{21}|^2 + |s_{11}|^2) \end{bmatrix}
$$

$$
= \begin{bmatrix} q_{11} & -q_{21} \\ -q_{21}^* & q_{11} \end{bmatrix} \tag{5}
$$

Any matrix which is equal to its complex conjugate transpose is known as a Hermitian matrix. Since

$$q_{ik} = q_{ki}^* \quad \text{and} \quad q_{kk} = q_{kk}^* \tag{6}$$

therefore the dissipation matrix is a Hermitian matrix. One of the properties of a Hermitian matrix is that it is positive definite.

The eigenvalue equation for the two-port dissipation matrix of (5) is

$$[Q] \, [X_i] = q_i \, [X_i] \qquad i = 1, 2 \tag{7a}$$

$$q_i = F(s_i) = 1 - s_i^* s_i \tag{7b}$$

or

$$q_1 = 1 - s_1^* s_1 \tag{7c}$$

$$q_2 = 1 - s_2^* s_2 \tag{7d}$$

The eigenvalues and eigenvectors of the real scattering matrix [S] are given by Eqs. (9-3), (9-7), and (9-9).

$$s_1 = s_{11} + s_{21} \tag{8a}$$

$$s_2 = s_{11} - s_{21} \tag{8b}$$

$$[X_1] = \frac{1}{\sqrt{2}} \, [\, 1 \quad 1 \,]^t \tag{8c}$$

$$[X_2] = \frac{1}{\sqrt{2}} \, [\, 1 \quad -1 \,]^t \tag{8d}$$

(c) The corresponding dissipation matrix, which is also real, may be diagonalized as follows.

$$[X]^{-1}[Q][X] = [X]^t[Q][X] = \frac{1}{2}\begin{bmatrix} 1 & 1 \\ 1 & -1 \end{bmatrix}\begin{bmatrix} q_{11} & -q_{21} \\ -q_{21} & q_{11} \end{bmatrix}\begin{bmatrix} 1 & 1 \\ 1 & -1 \end{bmatrix}$$

$$= \frac{1}{2}\begin{bmatrix} 2(q_{11}-q_{21}) & 0 \\ 0 & 2(q_{11}+q_{21}) \end{bmatrix}$$

$$= \text{diag}\,[\, q_1 \quad q_2\,] \tag{9a}$$

$$[X] = [\,[x_1] \quad [x_2]\,] \tag{9b}$$

$$q_1 \equiv q_{11} + q_{21} \tag{9c}$$

$$q_2 \equiv q_{11} - q_{21} \tag{9d}$$

The eigenvalues of $[Q]$ are therefore q_1 and q_2.

The elements of the dissipation matrix can be expressed in terms of the eigenvalues.

$$q_{11} = \frac{1}{2}\,(\,q_1 + q_2\,) \tag{10a}$$

$$q_{21} = \frac{1}{2}\,(\,q_1 - q_2\,) \tag{10b}$$

(c) The dissipation power is related to the incident parameters, (13-1b).

$$P = \frac{1}{2}\,[a^*]^t\,[Q]\,[a] \tag{11}$$

For a two-port device, one has

$$P = \frac{1}{2}\,[\,a_1^* \quad a_2^*\,]\begin{bmatrix} q_{11} & -q_{21} \\ -q_{21} & q_{11} \end{bmatrix}\begin{bmatrix} a_1 \\ a_2 \end{bmatrix}$$

$$= \frac{1}{2}\,(\,q_{11}a_1a_1^* - q_{21}a_1^*a_2 + q_{11}a_2a_2^* - q_{21}a_1a_2^*\,) \tag{12}$$

When the device is excited only at one of the ports, the dissipation depends on q_{11} and the incident wave at the input port. If more than one port is excited simultaneously, then the loss of the device depends on q_{11}, q_{21}, and their incident parameters and their phases.

(1) When $a_1 = a_o$ and $a_2 = 0$, (12) becomes

$$P = \frac{1}{2} q_{11} a_1^2 = \frac{1}{2} q_{11} a_o^2 \tag{13}$$

(2) When $a_1 = 0$ and $a_2 = a_o$, then

$$P = \frac{1}{2} q_{11} a_2^2 = \frac{1}{2} q_{11} a_o^2 \tag{14}$$

(3) When $a_1 = a_2 = a_o$, this is the case when both ports are excited in phase or the normalized eigenvector has in phase components.

$$P = \frac{1}{2} \{ q_{11} (a_1^2 + a_2^2) - q_{21} (a_1^* a_2 + a_1 a_2^*) \}$$

$$= a_o^2 (q_{11} - q_{21}) = a_o^2 q_2 \tag{15}$$

(4) When $a_1 = a_o = -a_2$, this is the case when ports are excited in opposite phase or excited by eigenvectors with oppositely phased components.

$$P = \frac{1}{2} \{ q_{11} (a_o^2 + a_o^2) + q_{21} (a_o^2 + a_o^2) \}$$

$$= a_o^2 (q_{11} + q_{21}) = a_o^2 q_1 \tag{16}$$

15. Problems

1. Determine the scattering matrix for an ideal transformer with a ratio of n:1. The normalizing resistance is $R_{o1} = 1 = R_{o2}$.

2. Determine the scattering matrix of an impedance element connected in series with a uniform lossless line with a characteristic impedance Z_o, Fig. 15-1.

3. Determine the eigenvalues and eigenvectors for the matrix

$$B = \begin{bmatrix} 2 & -2 & 3 \\ 1 & 1 & 1 \\ 1 & 3 & -1 \end{bmatrix}$$

4. Determine the eigenvalues and eigenvectors for the matrix

$$C = \begin{bmatrix} 0 & 1 & 0 \\ 0 & 0 & 1 \\ 1 & -3 & 3 \end{bmatrix}$$

5. Diagonalize the matrix given in Problem 3.

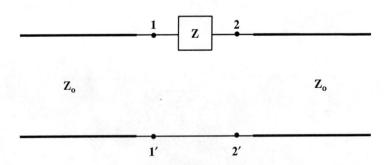

Figure 15-1: A series element as a two-port network

Chapter IV Immittance Matrices

1. Introduction

Reactive microwave elements are utilized in impedance matching and for construction of other devices such as filters. These reactive elements can be obtained by the introduction of physical discontinuities in an otherwise uniform waveguide. The discontinuity is designed to perturb predominantly either the electric or the magnetic field within the structure depending upon the type of reactance desired.

Besides a scattering description of such discontinuities, an immittance matrix description is also very useful since it provides information directly related to the circuit elements.

These matrices - scattering, impedance, and admittance - together with the relations between them and their eigenvalues and eigenvectors will be investigated.

A chosen scattering matrix for a given device may not have an impedance or admittance matrix, i.e., their elements become infinitely large. This situation is determined by the elements of the scattering matrix and it can be remedied by shifting the reference terminal planes.

2. Impedance Matrix

(a) When currents are chosen to be the independent variables, the current-voltage relation for an m-port device is

$$[V] = [Z][I] \tag{1a}$$

where

$$[V] = [\, v_1 \quad v_2 \quad \cdots \quad v_m \,]^t \tag{1b}$$

$$[I] = [\, i_1 \quad i_2 \quad \cdots \quad i_m \,]^t \tag{1c}$$

$$[Z] = \begin{bmatrix} z_{11} & z_{12} & \cdots & z_{1m} \\ z_{21} & z_{22} & \cdots & z_{2m} \\ \cdots & \cdots & \cdots & \cdots \\ z_{m1} & z_{m2} & \cdots & z_{mm} \end{bmatrix} \tag{1d}$$

where v_k is the voltage across the kth port and i_k is the current entering the kth port, and $[Z]$ is the impedance matrix.

(b) For a two-port device, (1) becomes

$$[V] = [\, v_1 \quad v_2 \,]^t \tag{2a}$$

$$[I] = [\, i_1 \quad i_2 \,]^t \tag{2b}$$

$$[Z] = \begin{bmatrix} z_{11} & z_{12} \\ z_{21} & z_{22} \end{bmatrix} \tag{2c}$$

Equation (1a) can be expanded for a two-port device as

$$v_1 = z_{11}i_1 + z_{12}i_2 \tag{3a}$$

$$v_2 = z_{21}i_1 + z_{22}i_2 \tag{3b}$$

The elements of the impedance matrix are defined as follows.

$$z_{11} = \left. \frac{v_1}{i_1} \right]_{i_2=0} = \text{open--circuit input impedance at port 1} \tag{4a}$$

$$z_{21} = \left. \frac{v_2}{i_1} \right]_{i_2=0} = \text{open--circuit forward transfer impedance} \tag{4b}$$

$$z_{12} = \left. \frac{v_1}{i_2} \right|_{i_1=0} = \text{open-circuit reverse transfer impedance} \qquad (4c)$$

$$z_{22} = \left. \frac{v_2}{i_2} \right|_{i_1=0} = \text{open-circuit output impedance at port 2} \qquad (4d)$$

A symmetrical two-port device is defined by $z_{11} = z_{22}$ and $z_{12} = z_{21}$. A reciprocal two-port device satisfies the condition $z_{12} = z_{21}$.

(c) The eigenvalue equation of the impedance matrix is

$$[Z] [e_i^z] = \lambda_i^z [e_i^z]$$

where $[e_i^z]$ is the eigenvector corresponding to the eigenvalue λ_i^z of the impedance matrix $[Z]$. In general,

$$[Z] [E^z] = [\lambda^z] [E^z] \qquad (5a)$$

$$[E^z] = [\ [e_1^z] \quad [e_2^z] \quad \cdots \quad [e_m^z]\ ] \qquad (5b)$$

$$[\lambda^z] = \text{diag} [\ \lambda_1^z \quad \lambda_2^z \quad \cdots \quad \lambda_m^z\ ] \qquad (5c)$$

The above equation can be written as

$$(\ [Z] - [\lambda^z]\ ) [E^z] = 0 \qquad (5d)$$

and its characteristic equation is

$$\det \left| [Z] - [\lambda^z] \right| = 0 \qquad (5e)$$

For symmetrical and reciprocal devices, $z_{11} = z_{22}$ and $z_{12} = z_{21}$, one has

$$\det \begin{bmatrix} z_{11} - \lambda^z & z_{21} \\ z_{21} & z_{11} - \lambda^z \end{bmatrix} = 0 \qquad (6a)$$

and the eigenvalues are

$$\lambda_1^z = z_{11} + z_{21} \qquad (6b)$$

$$\lambda_2^z = z_{11} - z_{21} \qquad (6c)$$

The impedance matrix elements can be obtained from (6).

$$z_{11} = \frac{1}{2} (\lambda_1^z + \lambda_2^z)$$ (7a)

$$z_{21} = \frac{1}{2} (\lambda_1^z - \lambda_2^z)$$ (7b)

(d) The eigenvector $[e_1^z]$ is determined as follows: for $\lambda = \lambda_1^z$,

$$\begin{bmatrix} z_{11} & z_{21} \\ z_{21} & z_{11} \end{bmatrix} \begin{bmatrix} e_{11} \\ e_{12} \end{bmatrix} = \lambda_1^z \begin{bmatrix} e_{11} \\ e_{12} \end{bmatrix} \qquad \text{and} \qquad [e_1^z] \equiv [\, e_{11} \quad e_{12} \,]^t$$

$$z_{11}e_{11} + z_{21}e_{12} = \lambda_1^z e_{11} = (z_{11} + z_{21}) e_{11}$$

$$z_{21}e_{11} + z_{11}e_{12} = \lambda_1^z e_{12}$$

or

$$e_{11} = e_{12}$$

Therefore,

$$[e_1^z] = e_{11} [\, 1 \quad 1 \,]^t$$

The normalized eigenvector is

$$[e_1^z] = \frac{1}{\sqrt{2}} [\, 1 \quad 1 \,]^t$$ (8a)

Similarly, the eigenvector $[e_2^z]$ $(\lambda = \lambda_2^z)$ is given by

$$\begin{bmatrix} z_{11} & z_{21} \\ z_{21} & z_{11} \end{bmatrix} \begin{bmatrix} e_{21} \\ e_{22} \end{bmatrix} = \lambda_2^z \begin{bmatrix} e_{21} \\ e_{22} \end{bmatrix} \qquad \text{and} \qquad [e_2^z] \equiv [\, e_{21} \quad e_{22} \,]^t$$

or

$$e_{22} = \frac{(z_{11} - \lambda_2^z) e_{21}}{-z_{21}} = -e_{21}$$

and

$$[e_2^z] = e_{21} [\, 1 \quad -1 \,]^t$$

The normalized eigenvector is

$$[e_2^z] = \frac{1}{\sqrt{2}} [\, 1 \quad -1 \,]^t$$ (8b)

(e) The impedance matrix can be diagonalized by the following process. Construct the matrix of eigenvectors of the $[Z]$ matrix.

$$[E^z] = [[e_1^z] \quad [e_2^z]] = \frac{1}{\sqrt{2}} \begin{bmatrix} 1 & 1 \\ 1 & -1 \end{bmatrix} \tag{9}$$

Then the diagonalized $[Z]$ is given by

$$[D^z] = [E^z]^t [Z] [E^z]$$

$$= \frac{1}{\sqrt{2}} \begin{bmatrix} 1 & 1 \\ 1 & -1 \end{bmatrix} \begin{bmatrix} z_{11} & z_{21} \\ z_{21} & z_{11} \end{bmatrix} \frac{1}{\sqrt{2}} \begin{bmatrix} 1 & 1 \\ 1 & -1 \end{bmatrix}$$

$$= \begin{bmatrix} z_{11} + z_{21} & 0 \\ 0 & z_{11} - z_{21} \end{bmatrix}$$

$$= \begin{bmatrix} \lambda_1^z & 0 \\ 0 & \lambda_2^z \end{bmatrix} \equiv \mathrm{diag} \, [\, \lambda_1^z \quad \lambda_2^z \,] \tag{10}$$

(f) This two-port device can be represented by an equivalent T network (Fig. 2-1). The circuit elements, Z_A and Z_C, can be determined by (4).

The input impedance across terminals 1–1′ under the condition $i_2 = 0$ is

$$Z_{in} = \left[\frac{v_1}{i_1} \right]_{i_2 = 0} = Z_A + Z_C = z_{11} \tag{11a}$$

The forward transfer impedance with $i_2 = 0$ is

$$Z_f = \left[\frac{v_2}{i_1} \right]_{i_2 = 0} = \frac{i_1 Z_C}{i_1} = Z_C = z_{21} \tag{11b}$$

Therefore,

$$Z_A = z_{11} - Z_C = z_{11} - z_{21} \tag{12a}$$

$$Z_C = z_{21} \tag{12b}$$

and the circuit is as shown in Fig. 2-2.

The circuit elements can be expressed in terms of the eigenvalues by substituting (6c) and (7b) into (12)

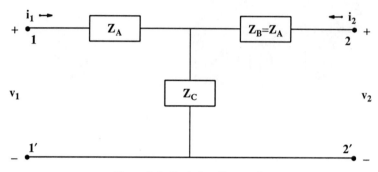

Figure 2-1: Equivalent T-network

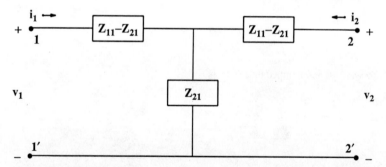

Figure 2-2: Equivalent T-network; elements expressed in terms of elements of impedance matrix

$$Z_A = \lambda_2^z \tag{13a}$$

$$Z_C = \frac{1}{2}(\lambda_1^z - \lambda_2^z) \tag{13b}$$

This circuit with elements expressed in terms of eigenvalues is shown in Fig. 2-3.

(g) Bisecting the circuit in Fig. 2-2 yields Fig. 2-4. Then

$$Z_{oc}^z = z_{11} + z_{21} = \lambda_1^z \tag{14a}$$

$$Z_{sc}^z = z_{11} - z_{21} = \lambda_2^z \tag{14b}$$

Bisecting the circuit in Fig. 2-3 yields Fig. 2-5. Hence

$$Z_{oc}^z = \lambda_1^z \tag{15a}$$

$$Z_{sc}^z = \lambda_2^z \tag{15b}$$

Equations (14a) and (15a) imply that the input impedance of the equivalent cirucit when excited by eigenvector $[e_1^z]$ (in-phase sources) is equal to the eigenvalue λ_1^z.

Similarly, (14b) and (15b) means that the input impedance of the circuit excited by eigenvector $[e_2^z]$ (oppositely phased sources) is equal to the eigenvalue λ_2^z.

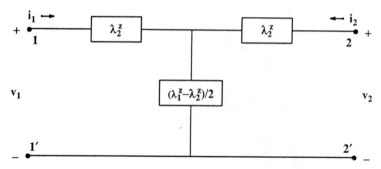

Figure 2-3: Equivalent T-network; elements expressed in terms of eigenvalues

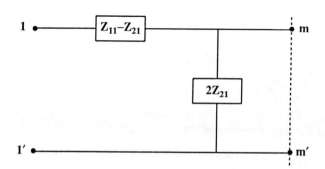

Figure 2-4: Eigen-network of Figure 2-2

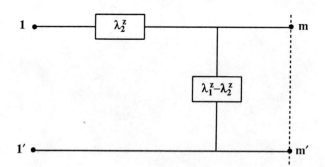

Figure 2-5: Eigen-network of Figure 2-3

3. Admittance Matrix

(a) When voltages are chosen as the independent variables, then one has

$$[I] = [Y][V] \tag{1a}$$

$$[I] = [i_1 \quad i_2 \quad \cdots \quad i_m]^t \tag{1b}$$

$$[V] = [v_1 \quad v_2 \quad \cdots \quad v_m]^t \tag{1c}$$

$$[Y] = \begin{bmatrix} y_{11} & y_{12} & \cdots & y_{1m} \\ y_{21} & y_{22} & \cdots & y_{2m} \\ \cdots & \cdots & \cdots & \cdots \\ y_{m1} & y_{m2} & \cdots & y_{mm} \end{bmatrix} \tag{1d}$$

(b) For a two-port device, one has

$$i_1 = y_{11}v_1 + y_{12}v_2 \tag{2a}$$

$$i_2 = y_{21}v_1 + y_{22}v_2 \tag{2b}$$

The admittance elements are defined as follows:

$$y_{11} = \frac{i_1}{v_1}\bigg]_{v_2 = 0} = \text{short–circuit input admittance at port 1} \tag{3a}$$

$$y_{21} = \frac{i_2}{v_1}\bigg]_{v_2 = 0} = \text{short–circuit forward transfer admittance} \tag{3b}$$

$$y_{12} = \frac{i_1}{v_2}\bigg]_{v_1 = 0} = \text{shrot–circuit reverse transfer admittance} \tag{3c}$$

$$y_{22} = \frac{i_2}{v_2}\bigg]_{v_1 = 0} = \text{short–circuit output admittance at port 2} \tag{3d}$$

(c) For symmetrical and reciprocal devices, one has $y_{11} = y_{22}$ and $y_{21} = y_{12}$. Then

$$[Y] = \begin{bmatrix} y_{11} & y_{21} \\ y_{21} & y_{11} \end{bmatrix} \tag{4}$$

The corresponding eigenvalue equation is

$$[Y] [e_k^y] = \lambda_k^y [e_k^y] \tag{5}$$

or

$$[Y] [E^y] = [\lambda^y] [E^y] \tag{6a}$$

$$[E^y] = [[e_1^y] \quad [e_2^y] \quad \cdots \quad [e_m^y]]^t \tag{6b}$$

$$[\lambda^y] = \text{diag} [\lambda_1^y \quad \lambda_2^y \quad \cdots \quad \lambda_m^y] \tag{6c}$$

Equation (6a) can be rearranged as

$$([Y] - [\lambda^y]) [E^y] = 0 \tag{7}$$

and the corresponding characteristic equation is

$$\det \left| [Y] - [\lambda^y] \right| = 0 \tag{8a}$$

or

$$\det \begin{bmatrix} y_{11} - \lambda^y & y_{21} \\ y_{21} & y_{11} - \lambda^y \end{bmatrix} = 0 \tag{8b}$$

The eigenvalues are

$$\lambda_1^y = y_{11} + y_{21} \tag{9a}$$

$$\lambda_2^y = y_{11} - y_{21} \tag{9b}$$

Conversely, the elements of the admittance matrix are

$$y_{11} = \frac{1}{2} (\lambda_1^y + \lambda_2^y) \tag{10a}$$

$$y_{21} = \frac{1}{2} (\lambda_1^y - \lambda_2^y) \tag{10b}$$

(d) The normalized eigenvectors of the admittance matrix can be obtained by the same procedure as in Section 2(d).

$$[e_1^y] = \frac{1}{\sqrt{2}} \begin{bmatrix} 1 & 1 \end{bmatrix}^t \tag{11a}$$

$$[e_2^y] = \frac{1}{\sqrt{2}} \begin{bmatrix} 1 & -1 \end{bmatrix}^t \tag{11b}$$

(e) The admittance matrix can be diagonalized as follows.

$$[D^y] = [E^y]^t [Y] [E^y] = \frac{1}{2} \begin{bmatrix} 1 & 1 \\ 1 & -1 \end{bmatrix} \begin{bmatrix} y_{11} & y_{21} \\ y_{21} & y_{11} \end{bmatrix} \begin{bmatrix} 1 & 1 \\ 1 & -1 \end{bmatrix}$$

$$= \begin{bmatrix} \lambda_1^y & 0 \\ 0 & \lambda_2^y \end{bmatrix} = \text{diag} \begin{bmatrix} \lambda_1^y & \lambda_2^y \end{bmatrix} \tag{12}$$

(f) The equivalent circuit of a [Y] matrix can be represented as a π network, Fig. 3-1. The input admittance is given by

$$Y_{in} = \frac{i_1}{v_1} \bigg]_{v_2 = 0} = Y_A + Y_C = y_{11} \tag{13a}$$

The forward transfer admittance is

$$Y_f = \frac{i_2}{v_1} \bigg]_{v_2 = 0} = -Y_C = y_{21} \tag{13b}$$

Thus, the admittance elements of the π network are

$$Y_A = y_{11} - Y_C = y_{11} + y_{21} \tag{14a}$$

$$Y_C = -y_{21} \tag{14b}$$

The equivalent circuit in terms of the Y-matrix elements is as shown in Fig. 3-2. The substitution of (9) into (14) yields

$$Y_A = \lambda_1^y \tag{15a}$$

$$Y_C = \frac{-1}{2} (\lambda_1^y - \lambda_2^y) \tag{15b}$$

The equivalent circuit in terms of the eigenvalues is shown in Fig. 3-3.

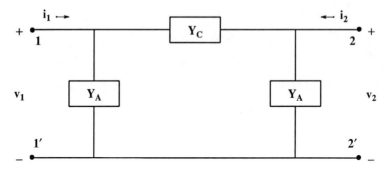

Figure 3-1: Equivalent π-network

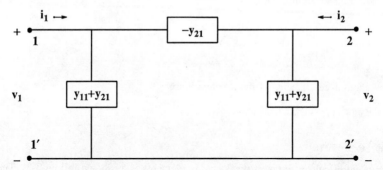

Figure 3-2: Equivalent π-network; elements expressed in terms of elements of admittance matrix

(g) Figure 3-3 can be rearranged as shown in Fig. 3-4. The bisection of Fig. 3-4 produces Fig. 3-5. The input admittances with terminals m–m' open-circuited and short-circuited are

$$Y_{oc} = \lambda_1^y \tag{16a}$$

$$Y_{sc} = \lambda_2^y \tag{16b}$$

The eigenvalue λ^y can be determined from the input admittance of the bisected equivalent circuit with its terminals at the mid-plane open-circuited and short-circuited, respectively.

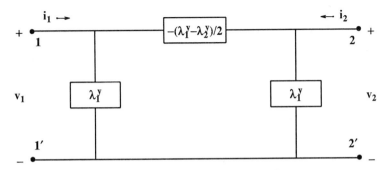

Figure 3-3: Equivalent π-network; elements expressed in terms of eigenvalues

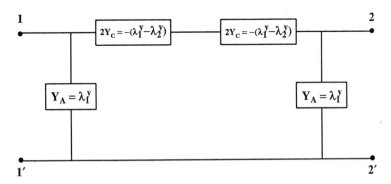

Figure 3-4: Equivalent π-network; bisectable format

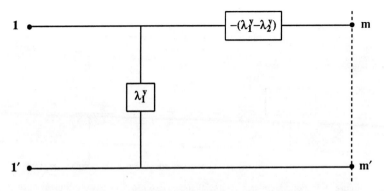

Figure 3-5: Eigen-network

3.1. Example: Admittance of a Waveguide Terminated by Reactances

A uniform waveguide of length L is terminated by irises at both ends (Fig. 3.1-1a). Determine the admittance matrix for this device.

Solution:

The equivalent π-network is shown in Fig. 3.1b. The eigen-network is shown in Fig. 3.1-1c and the eigenvalues are found to be

$$\lambda_1^y = Y_{oc} = jB + Y_o \tanh \frac{\theta}{2}$$

$$\lambda_2^y = Y_{oc} = jB + Y_o \coth \frac{\theta}{2}$$

where $\theta \equiv \beta L$, and L is the length of the waveguide. The admittance elements are

$$y_{11} = \frac{1}{2} (\lambda_1^y + \lambda_2^y) = jB + \frac{Y_o}{2} (\tanh \frac{\theta}{2} + \coth \frac{\theta}{2}) = jB + Y_o \coth \theta$$

$$y_{21} = \frac{1}{2} (\lambda_1^y - \lambda_2^y) = \frac{Y_o}{2} (\tanh \frac{\theta}{2} - \coth \frac{\theta}{2}) = - \frac{Y_o}{\sinh \frac{\theta}{2}}$$

Since

$$\tanh \frac{\theta}{2} + \coth \frac{\theta}{2} = \frac{e^{\frac{\theta}{2}} - e^{-\frac{\theta}{2}}}{e^{\frac{\theta}{2}} + e^{-\frac{\theta}{2}}} + \frac{e^{\frac{\theta}{2}} + e^{-\frac{\theta}{2}}}{e^{\frac{\theta}{2}} - e^{-\frac{\theta}{2}}}$$

$$= \frac{(e^{\theta} - 2 + e^{\theta}) + (e^{\theta} + 2 + e^{-\theta})}{e^{\theta} - e^{-\theta}} = 2 \coth \theta$$

and

$$\tanh \frac{\theta}{2} - \coth \frac{\theta}{2} = \frac{\{(e^{\theta} - 2 + e^{-\theta}) - (e^{\theta} + 2 + e^{-\theta})\}}{e^{\theta} - e^{-\theta}}$$

$$= \frac{-2}{\sinh \theta}$$

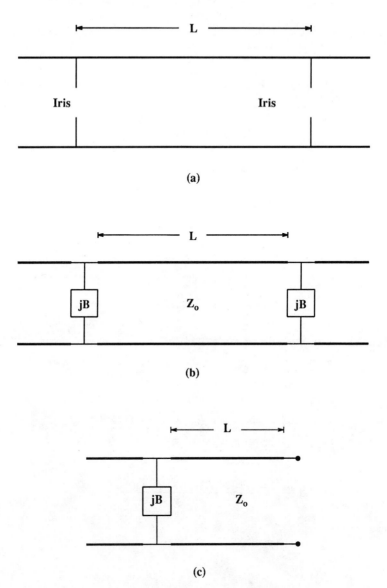

Figure 3.1-1: Uniform guide terminated by irises and its equivalent circuit and eigen-network

4. Eigen-networks

(a) It has been shown that the eigenvalues of a symmetrical and reciprocal network can be obtained from the input immittances of the bisected half-section with terminals at the mid-plane open-circuited and short-circuited. The circuits arranged in such manner are known as the eigen-networks.

This is due to the fact that when in-phase sources are applied to a symmetrical and reciprocal two-port device, currents of equal magnitude but oppositely directed will be created at the plane of symmetry. Consequently, there will be no net flow of current across the mid-plane, and hence the mid-plane is equivalent to an open circuit.

Similarly, if oppositely phased sources are applied to the two-port device, voltages of equal magnitude with opposite polarities will be created at the mid-plane. These voltages will cancel and a zero potential across mid-plane terminals will result. Thus, the mid-plane appears to be a short circuit.

(b) In impedance representation, the equivalent T network is generally used. The eigenvalues are obtained as shown in Fig. 4-1.

(c) In admittance representation, the equivalent π network is used. The eigenvalues are obtained as shown in Fig. 4-2.

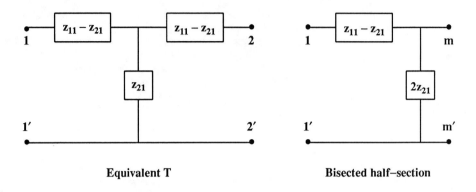

Equivalent T　　　　　　　　　　**Bisected half–section**

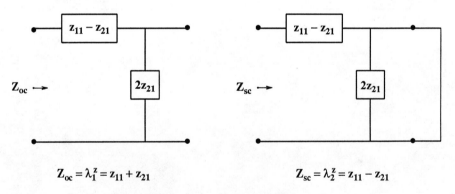

$$Z_{oc} = \lambda_1^z = z_{11} + z_{21}$$　　　　　$$Z_{sc} = \lambda_2^z = z_{11} - z_{21}$$

Figure 4-1: Equivalent Y-network and eigenvalues

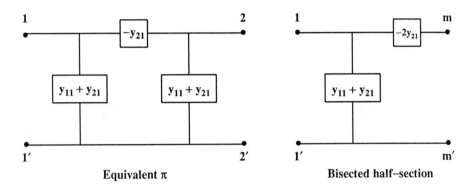

Equivalent π   **Bisected half–section**

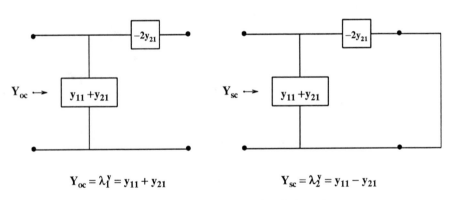

$$Y_{oc} = \lambda_1^y = y_{11} + y_{21} \qquad\qquad Y_{sc} = \lambda_2^y = y_{11} - y_{21}$$

Figure 4-2: Equivalent π-network and eigenvalues

4.1. Example: Scattering Matrix of a Shunt Impedance

Determine the scattering matrix of a shunt impedance element Z (Fig. 4.1-1).

Solution:

Figure 4.1-1 shows the arrangement of a lumped shunt impedance Z and its bisected section. The eigen-networks are shown in Fig. 4.1-2.

The scattering eigenvalue λ_1^s corresponding to in-phase excitation is given by [see Section III-9(e)]

$$\lambda_1^s = \frac{Z_1 - Z_0}{Z_1 + Z_0} = \frac{2Z - Z_0}{2Z + Z_0} \tag{1a}$$

where Z_1 is the open-circuit input impedance of the bisected section. The scattering eigenvalue λ_2^s corresponding to the opposite-phase excitation is [see Section III-9(g)]

$$\lambda_2^s = \frac{Z_2 - Z_0}{Z_2 + Z_0} = -1 \tag{1b}$$

where Z_2 is the short-circuit input impedance of the bisected section. Then

$$s_{11} = \frac{1}{2}(\lambda_1^s + \lambda_2^s) = \frac{-Z_0}{2Z + Z_0} \tag{2a}$$

$$s_{12} = \frac{1}{2}(\lambda_1^s - \lambda_2^s) = \frac{2Z}{2Z + Z_0} \tag{2b}$$

The impedance Z can be obtained in terms of the scattering elements.

$$\frac{Z}{Z_0} = -\frac{1}{2}\frac{\text{Eq. (2b)}}{\text{Eq. (2a)}} = -\frac{1}{2}\frac{s_{12}}{s_{11}} = -\frac{1}{2}\frac{\lambda_1^s - \lambda_2^s}{\lambda_1^s + \lambda_2^s} \tag{3}$$

When the scattering elements for a structure are known, then the structure can be represented by an equivalent shunt impedance given by (3).

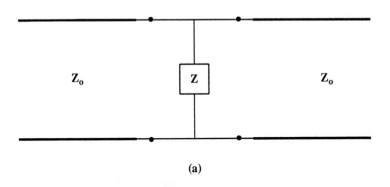

(a)

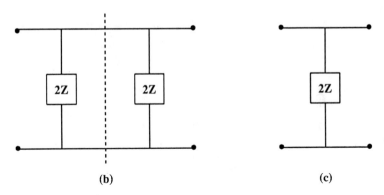

(b)   **(c)**

Figure 4.1-1: Shunt impedance element as a two-port junction

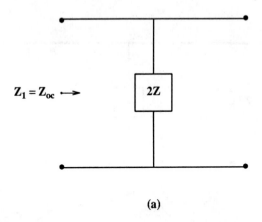

(a)

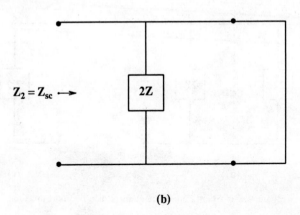

(b)

Figure 4.1-2: Eigen-networks for a shunt impedance element

4.2. Example: Scattering Matrix of a Section of Uniform Transmission Line

Determine the scattering matrix for a section of uniform transmission line with an electrical length $\theta = \beta L$, Fig. 4.2-1a.

Solution:

The in-phase eigen-network for the section of transmission line is the line of length $\theta/2$ terminated in an open circuit (Fig. 4.2-1b). The input impedance of this eigen-network is

$$Z_{oc} = Z_o \coth \frac{\theta}{2} \tag{1}$$

The oppositely phased eigen-network is the line of length $\theta/2$ terminated by a short circuit, Fig. 4.2-1c. The corresponding input impedance is

$$Z_{sc} = Z_o \tanh \frac{\theta}{2} \tag{2}$$

The scattering eigenvalues are therefore

$$\lambda_1^s = \frac{Z_{oc} - Z_o}{Z_{oc} + Z_o} = e^{-\theta} \tag{3a}$$

$$\lambda_2^s = \frac{Z_{sc} - Z_o}{Z_{sc} + Z_o} = -e^{-\theta} \tag{3b}$$

The scattering coefficients are therefore

$$s_{11} = s_{22} = \frac{\lambda_1^s + \lambda_2^s}{2} = 0 \tag{4a}$$

$$s_{21} = s_{12} = \frac{\lambda_1^s - \lambda_2^s}{2} = e^{-\theta} \tag{4b}$$

The scattering matrix is

$$[S] = \begin{bmatrix} 0 & e^{-\theta} \\ e^{-\theta} & 0 \end{bmatrix} \tag{5}$$

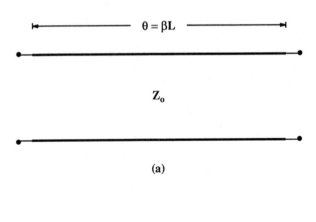

(a)

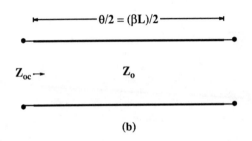

(b)

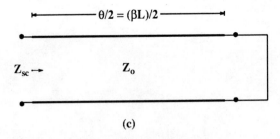

(c)

Figure 4.2-1: A section of uniform transmission line

5. Relations Between [S], [Z], and [Y]

(a) A device may be described either by its scattering or immittance matrix. In practice, it is necessary to relate one description to another, and relations between these matrices are needed.

From the definition of incident and reflected parameters, (III.3-7),

$$[a] = \frac{1}{2} ([V] + [I]) \tag{1a}$$

$$[b] = \frac{1}{2} ([V] - [I]) \tag{1b}$$

where

$$[a] = [\, a_1 \quad a_2 \quad \cdots \quad a_m \,]^t \tag{1c}$$

$$[b] = [\, b_1 \quad b_2 \quad \cdots \quad b_m \,]^t \tag{1d}$$

$$[V] = [\, \hat{v}_1 \quad \hat{v}_2 \quad \cdots \quad \hat{v}_m \,]^t \tag{1e}$$

$$[I] = [\, \hat{i}_1 \quad \hat{i}_2 \quad \cdots \quad \hat{i}_m \,]^t \tag{1f}$$

It is to be noted that [V] and [I] are the normalized voltage matrix and the normalized current matrix, respectively, with the subscript n omitted, i.e.,

$$\hat{v}_k \equiv v_{kn} = \frac{v_k}{\sqrt{R_o}}$$

$$\hat{i}_k \equiv i_{kn} = i_k \sqrt{R_o}$$

where v_k (or i_k) is the actual voltage (or current) at port k and R_o is the normalizing resistance.

The normalized voltage and normalized current matrices can be expressed in terms of the incident and reflected parameters.

$$[V] = [a] + [b] \tag{2a}$$

$$[I] = [a] - [b] \tag{2b}$$

The incident and reflected parameters are related by the scattering matrix.

$$[b] = [S] [a] \tag{3}$$

The substitution of (3) into (2) produces

$$[V] = ([u] + [S]) [a] \tag{4a}$$

$$[I] = ([u] - [S]) [a] \tag{4b}$$

(b) In the impedance representation, one has

$$[V] = [Z] [I] \tag{5}$$

Then

$$[V] = [Z] [I] = ([u] + [S]) [a] \tag{6}$$

The substitution of (4b) into (6) yields

$$[Z] ([u] - [S]) [a] = ([u] + [S]) [a] \tag{7}$$

or

$$[Z] = ([u] + [S]) ([u] - [S])^{-1} \tag{8}$$

This is a relation between $[Z]$ and $[S]$. Equation (8) exists provided $([u] - [S])$ is non-singular, i.e., the determinant of $([u] - [S])$ is nonzero. The corresponding relation between the eigenvalues of $[Z]$ and $[S]$ martrices is given by the Cayley-Hamilton relation [Eq. (III.8-6)].

$$f(\lambda^a) = f([A]) \tag{9}$$

Therefore

$$\lambda_i^z = \frac{(1 + \lambda_i^s)}{(1 - \lambda_i^s)} \tag{10}$$

where λ_i^z and λ_i^s are eigenvalues of the $[Z]$ and $[S]$ matrices, respectively.

(c) In the admittance representation one has

$$[I] = [Y] [V] \tag{11}$$

Then (4b) becomes

$$[I] = [Y] [V] = ([u] - [S]) [a] \tag{12}$$

The substitution of (4a) into (12) yields

$$[Y] ([u] + [S]) [a] = ([u] - [S]) [a]$$

or

$$[Y] = ([u] - [S]) ([u] + [S])^{-1} \tag{13}$$

Equation (13) exists provided the determinant of ([u] + [S]) is nonvanishing.
The corresponding relation for the eigenvalues is

$$\lambda_i^y = \frac{1 - \lambda_i^s}{1 + \lambda_i^s} \tag{14}$$

5.1. Example: Scattering Matrix of a Two-Port Device

The scattering matrix of a two-port device is given by

$$[S] = \begin{bmatrix} s_{11} & s_{21} \\ s_{21} & s_{11} \end{bmatrix}$$

subject to the condition

$$s_{11} - s_{21} = -1$$

Determine the equivalent circuit for this device.

Solution:

The impedance matrix is given by [Eq. (5-8)]

$$[Z] = ([u] + [S]) ([u] - [S])^{-1} \tag{1}$$

Let

$$[F] \equiv [u] - [S] = \begin{bmatrix} 1 - s_{11} & -s_{21} \\ -s_{21} & 1 - s_{11} \end{bmatrix}$$

The determinant of [F] is given by

$$\Delta_F = (1 - s_{11})^2 - s_{12}^2$$

The cofactor of [F] is

$$[F]_{cofactor} = \begin{bmatrix} 1 - s_{11} & s_{21} \\ s_{21} & 1 - s_{11} \end{bmatrix}$$

The adjoint of [F] is

$$Adj\ [F] = \begin{bmatrix} 1 - s_{11} & s_{21} \\ s_{21} & 1 - s_{11} \end{bmatrix}$$

The inverse of [F] is

$$[F]^{-1} = \frac{Adj\ [F]}{\Delta_F} = \frac{1}{\Delta_F} \begin{bmatrix} 1 - s_{11} & s_{21} \\ s_{21} & 1 - s_{11} \end{bmatrix} \tag{2}$$

The substitution of (2) into (1) yields

$$[Z] = \frac{1}{\Delta_F} \begin{bmatrix} 1 + s_{11} & s_{21} \\ s_{21} & 1 + s_{11} \end{bmatrix} \begin{bmatrix} 1 - s_{11} & s_{21} \\ s_{21} & 1 - s_{11} \end{bmatrix}$$

$$= \frac{1}{\Delta_F} \begin{bmatrix} 1 - s_{11}^2 + s_{21}^2 & 2s_{21} \\ 2s_{21} & 1 - s_{11}^2 + s_{21}^2 \end{bmatrix} \tag{3}$$

For $s_{11} - s_{21} = -1$ or $s_{21} = 1 + s_{11}$, then

$$\Delta_F = (1 - s_{11})^2 - (1 + s_{11})^2 = -4s_{11}$$

and

$$1 - s_{11}^2 + s_{21}^2 = 1 - s_{11}^2 + (1 + s_{11})^2 = 2(1 + s_{11})$$

Then the impedance matrix becomes

$$[Z] = \frac{1}{-4s_{11}} \begin{bmatrix} 2(1 + s_{11}) & 2(1 + s_{11}) \\ 2(1 + s_{11}) & 2(1 + s_{11}) \end{bmatrix} = \begin{bmatrix} Z_e & Z_e \\ Z_e & Z_e \end{bmatrix}$$

where

$$Z_e \equiv \frac{1 + s_{11}}{-2s_{11}} \tag{4}$$

and

$$s_{11} = -\frac{1}{1 + 2Z_e} \tag{5}$$

The equivalent circuit for the above scattering matrix is as shown in Fig. 5.1-1.

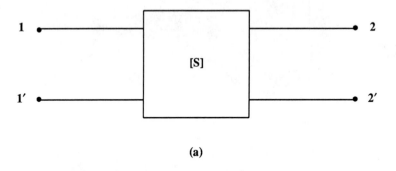

(a)

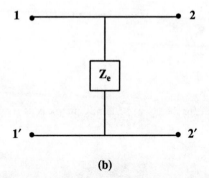

(b)

Figure 5.1-1: Equivalent circuit of a symmetrical two-port junction which satisfies $s_{11} - s_{21} = -1$

6. Problems

1. The scattering matrix of a lossless four-port structure is

$$[S] = \frac{-j}{\sqrt{2}} \begin{bmatrix} 0 & 0 & 1 & 1 \\ 0 & 0 & 1 & -1 \\ 1 & 1 & 0 & 0 \\ 1 & -1 & 0 & 0 \end{bmatrix}$$

Determine the corresponding admittance matrix.

2. Show that the normalized admittance matrix of the lossless structure given in Problem 1 is related to its scattering matrix as

$$[\hat{y}] = -[S]$$

3. Show that the normalized admittance matrix of a quarter-wavelength transmission line of characteristic admittance Y_{o1} is

$$[\hat{y}] = \frac{Y_{o1}}{Y_o} \begin{bmatrix} 0 & j \\ j & 0 \end{bmatrix}$$

4. Show that the normalized admittance matrix of a three-quarter-wavelength lossless line is

$$[\hat{y}] = \frac{Y_{o1}}{Y_o} \begin{bmatrix} 0 & -j \\ -j & 0 \end{bmatrix}$$

where Y_{o1} is the characteristic admittance of the line.

5. Show that the normalized admittance matrix of a quarter-wavelength lossless line is

$$[\hat{z}] = -j \frac{Z_{o1}}{Z_o} \begin{bmatrix} 0 & 1 \\ 1 & 0 \end{bmatrix}$$

where Z_{o1} is the characteristic impedance of the line.

Chapter V   Symmetrical Devices

1. Introduction

Microwave devices which are symmetrical in physical space have many special properties that are desirable in practical applications. Moreover, those properties may be predicted by intuitive arguments.

The characteristics of a symmetrical device will remain unchanged by a symmetry operation. For example, in Fig. 1-1, the symmetry operation corresponding to this structure is a rotation of the structure by 180° about the axis of symmetry a–a′. This operation interchanges the positions of port 1 and port 2. The exterior appearance of the structure remains unchanged after the symmetry operation, except for the labels of the ports. This structure is known to be invariant when subjected to symmetry operation.

For a device which has only two planes of symmetry, these planes intersect perpendicularly. This intersection is an axis of symmetry. For a device with more than two planes of symmetry, these planes may not all intersect normally.

Since the choice of coordinate system is entirely arbitrary in Maxwell's equations, therefore Maxwell's equations are independent of coordinates. A rotation of the coordinate frame and field quantity to a new situation should produce no changes in Maxwell's equations. This rotation is a symmetry operation.

Even though Maxwell's equations are invariant under a symmetry operation, however, its solution may not necessarily be so. For example, a traveling wave in one direction can be transformed by reflection into a wave propagating in the opposite direction. But a standing wave will be reflected to itself at its nodal plane. A solution of the latter type is said to be invariant under the symmetry operation.

Symmetrical devices will be examined by searching for symmetrical solutions of Maxwell's equations with the associated boundary conditions of the device. A general solution is a linear superposition of these symmetrical solutions.

A useful method for obtaining the properties of symmetrical devices is found in the theory of eigenvalue equations.

Some of the topics discussed in this chapter can be found in Montgomery, Dicke and Purcell [C-5] and Altman [C-1] at an advanced level.

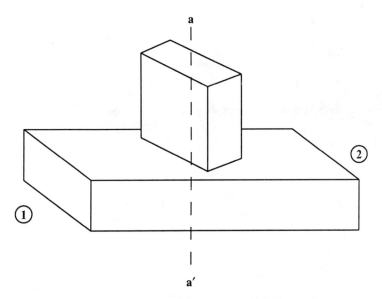

Figure 1-1: A junction with one plane of symmetry

2. Reflection Operation

The choice of the directions in space for a particular coordinate system is entirely arbitrary. It should be possible to employ a transformation of the geometrical axes and a conversion of the field quantities such that the new coordinates and the new field quantities satisfy Maxwell's equations. When such a case exists, Maxwell's equations are said to be invariant under the transformation.

The reflection operation in the yz-plane transforms x into −x, and the other two coordinates remain unchanged (Fig. 2-1). The geometrical conversions are:

$$x \rightarrow x' \equiv -x \tag{1a}$$

$$y \rightarrow y' \equiv y \tag{1b}$$

$$z \rightarrow z' \equiv z \tag{1c}$$

The unprimed and the primed quantities will be used to denote the original and the transformed quantities, respectively.

One conversion of the field components which leaves Maxwell's equations invariant is as listed below.

$$E_x \rightarrow E_x' = -E_x, \qquad H_x \rightarrow H_x' = H_x, \qquad J_x \rightarrow J_x' = -J_x \tag{2a}$$

$$E_y \rightarrow E_y' = E_y, \qquad H_y \rightarrow H_y' = -H_y, \qquad J_y \rightarrow J_y' = J_y \tag{2b}$$

$$E_z \rightarrow E_z' = E_z, \qquad H_z \rightarrow H_z' = -H_z, \qquad J_z \rightarrow J_z' = J_z \tag{2c}$$

$$\rho \rightarrow \rho' = \rho \tag{2d}$$

$$\omega \rightarrow \omega' = \omega \tag{2e}$$

where $E = E(x, y, z)$ and $E' = E(x', y', z')$. Maxwell's equations are invariant under the above transformations, (1) and (2). This can be verified by direct substitution of (1) and (2) into Maxwell's equations.

The reflection operation can be applied to electromagnetic fields only if the E-field alone or the H-field alone is considered; the Poynting theorem is then used to determine the associated field which is not considered in the reflection operation. This is because fields reflected with respect to a plane of symmetry are not solutions of Maxwell's equations. If the phase relation of the E-field and the H-field in the structure is such that power traveling to the left, the reflection of those fields will have the identical phase relation and will give rise to power travels to the left instead of to the right.

The reflection transformation on the field components that leaves Maxwell's equations invariant is obtained as follows. It is to be noted that the propagating directions of power in the plane of symmetry should remain unchanged before and after the reflection operation. The propagating directions of power normal to the plane of symmetry will be oppositely directed before and after the transformation.

$$x \to x' = -x$$

$$y \to y' = y$$

$$z \to z' = z$$

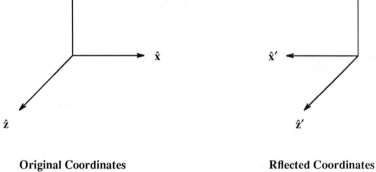

Original Coordinates   **Rflected Coordinates**

Right–hand System   **Left–hand System**

Figure 2-1: Reflection operation in the yz-plane

H_x': $\vec{P}_z' = \vec{E}_{y'} \times \vec{H}_x' = \hat{y}'E_{yo} \times \hat{x}'H_{xo} = \hat{z}'P_z$

by virtue of Fig. 2.1. In terms of the original coordinates, one has

$$\hat{z}P_z = \hat{y}E_{yo} \times (-\hat{x})\, H_{xo} = \hat{z}E_{yo}\, H_{xo,} \qquad \text{hence} \qquad \vec{H}_x' = \hat{x}'H_{xo} \tag{3a}$$

$\vec{H}_y'$: $\vec{P}_z' = \vec{E}_x' \times \vec{H}_y' = \hat{x}'(-E_{xo}) \times \hat{y}'H_{yo} = -\hat{z}'P_z$

$$-\hat{z}P_z = -\hat{x}(-E_{xo}) \times \hat{y}H_{yo} = -\hat{z}E_{xo}(-H_{yo}), \qquad \text{hence} \qquad \vec{H}_y' = \hat{y}'(-H_{yo}) \tag{3b}$$

$\vec{H}_z'$: $\vec{P}_y' = \vec{E}_x' \times \vec{H}_z' = \hat{y}'P_y$

$$\hat{y}P_y = (-\hat{x})\,(-E_{xo}) \times \hat{z}H_{zo} = \hat{y}E_{xo}(-H_{zo}), \qquad \text{hence} \qquad \vec{H}_z = \hat{z}'(-H_{zo}) \tag{3c}$$

The remaining components of the Poynting vector can be shown to be consistent with the above transformations by direct expansion.

The relation between the original coordinates (x, y, z) and the new coordinates (x', y', z') is given in Fig. 2-1. The invariance of Maxwell's equations implies that this newly transformed field is also a solution. This newly transformed field will satisfy the boundary conditions of the device only if the geometry of the device is invariant under the same transformation.

The transformations (1) and (2) will be denoted by the operator $[F_x]$. $[F_x]$ will operate on the field and coordinate and change them to their respective values as indicated in (1) and (2). Thus

$$[F_x] \cdot \hat{x} = \hat{x}' = -\hat{x} \tag{4a}$$

$$[F_x] \cdot \hat{x}E_x(x, y, z) = \hat{x}'E_x(x', y', z') = -\hat{x}E_x(-x, y, z) \tag{4b}$$

For a constant k, one has

$$[F_x] \cdot (k\vec{E}) = k\,[F_x] \cdot \vec{E} \tag{5}$$

Also

$$[F_x] \cdot (\vec{A} + \vec{B}) = [F_x] \cdot \vec{A} + [F_x] \cdot \vec{B} \tag{6}$$

A linear operator is an operator which satisfies both conditions (5) and (6). Hence the reflection operator is a linear operator.

Information concerning general properties of a device can be obtained from the solutions of Maxwell's equations that are invariant (except for a possible change in phase - which can be compensated by a change in zero reference) under the symmetry transformation.

If $\vec{E}_y$ is a symmetrical solution, then

$$[F_x] \cdot \vec{E}_y = k_f \vec{E}_y \tag{7}$$

where k_f is a scalar factor.

Operating on (7) by $[F_x]$ again, one has

$$[F_x] \cdot ([F_x] \cdot \vec{E}_y) = k_f [F_x] \cdot \vec{E}_y = k_f^2 \vec{E}_y \tag{8}$$

The reflection of a field twice returns it back to its original state; therefore

$$[F_x]^2 \cdot \vec{E}_y = \vec{E}_y = k_f^2 \vec{E}_y \tag{9}$$

Hence

$$k_f^2 = 1 \qquad \text{or} \qquad k_f = \pm 1 \tag{10}$$

With $k_f = +1$, solutions are even functions of x and when $k_f = -1$, the solution is known to be odd. Conventionally, a solution is considered to be even (or odd) when both components of the E-field in the plane of symmetry, i.e., $\vec{E}_y$ and $\vec{E}_z$ in the case of $[F_x]$, are even (or odd) functions of x. When $\vec{E}_y$ and $\vec{E}_z$ are even functions of x, $\vec{E}_x$ is an odd function and conversely.

Using this convention, the even and odd solutions are:

Even solution: $\vec{E}$-field in the plane of symmetry is even.

$$E_x(x, y, z) = -E_x(-x, y, z) \qquad H_x(x, y, z) = H_x(-x, y, z) \tag{11a}$$

$$E_y(x, y, z) = E_y(-x, y, z) \qquad H_y(x, y, z) = -H_y(-x, y, z) \tag{11b}$$

$$E_z(x, y, z) = E_z(-x, y, z) \qquad H_z(x, y, z) = -H_z(-x, y, z) \tag{11c}$$

Odd solution: $\vec{E}$-field in the plane of symmetry is odd.

$$E_x(x, y, z) = E_x(-x, y, z) \qquad H_x(x, y, z) = -H_x(-x, y, z) \tag{11d}$$

$$E_y(x, y, z) = -E_y(-x, y, z) \qquad H_y(x, y, z) = H_y(-x, y, z) \tag{11e}$$

$$E_z(x, y, z) = -E_z(-x, y, z) \qquad H_z(x, y, z) = H_z(-x, y, z) \tag{11f}$$

For solutions to be continuous at the plane of symmetry, i.e., a perfect electric wall does not exist at the plane of symmetry, then setting x = 0 in (11) yields

Even solution:

$$E_x(0, y, z) = -E_x(0, y, z) \quad \text{or} \quad E_x(0, y, z) = 0 \tag{12a}$$

$$E_y(0, y, z) = E_y(0, y, z) \tag{12b}$$

$$E_z(0, y, z) = E_z(0, y, z) \tag{12c}$$

$$H_x(0, y, z) = H_x(0, y, z) \tag{12d}$$

$$H_y(0, y, z) = -H_y(0, y, z) \quad \text{or} \quad H_y(0, y, z) = 0 \tag{12e}$$

$$H_z(0, y, z) = -H_z(0, y, z) \quad \text{or} \quad H_z(0, y, z) = 0 \tag{12f}$$

Odd solution:

$$E_x(0, y, z) = E_x(0, y, z) \tag{12g}$$

$$E_y(0, y, z) = -E_y(0, y, z) \quad \text{or} \quad E_y(0, y, z) = 0 \tag{12h}$$

$$E_z(0, y, z) = -E_z(0, y, z) \quad \text{or} \quad E_z(0, y, z) = 0 \tag{12i}$$

$$H_x(0, y, z) = -H_x(0, y, z) \quad \text{or} \quad H_x(0, y, z) = 0 \tag{12j}$$

$$H_y(0, y, z) = H_y(0, y, z) \tag{12k}$$

$$H_z(0, y, z) = H_z(0, y, z) \tag{12l}$$

Equations (12h) to (12j) imply that the odd solution at the plane of symmetry satisfies the condition of a perfect electric wall. In other words, the field distribution will not be disturbed if the plane of symmetry is replaced by a perfect electric conductor.

Similarly, Eqs.(12a), (12e), and (12f) show that the even solution will not be disturbed if the plane of symmetry is replaced by a perfect magnetic conductor.

3. Symmetry Operations

(a) The reflection operators $[F_y]$ and $[F_z]$ for reflections with respect to the xz-plane and the xy-plane, respectively, can be defined similarly to $[F_x]$ (Fig. 3-1).

These reflection operators may be applied successively. A reflection $[F_x]$ followed by a reflection $[F_y]$ is equivalent to a rotation of $180°$ about the z-axis (Fig. 3-2).

(b) The rotation operator $[R_z]$ represents the rotation of $180°$ about the z-axis (in accordance with the right-hand rule). Therefore, one has, from Fig. 3-2d,

$$[R_z] = [F_x] \cdot [F_y] = [F_y] \cdot [F_x] \tag{1a}$$

The order of reflection operators is interchangeable without affecting the final result.

Similarly,

$$[R_x] = [F_y] \cdot [F_z] = [F_z] \cdot [F_y] \tag{1b}$$

$$[R_y] = [F_x] \cdot [F_z] = [F_z] \cdot [F_x] \tag{1c}$$

(c) A reflection with respect to the origin, $[P]$, is obtained by reflections $[F_x]$, $[F_y]$, and $[F_z]$ successively.

$$[P] = [F_x] \cdot [F_y] \cdot [F_z] \tag{2}$$

(d) The identity operator, $[I]$, represents the operation which leaves the coordinates unchanged. The reflection operator with respect to any plane twice is one such operator.

$$[F_x] \cdot [F_x] = [I] \tag{3}$$

(e) The results of a multiple operations are tabulated in Fig. 3-3 and these results can be easily verified by a similar procedure as shown above.

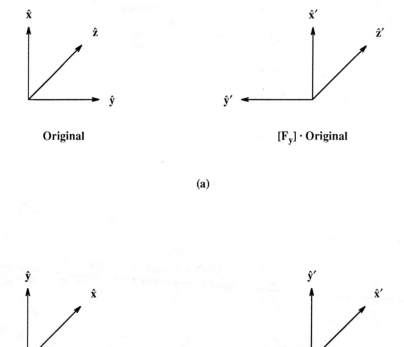

(a)

(b)

Figure 3-1: The reflection operators $[F_y]$ and $[F_z]$

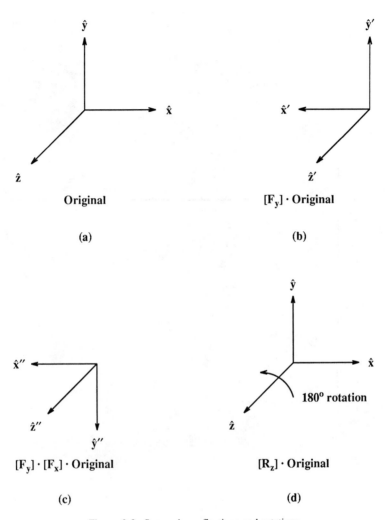

Figure 3-2: Successive reflections and rotations

	I	R_x	R_y	R_z	F_x	F_y	F_z	P
I	I	R_x	R_y	R_z	F_x	F_y	F_z	P
R_x	R_x	I	R_z	R_y	P	F_z	F_y	F_x
R_y	R_y	R_z	I	R_x	F_z	P	F_x	F_y
R_z	R_z	R_y	R_x	I	F_y	F_x	P	F_z
F_x	F_x	P	F_z	F_y	I	R_z	R_y	R_x
F_y	F_y	F_z	P	F_x	R_z	I	R_x	R_y
F_z	F_z	F_y	F_x	P	R_y	R_z	I	R_z
P	P	F_x	F_y	F_z	R_x	R_y	R_z	I

Figure 3-3: Multiplication table for operators

4. Symmetry Matrix

(a) The reflection, rotation, etc., operators discussed in the previous section may be grouped under the title of symmetry operators. For multiple-port devices, these operators, in general, have the form of a matrix and will be denoted by [G].

$$[G] [a] = [a'] \qquad \text{and} \qquad [G] [b] = [b'] \tag{1}$$

where [a] and [b] are the original solutions of a symmetrical device characterized by the symmetry operator matrix [G]; [a'] and [b'] are the corresponding transformed solutions which can exist and satisfy the same boundary conditions.

(b) The symmetry operator [G] should generally be determined by inspection. The coefficient of [G] should be either 0 or ± 1 (see Section 3) since the transformed field [a'] is no more than an interchange of the components of the original field [a].

[G] should have one and only one nonzero coefficient per row; otherwise, the field at one port would be exchanged with the sum or the difference of the fields from more than one port or would be left out completely.

There should be one and only one nonzero coefficient per column of [G]; otherwise, the field of some ports will be operated on more than once.

In view of these restrictions, the [G] matrix has only one nonzero coefficient per row and per column.

(c) By virtue of the fact that [G] has only one nonzero coefficient per row and per column, it can be shown that [G] is an orthogonal matrix. Let

$$[G] = \begin{bmatrix} g_{11} & g_{12} & \cdots & g_{1m} \\ g_{21} & g_{22} & \cdots & g_{2m} \\ \cdots & \cdots & \cdots & \cdots \\ g_{m1} & g_{m2} & \cdots & g_{mm} \end{bmatrix}$$

then

$$[G]^t [G] = \begin{bmatrix} g_{11} & g_{21} & \cdots & g_{m1} \\ g_{12} & g_{22} & \cdots & g_{m2} \\ \cdots & \cdots & \cdots & \cdots \\ g_{1m} & g_{2m} & \cdots & g_{mm} \end{bmatrix} \begin{bmatrix} g_{11} & g_{12} & \cdots & g_{1m} \\ g_{21} & g_{22} & \cdots & g_{2m} \\ \cdots & \cdots & \cdots & \cdots \\ g_{m1} & g_{m2} & \cdots & g_{mm} \end{bmatrix}$$

$$= \begin{bmatrix} \sum_{i=1}^{m} g_{i1} g_{i1} & \cdots & \sum_{i=1}^{m} g_{i1} g_{im} \\ \cdots & \cdots & \cdots \\ \sum_{i=1}^{m} g_{im} g_{i1} & \cdots & \sum_{i=1}^{m} g_{im} g_{im} \end{bmatrix}$$

$$[G]^t [G] = \begin{bmatrix} 1 & 0 & \cdots & 0 \\ 0 & 1 & \cdots & 0 \\ \multicolumn{4}{c}{\cdots\cdots\cdots\cdots} \\ 0 & 0 & \cdots & 1 \end{bmatrix} \tag{2}$$

since

$$\sum_{i=1}^{m} g_{ik}g_{ik} = 1$$

and

$$\sum_{i=1}^{m} g_{ik}g_{ij} = 0 \qquad \text{for} \qquad j \neq k$$

But

$$[G]^{-1}[G] = [u] \tag{3}$$

Therefore,

$$[G]^t = [G]^{-1} \tag{4}$$

The transpose of an orthogonal matrix is equal to its inverse.

(d) Since both [a] and [a'] are solutions, they both satisfy the eigenvalue equation.

$$[S] [a] = [b] \qquad \text{and} \qquad [S] [a'] = [b'] \tag{5}$$

Then from (1), one has

$$[S] [a'] = [S] ([G] [a]) = [S] [G] [a] \tag{6a}$$

$$[b'] = [G] [b] = [G] [S] [a] \tag{6b}$$

Since (6a) equals (6b), one then has

$$[S] [G] [a] = [G] [S] [a] \tag{7}$$

Equation (7) implies that the scattering matrix and the symmetry operator matrix commute. Commuting matrices are similar matrices and they have identical eigenvectors.

$$[G] [S] = [S] [G] \tag{8}$$

5. Commutable Matrices

(a) Given a diagonalizable mth-order matrix [A], it is desired to find another diagonalizable matrix [B] such that [A] and [B] commute, i.e.,

$$[A] [B] = [B] [A] \tag{1}$$

Since [B] is diagonalizable, it can be expressed as

$$[B] = [E] [D_B] [E]^{-1} \tag{2a}$$

where [E] is a matrix of eigenvectors of [B], and

$$[D_B] = [E]^{-1} [B] [E] = \text{diag} [\beta_1 \quad \beta_2 \quad \cdots \quad \beta_m] \tag{2b}$$

where β_i is the ith eigenvalue of [B]. Then (1) becomes

$$[A] [E] [D_B] [E]^{-1} = [E] [D_B] [E]^{-1} [A] \tag{3}$$

Pre-multiplying (3) by $[E]^{-1}$ and post-multiplying it by [E] yields

$$[E]^{-1} [A] [E] [D_B] = [D_B] [E]^{-1} [A] [E] \tag{4}$$

If [E] is also a matrix of eigenvectors of [A], then

$$[E]^{-1} [A] [E] = [D_A] = \text{diag} [\alpha_1 \quad \alpha_2 \quad \cdots \quad \alpha_m] \tag{5}$$

where α_i is the ith eigenvalue of [A], since [A] is diagonalizable. Equation (4) then reduces to

$$[D_A] [D_B] = [D_B] [D_A] \tag{6}$$

Equation (6) is always true since both $[D_A]$ and $[D_B]$ are diagonal matrices.

Thus, if [A] is a given diagonalizable matrix, then another diagonalizable matrix [B] commutes with [A] if both matrices have the identical matrix of eigenvectors, i.e., provided they have identical eigenvectors.

To find the matrix [B]:

(1) Find the eigenvectors of [A].

(2) Form the matrix of eigenvectors of [A], i.e., the matrix [E].

(3) Choose a diagonal matrix of parameters β_m for the diagonal matrix.

$$[D_B] = \text{diag} [\beta_1 \quad \beta_2 \quad \cdots \quad \beta_m]$$

(4) The desired matrix [B] is then given by

$$[B] = [E] [D_B] [E]^{-1} = [B(\beta_1, \beta_2, \ldots, \beta_m)]$$

(b) For symmetric m × m matrices [A] and [B], the product [A] [B] is also symmetric if [A] and [B] commute.

If ([A] [B]) is symmetric, this implies

$$[A] [B] = ([A] [B])^t = [B]^t [A]^t = [B] [A] \tag{7}$$

by symmetry.

5.1. Determination of Commutable Matrix

Determine a matrix [B] which commutes with the matrix

$$[A] = \begin{bmatrix} 4 & 1 \\ 2 & 3 \end{bmatrix} \tag{1}$$

and the eigenvalues of [B] are β_1 and β_2.

Solution:

The eigenvalues of [A] are determined from

$$\det \left| [A] - \lambda [u] \right| = 0 \tag{2}$$

and are found to be $\lambda_1 = 2$ and $\lambda_2 = 5$.

The eigenvectors of [A] are determined from

$$[A] [e_i] = \lambda_i [x_i] \qquad i = 1, 2 \tag{3}$$

The matrix of eigenvectors of [A] is found to be

$$[E] = [\; e_1 \quad e_2 \;] = \begin{bmatrix} 1 & 1 \\ -2 & 1 \end{bmatrix} \tag{4}$$

The inverse of [E] is

$$[E]^{-1} = \frac{\text{Adj } [E]}{\Delta_E} = \frac{1}{3} \begin{bmatrix} 1 & -1 \\ 2 & 1 \end{bmatrix} \tag{5a}$$

$$\text{Adj } [E] = \begin{bmatrix} 1 & -1 \\ 2 & 1 \end{bmatrix} \qquad \text{and} \qquad \Delta_E = 3 \tag{5b}$$

The [B] matrix is then given by

$$[B] = [E] [D_B] [E]^{-1} = \begin{bmatrix} 1 & 1 \\ -2 & 1 \end{bmatrix} \begin{bmatrix} \beta_1 & 0 \\ 0 & \beta_2 \end{bmatrix} \frac{1}{3} \begin{bmatrix} 1 & -1 \\ 2 & 1 \end{bmatrix}$$

$$= \frac{1}{3} \begin{bmatrix} \beta_1 + 2\beta_2 & \beta_2 - \beta_1 \\ 2(\beta_2 - \beta_1) & 2\beta_1 + \beta_2 \end{bmatrix} \tag{6}$$

If $\beta_1 = \lambda_1 = 2$ and $\beta_2 = \lambda_2 = 5$, then

$$[B] = \begin{bmatrix} 4 & 1 \\ 2 & 3 \end{bmatrix} \tag{7}$$

6. Properties of Commutable Matrices

The commutable matrices [G] and [S] are defined by

$$[G] [S] = [S] [G] \tag{1}$$

Let the eigenvalue equation for a square matrix [G] be

$$[G] [g_i] = \lambda_i^g [g_i] \tag{2}$$

where λ_i^g is the eigenvalue and $[g_i]$ is the corresponding eigenvector of [G].

(a) Let $[E^g]$ be the eigenvector matrix of [G].

$$[E^g] = [\; [g_1] \quad [g_2] \quad \cdots \quad [g_n] \;] = \begin{bmatrix} g_{11} & g_{21} & \cdots & g_{n1} \\ g_{12} & g_{22} & \cdots & g_{n2} \\ \cdots & \cdots & \cdots & \cdots \\ g_{1n} & g_{2n} & \cdots & g_{nn} \end{bmatrix} \tag{3}$$

Then (2) becomes

$$[G] [E^g] = [E^g] [D_G] \tag{4a}$$

$$[D_G] = \text{diag} [\; \lambda_1^g \quad \lambda_2^g \quad \cdots \quad \lambda_n^g \;] \tag{4b}$$

Equation (4) is simply the general expression of (2) including all eigenvectors of [G].

Equation (4) can be rearranged as

$$[D_G] = [E^g]^{-1} [G] [E^g] \tag{5a}$$

or

$$[G] = [E^g] [D_G] [E^g]^{-1} \tag{5b}$$

Equation (5b) expresses [G] in terms of its eigenvalues and eigenvectors.

(b) For distinct eigenvalues, both [G] and [S] can be reduced to their diagonal forms.

$$[E^g]^{-1} [G] [E^g] = [D_G] \quad \rightarrow \quad [G] = [E^g] [D_G] [E^g]^{-1} \tag{6a}$$

$$[E^g]^{-1} [S] [E^g] = [D_S] \quad \rightarrow \quad [S] = [E^g] [D_S] [E^g]^{-1} \tag{6b}$$

With (1) and (6), one has

$$([E^g] [D_G] [E^g]^{-1}) [S] = [S] ([E^g] [D_G] [E^g]^{-1})$$

$$[D_G] [E^g]^{-1} [S] [E^g] = [E^g]^{-1} [S] [E^g] [D_G]$$

or

$$[D_G] [D_S] = [D_S] [D_G] \tag{7}$$

(c) If the characteristic equation of [G] has multiple roots,

$$\lambda_1^g = \lambda_2^g = \cdots = \lambda_p^g \tag{8}$$

then λ_p^g is know to be a degenerate eigenvalue. The eigenvalue equations are

$$[G] [g_1] = \lambda_1^g [g_1] = \lambda_p^g [g_1] \tag{9a}$$

$$[G] [g_2] = \lambda_2^g [g_2] = \lambda_p^g [g_2] \tag{9b}$$

. . . .

$$[G] [g_p] = \lambda_p^g [g_p] \tag{9p}$$

if the determinant $\left| [G] - \lambda_p^g [u] \right|$ is multiple degenerate, i.e., with linearly related rows and columns.

When $\left| [G] - \lambda_p^g [u] \right|$ is simple degenerate, i.e., the determinant is equal to zero but has no related rows or columns, then

$$[g_1] = [g_2] = \cdots = [g_p] \tag{10}$$

If the equations in (9) are each multiplied by a different arbitrary constant and then added up, one obtains a new eigenvalue equation.

$$[G] (c_1 [g_1] + c_2 [g_2] + \cdots + c_p [g_p])$$

$$= \lambda_p^g (c_1 [g_1] + c_2 [g_2] + \cdots + c^p [g^p]) \tag{11}$$

Hence, the vector

$$c_1 [g_1] + c_2 [g_2] + \cdots + c^p [g^p] \tag{12}$$

is also an eigenvector associated with λ_p^g. Since the c_i's are arbitrarily chosen constants, there are infinite numbers of eigenvectors associated with λ_p^g.

(d) Since [G] and [S] commute, the eigenvector $[g_i]$ is also an eigenvector of [S]. From the eigenvalue equation

$$[G] [g_i] = \lambda_i^g [g_i] \tag{13}$$

one has

$$[S] ([G] [g_i]) = [S] (\lambda_i^g [g_i]) = \lambda_i^g [S] [g_i]$$

or

$$[G] \{ [S] [g_i] \} = \lambda_1^g \{ [S] [g_i] \} \tag{14}$$

Equation (14) implies that $\{[S] [g_i]\}$ is an eigenvector of $[G]$ with corresponding eigenvalue λ_1^g. But the eigenvector corresponding to λ_1^g is completely determined by (13) to within a constant multiplier if λ_1^g is nondegenerate. It follows that $[S][g_i]$ can differ from $[g_i]$ only by a constant, i.e.,

$$[S] [g_i] = \text{constant} \times [g_i] = \lambda_i^s [g_i] \tag{15}$$

Equation (15) has the form of an eigenvalue equation, and consequently the constant can be interpreted as the eigenvalue of $[S]$.

If λ_1^g is degenerate, $[S][g_i]$ is still an eigenvector of $[G]$, but it could assume a totally different form from $[g_i]$, i.e., it could have the form of (12) - a linear combination of other eigenvectors of $[G]$ corresponding to λ_1^g.

7. Symmetrical Two-Port Junction

(a) Figure 7-1 shows a symmetrical two-port junction with symmetrical reference plane. The reflection with respect to the plane of symmetry simply exchanges the fields at port 1 with those at port 2. The symmetry operator for this case is defined by

$$[G] [a] = [a']$$ (1a)

$$\begin{bmatrix} g_{11} & g_{12} \\ g_{21} & g_{22} \end{bmatrix} \begin{bmatrix} a_1 \\ a_2 \end{bmatrix} = \begin{bmatrix} a_1' \\ a_2' \end{bmatrix} = \begin{bmatrix} a_2 \\ a_1 \end{bmatrix}$$ (1b)

or

$$g_{11}a_1 + g_{12}a_2 = a_2 \quad \rightarrow \quad g_{11} = 0 \quad \text{and} \quad g_{12} = 1$$

$$g_{21}a_1 + g_{22}a_2 = a_1 \quad \rightarrow \quad g_{21} = 1 \quad \text{and} \quad g_{22} = 0$$

That is,

$$[G] = \begin{bmatrix} 0 & 1 \\ 1 & 0 \end{bmatrix}$$ (2)

(b) The eigenvalue equation is

$$[G] [g_i] = \lambda_i^g [g_i]$$ (3a)

$$[g_i] = [\, g_{i1} \quad g_{i2} \,]$$ (3b)

Since a sequence of two reflections with respect to the same plane will return to its original form, therefore

$$[G] ([G] [g]) = [G] (\lambda^g [g]) = \lambda^g ([G] [g]) = (\lambda^g)^2 [g] = [g]$$ (4)

That is,

$$(\lambda^g)^2 = 1$$

or

$$\lambda_1^g = 1 \quad \text{and} \quad \lambda_2^g = -1$$ (5)

(c) The eigenvectors of $[G]$ are determined as follows: for $\lambda_1^g = 1$,

$$[G] [g_1] = \lambda_1^g [g_1] \quad \rightarrow \quad \begin{bmatrix} 0 & 1 \\ 1 & 0 \end{bmatrix} \begin{bmatrix} g_{11} \\ g_{12} \end{bmatrix} = \begin{bmatrix} g_{11} \\ g_{12} \end{bmatrix}$$

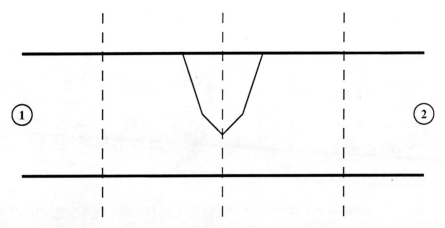

Figure 7-1: A symmetrical two-port junction

or

$$g_{12} = g_{11}$$

and

$$[g_1] = g_{11} [1 \quad 1]^t = \frac{1}{\sqrt{2}} [1 \quad 1]^t \tag{6a}$$

For $\lambda_2^g = -1$,

$$\begin{bmatrix} 0 & 1 \\ 1 & 0 \end{bmatrix} \begin{bmatrix} g_{21} \\ g_{22} \end{bmatrix} = - \begin{bmatrix} g_{21} \\ g_{22} \end{bmatrix}$$

$$g_{22} = -g_{21}$$

and

$$[g_2] = g_{21} [1 \quad -1]^t = \frac{1}{\sqrt{2}} [1 \quad -1]^t \tag{6b}$$

(d) The eigenvector matrix is

$$[E^g] = [[g_1] \quad [g_2]] = \frac{1}{\sqrt{2}} \begin{bmatrix} 1 & 1 \\ 1 & -1 \end{bmatrix} \tag{7}$$

The matrix $[E^g]$ is orthonormal.

$$[E^g]^t [E^g] = \frac{1}{2} \begin{bmatrix} 1 & 1 \\ 1 & -1 \end{bmatrix} \begin{bmatrix} 1 & 1 \\ 1 & -1 \end{bmatrix} = \begin{bmatrix} 1 & 0 \\ 0 & 1 \end{bmatrix} \tag{8}$$

Therefore,

$$[E^g]^t [E^g] = [E^g]^{-1} [E^g] = [u]$$

or

$$[E^g]^{-1} = [E^g]^t \tag{9}$$

(e) The scattering matrix $[S]$ commutes with the $[G]$ matrix, and they have identical eigenvectors. For nondegenerate eigenvalues, both $[S]$ and $[G]$ can be diagonalized by the $[E^g]$ matrix.

$$[E^g]^{-1} [G] [E^g] = [D_G] \tag{10a}$$

$$[E^g]^{-1} [S] [E^g] = [D_S] \tag{10b}$$

where $[D_G]$ and $[D_S]$ are the diagonalized $[G]$ and $[S]$ matrices, respectively.

The scattering matrix [S] can be expressed in terms of its eigenvalues by (10b).

$$[S] = [E^g] [D_S] [E^g]^{-1} = \frac{1}{2} \begin{bmatrix} 1 & 1 \\ 1 & -1 \end{bmatrix} \begin{bmatrix} s_1 & 0 \\ 0 & s_2 \end{bmatrix} \begin{bmatrix} 1 & 1 \\ 1 & -1 \end{bmatrix}$$

$$[S] = \frac{1}{2} \begin{bmatrix} s_1 + s_2 & s_1 - s_2 \\ s_1 - s_2 & s_1 + s_2 \end{bmatrix} = \begin{bmatrix} s_{11} & s_{12} \\ s_{12} & s_{11} \end{bmatrix}$$

where s_1 and s_2 are the eigenvalues of [S]. The simpler notation s_1 and s_2 is used instead of λ_1^s and λ_2^s, respectively. The latter notation will be used should confusion arise. Therefore,

$$s_{11} = \frac{1}{2} (s_1 + s_2) \tag{11a}$$

$$s_{12} = \frac{1}{2} (s_1 - s_2) \tag{11b}$$

(f) For a lossless junction, the scattering matrix satisfies the unitary condition, i.e.,

$$[S^*]^t [S] = [u] \tag{12a}$$

$$\begin{bmatrix} s_{11}^* & s_{12}^* \\ s_{12}^* & s_{11}^* \end{bmatrix} \begin{bmatrix} s_{11} & s_{12} \\ s_{12} & s_{11} \end{bmatrix} = \begin{bmatrix} 1 & 0 \\ 0 & 1 \end{bmatrix}$$

The expansion of the above relation yields

$$|s_{11}|^2 + |s_{12}|^2 = 1 \tag{12b}$$

$$s_{11}^* s_{12} + s_{12}^* s_{11} = 0 \tag{12c}$$

Let the eigenvalues of [S] be

$$s_1 = \alpha e^{j\theta} \quad \text{and} \quad s_2 = \beta e^{j\phi} \tag{13}$$

Then (11) becomes

$$s_{11} = \frac{1}{2} (\alpha e^{j\theta} + \beta e^{j\phi}) \tag{14a}$$

$$s_{12} = \frac{1}{2} (\alpha e^{j\theta} - \beta e^{j\phi}) \tag{14b}$$

and

$$|s_{11}|^2 = s_{11}s_{11}^* = \frac{1}{4} [\alpha^2 + \beta^2 + \alpha\beta e^{j(\theta-\phi)} + \alpha\beta e^{j(\phi-\theta)}] \qquad (15a)$$

$$|s_{12}|^2 = \frac{1}{4} [\alpha^2 + \beta^2 - \alpha\beta e^{j(\theta-\phi)} - \alpha\beta e^{j(\phi-\theta)}] \qquad (15b)$$

The substitution of (15) into (12b) yields

$$|s_{11}|^2 + |s_{12}|^2 = \frac{1}{2} (\alpha^2 + \beta^2) = 1$$

or

$$\alpha^2 + \beta^2 = 2 \qquad (16)$$

From (14)

$$s_{11}^* s_{12} = \frac{1}{4} (\alpha e^{-j\theta} + \beta e^{-j\phi})(\alpha e^{j\theta} - \beta e^{j\phi})$$

$$= \frac{1}{4} [(\alpha^2 - \beta^2 - 2j\alpha\beta \sin(\phi - \theta)] = (s_{12}^* s_{11})^* \qquad (17)$$

Equation (12c) then becomes

$$\alpha^2 - \beta^2 = 0 \qquad \text{or} \qquad \alpha^2 = \beta^2 \qquad (18)$$

Combining (16) and (18) gives

$$\alpha^2 = 1 = \beta^2 \qquad \text{or} \qquad |\alpha| = |\beta| = 1 \qquad (19a)$$

and

$$|s_1| = |s_2| = 1 \qquad (19b)$$

The eigenvalues of a lossless device have unit magnitude.

(g) It can be shown that s_{11} and s_{12} are in quadrature for a lossless device. Let

$$s_1 = \alpha e^{j\theta} = 1 e^{j\theta}$$

$$s_2 = \beta e^{j\phi} = 1 e^{j\phi}$$

To simplify the proof, let s_1 be chosen as the reference, $\theta = 0°$, and ϕ is now measured with respect to s_1. Thus

$$s_1 = 1 \qquad \text{and} \qquad s_2 = e^{j\phi} \qquad (20)$$

Then

$$s_{11} = \frac{1}{2} (s_1 + s_2) = \frac{1}{2} (1 + e^{j\phi})$$

$$= \frac{1}{2} (1 + \cos \phi + j \sin \phi) \equiv Ae^{j\theta_1} \tag{21a}$$

$$s_{21} = \frac{1}{2} (s_1 - s_2) = \frac{1}{2} (1 - \cos \phi - j \sin \phi) \equiv Be^{j\theta_2} \tag{21b}$$

where

$$A = \frac{1}{2} \sqrt{ (1 + \cos \phi)^2 + \sin^2 \phi } = \frac{1}{\sqrt{2}} \sqrt{1 + \cos \phi} \tag{21c}$$

$$B = \frac{1}{\sqrt{2}} \sqrt{1 - \cos \phi} \tag{21d}$$

$$\theta_1 = \tan^{-1} \left[\frac{\sin \phi}{1 + \cos \phi} \right] \tag{21e}$$

$$\theta_2 = \tan^{-1} \left[\frac{-\sin \phi}{1 - \cos \phi} \right] \tag{21f}$$

The angle between s_{11} and s_{12} is

$$\theta_1 - \theta_2 = \tan^{-1} \left[\frac{\sin \phi}{1 + \cos \phi} \right] - \tan^{-1} \left[\frac{-\sin \phi}{1 - \cos \phi} \right]$$

$$\tan (\theta_1 - \theta_2) = \left[\frac{\sin \phi}{1 + \cos \phi} - \frac{-\sin \phi}{1 - \cos \phi} \right] \left[1 + \frac{\sin \phi}{1 + \cos \phi} \times \frac{-\sin \phi}{1 - \cos \phi} \right]^{-1}$$

$$= \frac{2 \sin \phi}{1 - 1} = \infty$$

Hence,

$$\theta_1 - \theta_2 = \tan^{-1} \infty = \frac{\pi}{2} \tag{22}$$

Therefore, s_{11} and s_{12} are spaced $90°$ apart.

When s_{11} and s_{12} are known, then the two-port device can be represented by an equivalent shunt impedance given by (IV.4.1-3).

$$\frac{Z}{Z_o} = \frac{-s_{12}}{2s_{11}} = \frac{-(s_1 - s_2)}{2(s_1 + s_2)} \tag{23}$$

(h) When the obstacle is thin, the entire obstacle may be considered to be coincident with the plane of symmetry.

The oppositely phased excitation, $[g_2]$, Eq. (6b), sees a perfect short circuit at the plane of symmetry, i.e., $s_2 = -1$. Consequently, the presence of a thin obstacle at the plane of symmetry will not make any difference to $[g_2]$. In other words, $[g_2]$ will not see the thin obstacle. Thus, the eigenvalue s_2 is a function of the reference planes only and is always equal to -1 at the plane of symmetry and at every half-wavelength away.

The in-phase excitation, $[g_1]$, will see the obstacle, and the fields associated with $[g_1]$ will be reflected by the obstacle. Without the obstacle, the plane of symmetry is a magnetic wall and the fields are completely reflected, $s_1 = +1$. If the obstacle is a short circuit, then the plane of symmetry provides a null and $s_1 = -1$. For the intermediate case, $s_1 = e^{j\theta}$, where the phase angle depends on the nature of the obstacle.

The thin obstacle can be represented by an equivalent shunt impedance as given by (23), with $s_2 = -1$.

$$\hat{Z}_e = \frac{Z}{Z_o} = \frac{1 + s_1}{2(1 - s_1)} \tag{24}$$

(i) The plane of symmetry is always a null, an electric wall, to eigenvector $[g_2]$ and $s_2 = -1$, irrespective of the thickness of the obstacle.

A thick obstacle may still be represented by a shunt impedance at the plane of symmetry, (24).

However, for a thick obstacle, the electric wall will be extended longitudinally. The effective null plan for $s_2 = -1$ lies somewhere between the plane of symmetry and the outermost edges of the obstacle (Fig. 7-2). This is shown schematically in Fig. 7-2b, where the obstacle is replaced by a shunt impedance Z_e and a short piece of line of length L on each side.

For $L \ll \lambda_g$, the wavelength of the guide, the equivalent circuit can be obtained as follows. At the physical plane of symmetry, where, $s_2 = -1$, the corresponding eigenvalue z_2 for the impedance matrix is [Eq. (IV.5-10)]

$$z_2 = \frac{1 + s_2}{1 - s_2} \tag{25a}$$

and

$$z_2(s_2 = -1) = 0 \tag{25b}$$

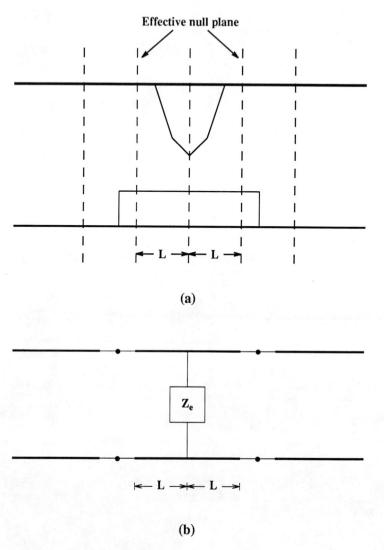

Figure 7-2: A thick obstacle and its equivalent circuit

At a distance $-L$ away from the plane of symmetry, the normalized impedance, $Z_2/Z_o = z_2$, is

$$z_2(-L) = -j \tan \frac{2\pi}{\lambda_g} L \approx -j \frac{2\pi}{\lambda_g} L \qquad \text{for } L \ll \lambda_g \qquad (26)$$

The impedance matrix is given by (IV.2-10).

$$[Z_n] = \frac{1}{2} \begin{bmatrix} z_1 + z_2 & z_1 - z_2 \\ z_1 - z_2 & z_1 + z_2 \end{bmatrix} \qquad (27a)$$

$$= \begin{bmatrix} z_{11} & z_{12} \\ z_{12} & z_{11} \end{bmatrix} \qquad (27b)$$

where

$$z_{11} = \frac{1}{2} (z_1 + z_2) \qquad (27c)$$

$$z_{12} = \frac{1}{2} (z_1 - z_2) \qquad (27d)$$

or

$$z_{11} - z_{12} = z_2 \qquad (27e)$$

$$z_{12} = \frac{1}{2} (z_1 - z_2) \qquad (27f)$$

where z_1 is the normalized impedance of the in-phase eigen-network. Both z_1 and z_2 are taken at the plane of symmetry. The equivalent T network is shown in Fig. 7-3.

(j) For a lossless, symmetrical but nonreciprocal two-port device, the symmetry matrix [G], Eq. (2), must be modified to take into account the phase differences.

$$[G] = \begin{bmatrix} 0 & e^{j\theta} \\ e^{-j\theta} & 0 \end{bmatrix} \qquad (28)$$

8. H-plane T-junction

(a) Figure 8-1 shows a T-junction in the H-plane. It has two planes of symmetry, P_A and P_B. P_A is a surface parallel to the yz-plane and divides the junction symmetrically with port 1 and port 2 on the opposite sides of this surface. P_B is a surface parallel to the xz-plane which slices the junction into two identical T's along the surface of the paper. P_A and P_B intersect perpendicularly, and their intersection J is an axis of symmetry. The device is assumed to be operated at its dominant TE_{10} mode.

(b) The reflection operation with respect to plane P_A interchanges the E-field at port 1 with that at port 2, and the fields at port 3 remain unchanged.

The symmetry matrix [G] is obtained from (4-1).

$$[G] [E] = [E'] \tag{1a}$$

where

$$[G] = \begin{bmatrix} g_{11} & g_{12} & g_{13} \\ g_{21} & g_{22} & g_{23} \\ g_{31} & g_{32} & g_{33} \end{bmatrix} \tag{1b}$$

$$[E] = [E_1 \quad E_2 \quad E_3]^t = \text{original field} \tag{1c}$$

$$[E'] = [E_1' \quad E_2' \quad E_3']^t = \text{transformed field} \tag{1d}$$

The expansion of (1a) yields

$$g_{11}E_1 + g_{12}E_2 + g_{13}E_3 = E_1' \tag{2a}$$

$$g_{21}E_1 + g_{22}E_2 + g_{23}E_3 = E_2' \tag{2b}$$

$$g_{31}E_1 + g_{32}E_2 + g_{33}E_3 = E_3' \tag{2c}$$

The result of reflection with respect to P_A is

$$E_1' = E_2 ; \qquad E_2' = E_1 ; \qquad E_3' = E_3 \tag{3a)-(3c}$$

The substitution of (3a) into (2a) gives

$$g_{12} = 1 \qquad \text{and} \qquad g_{11} = g_{13} = 0 \tag{4a}$$

since there can be only one nonzero element in any row of the symmetry matrix.

The other elements are determined similarly from (2b) and (3b) and from (2c) and (3c), respectively.

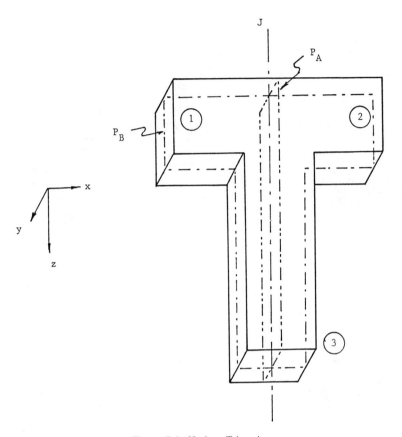

Figure 8-1: H-plane T-junction

$$g_{21} = 1 \quad \text{and} \quad g_{22} = g_{23} = 0 \tag{4b}$$

$$g_{33} = 1 \quad \text{and} \quad g_{31} = g_{32} = 0 \tag{4c}$$

Hence, the symmetry matrix for the T-junction is

$$[G_A] = \begin{bmatrix} 0 & 1 & 0 \\ 1 & 0 & 0 \\ 0 & 0 & 1 \end{bmatrix} \tag{5}$$

(c) The reflection with respect to the P_B-plane will transform the fields with a negative sign at the same port. This operation will have little contribution in problem solution.

(d) The rotation with respect to the J-axis will interchange the fields at port 1 with those at port 2, and in addition, the signs of the fields are reversed. The fields at port 3 also have an opposite sign.

$$[G_J] = \begin{bmatrix} 0 & -1 & 0 \\ -1 & 0 & 0 \\ 0 & 0 & -1 \end{bmatrix} \tag{6}$$

(e) For all practical purpose, the symmetry matrix for the T-junction is given by $[G] = [G_A]$ or

$$[G] = \begin{bmatrix} 0 & 1 & 0 \\ 1 & 0 & 0 \\ 0 & 0 & 1 \end{bmatrix} \tag{7}$$

(f) The scattering matrix of the T-junction can be obtained from the symmetry matrix by virtue of the commuting property between these matrices, (4-8).

$$[G] [S] = [S] [G]$$

or

$$[S] = [G]^{-1} [S] [G] = [G]^t [S] [G] \tag{8}$$

since the symmetry matrix is an orthogonal matrix, and thus, from (4-4), $[G]^t = [G]^{-1}$.

The scattering matrix for a three-port device has the general form

$$[S] = \begin{bmatrix} s_{11} & s_{12} & s_{13} \\ s_{21} & s_{22} & s_{23} \\ s_{31} & s_{32} & s_{33} \end{bmatrix} \tag{9}$$

Then (8) becomes

$$\begin{bmatrix} s_{11} & s_{12} & s_{13} \\ s_{21} & s_{22} & s_{23} \\ s_{31} & s_{32} & s_{33} \end{bmatrix} = \begin{bmatrix} 0 & 1 & 0 \\ 1 & 0 & 0 \\ 0 & 0 & 1 \end{bmatrix} \begin{bmatrix} s_{11} & s_{12} & s_{13} \\ s_{21} & s_{22} & s_{23} \\ s_{31} & s_{32} & s_{33} \end{bmatrix} \begin{bmatrix} 0 & 1 & 0 \\ 1 & 0 & 0 \\ 0 & 0 & 1 \end{bmatrix}$$

$$= \begin{bmatrix} 0 & 1 & 0 \\ 1 & 0 & 0 \\ 0 & 0 & 1 \end{bmatrix} \begin{bmatrix} s_{12} & s_{11} & s_{13} \\ s_{22} & s_{21} & s_{23} \\ s_{32} & s_{31} & s_{33} \end{bmatrix} = \begin{bmatrix} s_{22} & s_{21} & s_{23} \\ s_{12} & s_{11} & s_{13} \\ s_{32} & s_{31} & s_{33} \end{bmatrix} \tag{10}$$

It is apparent that

$$s_{11} = s_{22} \equiv a \tag{11a}$$

$$s_{12} = s_{21} \equiv b \tag{11b}$$

$$s_{13} = s_{31} = s_{23} = s_{32} \equiv c \tag{11c}$$

$$s_{33} \equiv d \tag{11d}$$

since $s_{13} = s_{31}$ for a reciprocal network. Equation (10) can be expressed as

$$[S] = \begin{bmatrix} a & b & c \\ b & a & c \\ c & c & d \end{bmatrix} \tag{11e}$$

The number of scattering coefficients is reduced to four by applying considerations of symmetry and isotropy.

(g) When the junction is lossless, the scattering matrix satisfies the unitary condition, (III.13-2).

$$[S^*]^t [S] = [u] = \begin{bmatrix} a^* & b^* & c^* \\ b^* & a^* & c^* \\ c^* & c^* & d^* \end{bmatrix} \begin{bmatrix} a & b & c \\ b & a & c \\ c & c & d \end{bmatrix} \tag{12a}$$

The expansion of the above relation yields the following result.

(1, 1), (2, 2) term: $\qquad$ $|a|^2 + |b|^2 + |c|^2 = 1$ $\qquad$ (12b)

(3, 3) term: $\qquad$ $2|c|^2 + |d|^2 = 1$ $\qquad$ (12c)

(2, 1), (1, 2)* term: $\qquad$ $ab^* + a^*b + |c|^2 = 0$ $\qquad$ (12d)

(3, 1), (1, 3)*, ((2, 3)*, (3, 2) term: $\quad ac^* + bc^* + d^*c = 0$ $\qquad$ (12e)

(h) Much information may be deduced from Eqs. (11) and (12). Specific properties of the junction can only be determined by solving the scattering coefficients in terms of the eigenvalues. Some such properties are the minimum VSWR possible in line 1-2 for a given coupling between port 1 and port 3, or the possibility that arm 1 may be decoupled from arm 2 by placing a short in arm 3, etc. The characteristic equation for matrix [G] is

$$\det \left| [G] - \lambda [u] \right| = 0 \qquad (13a)$$

where $\lambda \equiv \lambda^g$ (the superscript g has been omitted for simplicity and will be reinserted in the final result).

$$\det \begin{bmatrix} -\lambda & 1 & 0 \\ 1 & -\lambda & 0 \\ 0 & 0 & 1-\lambda \end{bmatrix} = -\lambda\,[-\lambda\,(1-\lambda)] - (1-\lambda) = 0$$

$$f(\lambda) = -\lambda^3 + \lambda^2 + \lambda - 1 = (\lambda^2 - 1)(1 - \lambda) \qquad (13b)$$

$$= (\lambda + 1)(\lambda - 1)(1 - \lambda) \qquad (13c)$$

The eigenvalues are

$$\lambda_1^g = -1, \qquad \lambda_2^g = 1, \qquad \lambda_3^g = 1 \qquad (14)$$

(i) The eigenvectors may be found as follows. The characteristic polynomial is given by (13b).

$$f(\lambda) = -\lambda^3 + \lambda^2 + \lambda - 1 \qquad (15)$$

where $\lambda \equiv \lambda^g$.

Construct the function $[F(\lambda)]$ [Eq. (III.7-5)].

$$[F(\lambda)] = \frac{[u]\,f(\lambda) - f([G])}{\lambda[u] - [G]} \tag{16}$$

where

$$f([G]) = -[G]^3 + [G]^2 + [G] - [u] \tag{17}$$

$$[F(\lambda)] = \frac{-([u]\,\lambda^3 - [G]^3) + ([u]\,\lambda^2 - [G]^2) + ([u]\,\lambda - [G])}{\lambda[u] - [G]}$$

$$= -([u]\,\lambda^2 + \lambda[G] + [G]^2) + ([u]\,\lambda + [G]) + [u]$$

$$= [u]\,(-\lambda^2 + \lambda + 1) - [G]^2 + (1 - \lambda)[G] \tag{18}$$

$$[G]^2 = [G]\,[G] = [u] \tag{19}$$

since the symmetry matrix is orthonormal. The last two terms on the right of (18) are equal to

$$(1-\lambda)[G] - [G]^2 = (1-\lambda) \begin{bmatrix} 0 & 1 & 0 \\ 1 & 0 & 0 \\ 0 & 0 & 1 \end{bmatrix} - [u]$$

$$= \begin{bmatrix} -1 & 1-\lambda & 0 \\ 1-\lambda & -1 & 0 \\ 0 & 0 & -\lambda \end{bmatrix} \tag{20}$$

The substitution of (20) into (18) yields

$$[F(\lambda)] = \begin{bmatrix} -\lambda^2 + \lambda & 1-\lambda & 0 \\ 1-\lambda & -\lambda^2 + \lambda & 0 \\ 0 & 0 & -\lambda^2 + 1 \end{bmatrix} \tag{21}$$

At $\lambda = -1$, one has

$$[F(\lambda = -1)] = \begin{bmatrix} -2 & 2 & 0 \\ 2 & -2 & 0 \\ 0 & 0 & 0 \end{bmatrix} \tag{22}$$

Equation (22) has only one independent column, the normalized eigenvector is

$$[x_1] = \frac{1}{\sqrt{2}} [\, 1 \quad -1 \quad 0 \,]^t \tag{23}$$

At $\lambda = 1$, Eq. (21) becomes

$$[F(\lambda = 1)] = \begin{bmatrix} 0 & 0 & 0 \\ 0 & 0 & 0 \\ 0 & 0 & 0 \end{bmatrix} \tag{24}$$

Equation (24) provides a trivial eigenvector. The first derivative of $[F(\lambda)]$ with respect to λ gives

$$[F(\lambda)]^{(1)} \equiv \frac{d[F]}{d\lambda} = \begin{bmatrix} -2\lambda + 1 & -1 & 0 \\ -1 & -2\lambda + 1 & 0 \\ 0 & 0 & -2\lambda \end{bmatrix}$$

$$[F(\lambda=1)]^{(1)} = \begin{bmatrix} -1 & -1 & 0 \\ -1 & -1 & 0 \\ 0 & 0 & -2 \end{bmatrix} \tag{25}$$

The normalized eigenvectors are

$$[x_2] = \frac{1}{\sqrt{2}} [\, 1 \quad 1 \quad 0 \,]^t \tag{26}$$

$$[x_3] = [\, 0 \quad 0 \quad 1 \,]^t \tag{27}$$

The eigenvector matrix $[E^g]$ for the symmetry matrix is

$$[E^g] = [\ [x_1] \quad [x_2] \quad [x_3]\ ]^t = \begin{bmatrix} \dfrac{1}{\sqrt{2}} & \dfrac{1}{\sqrt{2}} & 0 \\ \dfrac{-1}{\sqrt{2}} & \dfrac{1}{\sqrt{2}} & 0 \\ 0 & 0 & 1 \end{bmatrix} \tag{28}$$

As a check, the eigenvector matrix is used to diagonalize the symmetry matrix $[G]$ and is shown to be

$$[E^g]^t [G] [E^g] = \text{diag} [\, -1 \quad 1 \quad 1 \,] \tag{29}$$

The eigenvectors given by Eqs. (23), (26), and (27) are one possible set. In this choice, port 3 is not coupled to the other port; hence it is not useful for the general problem. This is also because there is only one symmetry matrix, and this does not supply sufficient conditions.

(j) The scattering matrix can be diagonalized.

$$[D_S] = [E^g]^{-1} [S] [E^g] = [E^g]^t [S] [E^g]$$

or

$$[S] = [E^g] [D_S] [E^g]^t$$

$$= \begin{bmatrix} \frac{1}{\sqrt{2}} & \frac{1}{\sqrt{2}} & 0 \\ \frac{-1}{\sqrt{2}} & \frac{1}{\sqrt{2}} & 0 \\ 0 & 0 & 1 \end{bmatrix} \begin{bmatrix} s_1 & 0 & 0 \\ 0 & s_2 & 0 \\ 0 & 0 & s_3 \end{bmatrix} \begin{bmatrix} \frac{1}{\sqrt{2}} & \frac{-1}{\sqrt{2}} & 0 \\ \frac{1}{\sqrt{2}} & \frac{1}{\sqrt{2}} & 0 \\ 0 & 0 & 1 \end{bmatrix}$$

$$= \begin{bmatrix} \frac{1}{2} (s_1 + s_2) & \frac{1}{2} (-s_1 + s_2) & 0 \\ \frac{1}{2} (-s_1 + s_2) & \frac{1}{2} (s_1 + s_2) & 0 \\ 0 & 0 & s_3 \end{bmatrix}$$

(k) An alternative approach to finding the engenvectors is as follows. Let the eigenvectors be

$$[x_1] = [1 \quad -1 \quad 0]^t \tag{30a}$$

$$[x_2] = [x_{21} \quad x_{22} \quad x_{23}]^t \tag{30b}$$

$$[x_3] = [x_{31} \quad x_{32} \quad x_{33}]^t \tag{30c}$$

Apply the orthogonal conditions.

$$[x_1]^t [x_2] = 0 = [1 \quad -1 \quad 0] [x_{21} \quad x_{22} \quad x_{23}]^t$$

$$= x_{21} - x_{22}$$

or

$$x_{22} = x_{21} \tag{31a}$$

Orthogonality between $[x_1]$ and $[x_3]$ implies:

$$[x_1]^t [x_3] = [\; 1 \quad -1 \quad 0 \;] [\; x_{31} \quad x_{32} \quad x_{33} \;]^t = x_{31} - x_{32} = 0$$

or

$$x_{32} = x_{31} \tag{31b}$$

and the orthogonality between $[x_2]$ and $[x_3]$ implies:

$$[x_2]^t [x_3] = x_{21} x_{31} + x_{21} x_{31} + x_{23} x_{33} = 0$$

or

$$2x_{21}x_{31} + x_{23}x_{33} = 0 \tag{31c}$$

Since there are more unknowns then equations, some arbitrariness exists in the solution. Equation (31c) can be arranged as

$$\frac{-x_{23}}{\sqrt{2}\, x_{21}} \frac{x_{33}}{\sqrt{2}\, x_{31}} = 1 \tag{32}$$

The eigenvectors $[x_2]$ and $[x_3]$ can be expressed as

$$[x_2] = x_{21} \left[\; 1 \quad 1 \quad \frac{x_{23}}{x_{21}} \;\right]^t \tag{33a}$$

$$[x_3] = x_{31} \left[\; 1 \quad 1 \quad \frac{x_{33}}{x_{32}} \;\right]^t \tag{33b}$$

Equation (32) can be satisfied by the following choice.

$$\frac{x_{33}}{\sqrt{2}\, x_{31}} = \left[\frac{-x_{23}}{\sqrt{2}\, x_{21}}\right]^{-1} \equiv -k^{-1} \tag{34}$$

That is,

$$\frac{x_{23}}{x_{21}} = \sqrt{2}\, k \quad \text{and} \quad \frac{x_{33}}{x_{31}} = -\sqrt{2}\, k^{-1} \tag{35}$$

Equation (33) then becomes

$$[x_2] = x_{21} [1 \quad 1 \quad \sqrt{2} \, k]^t \tag{36a}$$

$$[x_3] = x_{31} \left[1 \quad 1 \quad \frac{-\sqrt{2}}{k} \right]^t \tag{36b}$$

The normalized eigenvectors are:

$$[x_1] = \frac{1}{\sqrt{2}} [1 \quad -1 \quad 0]^t \tag{37a}$$

$$|x_2| = |x_{21}| \sqrt{1 + 1 + 2 k^2}$$

$$= \sqrt{2} \sqrt{1 + k^2} \, |x_{21}| = 1$$

or

$$|x_{21}| = \frac{1}{\sqrt{2 (1 + k^2)}}$$

and

$$[x_2] = \left[\frac{1}{\sqrt{2(1 + k^2)}} \quad \frac{1}{\sqrt{2(1 + k^2)}} \quad \frac{k}{\sqrt{1 + k^2}} \right]^t \tag{37b}$$

The magnitude of $[x_3]$ is

$$|x_3| = |x_{31}| \sqrt{1 + 1 + \frac{2}{k^2}} = 1$$

$$|x_{31}| = \frac{k}{\sqrt{2(1 + k^2)}}$$

and

$$[x_3] = \left[\frac{k}{\sqrt{2(1 + k^2)}} \quad \frac{k}{\sqrt{2(1 + k^2)}} \quad \frac{-1}{\sqrt{1 + k^2}} \right]^t \tag{37c}$$

The eigenvector matrix is

$$[E^g] = [[x_1] \quad [x_2] \quad [x_3]]^t$$

$$= \frac{1}{\sqrt{2(1 + k^2)}} \begin{bmatrix} \sqrt{1 + k^2} & 1 & k \\ -\sqrt{1 + k^2} & 1 & k \\ 0 & \sqrt{2}\,k & -\sqrt{2} \end{bmatrix} \tag{38}$$

The eigenvector matrix is orthonormal as can be shown by direct expansion of the relation

$$[E^g]^t [E^g] = [u]$$

Hence,

$$[E^g]^t = [E^g]^{-1} \tag{39}$$

(l) The scattering matrix is given by

$$[S] = [E^g] [D_S] [E^g]^{-1} = [E^g] [D_S] [E^g]^t \tag{40a}$$

$$[D_S] = \text{diag} [s_1 \quad s_2 \quad s_3] \tag{40b}$$

$$[D_S] [E^g]^t = \frac{1}{\sqrt{2(1 + k^2)}} \begin{bmatrix} s_1 & 0 & 0 \\ 0 & s_2 & 0 \\ 0 & 0 & s_3 \end{bmatrix} \begin{bmatrix} \sqrt{1 + k^2} & -\sqrt{1 + k^2} & 0 \\ 1 & 1 & k\sqrt{2} \\ k & k & -\sqrt{2} \end{bmatrix}$$

$$= \frac{1}{\sqrt{2(1 + k^2)}} \begin{bmatrix} s_1\sqrt{1 + k^2} & -s_1\sqrt{1 + k^2} & 0 \\ s_2 & s_2 & \sqrt{2}\,ks_2 \\ ks_3 & ks_3 & -\sqrt{2}\,s_3 \end{bmatrix} \tag{41a}$$

and

$$[S] = [E^g] \ [\text{Eq.(41a)}]$$

$$[S] = \frac{1}{2(1 + k^2)} \times$$

$$\begin{bmatrix} s_1(1 + k^2) + s_2 + k^2 s_3 & -s_1(1 + k^2) + s_2 + k^2 s_3 & \sqrt{2}\, k(s_2 - s_3) \\ -s_1(1 + k^2) + s_2 + k^2 s_3 & s_1(1 + k^2) + s_2 + k^2 s_3 & \sqrt{2}\, k(s_2 - s_3) \\ \sqrt{2}\, k(s_2 - s_3) & \sqrt{2}\, k(s_2 - s_3) & 2(k^2 s_2 + s_3) \end{bmatrix} \qquad (41b)$$

but [S] is also given by (11e), i.e.,

$$[S] = \begin{bmatrix} a & b & c \\ b & a & c \\ c & c & d \end{bmatrix} \qquad (42)$$

Therefore, the elements of the scattering matrix are

$$a = \frac{1}{2(1 + k^2)} \{ (1 + k^2)s_1 + s_2 + k^2 s_3 \} \qquad (43a)$$

$$b = \frac{1}{2(1 + k^2)} \{ -(1 + k^2)s_1 + s_2 + k^2 s_3 \} \qquad (43b)$$

$$c = \frac{k}{\sqrt{2}\,(1 + k^2)} \, (s_2 - s_3) \qquad (43c)$$

$$d = \frac{k^2 s_2 + s_3}{1 + k^2} \qquad (43d)$$

(m) It proves convenient to select the reference plane at port 3 such that

$$\sqrt{2}\, c + d = 1 \qquad (44)$$

which is equivalent to (12c). This can be visualized if $\sqrt{2}c$ and d are the mutually perpendicular sides of a right triangle and its hypotenuse has an amplitude of unity. The substitution of (43c) and (43d) into (44) yields

$$\frac{k(s_2 - s_3)}{1 + k^2} + \frac{k^2 s_2 + s_3}{1 + k^2} = 1$$

or

$$\frac{1}{1 + k^2} \{ (k + k^2)s_2 + (1 - k)s_3 \} = 1 \tag{45}$$

Equation (45) can be satisfied if

$$k = 1 \qquad \text{and} \qquad s_2 = 1 \tag{46a}$$

or

$$k = -1 \qquad \text{and} \qquad s_3 = 1 \tag{46b}$$

The substitution of (46a) into (43) gives

$$a = \frac{1}{4} (1 + 2s_1 + s_3) \tag{47a}$$

$$b = \frac{1}{4} (1 - 2s_1 + s_3) \tag{47b}$$

$$c = \frac{1}{2\sqrt{2}} (1 - s_3) \tag{47c}$$

$$d = \frac{1}{2} (1 + s_3) \tag{47d}$$

The substitution of (46b) into (43) produces the same set of relations as (47) with the exception that s_3 is replaced by s_2.

Equation (47) provides the following relations.

$$2a + d = 1 + s_1 + s_3 \tag{48a}$$

$$\sqrt{2}\, c + d = 1 \tag{48b}$$

$$a + b = d \tag{48c}$$

(n) When port 1 and port 2 are matched, Eqs. (42) and (47a) give

$$a = 0 = \frac{1}{4} (1 + 2s_1 + s_3) \tag{49}$$

Since $|s_1| = |s_3| = 1$, for a lossless junction, Eq. (49) can be fulfilled if

$$s_1 = -1 \quad \text{and} \quad s_3 = 1 \tag{50}$$

with respect to the reference plane previously selected, $s_2 = 1$ [Eq. (46a)]. Then (47) becomes

$$a = 0 = c \quad \text{and} \quad b = 1 = d \tag{51}$$

and the scattering matrix is [Eq. (42)]

$$[S] = \begin{bmatrix} 0 & 1 & 0 \\ 1 & 0 & 0 \\ 0 & 0 & 1 \end{bmatrix} \tag{52}$$

This implies that port 3 is completely decoupled from the other two ports. Since $s_{33} = 1$, hence port 3 appears as a perfect reflector such as an electric wall somewhere inside port 3.

Equation (52) can also be obtained from the unitary condition by setting $a = 0$ in (12).

$$|b|^2 + |c|^2 = 1 \tag{53a}$$

$$2|c|^2 + |d|^2 = 1 \tag{53b}$$

$$|c|^2 = 0 \tag{53c}$$

$$bc^* + d^*c = 0 \tag{53d}$$

Consequently,

$$c = 0 \quad b = 1 \quad d = 1 \tag{54}$$

(o) When port 3 is matched, (43d) becomes

$$d = 0 = \frac{k^2 s_2 + s_3}{1 + k^2} \tag{55}$$

This can hold if

$$s_3 = -1 \tag{56}$$

since from (46) one has

$$k = 1 \qquad \text{and} \qquad s_2 = 1$$

Then (43) becomes

$$a = \frac{s_1}{2}, \qquad b = \frac{-s_1}{2}, \qquad c = \frac{1}{\sqrt{2}}, \qquad d = 0 \tag{57}$$

The scattering matrix becomes

$$[S] = \frac{1}{2} \begin{bmatrix} s_1 & -s_1 & \sqrt{2} \\ -s_1 & s_1 & \sqrt{2} \\ \sqrt{2} & \sqrt{2} & 0 \end{bmatrix} \tag{58}$$

For a lossless junction, s_1 has an amplitude of unity, and its phase can be adjusted by appropriate choice of the reference plane. Let

$$s_1 = e^{j2\theta} \tag{59}$$

Adjust the reference planes for port 1 and port 2 by moving them symmetrically, such that $s_1 = -1 = e^{j\pi}$. The distance moved is

$$\pi - 2\theta = 2(\pi/2 - \theta)$$

That is, the reference plane at port 3 is moved by an angle $-(\pi/2 - \theta)$ so that c remains unchanged. Then (58) becomes

$$[S] = \frac{1}{2} \begin{bmatrix} -1 & 1 & \sqrt{2} \\ 1 & -1 & \sqrt{2} \\ \sqrt{2} & \sqrt{2} & 0 \end{bmatrix} = \begin{bmatrix} \dfrac{-1}{2} & \dfrac{1}{2} & \dfrac{1}{\sqrt{2}} \\ \dfrac{1}{2} & \dfrac{-1}{2} & \dfrac{1}{\sqrt{2}} \\ \dfrac{1}{\sqrt{2}} & \dfrac{1}{\sqrt{2}} & 0 \end{bmatrix} \tag{60}$$

Equation (60) implies that when a wave is incident onto port 3, it will split equally between port 1 and port 2. When a wave is incident onto port 1, one-quarter of the power is reflected back to arm 1, one-quarter of the power is transmitted to port 2, and one-half of the power is delivered to port 3.

When the three-port device is excited by in-phase signals to port 1 and port 2, i.e.,

$$[a] = [\, e_1 \quad e_1 \quad 0 \,]^t \tag{61}$$

the scattered wave is given by

$$[b] = [S][a]$$

$$= \frac{1}{2} \begin{bmatrix} -1 & 1 & \sqrt{2} \\ 1 & -1 & \sqrt{2} \\ \sqrt{2} & \sqrt{2} & 0 \end{bmatrix} \begin{bmatrix} e_1 \\ e_1 \\ 0 \end{bmatrix}$$

$$= [0 \quad 0 \quad \sqrt{2}\, e_1]^t \tag{62}$$

In other words, when port 1 and port 2 are excited by in-phase signals, all signal will be transmitted to port 3 only.

(p) When port 3 is terminated by an impedance which is characterized by the reflection coefficient ρ_3, and port 1 is excited by an incident wave A_1, then one has

$$[S][a] = \begin{bmatrix} a & b & c \\ b & a & c \\ c & c & d \end{bmatrix} \begin{bmatrix} A_1 \\ 0 \\ \rho_3 B_3 \end{bmatrix} = \begin{bmatrix} B_1 \\ B_2 \\ B_3 \end{bmatrix} \tag{63}$$

where $\rho_3 B_3$ is the wave reflected by the load. Equation (63) can be expanded as follows.

$$aA_1 = B_1 + 0 - c\rho_3 B_3 \tag{64a}$$

$$bA_1 = 0 + B_2 - c\rho_3 B_3 \tag{64b}$$

$$cA_1 = 0 + 0 + B_3(1 - d\rho_3) \tag{64c}$$

The B's can be determined by Cramer's rule.

$$B_1 = \left[a + \frac{c^2 \rho_3}{1 - d\rho_3} \right] A_1 \tag{65a}$$

$$B_2 = \left[b + \frac{c^2 \rho_3}{1 - d\rho_3} \right] A_1 \tag{65b}$$

$$B_3 = \frac{c}{1 - d\rho_3} A_1 \tag{65c}$$

The present three-port device with port 3 arbitrarily terminated can be considered as a symmetrical two-port device (symmetric by virtue of the original [S] matrix). One can construct a reduced 2×2 matrix, $[S_R]$, from (65).

$$[S_R][a] = \begin{bmatrix} s_{11} & s_{21} \\ s_{21} & s_{11} \end{bmatrix} \begin{bmatrix} A_1 \\ 0 \end{bmatrix} = \begin{bmatrix} \left(a + \dfrac{c^2 \rho_3}{1 - d\rho_3} \right) A_1 \\[4mm] \left(b + \dfrac{c^2 \rho_3}{1 - d\rho_3} \right) A_1 \end{bmatrix}$$

$$= \begin{bmatrix} (a + a_1)A_1 \\ (b + a_1)A_1 \end{bmatrix} = \begin{bmatrix} s_{11}A_1 \\ s_{21}A_1 \end{bmatrix} = \begin{bmatrix} B_1 \\ B_2 \end{bmatrix} = [b] \tag{66a}$$

$$a_1 \equiv \frac{c^2 \rho_3}{1 - d\rho_3} \tag{66b}$$

Therefore,

$$s_{11} = a + a_1 \qquad s_{21} = b + a_1 \tag{66}$$

The reduced scattering matrix is therefore

$$[S_R] = \begin{bmatrix} a + a_1 & b + a_1 \\[2mm] b + a_1 & a + a_1 \end{bmatrix} \tag{67}$$

8.1. Shifting Reference Planes in a Scattering Matrix

A scattering matrix is defined for a given set of reference planes. The phase of scattering coefficients will vary when reference planes are changed. This is not valid for the immitance matrices because the voltages and currents are functions of position, in both magnitude and phase. However, the incident and reflected wave parameters, a_i and b_i, remain independent of position in a lossless guide, i.e., constant in magnitude. This is because the basis of normalization for a_i and b_i is power.

Consider a two-port device attached to a lossless transmission line as shown in Fig. 8.1-1. The relations between incident and reflected parameters, [a] and [b], with respect to the original reference planes A and B are

$$b_1 = s_{11}a_1 + s_{12}a_2 \tag{1a}$$

$$b_2 = s_{21}a_1 + s_{22}a_2 \tag{1b}$$

The relation between parameters [a′] and [b′] with respect to the new reference planes A′ and B′ are

$$b_1' = s_{11}'a_1' + s_{12}'a_2' \tag{2a}$$

$$b_2' = s_{21}'a_1' + s_{22}'a_2' \tag{2b}$$

From the theory of wave propagation, the corresponding parameters at the original and the new reference planes are related as follows.

$$a_1' = a_1 e^{-j\theta_1}, \qquad b_1' = b_1 e^{j\theta_1} \tag{3a}$$

$$a_2' = a_2 e^{-j\theta_2}, \qquad b_2' = b_2 e^{j\theta_2} \tag{3b}$$

The substitution of (3) into (2) yields

$$b_1 e^{j\theta_1} = s_{11}'a_1 e^{-j\theta_1} + s_{12}'a_2 e^{-j\theta_2}$$

or

$$b_1 = s_{11}'e^{-j2\theta_1} a_1 + s_{12}'e^{-j(\theta_1 + \theta_2)}a_2 \tag{4a}$$

and

$$b_2 e^{j\theta_2} = s_{21}'a_1 e^{-j\theta_1} + s_{22}'a_2 e^{-j\theta_2}$$

or

$$b_2 = s_{21}'e^{-j(\theta_1 + \theta_2)} + s_{22}'e^{-j2\theta_2}a_2 \tag{4b}$$

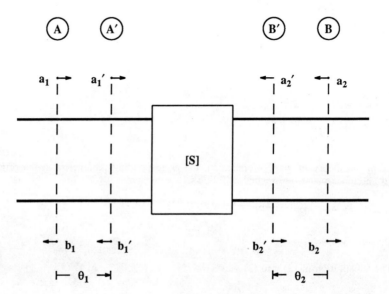

Figure 8.1-1: A two-port junction

A direct comparison of (1) with (4) produces

$$s_{11} = s_{11}'e^{-j2\theta_1} \qquad \text{or} \qquad s_{11}' = s_{11}e^{j2\theta_1} \tag{5a}$$

$$s_{12} = s_{12}'e^{-j(\theta_1 + \theta_2)} \qquad \text{or} \qquad s_{12}' = s_{12}e^{j(\theta_1 + \theta_2)} \tag{5b}$$

$$s_{21} = s_{21}'e^{-j(\theta_1 + \theta_2)} \qquad \text{or} \qquad s_{21}' = s_{21}e^{j(\theta_1 + \theta_2)} \tag{5c}$$

$$s_{22} = s_{22}'e^{-j2\theta_2} \qquad \text{or} \qquad s_{22}' = s_{22}e^{j2\theta_2} \tag{5d}$$

One can show, in general, for a p-port structure,

$$s_{mn}' = s_{mn}e^{j(\theta_m + \theta_n)} \qquad m, n = 1, 2, \ldots, p \tag{6}$$

where θ_k, $k = 1, 2, \ldots, p$, is the electrical length of the shift in position toward the junction. Equation (6) implies that elements of the scattering matrix can be selected to be real or imaginary by an appropriate choice of reference planes. This is a very useful concept in the application of the scattering matrix to practical structures.

8.2. Example: E-Plane T-Junction

An E-plane symmetrical T-junction is shown in Fig. 8.2-1. The junction is filled with isotropic medium. Determine the following:

(a) The significant symmetry matrix.

(b) The number of independent scattering coefficients.

(c) For a lossless junction, find the relations between various scattering coefficients.

(d) Eigenvalues and eigenvectors of the [G] matrix.

(e) The scattering matrix in terms of its eigenvalues.

(f) The scattering matrix if $\sqrt{2}\,c + d = 1$.

(g) The necessary conditions for port 1 and port 2 to be matched simultaneously.

(h) The condition for a matched port 3.

(i) The reduced scattering matrix when port 3 is terminated by a normalized impedance.

Solution:

(a) There are three significant symmetry matrices, $[G_A]$, $[G_B]$, and $[G_J]$. By the same argument as for the H-plane T-junction, the significant symmetry matrix is $[G_A]$ and it is given by

$$[G_A] = \begin{bmatrix} 0 & 1 & 0 \\ 1 & 0 & 0 \\ 0 & 0 & -1 \end{bmatrix} \tag{1}$$

The negative sign in the g_{33} element is due to the change of sign of the E-field at port 3 on reflection. $[G_A]$ is symmetric and orthogonal.

(b) By virtue of the commuting property between the scattering and symmetry matrices, one has

$$[S]\,[G] = [G]\,[S] \qquad \text{or} \qquad [S] = [G]\,[S]\,[G]^{-1} = [G]\,[S]\,[G]^t = [G]\,[S]\,[G]$$

$$[S] = \begin{bmatrix} 0 & 1 & 0 \\ 1 & 0 & 0 \\ 0 & 0 & -1 \end{bmatrix} \begin{bmatrix} s_{11} & s_{12} & s_{13} \\ s_{12} & s_{22} & s_{23} \\ s_{13} & s_{23} & s_{33} \end{bmatrix} \begin{bmatrix} 0 & 1 & 0 \\ 1 & 0 & 0 \\ 0 & 0 & -1 \end{bmatrix} \tag{2a}$$

The expansion of (2a) yields

$$\begin{bmatrix} s_{11} & s_{12} & s_{13} \\ s_{12} & s_{22} & s_{23} \\ s_{13} & s_{23} & s_{33} \end{bmatrix} = \begin{bmatrix} s_{22} & s_{12} & -s_{23} \\ s_{12} & s_{11} & -s_{13} \\ -s_{23} & -s_{13} & s_{33} \end{bmatrix} \tag{2b}$$

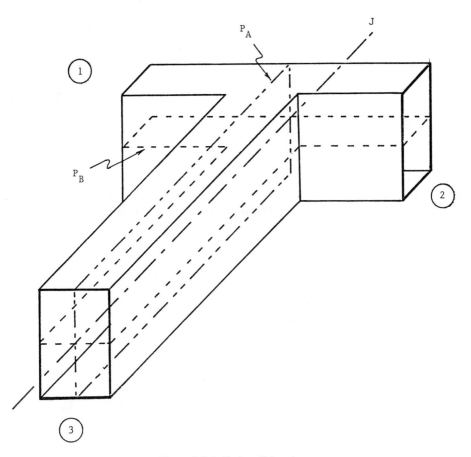

Figure 8.2-1: E-plane T-junction

Equation (2b) produces the following relations.

$$s_{11} = s_{22} \equiv a \tag{3a}$$

$$s_{12} \equiv b \tag{3b}$$

$$s_{13} = -s_{23} \equiv c \tag{3c}$$

$$s_{33} \equiv d \tag{3d}$$

The scattering matrix is therefore

$$[S] = \begin{bmatrix} a & b & c \\ b & a & -c \\ c & -c & d \end{bmatrix} \tag{4}$$

(c) For a lossless T-junction, the [S] matrix satisfies the unitary condition.

$$[S^*]^t [S] = [u] = \begin{bmatrix} a^* & b^* & c^* \\ b^* & a^* & -c^* \\ c^* & -c^* & d^* \end{bmatrix} \begin{bmatrix} a & b & c \\ b & a & -c \\ c & -c & d \end{bmatrix} \tag{5a}$$

The expansion of the above expression yields,

$$(1, 1) \quad |a|^2 + |b|^2 + |c|^2 = 1 \tag{5b}$$

$$(3, 3) \quad 2|c|^2 + |d|^2 = 1 \tag{5c}$$

$$(2, 1) \quad b^*a + a^*b - |c|^2 = 0 \tag{5d}$$

$$(3, 1) \quad c^*a - c^*b + d^*c = 0 \tag{5e}$$

(d) The characteristic equation is

$$\det | [G] - \lambda [u] | = 0 = \det \begin{bmatrix} -\lambda & 1 & 0 \\ 1 & -\lambda & 0 \\ 0 & 0 & -1 - \lambda \end{bmatrix}$$

$$= -\lambda^2 (1 + \lambda) + 1 + \lambda = (1 + \lambda)^2 (1 - \lambda) \tag{6a}$$

The eigenvalues are

$$\lambda_1 = 1 \qquad \lambda_2 = -1 \qquad \lambda_3 = -1 \tag{6b}$$

The eigenvector is determined from the following relation.

$$[G] [x_i] = \lambda_i [x_i] \tag{7}$$

For $\lambda = \lambda_1 = 1$, one has

$$\begin{bmatrix} 0 & 1 & 0 \\ 1 & 0 & 0 \\ 0 & 0 & -1 \end{bmatrix} \begin{bmatrix} x_{11} \\ x_{12} \\ x_{13} \end{bmatrix} = \begin{bmatrix} x_{11} \\ x_{12} \\ x_{13} \end{bmatrix}$$

The expansion of the above expression yields

$$x_{12} = x_{11} \qquad \text{and} \qquad x_{13} = -x_{13} \quad \text{or} \quad x_{13} = 0 \tag{8a}$$

The eigenvector for λ_1 is

$$[x_1] = [\ x_{11} \quad x_{11} \quad 0\ ]^t = \frac{1}{\sqrt{2}} [\ 1 \quad 1 \quad 0\ ]^t \tag{8b}$$

The other eigenvectors will be determined by the property of orthogonality. Let

$$[x_2] = [\ x_{21} \quad x_{22} \quad x_{23}\ ]^t \qquad [x_3] = [\ x_{31} \quad x_{32} \quad x_{33}\ ]^t \tag{9}$$

The orthogonality between $[x_1]$ and $[x_2]$ gives

$$[x_1]^t [x_2] = 0 = \frac{1}{\sqrt{2}} [\ 1 \quad 1 \quad 0\ ] \begin{bmatrix} x_{21} \\ x_{22} \\ x_{23} \end{bmatrix} = \frac{1}{\sqrt{2}} (\ x_{21} + x_{22}\ )$$

or

$$x_{22} = -x_{21} \tag{10a}$$

$$[x_2] = x_{21} \begin{bmatrix} 1 & -1 & \dfrac{x_{23}}{x_{21}} \end{bmatrix}^t \tag{10b}$$

The orthogonality between $[x_1]$ and $[x_3]$ yields

$$\frac{1}{\sqrt{2}} \begin{bmatrix} 1 & 1 & 0 \end{bmatrix} \begin{bmatrix} x_{31} \\ x_{32} \\ x_{33} \end{bmatrix} = \frac{1}{\sqrt{2}} (x_{31} + x_{32}) = 0 \quad \text{or} \quad x_{32} = -x_{31}$$

The eigenvector correspondng to λ_3 is

$$[x_3] = x_{31} \begin{bmatrix} 1 & -1 & \dfrac{x_{33}}{x_{31}} \end{bmatrix}^t \tag{11}$$

The orthogonality between $[x_2]$ and $[x_3]$ provides the following relation.

$$[x_2]^t [x_3] = x_{21} \begin{bmatrix} 1 & -1 & \dfrac{x_{23}}{x_{21}} \end{bmatrix} x_{31} \begin{bmatrix} 1 \\ -1 \\ \dfrac{x_{33}}{x_{31}} \end{bmatrix}$$

$$= x_{21} x_{31} \begin{bmatrix} 1 + 1 + \dfrac{x_{23}}{x_{21}} \dfrac{x_{33}}{x_{31}} \end{bmatrix} = 0 \tag{12a}$$

$$\frac{x_{23}}{x_{21}} \frac{x_{33}}{x_{31}} = -2 \quad \text{or} \quad \frac{x_{23}}{-\sqrt{2}\, x_{21}} \frac{x_{33}}{\sqrt{2}\, x_{31}} = 1 \tag{12b}$$

Let

$$k = \frac{x_{33}}{\sqrt{2}\, x_{31}} \quad \text{or} \quad \frac{x_{33}}{x_{31}} = \sqrt{2}\, k \tag{13a}$$

$$\frac{1}{k} = \frac{x_{23}}{-\sqrt{2}\, x_{21}} \quad \text{or} \quad \frac{x_{23}}{x_{21}} = -\frac{\sqrt{2}}{k} \tag{13b}$$

then

$$[x_2] = x_{21} \begin{bmatrix} 1 & -1 & \dfrac{-\sqrt{2}}{k} \end{bmatrix}^t = \frac{k}{\sqrt{2}\,(1+k^2)} \begin{bmatrix} 1 & -1 & \dfrac{-\sqrt{2}}{k} \end{bmatrix}^t \tag{14a}$$

$$[x_3] = x_{31} \begin{bmatrix} 1 & -1 & \sqrt{2}\, k \end{bmatrix}^t = \frac{1}{\sqrt{2}\,(1+k^2)} \begin{bmatrix} 1 & -1 & \sqrt{2}\, k \end{bmatrix}^t \tag{14b}$$

The matrix of eigenvectors is

$$[E^g] = \frac{1}{\sqrt{2(1+k^2)}} \begin{bmatrix} \sqrt{1+k^2} & k & 1 \\ \sqrt{1+k^2} & -k & -1 \\ 0 & -\sqrt{2} & k\sqrt{2} \end{bmatrix} \tag{15}$$

(e) The scattering matrix can be expressed in terms of its eigenvalues by means of the following relation.

$$[S] = [E^g][S_d][E^g]^{-1} \tag{16}$$

The eigenvector matrix $[E^g]$ is orthonormal, i.e.,

$$[E^g]^t[E^g] = [u] \tag{17a}$$

hence,

$$[E^g]^t = [E^g]^{-1} \tag{17b}$$

Then, the expansion of (15) yields

$$[S] = [E^g][S_d][E^g]^t$$

$$= \frac{1}{\sqrt{2(1+k^2)}}[E^g] \begin{bmatrix} s_1 & 0 & 0 \\ 0 & s_2 & 0 \\ 0 & 0 & s_3 \end{bmatrix} \begin{bmatrix} \sqrt{1+k^2} & \sqrt{1+k^2} & 0 \\ k & -k & -\sqrt{2} \\ 1 & -1 & \sqrt{2}k \end{bmatrix}$$

$$= \frac{1}{\sqrt{2(1+k^2)}}[E^g] \begin{bmatrix} s_1\sqrt{1+k^2} & s_1\sqrt{1+k^2} & 0 \\ s_2k & -s_2k & -\sqrt{2}s_2 \\ s_3 & -s_3 & \sqrt{2}ks_3 \end{bmatrix}$$

$$[S] = \frac{1}{2(1+k^2)} \times$$

$$\begin{bmatrix} (1+k^2)s_1 + k^2s_2 + s_3 & (1+k^2)s_1 - k^2s_2 - s_3 & \sqrt{2}k(s_3 - s_2) \\ (1+k^2)s_1 - k^2s_2 - s_3 & (1+k^2)s_1 + k^2s_2 + s_3 & \sqrt{2}k(s_2 - s_3) \\ \sqrt{2}k(s_3 - s_2) & \sqrt{2}k(s_2 - s_3) & 2(s_2 + ks_3) \end{bmatrix} \tag{18}$$

A direct comparison between (18) and (4) shows that

$$a = \frac{1}{2(1 + k^2)} [(1 + k^2)s_1 + k^2 s_2 + s_3] \tag{19a}$$

$$b = \frac{1}{2(1 + k^2)} [(1 + k^2)s_1 - k^2 s_2 - s_3] \tag{19b}$$

$$c = \frac{k}{\sqrt{2} (1 + k^2)} (s_3 - s_2) \tag{19c}$$

$$d = \frac{1}{(1 + k^2)} (s_2 + ks_3) \tag{19d}$$

(f) To satisfy the condition

$$\sqrt{2} c + d = 1 \tag{20}$$

one must have

$$\frac{k}{1 + k^2} (s_3 - s_2) + \frac{1}{1 + k^2} (s_2 + ks_3) = 1$$

or

$$\frac{1}{1 + k^2} [(1 - k)s_2 + 2k_3] = 1 \tag{21}$$

Equation (21) can be realized if

$$k = 1 \qquad \text{and} \qquad s_3 = 1 \tag{22}$$

Then (19) becomes

$$a = \frac{1}{4} (2s_1 + s_2 + 1) \tag{23a}$$

$$b = \frac{1}{4} (2s_1 - s_2 - 1) \tag{23b}$$

$$c = \frac{1}{2\sqrt{2}} (1 - s_2) \tag{23c}$$

$$d = \frac{1}{2} (s_2 + 1) \tag{23d}$$

(g) Port 1 and port 2 will be matched if [Eq. (19a)]

$$a = 0 = \frac{1}{2(1 + k^2)} [(1 + k^2)s_1 + k^2 s_2 + s_3] \tag{24}$$

Subject to the condition

$$\sqrt{2} c + d = 1 \quad \text{or} \quad k = 1 \quad \text{and} \quad s_3 = 1 \tag{25}$$

Eq. (24) becomes

$$a = \frac{1}{4} (2s_1 + s_2 + 1) = 0 \tag{26}$$

Since for a lossless structure, by virtue of the unitary property, (Section 7(f)),

$$| s_1 | = 1 \quad \text{and} \quad | s_2 | = 1 \tag{27}$$

and Eq. (27) would be satisfied if

$$s_1 = -1 \quad \text{and} \quad s_2 = 1 \tag{28}$$

Then, (23) yields

$$a = 0 \qquad b = -1 \qquad c = 0 \qquad d = 1 \tag{29a}$$

and the scattering matrix becomes

$$[S] = \begin{bmatrix} 0 & -1 & 0 \\ -1 & 0 & 0 \\ 0 & 0 & 1 \end{bmatrix} \tag{29b}$$

Port 3 is completely decoupled from arm 1 and arm 2 when port 1 and port 2 are matched simultaneously. Since $s_{33} = 1$, a metallic wall must exist in arm 3.

Port 1 and port 2 would be mismatched if power is extracted from port 3.

(h) For port 3 to be matched, one must have, from (19d), subject to condition (20),

$$d = 0 = \frac{1}{2} (s_2 + 1) \quad \text{or} \quad s_2 = -1 \tag{30a}$$

Then, (23) becomes

$$a = \frac{s_1}{2} \qquad b = \frac{s_1}{2} \qquad c = \frac{1}{\sqrt{2}} \qquad d = 0$$

and the scattering matrix is

$$[s] = \frac{1}{2} \begin{bmatrix} s_1 & s_1 & \sqrt{2} \\ s_1 & s_1 & -\sqrt{2} \\ \sqrt{2} & -\sqrt{2} & 0 \end{bmatrix} \tag{30b}$$

In general, s_1 is complex, i.e.,

$$s_1 = e^{j2\theta_1} \tag{31}$$

The coefficient in (30b) can be made real by a shift in reference planes. Let reference planes at port 1 and port 2 be shifted by an electrical length of $-\theta_1$; then s_{11} and s_{22} will be changed by $e^{-j2\theta_1}$, and s_{12} and s_{13} by $e^{-j\theta_1}$, i.e.,

$$s_{11} = e^{j2\theta_1} e^{-j2\theta_1} = 1 = s_{22} \tag{32a}$$

$$s_{12} = e^{j2\theta_1} e^{-j2\theta_1} = 1 = s_{21} \tag{32b}$$

$$s_{13} = \sqrt{2} e^{j(\theta_1 - \theta_1)} = \sqrt{2} = s_{31} \tag{32c}$$

$$s_{23} = s_{32} = -\sqrt{2} \tag{32d}$$

The new scattering matrix is

$$[S] = \frac{1}{2} \begin{bmatrix} 1 & 1 & \sqrt{2} \\ 1 & 1 & -\sqrt{2} \\ \sqrt{2} & -\sqrt{2} & 0 \end{bmatrix} \tag{33}$$

When power is fed into port 1, one-quarter of the power is reflected back to port 1; one-quarter of the power is transmitted to port 2; and one-half of the power is transmitted to port 3.

If equal and oppositely phased signals are applied to the reference plane of port 1 and port 2, i.e.,

$$[a] = [\, a_0 \quad -a_0 \quad 0 \,]^t \tag{34}$$

then

$$[b] = [S][a] = \frac{1}{2} \begin{bmatrix} 1 & 1 & \sqrt{2} \\ 1 & 1 & -\sqrt{2} \\ \sqrt{2} & -\sqrt{2} & 0 \end{bmatrix} \begin{bmatrix} a_0 \\ -a_0 \\ 0 \end{bmatrix} = \begin{bmatrix} 0 \\ 0 \\ \sqrt{2}a_0 \end{bmatrix} \tag{35}$$

The total power is transmitted to port 3.

(i) Consider the case when port 3 is terminated by a normalized impedance $\hat{Z}$ and condition (20) is satisfied. Let port 2 be match terminated. For a signal A applied to port 1, the outputs are

$$\begin{bmatrix} a & b & c \\ b & a & -c \\ c & -c & d \end{bmatrix} \begin{bmatrix} A \\ 0 \\ \Gamma b_3 \end{bmatrix} = \begin{bmatrix} b_1 \\ b_2 \\ b_3 \end{bmatrix} \tag{36a}$$

$$\Gamma = \frac{\hat{Z} - 1}{\hat{Z} + 1} \tag{36b}$$

This can be rearranged as

$$aA + c\Gamma b_3 = b_1 \quad \text{or} \quad b_1 - c\Gamma b_3 = aA \tag{37a}$$

$$bA - c\Gamma b_3 = b_2 \quad \text{or} \quad b_2 + c\Gamma b_3 = bA \tag{37b}$$

$$cA + d\Gamma b_3 = b_3 \quad \text{or} \quad (1 - d\Gamma)b_3 = cA \tag{37c}$$

or

$$\begin{bmatrix} 1 & 0 & -c\Gamma \\ 0 & 1 & c\Gamma \\ 0 & 0 & 1 - d\Gamma \end{bmatrix} \begin{bmatrix} b_1 \\ b_2 \\ b_3 \end{bmatrix} = \begin{bmatrix} a \\ b \\ c \end{bmatrix} A \tag{38a}$$

$$\Delta = 1 - d\Gamma \tag{38b}$$

The reflected parameters, b_i, may be solved from (38).

$$b_1 = \frac{A}{\Delta} \begin{bmatrix} a & 0 & -c\Gamma \\ b & 1 & c\Gamma \\ c & 0 & 1 - d\Gamma \end{bmatrix} = \frac{A}{\Delta} [a(1 - d\Gamma) + c^2\Gamma] \tag{39a}$$

$$b_2 = \frac{A}{\Delta} [b(1 - d\Gamma) - c^2\Gamma] \tag{39b}$$

$$b_3 = \frac{Ac}{\Delta} \tag{39c}$$

The reduced scattering matrix, $[S_R]$, is therefore

$$[S_R] = \begin{bmatrix} s_{11} & s_{12} \\ s_{12} & s_{11} \end{bmatrix} \begin{bmatrix} A \\ 0 \end{bmatrix}$$

$$= \begin{bmatrix} b_1 \\ b_2 \end{bmatrix} = \frac{A}{\Delta} \begin{bmatrix} a(1-d\Gamma) - c^2\Gamma \\ B(1-d\Gamma) - c^2\Gamma \end{bmatrix} \tag{40a}$$

$$s_{11} = A \begin{bmatrix} a - \dfrac{c^2\Gamma}{\Delta} \end{bmatrix} \tag{40b}$$

$$s_{12} = A \begin{bmatrix} b - \dfrac{c^2\Gamma}{\Delta} \end{bmatrix} \tag{40c}$$

9. Symmetrical Y-Junction

(a) Figure 9-1a shows an H-plane symmetrical Y-junction, and its symmetries are illustrated in Fig. 9-1b. F_1, F_2, and F_3 are planes of symmetry for reflection, and these planes intersect along the three fold symmetry axis J. The structure is invariant under rotation by $120°$, R_1-rotation, and $240°$, R_2-rotation, about the J-axis.

The plane containing the axes of these three guides is also a plane of symmetry, and the intersections of this plane and other planes of symmetry, such as F_1, F_2, . . ., are axes of symmetry.

(b) The results of multiple operations among the operators R_1, R_2, F_1, F_2, F_3, as well as the unit operator I are summarized in Fig. 9-2.

It is to be noted that (Fig. 9-3)

$$F_3F_1 = R_2 \tag{1}$$

This implies that a reflection in the F_1-plane followed by a reflection in the F_3-plane is equivalent to the rotation R_2, a rotation about the J-axis by $240°$.

In general,

$$F_3F_1 \neq F_1F_3 \tag{2}$$

That is, these operators are noncommutative.

Any element of the group may be generated by one or more operators in R_1 and F_1. For example,

$$F_2 = F_1R_1^2 \tag{3}$$

Consequently, F_1 and R_1 can be called the generators of the group.

(c) The rotation operator R_1 rotates the junction by $120°$ about the J-axis. The voltages and currents at one terminal plane are replaced by those at another terminal plane. It is convenient to keep the terminal numbers fixed but the structure is allowed to transform. The R_1 rotation matrix is therefore

$$[R_1] = \begin{bmatrix} 0 & 0 & 1 \\ 1 & 0 & 0 \\ 0 & 1 & 0 \end{bmatrix} \tag{4a}$$

since

$$[R_1][e] = [e'] \tag{4b}$$

$$[e] = [\, e_1 \quad e_2 \quad e_3 \,]^t = \text{original field} \tag{4c}$$

$$[e'] = [\, e_3 \quad e_1 \quad e_2 \,]^t = \text{rotated field} \tag{4d}$$

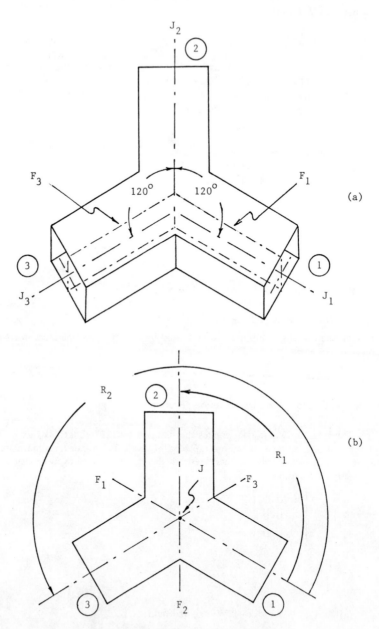

Figure 9-1: A symmetrical γ-junction

	I	R_1	R_2	F_1	F_2	F_3
I	I	R_1	R_2	F_1	F_2	F_3
R_1	R_1	R_2	I	F_2	F_3	F_1
R_2	R_2	I	R_1	F_3	F_1	F_2
F_1	F_1	F_3	F_2	I	R_2	R_1
F_2	F_2	F_1	F_3	R_1	I	R_2
F_3	F_3	F_2	F_1	R_2	R_1	I

Figure 9-2: Multiplication table for operators

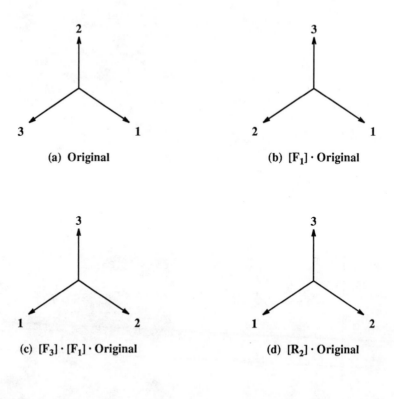

(a) Original

(b) $[F_1] \cdot$ Original

(c) $[F_3] \cdot [F_1] \cdot$ Original

(d) $[R_2] \cdot$ Original

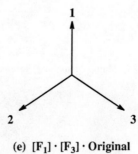

(e) $[F_1] \cdot [F_3] \cdot$ Original

Figure 9-3: Successive operations

(d) The terminal voltages and currents are transformed accordingly.

$$[R_1] [i] = [i'] \qquad \text{and} \qquad [R_1] [v] = [v'] \tag{5}$$

In the impedance representation, one has

$$[v] = [Z] [i] \tag{6}$$

The rotated voltages and currents also satisfy (6).

$$[v'] = [Z] [i'] \tag{7}$$

The substitution of (5) into (7) yields

$$[R_1] [v] = [Z] [R_1] [i] \tag{8}$$

Replacing [v] by (6), one has

$$[R_1] ([Z] [i]) = [Z] [R_1] [i] \qquad \text{or} \qquad ([R_1] [Z] - [Z] [R_1]) [i] = 0 \tag{9}$$

Since this relation is valid for any [i], and in general, [i] is nonzero, therefore,

$$[R_1] [Z] - [Z] [R_1] = 0 \qquad \text{or} \qquad [R_1] [Z] = [Z] [R_1] \tag{10}$$

The rotation matrix $[R_1]$ and the impedance matrix $[Z]$ are commutative. Similarly, it can be established that the rotation matrix is also commutative with the immittance and scattering matrices.

$$[R_1] [Y] = [Y] [R_1] \qquad \text{and} \qquad [R_1] [S] = [S] [R_1]$$

(e) The reflection matrix $[F_1]$ is given by

$$[F_1] = \begin{bmatrix} 1 & 0 & 0 \\ 0 & 0 & 1 \\ 0 & 1 & 0 \end{bmatrix} \tag{11}$$

Figure 9-1b indicates that a reflection with respect to the F_1-plane interchanges the field between port 2 and port 3, and the field at port 1 remains unchanged.

It can be shown by a procedure similar to (d) that the matrices $[F_1]$ and $[Z]$ commute.

$$[F_1] [Z] = [Z] [F_1] \tag{12}$$

(f) The commutation of the generator functions, $[F_1]$ and $[R_1]$, with the impedance matrix (also admittance and scattering matrices) automatically guarantees the commutation of every symmetry matrix. For example, the rotation $[R_1]$ operated on (10) yields

$$[R_1] ([R_1] [Z]) = [R_1] ([Z] [R_1])$$

or

$$[R_1]^2 [Z] = ([R_1] [Z]) [R_1] = ([Z] [R_1]) [R_1] = [Z] [R_1]^2 \qquad (13)$$

Next apply reflection $[F_1]$ to (13).

$$[F_1] ([R_1]^2 [Z]) = [F_1] ([Z] [R_1]^2) = ([F_1] [Z]) [R_1]^2$$

$$= ([Z] [F_1]) [R_1]^2 = [Z] ([F_1] [R_1]^2) \qquad (14)$$

But from (3), $[F_2] = [F_1] [R_1]^2$, so (14) becomes

$$[F_2] [Z] = [Z] [F_2] \qquad (15)$$

Thus, the reflection $[F_2]$ matrix commutes with the impedance matrix.

Any other symmetry operators can be shown in a similar way to be commutative with the impedance matrix. Thus, if the generator operators of the symmetry operator group commute with $[Z]$, then every symmetry operator of the group commutes with $[Z]$ and all conditions which symmetry imposes on $[Z]$ are fulfilled.

(g) The eigenvalue equation for $[R_1]$ is

$$[R_1] [x_i] = r_i [x_i] \qquad (16a)$$

$$[x_i] = [\, x_{i1} \quad x_{i2} \quad x_{i3} \,]^t \qquad (16b)$$

where $[x_i]$ is the eigenvector corresponding to the eigenvalue r_i of $[R_1]$.

Since $[R_1]$ is a rotation of $120°$, three successive $[R_1]$ operations will restore the original position, i.e.,

$$[R_1]^3 = [\, u \,] = \text{identity operator} \qquad (17)$$

Combining (17) with (16a) gives

$$[R_1]^3 [x_i] = r_i^3 [x_i] = [\, u \,] [x_i] \qquad (18)$$

Hence,

$$r_i^3 = 1 = e^{\pm j 2 n \pi} \qquad \text{or} \qquad r = e^{\pm j 2 n \pi / 3} \qquad n = 0, 1, 2, \ldots \qquad (19a)$$

That is,

$$n = 0 \qquad r_1 = 1 \tag{19b}$$

$$n = 1 \qquad r_2 = e^{j2\pi/3} = \frac{-1}{2} - j\frac{\sqrt{3}}{2} \equiv \alpha \tag{19c}$$

$$r_3 = e^{-j2\pi/3} = \frac{-1}{2} + j\frac{\sqrt{3}}{2} \equiv \alpha^* \tag{19d}$$

(h) For $r = r_1 = 1$, the eigenvalue equation is

$$[R_1][x_1] = r_1[x_1] \qquad \rightarrow \qquad \begin{bmatrix} 0 & 0 & 1 \\ 1 & 0 & 0 \\ 0 & 1 & 0 \end{bmatrix} \begin{bmatrix} x_{11} \\ x_{12} \\ x_{13} \end{bmatrix} = \begin{bmatrix} x_{11} \\ x_{12} \\ x_{13} \end{bmatrix}$$

or

$$x_{13} = x_{11}, \qquad x_{11} = x_{12}, \qquad x_{12} = x_{13}$$

Thus,

$$[x_1] = x_{11}[\,1 \quad 1 \quad 1\,]^t$$

The normalized eigenvector will be obtained at the end of this section. Presently, it is the format of the eigenvector that is of interest, and the constant factor will be neglected, i.e.,

$$[x_1] = [\,1 \quad 1 \quad 1\,]^t \tag{20}$$

For $r_2 = \alpha$, the eigenvalue equation is

$$\begin{bmatrix} 0 & 0 & 1 \\ 1 & 0 & 0 \\ 0 & 1 & 0 \end{bmatrix} \begin{bmatrix} x_{21} \\ x_{22} \\ x_{23} \end{bmatrix} = \alpha \begin{bmatrix} x_{21} \\ x_{22} \\ x_{23} \end{bmatrix}$$

which gives

$$x_{23} = \alpha x_{21}$$

$$x_{21} = \alpha x_{22} \qquad \text{or} \qquad x_{22} = \frac{x_{21}}{\alpha} = \alpha^* x_{21}$$

Hence,

$$[x_2] = x_{21}[\,1 \quad \alpha^* \quad \alpha\,]^t \qquad \text{or} \qquad [x_2] = [\,1 \quad \alpha^* \quad \alpha\,]^t \tag{21}$$

For $r_3 = \alpha^*$, the eigenvalue equation is

$$\begin{bmatrix} 0 & 0 & 1 \\ 1 & 0 & 0 \\ 0 & 1 & 0 \end{bmatrix} \begin{bmatrix} x_{31} \\ x_{32} \\ x_{33} \end{bmatrix} = \alpha^* \begin{bmatrix} x_{31} \\ x_{32} \\ x_{33} \end{bmatrix}$$

which yields

$$x_{33} = \alpha^* x_{31}$$

$$x_{31} = \alpha^* x_{32} \quad \text{or} \quad x_{32} = \alpha x_{31}$$

Therefore,

$$[x_3] = x_{31} [1 \quad \alpha \quad \alpha^*]^t \quad \text{or} \quad [x_3] = [1 \quad \alpha \quad \alpha^*]^t \quad (22)$$

Since $[R_1]$ commutes with $[Z]$, $[Y]$, and $[S]$ [Eq. (10)], these eigenvectors, (20) - (22), are also eigenvectors of $[Z]$, $[Y]$, and $[S]$ for nondegenerate eigenvalues. This can be verified by pre-multiplying (16a) by $[Z]$.

$$[Z] ([R_1] [x_i]) = [Z] (r_i [x_i]) = r_i ([Z] [x_i])$$

or

$$[R_1] ([Z] [x_i]) = r_i ([Z] [x_i])$$

since $[Z]$ and $[R_1]$ are commutative. The above expression implies that $([Z][x_i])$ is an eigenvector of $[R_1]$ corresponding to the nondegenerate eigenvalue r_i. Therefore, $([Z][x_i])$ can differ from $[x_i]$ by at most a constant factor z_i, i.e.,

$$[Z] [x_i] = z_i [x_i]$$

This is an eigenvalue equation which defines the eigenvector $[x_i]$ for the matrix $[Z]$ and the corresponding eigenvalue is z_i.

If the eigenvalue r_i is degenerate, then $([Z][x_i])$ is a linear combination of all linearly independent eigenvectors of the eigenvalue.

(i) The eigenvectors $[x_2]$ and $[x_3]$ are complex. It is convenient to choose eigenvectors such that $[Z]$ is real. One possibility is that if $[Z]$ has degenerate eigenvalues, then a linear combination of x_k's can be the desired eigenvectors. The eigenvalue equation for $[Z]$ is

$$[Z] \, [x_k] \; = \; z_k \, [x_k] \tag{23}$$

Applying the reflection $[F_1]$ to (23) yields

$$[F_1] \, (\, [Z] \, [x_k] \,) \; = \; z_k \, (\, [F_1] \, [x_k] \,) \quad \text{or} \quad [Z] \, (\, [F_1] \, [x_k] \,) \; = \; z_k \, (\, [F_1] \, [x_k] \,) \tag{24}$$

Hence, $([F_1][x_k])$ is an eigenvector of $[Z]$ corresponding to the eigenvalue z_k.

However,

$$[F_1] \, [x_2] \; = \; \begin{bmatrix} 1 & 0 & 0 \\ 0 & 0 & 1 \\ 0 & 1 & 0 \end{bmatrix} \begin{bmatrix} 1 \\ \alpha^* \\ \alpha \end{bmatrix} \; = \; \begin{bmatrix} 1 \\ \alpha \\ \alpha^* \end{bmatrix} \; = \; [x_3] \tag{25a}$$

$$[F_1] \, [x_3] \; = \; \begin{bmatrix} 1 & 0 & 0 \\ 0 & 0 & 1 \\ 0 & 1 & 0 \end{bmatrix} \begin{bmatrix} 1 \\ \alpha \\ \alpha^* \end{bmatrix} \; = \; \begin{bmatrix} 1 \\ \alpha^* \\ \alpha \end{bmatrix} \; = \; [x_2] \tag{25b}$$

The substitution of (25) into (24) yields

$$k \; = \; 2 \qquad [Z] \, [x_3] \; = \; z_2 \, [x_3] \tag{26a}$$

$$k \; = \; 3 \qquad [Z] \, [x_2] \; = \; z_3 \, [x_2] \tag{26b}$$

A direct comparison of (26) with (23) indicates that

$$z_2 \; = \; z_3 \tag{27}$$

That is, the $[Z]$ matrix has degenerate eigenvalues and hence the corresponding eigenvectors can be constructed by linear combination of eigenvectors $[x_2]$ and $[x_3]$.

Let the real normalized eigenvectors of $[Z]$ be represented by y_k; then,

$$[y_1] \; = \; [x_1] \; = \; \frac{1}{\sqrt{3}} \, [\, 1 \quad 1 \quad 1 \,]^t \tag{28a}$$

$$[y_2] \; = \; [x_2] + [x_3] \; = \; \frac{1}{\sqrt{6}} \, [\, 2 \quad -1 \quad -1 \,]^t \tag{28b}$$

$$[y_3] \; = \; j \, \frac{1}{\sqrt{3}} \, (\, [x_2] - [x_3] \,) \; = \; \frac{1}{\sqrt{2}} \, [\, 0 \quad 1 \quad -1 \,]^t \tag{28c}$$

(j) The eigenvalue equation of a Y-junction is given by

$$[S] \, [y_i] \; = \; s_i \, [y_i] \tag{29}$$

The normalized eigenvector matrix is

$$[Y] = [\ [y_1] \quad [y_2] \quad [y_3]\ ] = \frac{1}{\sqrt{6}} \begin{bmatrix} \sqrt{2} & 2 & 0 \\ \sqrt{2} & -1 & \sqrt{3} \\ \sqrt{2} & -1 & -\sqrt{3} \end{bmatrix} \tag{30}$$

Then

$$[S][Y] = [Y] \, diag [\ s_1 \quad s_2 \quad s_3\ ] \quad or \quad [S] = [Y] \, diag [\ s_1 \quad s_2 \quad s_3\ ] [Y]^{-1} \tag{31}$$

It can be shown by direct expansion that

$$[Y]^t [Y] = [u] \quad therefore \quad [Y]^t = [Y]^{-1} \tag{32}$$

Hence,

$$[S] = [Y] \, diag [\ s_1 \quad s_2 \quad s_3\ ] [Y]^t$$

$$= \frac{1}{6} \begin{bmatrix} \sqrt{2} & 2 & 0 \\ \sqrt{2} & -1 & \sqrt{3} \\ \sqrt{2} & -1 & -\sqrt{3} \end{bmatrix} \begin{bmatrix} s_1 & 0 & 0 \\ 0 & s_2 & 0 \\ 0 & 0 & s_3 \end{bmatrix} \begin{bmatrix} \sqrt{2} & \sqrt{2} & \sqrt{2} \\ 2 & -1 & -1 \\ 0 & \sqrt{3} & -\sqrt{3} \end{bmatrix}$$

$$= \frac{1}{6} \begin{bmatrix} 2(s_1 + 2s_2) & 2(s_1 - s_2) & 2(s_1 - s_2) \\ 2(s_1 - s_2) & 2(s_1 + 2s_2) & 2(s_1 - s_2) \\ 2(s_1 - s_2) & 2(s_1 - s_2) & 2(s_1 + 2s_2) \end{bmatrix}$$

$$= \frac{1}{3} \begin{bmatrix} \alpha & \beta & \beta \\ \beta & \alpha & \beta \\ \beta & \beta & \alpha \end{bmatrix} \tag{33a}$$

$$\alpha = \frac{1}{3} (s_1 + 2s_2) \tag{33b}$$

$$\beta = \frac{1}{3} (s_1 - s_2) \tag{33c}$$

(k) The scattering matrix can also be obtained from the rotation matrix. The scattering matrix is defined by

$$[S] [a] = [b] \tag{34}$$

The field parameters [a] and [b] are transformed according to

$$[R_1] [a] = [a'] \quad \text{and} \quad [R_1] [b] = [b'] \tag{35}$$

where the unprimed and primed fields are the original and transformed fields, respectively. The transformed fields also satisfy (34).

$$[S] [a'] = [b'] \tag{36}$$

The substitution of (35) into (36) yields

$$[S] ([R_1] [a]) = [R_1] [b] = [R_1] ([S] [a])$$

$$\{ [S] [R_1] - [R_1] [S] \} [a] = 0 \tag{37}$$

The scattering matrix commutes with the rotation matrix.

$$[S] [R_1] = [R_1] [S] \quad \text{or} \quad [S] = [R_1] [S] [R_1]^{-1} = [R_1] [S] [R_1]^t$$

For a reciprocal junction, one has

$$
\begin{bmatrix} s_{11} & s_{12} & s_{13} \\ s_{12} & s_{22} & s_{23} \\ s_{13} & s_{23} & s_{33} \end{bmatrix} =
\begin{bmatrix} 0 & 0 & 1 \\ 1 & 0 & 0 \\ 0 & 1 & 0 \end{bmatrix}
\begin{bmatrix} s_{11} & s_{12} & s_{13} \\ s_{12} & s_{22} & s_{23} \\ s_{13} & s_{23} & s_{33} \end{bmatrix}
\begin{bmatrix} 0 & 1 & 0 \\ 0 & 0 & 1 \\ 1 & 0 & 0 \end{bmatrix}
$$

$$
=
\begin{bmatrix} 0 & 0 & 1 \\ 1 & 0 & 0 \\ 0 & 1 & 0 \end{bmatrix}
\begin{bmatrix} s_{13} & s_{11} & s_{12} \\ s_{23} & s_{12} & s_{22} \\ s_{33} & s_{13} & s_{23} \end{bmatrix}
$$

$$
=
\begin{bmatrix} s_{33} & s_{13} & s_{23} \\ s_{13} & s_{11} & s_{12} \\ s_{23} & s_{12} & s_{22} \end{bmatrix} \tag{38}
$$

A direct comparison of the corresponding elements of the matrices on both sides of (38) yields

$$s_{11} = s_{22} = s_{33} \equiv \alpha \qquad \text{and} \qquad s_{12} = s_{13} = s_{23} \equiv \beta \tag{39a}$$

That is,

$$[S] = \begin{bmatrix} \alpha & \beta & \beta \\ \beta & \alpha & \beta \\ \beta & \beta & \alpha \end{bmatrix} \tag{39b}$$

(l) For a lossless junction, the scattering matrix satisfies the unitary condition.

$$[S^*]^t [S] = [u] = \begin{bmatrix} \alpha^* & \beta^* & \beta^* \\ \beta^* & \alpha^* & \beta^* \\ \beta^* & \beta^* & \alpha^* \end{bmatrix} \begin{bmatrix} \alpha & \beta & \beta \\ \beta & \alpha & \beta \\ \beta & \beta & \alpha \end{bmatrix} \tag{40a}$$

The expansion of the above relation produces the following results.

$$(1, 1) \text{ term:} \qquad |\alpha|^2 + 2|\beta|^2 = 1 \tag{40b}$$

$$(1, 2) \text{ term:} \qquad 2 \, \text{Re} \, (\alpha^*\beta) + |\beta|^2 = 0 \tag{40c}$$

(m) A special property of a Y-junction is that it is impossible to have all three ports matched. In order to accomplish this, one must have zero values for the diagonal elements in the scattering matrix, i.e., from (33),

$$\alpha = \frac{1}{3} \, (s_1 + 2s_2) = 0 \tag{41}$$

For a lossless junction, it is necessary that the eigenvalues of [S] have unit magnitude.

$$|s_i| = 1 \tag{42}$$

This is because a scattering matrix is similar to a diagonal matrix or a Jordan form depending upon whether it has distinct or multiple eigenvalues. In either case, the elements on the main diagonal of the similar matrix (diagonal of Jordan form) are the eigenvalues of the matrix. For lossless devices, all incident power must be totally scattered.

$$|s_{jj}| = 1 = |s_i|$$

Otherwise, the scattered power would be greater than or less than the incident power.

Equation (41) requires

$$s_1 = -2s_2 \tag{43}$$

which is impossible since $|s_1| = |s_2| = 1$ by (42).

(n) When port 3 is terminated by some impedance which is characterized by a reflection coefficient ρ_3 and port 1 is excited by an incident wave a_1, then the relation $[S][a] = [b]$ gives

$$\begin{bmatrix} \alpha & \beta & \beta \\ \beta & \alpha & \beta \\ \beta & \beta & \alpha \end{bmatrix} \begin{bmatrix} a_1 \\ 0 \\ \rho_3 b_3 \end{bmatrix} = \begin{bmatrix} b_1 \\ b_3 \\ b_3 \end{bmatrix} \tag{44}$$

The expansion of (44) yields

$$\alpha a_1 + \beta \rho_3 b_3 = b_1$$

$$\beta a_1 + \beta \rho_3 b_3 = b_2$$

$$\beta a_1 + \alpha \rho_3 b_3 = b_3$$

These relations can be expressed in matrix form.

$$\begin{bmatrix} 1 & 0 & -\beta_1 \\ 0 & 1 & -\beta_1 \\ 0 & 0 & 1-\beta_1 \end{bmatrix} \begin{bmatrix} b_1 \\ b_2 \\ b_3 \end{bmatrix} = a_1 \begin{bmatrix} \alpha \\ \beta \\ \beta \end{bmatrix} \tag{45a}$$

$$\beta_1 \equiv \beta \rho_3 \tag{45b}$$

The scattered wave can be determined by Cramer's rule.

$$\Delta = \det \begin{bmatrix} 1 & 0 & -\beta_1 \\ 0 & 1 & -\beta_1 \\ 0 & 0 & 1-\beta_1 \end{bmatrix} = 1 - \beta_1 \tag{46a}$$

$$b_1 = \frac{a_1}{\Delta} \begin{bmatrix} \alpha & 0 & -\beta_1 \\ \beta & 1 & -\beta_1 \\ \beta & 0 & 1-\beta_1 \end{bmatrix} = a_1 \left[\alpha + \frac{\beta_1 \beta}{\Delta} \right] \tag{46b}$$

$$b_2 = a_1 \beta \left[1 + \frac{\beta_1 \beta}{\Delta} \right] \tag{46c}$$

$$b_3 = \frac{a_1 \beta}{\Delta} \tag{46d}$$

Since port 3 is terminated, the junction may be considered as a two-port device and one can construct a 2×2 scattering matrix for the reciprocal device, such that $[S_R][a]=[b]$.

$$[S_R] [a] = \begin{bmatrix} s_{11} & s_{12} \\ s_{12} & s_{11} \end{bmatrix} \begin{bmatrix} a_1 \\ 0 \end{bmatrix} = \begin{bmatrix} b_1 \\ b_2 \end{bmatrix} = \begin{bmatrix} a_1 \left[\alpha + \dfrac{\beta\beta_1}{\Delta} \right] \\ a_1 \beta \left[1 + \dfrac{\beta\beta_1}{\Delta} \right] \end{bmatrix} \tag{47}$$

The elements of $[S_R]$ can be obtained by inspection.

$$s_{11} = \alpha + \frac{\beta\beta_1}{\Delta} \tag{48a}$$

$$s_{12} = \beta \left[1 + \frac{\beta\beta_1}{\Delta} \right] \tag{48b}$$

10. Problems

1. The combination of an E-plane T-junction and a H-plane T-junction is known as a Magic Tee; (Fig. 10-1).

 (a) Determine the planes of symmetry.

 (b) For a Magic Tee filled with isotropic dielectric, show that the scattering matrix is given by

$$[S] = \begin{bmatrix} s_{11} & s_{12} & s_{13} & s_{14} \\ s_{12} & s_{11} & s_{13} & -s_{14} \\ s_{13} & s_{13} & s_{33} & 0 \\ s_{14} & -s_{14} & 0 & s_{44} \end{bmatrix}$$

2. Can port 3 and port 4 of a Magic Tee be match-terminated simultaneously? If possible, what would be the scattering matrix of the structure with both of these ports matched?

3. If a lossless Magic Tee does not contain an exceptional high-energy storage element within its junction, show that the corresponding scattering matrix is

$$[S] = \frac{1}{\sqrt{2}} \begin{bmatrix} 0 & 0 & e^{j\theta} & e^{j\phi} \\ 0 & 0 & e^{j\theta} & -e^{j\phi} \\ e^{j\theta} & e^{j\theta} & 0 & 0 \\ e^{j\phi} & -e^{j\phi} & 0 & 0 \end{bmatrix}$$

where θ and ϕ are phase angles which depend upon the choice of reference plane.

4. The port 1 of a lossless Magic Tee is driven by a matched oscillator and produces a normalized incident wave a_1 at port 1; and other ports are match-terminated. Determine the scattered wave and power output of each port.

5. A lossless Magic Tee is driven by a matched source at port 4 and produces an incident wave a_4 at this port; all other ports are match-terminated. Determine the scattered waves and power outputs at all ports.

6. A lossless Magic Tee is driven by matched sources at port 1 and port 2 such that the incident waves are a_1 and a_2 at port 1 and port 2, respectively. The remaining ports are match-terminated. Determine the scattered wave and power output at each port.

7. A lossless Magic Tee is driven by a matched source at port 3 and produces an incident wave a_3 at this port. Ports 1, 2, and 4 are terminated by loads with reflection coefficients Γ_1, Γ_2, and Γ_4 at the corresponding reference planes, respectively. Determine the scattered wave and power output at each port.

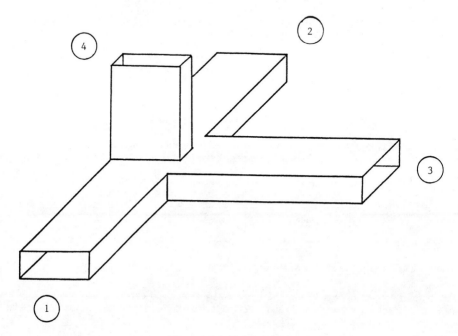

Figure 10-1: A magic Tee

CHAPTER VI Directional Couplers

1. Introduction

A directional coupler consists of two waveguides which are coupled through small holes such that the incident wave in one of the guides is partially transmitted to the other with certain directional characteristics. Such devices are used in monitoring the power in a system with little loss of energy to the main system.

The topics discussed in this and the next chapter are available in more advanced textbooks such as Altman [C-1], Collin [C-2] and Helszajn [D-2].

2. Directional Couplers

(a) A reciprocal four-port device is shown in Fig. 2-1. A reciprocal device is characterized by a scattering matrix which is symmetrical with respect to its main diagonal. A reciprocal four-port device is a directional coupler if the following conditions are satisfied.

(a.1)   The input and the output ports can be grouped into pairs

       (a.1.1)   port 1 and port 4

       (a.1.2)   port 2 and port 3

       such that the input ports (1 and 2) are isolated and the output ports (3 and 4) are isolated, i.e.,

$$s_{12} = 0 \qquad \text{and} \qquad s_{34} = 0 \tag{1a}$$

(a.2)   The input ports are matched.

$$s_{11} = 0 \qquad \text{and} \qquad s_{22} = 0 \tag{2}$$

       Since the device is reciprocal, one also has

$$s_{jk} = s_{kj} \tag{1b}$$

Thus, the scattering matrix of a four-port device is given by

$$[S] = \begin{bmatrix} s_{11} & s_{12} & s_{13} & s_{14} \\ s_{21} & s_{22} & s_{23} & s_{24} \\ s_{31} & s_{32} & s_{33} & s_{34} \\ s_{41} & s_{42} & s_{43} & s_{44} \end{bmatrix} \tag{3a}$$

and for a directional coupler, it becomes

$$[S] = \begin{bmatrix} 0 & 0 & s_{13} & s_{14} \\ 0 & 0 & s_{23} & s_{24} \\ s_{13} & s_{23} & s_{33} & 0 \\ s_{14} & s_{24} & 0 & s_{44} \end{bmatrix} \tag{3b}$$

Figure 2-1: A four-port device

(b) The scattering matrix of a lossless device satisfies the unitary condition.

$$[S^*]^t [S] = [u] \tag{4}$$

or

$$\begin{bmatrix} 0 & 0 & s_{13}^* & s_{14}^* \\ 0 & 0 & s_{23}^* & s_{24}^* \\ s_{13}^* & s_{23}^* & s_{33}^* & 0 \\ s_{14}^* & s_{24}^* & 0 & s_{44}^* \end{bmatrix} \begin{bmatrix} 0 & 0 & s_{13} & s_{14} \\ 0 & 0 & s_{23} & s_{24} \\ s_{13} & s_{23} & s_{33} & 0 \\ s_{14} & s_{24} & 0 & s_{44} \end{bmatrix} = [u] \tag{5}$$

Terms along the main diagonal of (5) are:

$$|s_{13}|^2 + |s_{14}|^2 = 1 \tag{6a}$$

$$|s_{23}|^2 + |s_{24}|^2 = 1 \tag{6b}$$

$$|s_{13}|^2 + |s_{23}|^2 + |s_{33}|^2 = 1 \tag{6c}$$

$$|s_{14}|^2 + |s_{24}|^2 + |s_{44}|^2 = 1 \tag{6d}$$

The subtraction of (6a) from (6c) produces

$$|s_{23}|^2 + |s_{33}|^2 - |s_{14}|^2 = 0 \tag{7a}$$

The subtraction of (6b) from (6d) yields

$$|s_{14}|^2 + |s_{44}|^2 - |s_{23}|^2 = 0 \tag{7b}$$

The addition of (7a) and (7b) gives

$$|s_{33}|^2 + |s_{44}|^2 = 0 \tag{8}$$

This implies that

$$|s_{33}| = 0 = |s_{44}| \tag{9}$$

since neither term in (8) can be negative. Thus, if port 1 and port 2 in a four-port directional coupler are matched, then port 3 and port 4 are also matched.

(c) Equation (6) can be simplified by using (9).

$$|s_{13}|^2 + |s_{14}|^2 = 1 \tag{10a}$$

$$|s_{23}|^2 + |s_{24}|^2 = 1 \tag{10b}$$

$$|s_{13}|^2 + |s_{23}|^2 = 1 \tag{10c}$$

$$|s_{14}|^2 + |s_{24}|^2 = 1 \tag{10d}$$

The difference between (10a) and (10c) is

$$|s_{14}|^2 - |s_{23}|^2 = 0 \quad \text{or} \quad |s_{14}| = |s_{23}| \tag{11a}$$

This means that the magnitude of the coupling between port 1 and port 4 is the same as that between port 2 and port 3. It is to be noted that (11a) does not specify the phase relation between s_{14} and s_{23}.

The difference between (10a) and (10d) is

$$|s_{13}| = |s_{24}| \tag{11b}$$

The magnitude of the coupling between port 1 and port 3 is the same as that between port 2 and port 4.

(d) If the four-port directional coupler has symmetry with respect to plane a-a' (Fig. 2-2), that is, port 1 is indistinguishable from port 2, and port 3 and port 4 are identical, then by symmetry, one has

$$s_{14} = s_{23} \quad \text{and} \quad s_{13} = s_{24} \tag{12}$$

(e) The scattering matrix for a symmetrical and reciprocal directional coupler is given by [Eqs. (3b) and (12)]

$$[S] = \begin{bmatrix} 0 & 0 & s_{13} & s_{14} \\ 0 & 0 & s_{14} & s_{13} \\ s_{13} & s_{14} & 0 & 0 \\ s_{14} & s_{13} & 0 & 0 \end{bmatrix} \quad \text{for symmetrical coupler} \tag{13}$$

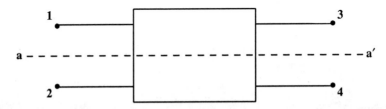

Figure 2-2: A symmetrical four-port device

Figure 2 - 2

The unitary condition $[S^{*T}] [S] = [u]$ yields

(1, 1) term: $\quad |s_{13}|^2 + |s_{14}|^2 = 1$ $\qquad\qquad$ (14a)

(1, 2) term: $\quad s_{13}^* s_{14} + s_{14}^* s_{13} = 0$ $\qquad\qquad$ (14b)

Equation (14a) specifies only the magnitudes of the scattering elements; therefore, a great deal of arbitrariness remains in the phase of each element of [S]. Let

$$s_{13} \equiv a e^{j\theta} \qquad \text{and} \qquad s_{14} \equiv b e^{j\phi} \qquad\qquad (15)$$

Then one has [Eq. (14)]

$$a^2 + b^2 = 1 \qquad\qquad (16a)$$

and

$$ab [e^{-j(\theta - \phi)} + e^{j(\theta - \phi)}] = 0 \qquad \text{or} \qquad 2ab \cos (\theta - \phi) = 0 \qquad\qquad (16b)$$

Since neither a nor b is zero, one must have

$$\cos (\theta - \phi) = \cos \left[\frac{\pm(2n + 1)\pi}{2} \right] = 0$$

or

$$\theta = \phi \pm \frac{(2n + 1)\pi}{2} \qquad\qquad n = 0, 1, 2, \ldots \qquad\qquad (17)$$

Since there are only two equations with four unknowns, there are two degrees of arbitrariness. One may choose

$$\phi = 0 \qquad \text{and} \qquad \theta = \frac{\pi}{2}$$

Then

$$s_{13} = ja \qquad \text{and} \qquad s_{14} = b \qquad\qquad (18)$$

Then the scattering matrix for the symmetrical and reciprocal directional coupler is, from (13),

$$[S] = \begin{bmatrix} 0 & 0 & ja & b \\ 0 & 0 & b & ja \\ ja & b & 0 & 0 \\ b & ja & 0 & 0 \end{bmatrix} \qquad \text{for symmetrical coupler} \qquad\qquad (19)$$

(f) If the directional coupler is not symmetrical with respect to plane a-a′, Fig. 2-2, i.e.,

$$s_{13} \neq s_{24} \qquad \text{and} \qquad s_{14} \neq s_{23}$$

one can choose the reference planes for each port to simplify the problem.

The substitution of (11a) into (10c) gives

$$|s_{13}|^2 + |s_{14}|^2 = 1 \tag{20}$$

The off-diagonal terms of (5) are

$$(3, 4) \text{ term:} \qquad s_{13}^* s_{14} + s_{23}^* s_{24} = 0 \tag{21a}$$

$$(4, 3) \text{ term:} \qquad s_{14}^* s_{13} + s_{24}^* s_{23} = 0 \tag{21b}$$

$$(1, 2) \text{ term:} \qquad s_{13}^* s_{23} + s_{14}^* s_{24} = 0 \tag{21c}$$

To satisfy (21), one can express the scattering elements in the following form [Eq. (11)].

$$s_{13} = ae^{j\phi_1} \qquad s_{24} = ae^{j\phi_2} \tag{22a}$$

$$s_{14} = be^{j\theta_1} \qquad s_{23} = be^{j\theta_2} \tag{22b}$$

The combination of (21b) and (22) yields

$$ab\, e^{j(\phi_1 - \theta_1)} + ab\, e^{j(\theta_2 - \phi_2)} = 0$$

or

$$e^{j(\phi_1 + \phi_2)} = -e^{j(\theta_1 + \theta_2)}$$

$$\phi_1 + \phi_2 = \theta_1 + \theta_2 + (2n+1)\pi \qquad n = 0, 1, 2, \ldots \tag{23}$$

Again there are more unknowns than specified relations. The phase angles can be chosen to simplify the problem by adjusting the reference plane for each port.

One possibility is to let

$$\theta_1 = \phi_1 = 0 \qquad \text{and} \qquad \phi_2 = \theta_2 = \frac{\pi}{2} \tag{24}$$

Then

$$s_{13} = a \qquad s_{24} = ja \tag{25a}$$

$$s_{14} = b \qquad s_{23} = jb \tag{25b}$$

The scattering matrix for an unsymmetrical directional coupler is [Eq. (3b)]

$$[S] = \begin{bmatrix} 0 & 0 & a & b \\ 0 & 0 & jb & ja \\ a & jb & 0 & 0 \\ b & ja & 0 & 0 \end{bmatrix} \qquad \text{for unsymmetrical coupler} \tag{26}$$

(g) A special class of directional couplers are the 3-db couplers. A 3-db coupler transmits power equally to both output ports. In this case, one has

$$|s_{13}|^2 = |s_{14}|^2 \tag{27}$$

Then (18) becomes

$$s_{13} = +ja \qquad s_{14} = a \tag{28}$$

and (14a) yields

$$2a^2 = 1 \qquad \text{or} \qquad a = \frac{1}{\sqrt{2}} \tag{29}$$

The scattering matrix for a 3-db coupler is therefore [Eq. (19)]

$$[S_{3\,db}] = \frac{1}{\sqrt{2}} \begin{bmatrix} 0 & 0 & j & 1 \\ 0 & 0 & 1 & j \\ j & 1 & 0 & 0 \\ 1 & j & 0 & 0 \end{bmatrix} \qquad \text{for symmetrical 3-db coupler} \tag{30}$$

3. Even- and Odd-Mode Theory

Characteristics of a device may be obtained from the eigenvalues of the scattering matrix. These eigenvalues are the reflection coefficients of the device associated with excitations of the device, i.e., its eigenvectors. For a symmetrical four-port device, these excitations reduce to exciting the isolated input ports with in-phase and oppositely phased sources separately.

Consider a symmetrical four-port device (Fig. 3-1). The scattering relation is

$$[b] = [S] [a] \tag{1a}$$

$$
\begin{bmatrix} b_1 \\ b_2 \\ b_3 \\ b_4 \end{bmatrix}
=
\begin{bmatrix}
s_{11} & s_{12} & s_{13} & s_{14} \\
s_{12} & s_{11} & s_{14} & s_{13} \\
s_{13} & s_{14} & s_{11} & s_{12} \\
s_{14} & s_{13} & s_{12} & s_{11}
\end{bmatrix}
\begin{bmatrix} a_1 \\ a_2 \\ a_3 \\ a_4 \end{bmatrix}
\tag{1b}
$$

since, by symmetry, one has

$$s_{11} = s_{22} = s_{33} = s_{44}$$

$$s_{12} = s_{21} = s_{34} = s_{43}$$

$$s_{13} = s_{31} = s_{24} = s_{42}$$

$$s_{14} = s_{41} = s_{23} = s_{32}$$

(a) Let port 1 and port 2 be excited by in-phase sources of equal magnitude (Fig. 3-2). This arrangement is known as even-mode excitation.

$$a_{1e} = a_{2e} = a_e \tag{2}$$

With even-mode excitation, the plane of symmetry is equivalent to an open circuit or a magnetic wall. The even-mode reflection parameters are

$$[b_e] = [S] [a_e] \tag{3a}$$

$$
\begin{bmatrix} b_{1e} \\ b_{2e} \\ b_{3e} \\ b_{4e} \end{bmatrix}
=
\begin{bmatrix}
s_{11} & s_{12} & s_{13} & s_{14} \\
s_{12} & s_{11} & s_{14} & s_{13} \\
s_{13} & s_{14} & s_{11} & s_{12} \\
s_{14} & s_{13} & s_{12} & s_{11}
\end{bmatrix}
\begin{bmatrix} a_e \\ a_e \\ 0 \\ 0 \end{bmatrix}
\tag{3b}
$$

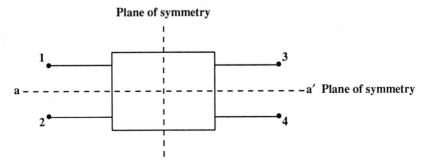

Figure 3-1: A symmetrical four-port device

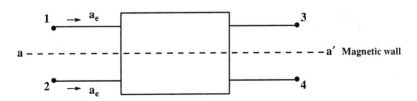

Figure 3-2: Even-mode excited four-port junction

or

$$b_{1e} = a_e (s_{11} + s_{12}) = b_{2e} \qquad (4a)$$

$$b_{3e} = a_e (s_{13} + s_{14}) = b_{4e} \qquad (4b)$$

The reflection and transmission coefficients are

$$\rho_{1e} = \frac{b_{1e}}{a_{1e}} = \frac{b_{1e}}{a_e} = s_{11} + s_{12} \qquad (5a)$$

$$\tau_{1e} = \frac{b_{3e}}{a_{1e}} = s_{13} + s_{14} \qquad (5b)$$

$$\rho_{2e} = \frac{b_{2e}}{a_{2e}} = s_{11} + s_{12} = \rho_{1e} \qquad (5c)$$

$$\tau_{4e} = \frac{b_{4e}}{a_{2e}} = s_{13} + s_{14} = \tau_{1e} \qquad (5d)$$

The transmission and reflection coefficients for port 1 and port 3 are identical to those for port 2 and port 4. Since these port pairs are isolated by a magnetic wall with even-mode excitations, it is thus permissible to use a single relation of a two-port network to describe the behavior of a four-port device, (5), i.e.,

$$\rho_e = s_{11} + s_{12} \qquad (6a)$$

$$\tau_e = s_{13} + s_{14} \qquad (6b)$$

These are the reflection and transmission coefficients for the even mode.

(b) The odd-mode relations are obtained by exciting the device with oppositely phased sources, Fig. 3-3. This is equivalent to setting

$$a_{1o} = -a_{2o} = a_o \qquad (7)$$

Then (1) becomes

$$[b_o] = [S] [a_o] \qquad (8a)$$

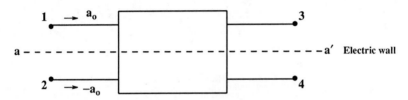

Figure 3-3: Odd-mode excited four-port junction

$$
\begin{bmatrix} b_{1o} \\ b_{2o} \\ b_{3o} \\ b_{4o} \end{bmatrix} = \begin{bmatrix} s_{11} & s_{12} & s_{13} & s_{14} \\ s_{12} & s_{11} & s_{14} & s_{13} \\ s_{13} & s_{14} & s_{11} & s_{12} \\ s_{14} & s_{13} & s_{12} & s_{11} \end{bmatrix} \begin{bmatrix} a_o \\ -a_o \\ 0 \\ 0 \end{bmatrix} \tag{8b}
$$

or

$$
b_{1o} = a_o (s_{11} - s_{12}) = -b_{2o} \tag{9a}
$$

$$
b_{3o} = a_o (s_{13} - s_{14}) = -b_{4o} \tag{9b}
$$

The reflection and transmission coefficients are

$$
\rho_{1o} = \frac{b_{1o}}{a_{1o}} = s_{11} - s_{12} \tag{10a}
$$

$$
\tau_{3o} = \frac{b_{3o}}{a_{1o}} = s_{13} - s_{14} \tag{10b}
$$

$$
\rho_{2o} = \frac{b_{2o}}{a_{2o}} = \frac{b_{2o}}{-a_o} = s_{11} - s_{12} = \rho_{1o} \tag{10c}
$$

$$
\tau_{4o} = \frac{b_{4o}}{-a_o} = s_{13} - s_{14} = \tau_{3o} \tag{10d}
$$

The equivalent two-port description for the odd-mode reflection and transmission coefficients is therefore

$$
\rho_o = s_{11} - s_{12} \tag{11a}
$$

$$
\tau_o = s_{13} - s_{14} \tag{11b}
$$

These are the reflection and transmission coefficients for the odd mode.

The elements of the scattering matrix can now be obtained from (6) and (11).

$$s_{11} = \frac{1}{2} (\rho_e + \rho_o) \tag{12a}$$

$$s_{12} = \frac{1}{2} (\rho_e - \rho_o) \tag{12b}$$

$$s_{13} = \frac{1}{2} (\tau_e + \tau_o) \tag{12c}$$

$$s_{14} = \frac{1}{2} (\tau_e - \tau_o) \tag{12d}$$

(c) The superposition of the even- and odd-mode excitations yields, for $a_e = a_o = a$, from (2) and (7),

$$a_1 = a_{1e} + a_{1o} = 2a \tag{13a}$$

$$a_2 = a_{2e} + a_{2o} = 0 \tag{13b}$$

and, from (4) and (9),

$$b_1 = b_{1e} + b_{1o} = 2s_{11}a \tag{14a}$$

$$b_2 = b_{2e} + b_{2o} = 2s_{12}a \tag{14b}$$

$$b_3 = b_{3e} + b_{3o} = 2s_{13}a \tag{14c}$$

$$b_4 = b_{4e} + b_{4o} = 2s_{14}a \tag{14d}$$

The substitution of (12) into (14) yields

$$b_1 = (\rho_e + \rho_o) a \tag{15a}$$

$$b_2 = (\rho_e - \rho_o) a \tag{15b}$$

$$b_3 = (\tau_e + \tau_o) a \tag{15c}$$

$$b_4 = (\tau_e - \tau_o) a \tag{15d}$$

This is shown schematically in Fig. 3-4.

(d) The properties of the device will be determined by the values of the reflection and transmission coefficients.

For an ideal symmetrical directional coupler, one has, from (2-1) and (2-2),

$$s_{11} = 0 \qquad \text{and} \qquad s_{12} = 0 \tag{16}$$

This requires [Eq. (12)]

$$\rho_e = \rho_o = 0 \tag{17}$$

(e) An alternative definition of a directional coupler is

$$s_{11} = 0 \qquad \text{and} \qquad s_{14} = 0 \tag{18}$$

i.e., port 1 and port 4 are isolated. This can be satisfied by requiring [Eq. (12)]

$$\rho_e = -\rho_o \qquad \text{and} \qquad \tau_e = \tau_o \tag{19}$$

Then (12) becomes

$$s_{11} = 0 \qquad s_{12} = \rho_e \qquad s_{13} = \tau_e \qquad s_{14} = 0 \tag{20}$$

The directional coupler specified by (20) is known as a backward wave coupler since the coupled wave emerges from port 2. When the coupled wave travels in the same direction as the wave in the main guide, the coupler is known as a co-directional coupler. If, on the other hand, the coupled wave travels in the opposite direction to the primary wave, it is known as the contra-directional coupler or the backward wave coupler.

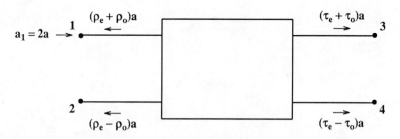

Figure 3-4: Arbitrarily excited four-port junction

4. Lorentz Reciprocity Theorem

This is one of the most useful theorems in the solution of electromagnetic problems. It shows the reciprocal properties of microwave circuits. It also demonstrates the identical transmitting and receiving characteristics of antennas. It can be used to establish the orthogonality properties of the modes in waveguides and cavities.

This theorem can be derived by considering a volume V bounded by a closed surface S (Fig. 4-1). Let a current source $\vec{J}_1$ create a field ($\vec{E}_1$, $\vec{H}_1$) and a second source $\vec{J}_2$ produce a field ($\vec{E}_2$, $\vec{H}_2$). Maxwell's equations for these fields are:

$$\nabla \times \vec{E}_1 = -j\omega\mu\vec{H}_1 \quad \text{and} \quad \nabla \times \vec{H}_1 = j\omega\epsilon\vec{E}_1 + \vec{J}_1 \tag{1}$$

$$\nabla \times \vec{E}_2 = -j\omega\mu\vec{H}_2 \quad \text{and} \quad \nabla \times \vec{H}_2 = j\omega\epsilon\vec{E}_2 + \vec{J}_2 \tag{2}$$

From the experience in the general theory on waveguides, it is found that the quantity of the form ($\vec{E}_a \times \vec{H}_b$) simplifies theoretical analysis. Form the following relations.

$$\nabla \cdot (\vec{E}_1 \times \vec{H}_2) = \vec{H}_2 \cdot \nabla \times \vec{E}_1 - \vec{E}_1 \cdot \nabla \times \vec{H}_2$$

$$= -j\omega\mu\vec{H}_1 \cdot \vec{H}_2 - (j\omega\epsilon\vec{E}_1 \cdot \vec{E}_2 + \vec{E}_1 \cdot \vec{J}_2) \tag{3a}$$

$$\nabla \cdot (\vec{E}_2 \times \vec{H}_1) = \vec{H}_1 \cdot \nabla \times \vec{E}_2 - \vec{E}_2 \cdot \nabla \times \vec{H}_1$$

$$= -j\omega\mu\vec{H}_1 \cdot \vec{H}_2 - (j\omega\epsilon\vec{E}_1 \cdot \vec{E}_2 + \vec{E}_2 \cdot \vec{J}_1) \tag{3b}$$

The subtraction of (3b) from (3a) yields

$$\nabla \cdot (\vec{E}_1 \times \vec{H}_2 - \vec{E}_2 \times \vec{H}_1) = \vec{J}_1 \cdot \vec{E}_2 - \vec{J}_2 \cdot \vec{E}_1 \tag{4}$$

Integrating both sides of (4) over the volume V yields

$$\oint_S [\vec{E}_1 \times \vec{H}_2 - \vec{E}_2 \times \vec{H}_1] \cdot \hat{n} \, da = \int_V [\vec{J}_1 \cdot \vec{E}_2 - \vec{J}_2 \cdot \vec{E}_1] \, dv \tag{5}$$

where $\hat{n}$ is the outward normal of the surface S. Equation (5) is the basic form of the Lorentz reciprocity theorem.

Under each of the following restrictions, the surface integral of (5) vanishes.

(i) If the surface S is a perfectly conducting surface, then

$$\hat{n} \times \vec{E}_1 = 0 = \hat{n} \times \vec{E}_2 \tag{6}$$

Consequently,

$$(\vec{E}_1 \times \vec{H}_2) \cdot \hat{n} = (\hat{n} \times \vec{E}_1) \cdot \vec{H}_2 = 0 \tag{7a}$$

$$(\vec{E}_2 \times \vec{H}_1) \cdot \hat{n} = (\hat{n} \times \vec{E}_2) \cdot \vec{H}_1 = 0 \tag{7b}$$

278

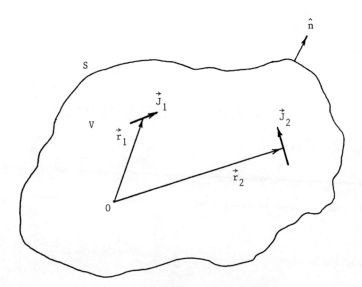

Figure 4-1: Arrangement of two sources

Therefore, the surface integral in (5) vanishes for this case.

(ii) If the surface S can be characterized by a surface impedance Z_s, which is defined by

$$Z_s \equiv \frac{\vec{E}_t}{\vec{J}_s} \tag{8a}$$

or

$$\vec{E}_t = Z_s \vec{J}_s \tag{8b}$$

$$\vec{E}_t = \text{tangential component of } \vec{E} \text{ on the surface} \tag{8c}$$

$$\vec{J}_s = \text{surface current density on the surface} \tag{8d}$$

with

$$\vec{J}_s = \hat{n} \times \vec{H} \tag{9}$$

one then has

$$\vec{E}_t = Z_s \hat{n} \times \vec{H} \tag{10}$$

The integrand of of the surface integral in (5) then becomes

$$(\hat{n} \times \vec{E}_1) \cdot \vec{H}_2 = Z_s \hat{n} \times (\hat{n} \times \vec{H}_1) \cdot \vec{H}_2$$

$$= Z_s \vec{H}_2 \cdot \hat{n} \times (\hat{n} \times \vec{H}_1)$$

$$= Z_s (\vec{H}_2 \times \hat{n}) \cdot (\hat{n} \times \vec{H}_1)$$

$$= -Z_s (\hat{n} \times \vec{H}_1) \cdot (\hat{n} \times \vec{H}_2) \tag{11a}$$

$$(\hat{n} \times \vec{E}_2) \cdot \vec{H}_1 = Z_s \hat{n} \times (\hat{n} \times \vec{H}_2) \cdot \vec{H}_1$$

$$= -Z_s (\hat{n} \times \vec{H}_1) \cdot (\hat{n} \times \vec{H}_2) = (\hat{n} \times \vec{E}_1) \cdot \vec{H}_2 \tag{11b}$$

or

$$(\hat{n} \times \vec{E}_1) \cdot \vec{H}_2 - (\hat{n} \times \vec{E}_2) \cdot \vec{H}_1 = 0 \tag{11c}$$

Therefore, when the surface can be characterized by a surface impedance, the surface integral of (5) vanishes.

(iii) When the surface S is a spherical surface with an infinitely large radius, then the field on this sphere is a spherical TEM radiation field, and

$$\vec{H} = \sqrt{\frac{\varepsilon}{\mu}} \; \hat{r} \times \vec{E} = Y\hat{r} \times \vec{E} \tag{12}$$

Then the integrand of (5) becomes, with $\hat{n} = \hat{r}$,

$$(\hat{n} \times \vec{E}_1) \cdot \vec{H}_2 - (\hat{n} \times \vec{E}_2) \cdot \vec{H}_1 = Y(\hat{r} \times \vec{E}_1) \cdot (\hat{r} \times \vec{E}_2) - Y(\hat{r} \times \vec{E}_2) \cdot (\hat{r} \times \vec{E}_1)$$

$$= 0 \tag{13}$$

Thus, the surface integral of (5) vanishes when the surface S is chosen to be a sphere with an infinite radius.

(iv) If the surface S encloses all sources for the field, the surface integral in (5) vanishes.

This can be shown by applying (5) to a volume which is internally bounded by the surface S and externally by S_∞, an infinitely large spherical surface (Fig. 4-2).

Since the volume V_0 contains no sources, therefore the right-hand side of (5) vanishes. One has

$$\int\limits_{S_0 = S + S_\infty} [\vec{E}_1 \times \vec{H}_2 - \vec{E}_2 \times \vec{H}_1] \cdot \hat{n}_0 \; da = 0 \tag{14}$$

where $\hat{n}_0$ is the unit outward normal of the surface S_0. Or

$$\int\limits_{S} [\vec{E}_1 \times \vec{H}_2 - \vec{E}_2 \times \vec{H}_1] \cdot \hat{n} \; da = \int\limits_{S_\infty} [\vec{E}_1 \times \vec{H}_2 - \vec{E}_2 \times \vec{H}_1] \cdot \hat{n}_0 \; da = 0 \tag{15}$$

The surface integral over the surface S_∞ vanishes by the reason given in part (iii).

This proves that the surface integral over the surface S vanishes if S encloses all sources.

When the surface integral in (5) vanishes, the Lorentz reciprocity theorem, (5), reduces to

$$\int\limits_{V} \vec{E}_1 \cdot \vec{J}_2 \; dv = \int\limits_{V} \vec{E}_2 \cdot \vec{J}_1 \; dv \tag{16}$$

For infinitesmal current elements, $\vec{J}_1$ and $\vec{J}_2$, then

$$\vec{E}_1(\vec{r}_2) \cdot \vec{J}_2(\vec{r}_2) = \vec{E}_2(\vec{r}_1) \cdot \vec{J}_1(\vec{r}_1) \tag{17}$$

Equation (17) implies that if $\vec{J}_1$ and $\vec{J}_2$ have identical amplitude, then the component of $\vec{E}_1$ (created by $\vec{J}_1$) in the direction of $\vec{J}_2$ is equal to the component of $\vec{E}_2$ (generated by $\vec{J}_2$) that is in parallel with $\vec{J}_1$.

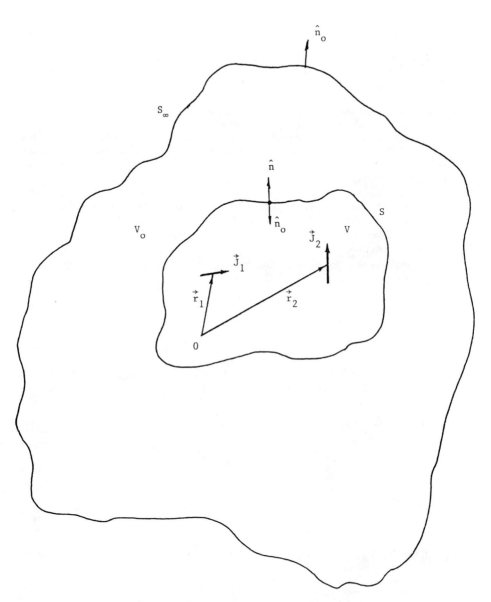

Figure 4-2: Arrangement of a volume without sources

5. Probe Coupling in a Waveguide

A coaxial-line feed to a rectangular waveguide is shown in Fig. 5-1. The probe is the extension of the inner conductor of the coaxial line into the guide. The position, L, of the short circuit and the length, d, of the probe can be adjusted to maximize the transfer of power from the coaxial line to the waveguide.

Any waveguide modes with nonzero E-field tangential to the probe will induce current on the probe. By the principle of reciprocity (Section 4), a probe carrying a current which is identical to that created by the waves in the guide will excite the same waveguide modes. Thus, to excite a particular mode, say, the TE_{10} mode, in a rectangular waveguide, the probe should be located at the position where the maximum E-field would exist for that mode. For the TE_{10} mode, this location is along the center of the broad face of the rectangular guide, [Section II.12.(b)].

The higher-order modes are also excited, and these are evanescent modes for guides designed for the propagation of a dominant mode. These evanescent modes are non-propagating and exist within a short distance from the probe. They store reactive energy which provides the reactive characteristics of the probe.

The section of short-circuited guide is used to supply the necessary reactance to cancel out the probe reactance.

A sinusoidal standing-wave current distribution is a good approximation for the current on a thin probe.

$$\vec{i} = \hat{y} I_0 \sin \gamma_0 (d - y) \qquad \text{at } x = \frac{a}{2}, \ 0 \le y \le d, \ z = 0 \tag{1a}$$

$$\gamma_0 = \omega \sqrt{\mu_0 \varepsilon_0} \tag{1b}$$

Consider an infinitely long waveguide, with a distribution of a current source $\vec{J}$ within the interval $z_a \le z \le z_b$ (Fig. 5-2). The fields produced by this source $\vec{J}$ can be expressed in terms of waveguide modes.

$$\vec{E}^+ = \sum_m A_m^+ (\vec{e}_m + \vec{e}_{zm}) e^{-j\beta_m z} \qquad z > z_b \tag{2a}$$

$$\vec{H}^+ = \sum_m A_m^+ (\vec{h}_m + \vec{h}_{zm}) e^{-j\beta_m z} \qquad z > z_b \tag{2b}$$

$$\vec{E}^- = \sum_m A_m^- (\vec{e}_m - \vec{e}_{zm}) e^{j\beta_m z} \qquad z < z_a \tag{2c}$$

$$\vec{H}^- = \sum_m A_m^- (-\vec{h}_m + \vec{h}_{zm}) e^{j\beta_m z} \qquad z < z_a \tag{2d}$$

where $A_m^{\pm}$ are constant amplitudes, and

$$\vec{e}_m = \vec{e}_m(x, y) = \hat{x} e_{xm} + \hat{y} e_{ym} \tag{2e}$$

$$\vec{e}_{zm} = \vec{e}_{zm}(x, y) = \hat{z} e_{zm} \tag{2f}$$

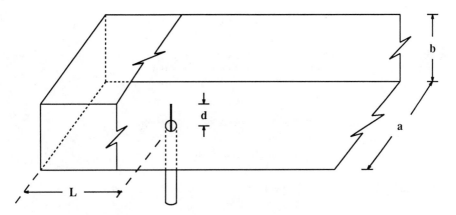

Figure 5-1: Coaxial feed for a rectangular guide

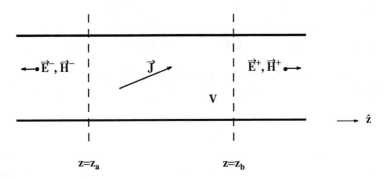

Figure 5-2: Source region and field region within a guide

$$\vec{h}_m = \vec{h}_m(x, y) = \hat{x} h_{xm} + \hat{y} h_{ym} \tag{2g}$$

$$\vec{h}_{zm} = \vec{h}_{zm}(x, y) = \hat{z} h_{zm} \tag{2h}$$

The sign convention is chosen such that transverse components of the $\vec{e}$-fields in the positive and negative traveling waves remain the same (see Section II.4), i.e.,

$$\vec{E}_t(x, y, z) = \vec{e}(x, y)\, e^{\pm j\beta z} \tag{3}$$

and the corresponding $\vec{h}$-field is determined by Poynting's theorem.

$$\frac{\vec{e}}{e} \times \frac{\vec{h}}{h} = \frac{\vec{S}}{S} = \text{direction of propagation} \tag{4}$$

The unknown magnitude $A_m^{\pm}$ will be determined by invoking the reciprocity theorem, (4-5).

$$\int_S (\vec{E}_1 \times \vec{H}_2 - \vec{E}_2 \times \vec{H}_1) \cdot \hat{n}\, da = \int_V (\vec{E}_2 \cdot \vec{J}_1 - \vec{E}_1 \cdot \vec{J}_2)\, dv \tag{5}$$

where the volume V is the volume bounded by the guide in the interval $z_a \le z \le z_b$ (Fig. 5-2).

Let $\vec{E}_1$ and $\vec{H}_1$ be the field radiated by the source $\vec{J} = \vec{J}_1$; this is the field expressed in terms of waveguide modes as given by (2). The fields $\vec{E}_2$ and $\vec{H}_2$, are chosen to be the kth normalized waveguide mode.

$$\vec{E}_2 \equiv \vec{E}_k^- = (\vec{e}_k - \vec{e}_{zk})\, e^{j\beta_k z} \tag{6a}$$

$$\vec{H}_2 \equiv \vec{H}_k^- = (-\vec{h}_k + \vec{h}_{zk})\, e^{j\beta_k z} \tag{6b}$$

For an infinitely long guide, $\vec{E}_2$ and $\vec{H}_2$ are excited by sources located at infinity, or $\vec{J}_2 = 0$ within the volume V. Then (5) gives

$$\oint_S (\vec{E}_1 \times \vec{H}_k^- - \vec{E}_k^- \times \vec{H}_1) \cdot \hat{n}\, da = \int_V \vec{E}_k^- \cdot \vec{J}\, dv \tag{7}$$

The surface S is made up of the guide walls and the cross-sectional surfaces S_0 of the guide at $z = z_a$ and at $z = z_b$. The integral over the guide walls is zero by virtue of the vanishing tangential $\vec{E}$-field on the conducting surface.

$$\hat{n} \times \vec{E}_1 = 0 \qquad\qquad \hat{n} \times \vec{E}_k^- = 0 \tag{8}$$

Then (7) becomes

$$\int_{S_{o1}+S_{o2}} (\vec{E}_1 \times \vec{H}_k^- - \vec{E}_k^- \times \vec{H}_1) \cdot \hat{n} \, da = \int_V \vec{E}_k^- \cdot \vec{J} \, dv \qquad (9)$$

where $S_{oi} \equiv S_o(z = z_i)$ with i = a,b. Because the modes are orthogonal, one has

$$\int_{S_o} \vec{E}_m^\pm \times \vec{H}_n^\pm \cdot \hat{n} \, da = 0 \qquad \text{for} \qquad m \neq n \qquad (10)$$

Hence, (9) reduces to

$$\int_{S_{oa}+S_{ob}} (\vec{E}_{1k} \times \vec{H}_k^- - \vec{E}_k^- \times \vec{H}_{1k}) \cdot \hat{n} \, da = \int_V \vec{E}_k^- \cdot \vec{J} \, dv \qquad (11)$$

Each term in (11) is expanded as follows.

$$[\vec{E}_{1k} \times \vec{H}_k^- - \vec{E}_k^- \times \vec{H}_{1k}]_{z=z_a} = [A_k^- (\vec{e}_k - \vec{e}_{zk}) e^{j\beta_k z_a}] \times [(-\vec{h}_k + \vec{h}_{zk}) e^{j\beta_k z_a}]$$

$$- [(\vec{e}_k - \vec{e}_{zk}) e^{j\beta_k z_a}] \times [A_k^- (-\vec{h}_k + \vec{h}_{zk}) e^{j\beta_k z_a}]$$

$$= 0 \qquad (12a)$$

$$[\vec{E}_{1k} \times H_k^- - \vec{E}_k^- \times H_{1k}]_{z=z_b} = [A_k^+ (\vec{e}_k + \vec{e}_{zk}) e^{-j\beta_k z_b}] \times [(-\vec{h}_k + \vec{h}_{zk}) e^{j\beta_k z_b}$$

$$- [(\vec{e}_k - \vec{e}_{zk}) e^{j\beta_k z_b}] \times [A_k^+ (\vec{h}_k + \vec{h}_{zk}) e^{-j\beta_k z_b}]$$

$$= A_k^+ [-\vec{e}_k \times \vec{h}_k + \vec{e}_k \times \vec{h}_{zk} - \vec{e}_{zk} \times \vec{h}_k + \vec{e}_{zk} \times \vec{h}_{zk}$$

$$- (\vec{e}_k \times \vec{h}_k + \vec{e}_k \times \vec{h}_{zk} - \vec{e}_{zk} \times \vec{h}_k - \vec{e}_{zk} \times \vec{h}_{zk})]$$

$$= -2A_k^+ \vec{e}_k \times \vec{h}_k \qquad (12b)$$

Therefore, (11) is given by

$$-2A_k^+ \int_{S_o(z_b)} (\vec{e}_k \times \vec{h}_k) \cdot \hat{z} \, da = \int_V \vec{E}_k^- \cdot \vec{J} \, dv \qquad (13)$$

or

$$A_k^+ = \frac{-1}{P_k} \int_V (\vec{e}_k - \vec{e}_{zk}) \, e^{j\beta_k z} \cdot \vec{J} \, dv \tag{14a}$$

$$P_k \equiv 2 \int_{S_o(z_b)} (\vec{e}_k \times \vec{h}_k) \cdot \hat{z} \, da \tag{14b}$$

The coefficient A_k^- can be similarly determined by choosing $\vec{E}_2$ and $\vec{H}_2$ as follows.

$$\vec{E}_2 \equiv \vec{E}_k^+ = (\vec{e}_k + \vec{e}_{zk}) \, e^{-j\beta_k z} \tag{15a}$$

$$\vec{H}_2 \equiv \vec{H}_k^+ = (\vec{h}_k + \vec{h}_{zk}) \, e^{-j\beta_k z} \tag{15b}$$

Then

$$A_k^- = \frac{-1}{P_k} \int_V \vec{E}_k^+ \cdot \vec{J} \, dv$$

$$= \frac{-1}{P_k} \int_V (\vec{e}_k + \vec{e}_{zk}) \, e^{-j\beta_k z} \cdot \vec{J} \, dv \tag{16}$$

Since the determination of all the modes is rather lengthy, only the dominant TE_{10} mode will be determined. The TE_{10} mode in a rectangular waveguide is given by [Section II.12(b)]

$$E_y = e_y e^{-j\beta z} = \sin \frac{\pi x}{a} \, e^{-j\beta z} \tag{17a}$$

$$H_x = h_x e^{-j\beta z} = -Y_w \sin \frac{\pi x}{a} \, e^{-j\beta z} \tag{17b}$$

where

$$Y_w = \frac{\beta_{10}}{\gamma_o Z_o} = \text{wave admittance of } TE_{10} \text{ mode} \tag{17c}$$

$$\gamma_o = \omega \sqrt{\mu_o \varepsilon_o} \tag{17d}$$

$$\beta_{mn} = \left\{ \gamma_o^2 - \left[\left(\frac{m\pi}{a} \right)^2 + \left(\frac{n\pi}{b} \right)^2 \right] \right\}^{\frac{1}{2}} \tag{17e}$$

$$\beta_{10} = \left\{ \gamma_o^2 - \left[\frac{\pi}{a} \right]^2 \right\}^{\frac{1}{2}} \tag{17f}$$

Then, with the probe located at $z = 0$, one has

$$P_{10} = 2 \int\limits_0^a \int\limits_0^b (\hat{y}E_y \times \hat{x}H_x) \cdot \hat{z} \, da$$

$$= 2 \int\limits_0^a \int\limits_0^b Y_w \sin^2 \frac{\pi x}{a} \, dx \, dy = abY_w \tag{18}$$

A probe situated at a distance L from a perfect short circuit, Fig. 5-3a, is equivalent to the original probe plus its image located at a distance 2L away with the short circuit removed; both probes are located within an infinitely long waveguide (Fig. 5-3b).

It will be considered that the field will only radiate into the region where $z > 0$. Then

$$E_y^+ = A^+ \sin \frac{\pi x}{a} e^{-j\beta z} \tag{19a}$$

$$H_x^+ = -A^+ Y_w \sin \frac{\pi x}{a} e^{-j\beta z} = -Y_w E_y^+ \tag{19b}$$

Equation (14a) then gives

$$A^+ = \frac{-1}{P_{10}} \left[\int \vec{e_2} e^{j\beta z} \cdot \vec{J} \, dv \mid_{z=0} - \int \vec{e_2} e^{j\beta z} \cdot \vec{J} \, dv \mid_{z=-2L} \right]$$

$$= \frac{-1}{abY_w} \left[\int\limits_0^d \sin \frac{\pi x}{a} \, \hat{y} \cdot \hat{y} I_o \sin \gamma_o(d - y) \, dy \right.$$

$$\left. - \int\limits_0^d \sin \frac{\pi x}{a} e^{-j2\beta L} I_o \sin \gamma_o(d - y) \, dy \right] \tag{20}$$

At $x = a/2$, $\sin \pi x/a = \sin \pi/2 = 1$; then

$$A^+ = \frac{-I_o}{abY_w} (1 - e^{-j2\beta L}) \int\limits_0^d \sin \gamma_o(d - y) \, dy$$

$$= \frac{-I_o}{ab\gamma_o Y_w} (1 - e^{-j2\beta L}) \cos \gamma_o(d - y) \Big|_0^d$$

$$= \frac{I_o}{ab\gamma_o Y_w} (1 - \cos \gamma_o d)(e^{-j2\beta L} - 1) \tag{21}$$

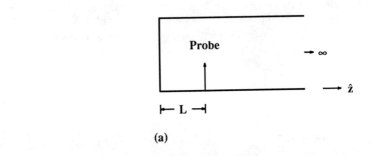

(a)

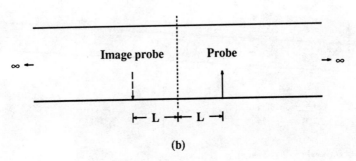

(b)

Figure 5-3: A probe in front of a conductor

The total transverse TE_{10} mode radiated by the probe is, for $z > 0$ [Eq. (19)],

$$E_y^+ = \frac{I_o Z_w}{ab\gamma_o} \left(e^{-j2\beta L} - 1 \right) \left(1 - \cos \gamma_o d \right) \sin \frac{\pi x}{a} e^{-j\beta z} \tag{22a}$$

$$H_x^+ = -Y_w E_y^+ \tag{22b}$$

The total radiated power is given by

$$P = \frac{Y_w}{2} \int_0^a \int_0^b |E_y^+|^2 \, dx \, dy$$

$$= \frac{I_o^2 Z_w}{4ab\gamma_o^2} \, | \, e^{-j2\beta L} - 1 \, |^2 \, (1 - \cos \gamma_o d)^2 \tag{23}$$

At the base of the probe, $y = 0$, the total current in the coaxial line is [Eq. (1)]

$$i \equiv I_o \sin \gamma_o d \tag{24}$$

Then the input impedance as seen by the coaxial line is given by the complex Poynting vector,

$$Z_{in} = \frac{P + jQ}{\dfrac{I \, I^*}{2}} = R_{in} + jX_{in} \tag{25a}$$

$$P = \text{real power radiated into the guide, Eq. (23)} \tag{25b}$$

$$Q = \text{reactive volt-amperes} = 2\omega (w_m - w_e)$$

$$= \text{reactive energy stored in the evanescent modes} \tag{25c}$$

With P given by (23), the input resistance is therefore determined.

$$R_{in} = \frac{2P}{I \, I^*} \tag{25d}$$

To find the input reactance X_{in}, one would have to evaluate all the evanescent modes. However, in practical applications, one usually adjusts experimentally the distance of the short circuit and the depth of penetration of the probe so that maximum power is obtained. For certain applications, the knowledge of the input reactance is not essential.

6. Radiation from Linear Current Elements

The radiation from a general linear current element can be more easily analyzed by sub-dividing the element into a transverse current element and an axial current element, Fig. 6-1. The radiation of each type of current element will be studied separately. The total radiation from a general current element is the sum of the contributions from each individual component.

(a) Transverse Current Element

For a transverse current element located at $z = 0$, (5-14) and (5-16) show that

$$A_k^+ = \frac{-1}{P_k} \int \vec{e}_k \cdot \vec{J} \, dL = A_k^- \tag{1a}$$

$$\vec{J} = \hat{x} J_x + \hat{y} J_y \tag{1b}$$

This implies that A_k^+ has a value proportional to $|\vec{J}|$, or the transverse current element gives rise to a voltage source in the transverse direction parallel to $\vec{J}$. In other words, the transverse current element is equivalent to a voltage source connected in parallel across an equivalent transmission line for the waveguide mode. The transverse current element radiates identical transverse electric fields next to the current element. This means that the voltage is continuous across the source region, $0^- \le z \le 0^+$. The corresponding radiated transverse magnetic field is discontinuous across the source region, and hence, the equivalent current is discontinuous across the equivalent voltage source. The equivalent circuit for this type of source is shown in Fig. 6-2. The ideal transformer provides the proper coupling between the generator and the line to provide the same amount of power as is radiated in the guide. The energy stored in the evanescent modes is denoted by the susceptance jB.

(b) Axial Current Element

For an axial current element, $\vec{J} = \hat{z} J_z$, situated at $z = 0$, (5-14) and (5-16) give

$$A_k^+ = \frac{1}{P_k} \int \vec{J} \cdot \vec{e}_{zk} e^{j\beta_k z} \, dz \tag{2a}$$

$$A_k^- = \frac{-1}{P_k} \int \vec{J} \cdot \vec{e}_{zk} e^{-j\beta_k z} \, dz \tag{2b}$$

$$\vec{J} = \hat{z} J_z \tag{2c}$$

Suppose the current is an even function of z, for $-L \le z \le L$, and since $\vec{e}_{zk} = \vec{e}_{zk}(x, y)$, the integral can be arranged as

$$\int \vec{J}(z) \cdot \vec{e}_{zk}(x, y) \, e^{j\beta_k z} \, dz = \int_{-L}^{L} \vec{J}(z) \cdot \vec{e}_{zk}(x, y) \, e^{j\beta_k z} \, dz$$

and

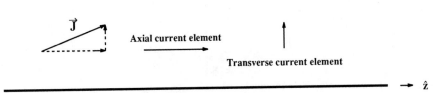

Figure 6-1: Linear current elements

Main guide

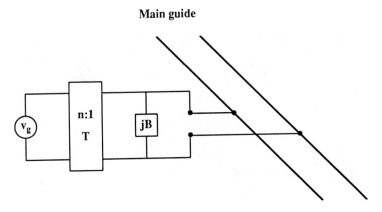

Figure 6-2: Equivalent circuit for transverse current element

$$A_k^+ = \frac{1}{P_k} \int_{-L}^{L} \vec{J}(z) \cdot \vec{e}_{zk}(x, y) \, e^{j\beta_k z} \, dz$$

$$= \frac{1}{P_k} \left[\int_{-L}^{0} \vec{J}(z) \cdot \vec{e}_{zk}(x, y) \, e^{j\beta_k z} \, dz + \int_{0}^{L} \vec{J}(z) \cdot \vec{e}_{zk}(x, y) \, e^{j\beta_k z} \, dz \right]$$

$$\equiv \frac{1}{P_k} [\, I_1 + I_2 \,] \tag{3}$$

Let $z = -z_1$. Then the integral I_1 becomes

$$I_1 = \int_{L}^{0} \vec{J}(-z_1) \cdot \vec{e}_{zk} e^{-j\beta_k z_1} \, d(-z_1)$$

$$= \int_{0}^{L} \vec{J}(z_1) \cdot \vec{e}_{zk} e^{-j\beta_k z_1} \, dz_1$$

$$= \int_{0}^{L} \vec{J}(z) \cdot \vec{e}_{zk} e^{-j\beta_k z} \, dz \tag{4}$$

where the dummy variable is changed back to z in the final expression. The substitution of (4) into (3) yields

$$A_k^+ = \frac{1}{P_k} \int_{0}^{L} \vec{J}(z) \cdot \vec{e}_{zk}(x, y) \left[e^{-j\beta_k z} + e^{j\beta_k z} \right] dz$$

$$= \frac{2}{P_k} \int_{0}^{L} \vec{J}(z) \cdot \vec{e}_{zk} \cos \beta_k z \, dz \tag{5a}$$

The coefficient A_k^- can be obtained similarly.

$$A_k^- = -A_k^+ \tag{5b}$$

In this case, the substitution of (5b) into (5-2) shows that the radiated transverse magnetic field is continuous across the source while the transverse electric field is discontinuous. This is equivalent to a voltage source connected in series with the main guide as shown in Fig. 6-3.

(c) Equivalent Dipole for a Linear Current Element

A linear current element can be represented by an equivalent oscillating electric dipole. From Maxwell's equations, one has

$$\nabla \times \vec{H} = j\omega\varepsilon\vec{E} + \vec{J}$$

$$= j\omega\,(\,\varepsilon_o\vec{E} + \vec{P}\,) + \vec{J}$$

$$= j\omega\varepsilon_o\vec{E} + j\omega\,(\,\vec{P} + \frac{\vec{J}}{j\omega}\,)$$

$$= j\omega\varepsilon_o\vec{E} + j\omega\,(\,\vec{P} + \vec{P}_J\,) \tag{6a}$$

where

$$\vec{P}_J \equiv \frac{\vec{J}}{j\omega} = \text{equivalent dipole for a current element} \tag{6b}$$

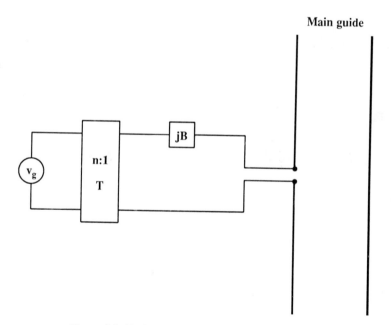

Main guide

Figure 6-3: Equivalent circuit for axial current element

7. Radiation from a Current Loop

(a) A current loop carrying a uniform current in a waveguide is illustrated in Fig. 7-1. The magnitude of the kth mode is given by (5-14a).

$$A_k^+ = \frac{-1}{P_k} \int_V \vec{E}_k^- \cdot \vec{J}\, dv$$

$$= \frac{-1}{P_k} \int_C \vec{E}_k^- \cdot \hat{t} I\, dL \tag{1a}$$

$$\vec{J}\, dv = \hat{t} I\, dL \tag{1b}$$

where $\hat{t}$ is the unit vector tangential to the current loop, and I is the constant amplitude of the current. By Stokes' theorem, (1a) becomes

$$A_k^+ = \frac{-I}{P_k} \int_S \nabla \times \vec{E}_k^- \cdot \hat{n}\, da \tag{2}$$

where S is the surface bounded by the current loop. Since

$$\nabla \times \vec{E}_k^- = -j\omega\mu_0 \vec{H}_k^- \tag{3}$$

therefore,

$$A_k^+ = \frac{j\omega I}{P_k} \int_S \vec{B}_k^- \cdot \hat{n}\, da \tag{4}$$

Similarly, from (5-16),

$$A_k^- = \frac{j\omega I}{P_k} \int_S \vec{B}_k^+ \cdot \hat{n}\, da \tag{5}$$

The magnitude of the excited kth mode is determined by the total magnetic flux of this mode passing through the loop.

(b) For a small loop when the flux of the kth mode is uniform over the area of the loop, one obtains

$$A_k^+ = \frac{j\omega I}{P_k} \vec{B}_k^- \cdot \int_S \hat{n}\, da = \frac{j\omega \vec{B}_k^-}{P_k} \cdot \hat{n} I S$$

$$= \frac{j\omega}{P_k} \vec{B}_k^- \cdot \vec{M} \tag{6a}$$

$$\vec{M} = \hat{n} I S = \text{magnetic dipole moment of the loop} \tag{6b}$$

Similarly, one obtains

$$A_k^- = \frac{j\omega}{P_k} \vec{B}_k^+ \cdot \vec{M} \qquad (6c)$$

The radiation of a small loop is equivalent to the radiation of a magnetic dipole.

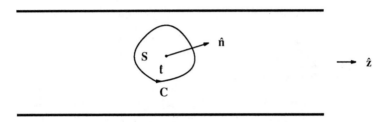

Figure 7-1: A current loop

8. Waveguide Coupling by an Aperture

The field coupled through a small aperture can be approximated by the radiation from the unperturbed field of (a) the normal component of the electric field, and (b) the tangential component of the magnetic field, at the aperture.

The normal component of the E-field at a conducting surface without the aperture is shown in Fig. 8-1a and the surface carries a negative surface charge density. When a hole is created on the conducting surface, some of the surface charges will be forced into the upper side of the surface and the distribution of the E-field is approximately sketched in Fig. 8-1b. The electric field of an ideal electric dipole has the shape shown in Fig. 8-1c. The similarity between Fig. 8-1b and the upper half of Fig. 8-1c suggests that the electric coupling of a small hole on a thin conducting surface may be approximated by an electric dipole oriented normal to the aperture and parallel to the unperturbed $\vec{E}$-field.

The tangential component of the $\vec{H}$-field along a smooth conducting plane is shown in Fig. 8-2a and this plane carries a surface current density. When a hole is placed on the plane, the surrface current density will redistribute and some will appear on the upper side of the conducting plane. The perturbed surface current density will generate magnetic field on both sides of the hole. The resultant field is sketched in Fig. 8-2b and may be interpreted as the leakage through the hole. The H-field of an ideal magnetic dipole, which is equivalent to an infinitesimal current loop, is shown in Fig. 8-2c. A direct comparison of Fig. 8-2b and Fig. 8-2c proposes that the magnetic coupling of a small hole on a thin conducting surface can be approximated by a magnetic dipole oriented in the plane of the aperture and parallel to the original unperturbed H-field.

The dipole moments of a small circular aperture may be related to the unperturbed field as follows:

$$P = -\varepsilon_0 \alpha_e (\hat{n} \cdot \vec{E}) \hat{n} \tag{1a}$$

$$M = -\alpha_m \vec{H}_{tangential} \tag{1b}$$

where $\hat{n}$ is the unit normal vector, and $\alpha_{e,m}$ are the polarizabilities introduced by H.A. Bethe (Physical Review, Vol. 66, pp. 163 -182).

$$\alpha_e = \frac{-2}{3} r_a^3 = \text{electric polarizability} \tag{2a}$$

$$\alpha_m = \frac{4}{3} r_a^3 = \text{magnetic polarizability} \tag{2b}$$

where r_a is the radius of the aperture.

The above expressions provide good approximation for $r_a \ll \lambda$.

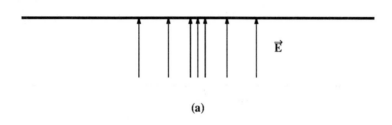

(a)

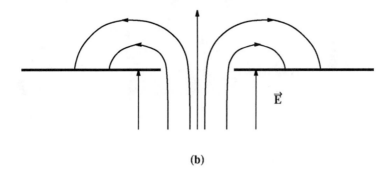

(b)

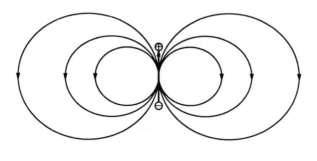

(c)

Figure 8-1: Distribution of E-field

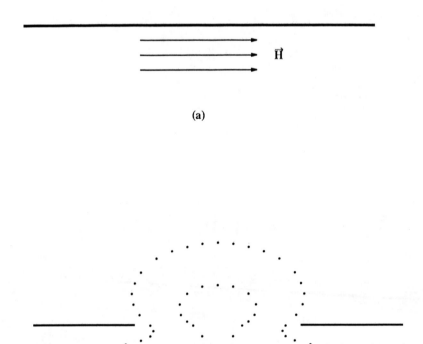

(a)

(b)

Figure 8-2: Distribution of H-field

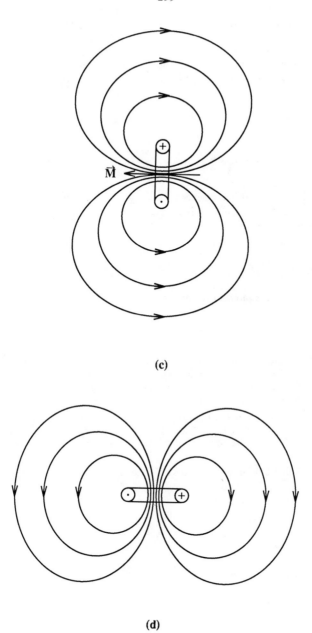

(c)

(d)

Figure 8-2: Distribution of H-field

9. Aperture in a Transverse Wall of a Waveguide

(a) A small circular aperture in a conducting surface transverse to the axis of a rectangular guide is shown in Fig. 9-1. The standing wave of the TE_{10} mode produced by an un-perturbed transverse conducting surface (without aperture) in the region $z < 0$ is (Sections II.4 and II.12)

$$E_y = E_o (e^{-j\beta z} - e^{j\beta z}) \sin \frac{\pi x}{a} \tag{1a}$$

$$H_x = - E_o Y_w (e^{-j\beta z} + e^{j\beta z}) \sin \frac{\pi x}{a} \tag{1b}$$

The z-component of the $\vec{H}$-field is not needed for the present problem and is omitted here.

Since the $\vec{E}$-field does not have any component normal to this aperture, there will be no induced electric dipole.

The total tangential $\vec{H}$-field at the aperture, $z = 0$ and $x = a/2$, is given by (1b).

$$H_x(z = 0) = -2E_o Y_w \tag{2}$$

The induced magnetic dipole moment is given by (8-1b).

$$\vec{M} = - \hat{x}\alpha_m H_x = \hat{x}\frac{8}{3}r_a^3 E_o Y_w \tag{3}$$

This equivalent magnetic dipole is shown in Fig. 9-2a. This magnetic dipole can be represented by one-half of a small current loop in the yz-plane as shown in Fig. 9-2b. The radiated field of this dipole in the presence of the conducting plane is equal to the field of this dipole plus that of its image with the conducting plane removed.

The image of the half circular current loop is simply the other half of the loop, thus the image of a magnetic dipole with respect to a conducting surface is another dipole $\vec{M}$. The effective magnetic dipole with the conducting surface removed is $2\vec{M}$ as shown in Fig. 9-2c.

(b) Let the radiated field into the region $z > 0$ be

$$E_y^+ = A^+ e^{-j\beta z} \sin \frac{\pi x}{a} \equiv A^+ e_y e^{-j\beta z} \tag{4a}$$

$$H_x^+ = - A^+ Y_w e^{-j\beta z} \sin \frac{\pi x}{a} \equiv A^+ h_x e^{-j\beta z} \tag{4b}$$

For $z < 0$, one has

$$E_y^- = A^- e^{j\beta z} \sin \frac{\pi x}{a} \equiv A^- e_y e^{j\beta z} \tag{4c}$$

$$H_x^- = A^- Y_w e^{j\beta z} \sin \frac{\pi x}{a} \equiv -A^- h_x e^{j\beta z} \tag{4d}$$

Then, Eq. (7-6) yields

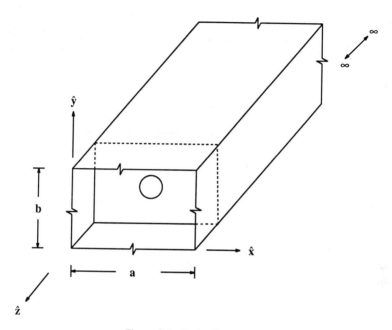

Figure 9-1: A circular aperture

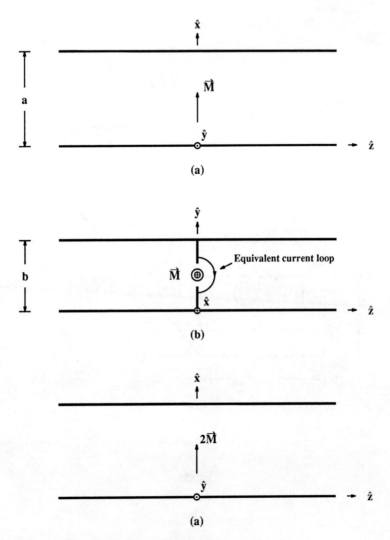

Figure 9-2: Magnetic moment of an aperture

$$A^+ = \frac{j\omega}{P_k} \vec{B}^- \cdot \vec{M} = \frac{j\omega}{P_{10}} 2MB_x^- \tag{5}$$

But $\vec{B}^-$ is the normalized unperturbed TE_{10} field,

$$\vec{B}^- = \hat{x}\mu_o h_x = \hat{x}\mu_o Y_w \sin \frac{\pi x}{a} \qquad x = \frac{a}{2}$$

$$= \hat{x}\mu_o Y_w \tag{6}$$

and from Eq. (5-18),

$$P_{10} = -2 \int_0^a \int_0^b e_y h_x \, dx \, dy = abY_w \tag{7}$$

Therefore,

$$A^+ = \frac{j\omega}{abY_w} (\mu_o Y_w) \frac{16}{3} r_a^3 E_o Y_w$$

$$= \frac{16}{3} r_a^3 \frac{j\omega\mu_o}{abZ_w} E_o \tag{8}$$

(c) The transmission coefficient of the aperture is [Eq. (8)]

$$\tau_a = \frac{A^+}{E_o} = j \frac{16}{3} r_a^3 \frac{k_o Z_o}{abZ_w}$$

$$= j \frac{16}{3} r_a^3 \frac{\beta}{ab} \tag{9a}$$

$$\omega\mu_o = \omega \sqrt{\mu_o \varepsilon_o} \sqrt{\frac{\mu_o}{\varepsilon_o}} = \gamma_o Z_o \tag{9b}$$

$$Z_w = \frac{\gamma_o}{\beta} Z_o \tag{9c}$$

(d) The field scattered into the input guide is calculated from (7-6c).

$$A^- = \frac{j\omega}{P_k} \vec{B}_k^- \cdot \vec{M} = A^+ \tag{10}$$

and (4c) gives

$$E_y^- = A^+ \sin \frac{\pi x}{a} \, e^{j\beta z} \qquad z < 0 \tag{11}$$

The total $\vec{E}$-field in the region $z < 0$ is the sum of the exciting field, (1a), and the scattered field, (11).

$$E_y(z < 0) = \left[E_0 e^{-j\beta z} + (A^+ - E_0) \, e^{j\beta z} \right] \sin \frac{\pi x}{a}$$

$$= E_0 \left[e^{-j\beta z} + \frac{A^+ - E_0}{E_0} \, e^{j\beta z} \right] \sin \frac{\pi x}{a} \tag{12}$$

Thus, the reflection coefficient produced by the presence of an aperture on a transverse conducting wall is, from (12) and (9),

$$\rho = \frac{A^+ - E_0}{E_0} = \frac{A^+}{E_0} - 1 = \tau_a - 1$$

$$= j \frac{16}{3} r_a^3 \frac{\beta}{ab} - 1 \tag{13}$$

The reflection coefficient produced by a normalized shunt susceptance jb_n in an infinite guide is (Fig. 9-3)

$$\rho = \frac{1 - y_n}{1 + y_n} = \frac{1 - (1 + jb_n)}{1 + (1 + jb_n)} = \frac{-jb_n}{2 + jb_n} = - \frac{1}{1 + \dfrac{2}{jb_n}}$$

$$\approx - (1 - \frac{2}{jb_n}) = -1 + j \frac{2}{b_n} \qquad \text{for} \quad jb_n \gg 1 \tag{14}$$

A comparison of (13) with (14) shows that the aperture on a transverse conducting surface is equivalent to a shunt susceptance given by

$$b_n = \frac{3ab}{8r_a^3 \beta} \tag{15}$$

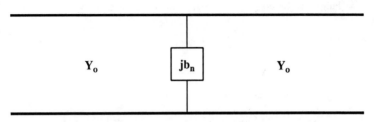

Figure 9-3: A shunt susceptance

10. Side-wall Coupler - Even-Odd Mode Theory

(a) Parameters of a side-wall directional coupler will be determined by the even- and the odd-mode techniques.

A side-wall coupler is shown in Fig. 10-1. The field distribution of even-mode excitation is shown in Fig. 10-2. This is simply the TE_{10} mode in two separate $b \times a$ rectangular waveguides. Since the tangential $\vec{E}$-field is zero at the plane of symmetry, an electric wall may be introduced at this plane.

The transmission coefficient from port 1 to port 3 is identical to the propagation factor in the simple guide, with ports 3 and 4 matched.

$$\tau_{13} = \tau_e = e^{-j\frac{2\pi L}{\lambda_{ge}}}$$

$$= \tau_{24} = e^{-j\beta_e L} \qquad \beta_e = \frac{2\pi}{\lambda_{ge}} \tag{1a}$$

where L is the length of the coupling section, λ_{ge} is the guided wavelength of the even mode, TE_{10}, and

$$\left[\frac{2\pi}{\lambda_{ge}}\right]^2 = \left[\frac{2\pi}{\lambda_o}\right]^2 - \left[\frac{\pi}{a}\right]^2 \tag{1b}$$

λ_o is the wavelength in free space.

(b) The field distribution of odd-mode excitation is shown in Fig. 10-3. This is identical to the TE_{20} mode in a rectangular guide with dimensions $b \times 2a$. Since the tangential $\vec{H}$-field is zero at the plane of symmetry, a magnetic wall may be inserted at this plane.

The transmission coefficient in this case is equal to the propagation factor for the TE_{20} mode, with output ports matched.

$$\tau_{13}' = \tau_o = e^{-j\frac{2\pi L}{\lambda_{go}}} = \tau_{24}' = e^{-j\beta_{go}L} \tag{2a}$$

$$\beta_{go} \equiv \frac{2\pi}{\lambda_{go}} \tag{2b}$$

λ_{go} is the guided wavelength of the TE_{20} mode, and

$$\left[\frac{2\pi}{\lambda_{go}}\right]^2 \equiv \left[\frac{2\pi}{\lambda_o}\right]^2 - \left[\frac{\pi}{2a}\right]^2 \tag{2c}$$

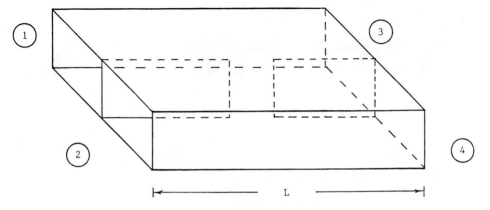

Figure 10-1: A side-wall directional coupler

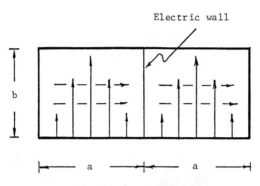

Even-mode excitation

Figure 10-2: Even-mode excitation

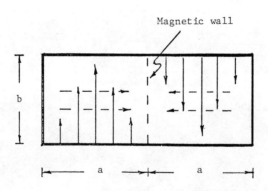

Odd-mode excitation

Figure 10-3: Odd-mode excitation

(c) The elements of the scattering matrix are given by (2-13) and (3-12).

$$s_{11} = 0 = s_{22} \qquad \text{by definition of a directional coupler} \tag{3a}$$

$$s_{13} = \frac{\tau_e + \tau_o}{2} = \frac{1}{2}\left(e^{-j\beta_e L} + e^{-j\beta_{go}L}\right) \tag{3b}$$

$$s_{14} \equiv \frac{\tau_e - \tau_o}{2} = \frac{1}{2}\left(e^{-j\beta_e L} - e^{-j\beta_{go}L}\right) \tag{3c}$$

Since

$$\frac{1}{2}\left(e^{-j\theta} \pm e^{-j\phi}\right) = \frac{1}{2}e^{-\frac{j\Delta_+}{2}}\left[e^{-\frac{j\Delta_-}{2}} \pm e^{\frac{j\Delta_-}{2}}\right]$$

$$= e^{-\frac{j\Delta_+}{2}}\cos\frac{\Delta_-}{2} \qquad \text{for plus sign} \tag{4a}$$

$$= -je^{-\frac{j\Delta_+}{2}}\sin\frac{\Delta_-}{2} \qquad \text{for minus sign} \tag{4b}$$

where

$$\Delta_+ \equiv \theta + \phi \qquad \Delta_- \equiv \theta - \phi \tag{4c}$$

then relations (3b) and (3c) can be arranged as follows.

$$s_{13} = e^{-j\frac{(\beta_e + \beta_{go})L}{2}}\cos\frac{(\beta_e - \beta_{go})L}{2} \tag{5a}$$

$$s_{14} = -je^{-j\frac{(\beta_e + \beta_{go})L}{2}}\sin\frac{(\beta_e - \beta_{go})L}{2} \tag{5b}$$

(d) The unitary conditions [Eq. (2-14)]

$$|s_{13}|^2 + |s_{14}|^2 = 1 \tag{6a}$$

$$s_{13}^* s_{14} + s_{14}^* s_{13} = 0 \tag{6b}$$

are satisfied by (5).

(e) Equation (5) implies that the power transfer between two waveguides is periodic. For a 3-db coupler,

$$s_{13} = s_{14} = 0.707 \tag{7}$$

This requires

$$(\beta_e - \beta_{go}) \frac{L}{2} = \frac{\pi}{4} \tag{8}$$

11. Side-wall Coupler - Eigenvalue Theory

Eigenvalue techniques may be used to analyze the side-wall couplers. The general scattering matrix for a directional coupler is a variation of (2-26) when setting $\theta_1 = \theta_2 = \pi/2$ and $\phi_1 = \phi_2 = 0$ [Eqs. (2-22) and (2-3b)].

$$[S] = \begin{bmatrix} 0 & 0 & a & jb \\ 0 & 0 & jb & a \\ a & jb & 0 & 0 \\ jb & a & 0 & 0 \end{bmatrix} \tag{1}$$

For a 3-db coupler with $a = b = \dfrac{1}{\sqrt{2}}$, (1) becomes

$$[S] = \frac{1}{\sqrt{2}} \begin{bmatrix} 0 & 0 & 1 & j \\ 0 & 0 & j & 1 \\ 1 & j & 0 & 0 \\ j & 1 & 0 & 0 \end{bmatrix} \tag{2}$$

and the eigenvalue equation is

$$\left| \ [S] - \lambda [u] \ \right| = 0$$

$$\begin{bmatrix} -\lambda & 0 & 0.707 & j0.707 \\ 0 & -\lambda & j0.707 & 0.707 \\ 0.707 & j0.707 & -\lambda & 0 \\ j0.707 & 0.707 & 0 & -\lambda \end{bmatrix} = 0 \tag{3}$$

The expansion of (3) yields

$$\left| \ [S] - \lambda [u] \ \right| \equiv D_1 + D_2 + D_3$$

where

$$D_1 \equiv -\lambda \begin{bmatrix} -\lambda & \dfrac{j}{\sqrt{2}} & \dfrac{1}{\sqrt{2}} \\ \dfrac{j}{\sqrt{2}} & -\lambda & 0 \\ \dfrac{1}{\sqrt{2}} & 0 & -\lambda \end{bmatrix}$$

$$D_1 = -\lambda \left[-\lambda\lambda^2 - \frac{j}{\sqrt{2}} \left(-j\frac{\lambda}{\sqrt{2}} \right) + \frac{1}{\sqrt{2}} \frac{\lambda}{\sqrt{2}} \right]$$

$$= \lambda^4 - \frac{\lambda^2}{2} + \frac{\lambda^2}{2} = \lambda^4$$

and

$$D_2 \equiv \frac{1}{\sqrt{2}} \begin{bmatrix} 0 & -\lambda & \frac{1}{\sqrt{2}} \\ \frac{1}{\sqrt{2}} & \frac{j}{\sqrt{2}} & 0 \\ \frac{j}{\sqrt{2}} & \frac{1}{\sqrt{2}} & -\lambda \end{bmatrix}$$

$$= \frac{1}{\sqrt{2}} \left[0 + \lambda \frac{-\lambda}{\sqrt{2}} + \frac{1}{\sqrt{2}} \left(\frac{1}{2} + \frac{1}{2} \right) \right]$$

$$= -\frac{\lambda^2}{2} + \frac{1}{2}$$

and

$$D_3 \equiv \frac{j}{\sqrt{2}} \begin{bmatrix} 0 & -\lambda & \frac{j}{\sqrt{2}} \\ \frac{1}{\sqrt{2}} & \frac{j}{\sqrt{2}} & -\lambda \\ \frac{j}{\sqrt{2}} & \frac{1}{\sqrt{2}} & 0 \end{bmatrix}$$

$$= \frac{-j}{\sqrt{2}} \left[\lambda j\frac{\lambda}{\sqrt{2}} + \frac{j}{\sqrt{2}} \left(\frac{1}{2} + \frac{1}{2} \right) \right]$$

$$= \frac{\lambda^2}{2} + \frac{1}{2}$$

Therefore,

$$| [S] - \lambda [u] | = \lambda^4 + 1 = 0 \tag{4a}$$

$$\lambda^4 = -1 = e^{j(2n+1)\pi}$$

$$\lambda = e^{j(2n+1)\frac{\pi}{4}} \qquad n = 0, 1, 2, \ldots \tag{4b}$$

The eigenvalues are

$$n = 0 \qquad \lambda_1 = e^{j\frac{\pi}{4}} \tag{4c}$$

$$n = 1 \qquad \lambda_2 = e^{j\frac{3\pi}{4}} = e^{j(1-\frac{1}{4})\pi} = -e^{-j\frac{\pi}{4}} \tag{4d}$$

$$n = 2 \qquad \lambda_3 = e^{j\frac{5\pi}{4}} = -e^{j\frac{\pi}{4}} \tag{4e}$$

$$n = 3 \qquad \lambda_4 = e^{j\frac{7\pi}{4}} = e^{-j\frac{\pi}{4}} \tag{4f}$$

The eigenvalue equation is

$$[S][x_i] = \lambda_i[x_i] \qquad i = 1, 2, 3, 4 \tag{5a}$$

For $\lambda = \lambda_1 = e^{j\frac{\pi}{4}}$,

$$\frac{1}{\sqrt{2}} \begin{bmatrix} 0 & 0 & 1 & j \\ 0 & 0 & j & 1 \\ 1 & j & 0 & 0 \\ j & 1 & 0 & 0 \end{bmatrix} \begin{bmatrix} x_{11} \\ x_{12} \\ x_{13} \\ x_{14} \end{bmatrix} = e^{j\frac{\pi}{4}} \begin{bmatrix} x_{11} \\ x_{12} \\ x_{13} \\ x_{14} \end{bmatrix} \tag{6a}$$

The expansion of the above relation yields

$$K_1(x_{13} + jx_{14}) = x_{11} \tag{6b}$$

$$K_1(jx_{13} + x_{14}) = x_{12} \tag{6c}$$

$$K_1(x_{11} + jx_{12}) = x_{13} \tag{6d}$$

$$K_1(jx_{11} + x_{12}) = x_{14} \tag{6e}$$

where

$$K_1 \equiv \frac{e^{-j\frac{\pi}{4}}}{\sqrt{2}}$$

(6f)

Multiplying (6c) by j and adding to (6b) yields

$$2jK_1x_{14} = x_{11} + jx_{12}$$

(7a)

From (6d), one has

$$x_{11} + jx_{12} = \frac{x_{13}}{K_1}$$

(7b)

The substitution of (7b) into (7a) yields

$$2jK_1^2x_{14} = x_{13}$$

Since $2jK_1^2 = 1$, one then has

$$x_{14} = x_{13}$$

(8)

Equations (8) and (6b) result in

$$K_1 (1 + j) x_{13} = x_{11} \qquad \text{or} \qquad x_{13} = x_{14} = x_{11}$$

(9)

since $K_1 (1 + j) = 1$.

Equations (8) and (6c) yield

$$K_1 (1 + j) x_{13} = x_{12} \qquad \text{or} \qquad x_{12} = x_{13}$$

(10)

Thus

$$x_{11} = x_{12} = x_{13} = x_{14}$$

(11)

and the eigenvector $[x_1]$ is

$$[x_1] = x_{11} [1 \quad 1 \quad 1 \quad 1]^t$$

(12)

For $\lambda = \lambda_2 = e^{-j\frac{\pi}{4}}$, the eigenvalue equation is

$$\frac{1}{\sqrt{2}} \begin{bmatrix} 0 & 0 & 1 & j \\ 0 & 0 & j & 1 \\ 1 & j & 0 & 0 \\ j & 1 & 0 & 0 \end{bmatrix} \begin{bmatrix} x_{21} \\ x_{22} \\ x_{23} \\ x_{24} \end{bmatrix} = e^{-j\frac{\pi}{4}} \begin{bmatrix} x_{21} \\ x_{22} \\ x_{23} \\ x_{24} \end{bmatrix} \tag{13}$$

The expansion of the above relation yields

$$K_2 (x_{23} + jx_{24}) = x_{21} \tag{14a}$$

$$K_2 (jx_{23} + x_{24}) = x_{22} \tag{14b}$$

$$K_2 (x_{21} + jx_{22}) = x_{23} \tag{14c}$$

$$K_2 (jx_{21} + x_{22}) = x_{24} \tag{14d}$$

where

$$K_2 \equiv \frac{e^{j\frac{\pi}{4}}}{\sqrt{2}} \tag{14e}$$

Equations (14a), (14b), and (14c) yield

$$2jK_2 x_{24} = x_{21} + jx_{22} = \frac{x_{23}}{K_2} \tag{15a}$$

with

$$2jK_2^2 = 2j \left[\frac{e^{j\frac{\pi}{4}}}{\sqrt{2}} \right]^2 = -1 \tag{15b}$$

one has

$$x_{24} = -x_{23} \tag{15c}$$

Then (14a) becomes

$$K_2 (1 - j) x_{23} = x_{21} \quad \text{or} \quad x_{23} = x_{21} = -x_{24} \tag{16a}$$

since

$$K_2 (1 - j) = (1 - j) \frac{1+j}{\sqrt{2}} \frac{1}{\sqrt{2}} = 1 \tag{16b}$$

Equation (14b) gives

$$K_2 (j - 1) x_{23} = x_{22} \quad \text{or} \quad -K_2 (1 - j) x_{23} = x_{22}$$

Thus

$$x_{22} = -x_{23} = -x_{21} = x_{24} \tag{17}$$

The eigenvector is

$$[x_2] = x_{21} [1 \quad -1 \quad 1 \quad -1]^t \tag{18}$$

The eigenvectors for λ_3 and λ_4 are

$$[x_3] = x_{31} [1 \quad -1 \quad -1 \quad 1]^t \tag{19}$$

$$[x_4] = x_{41} [1 \quad 1 \quad -1 \quad -1]^t \tag{20}$$

The normalized eigenvectors are therefore,

$$[e_1] = \frac{1}{\sqrt{4}} [1 \quad 1 \quad 1 \quad 1]^t \tag{21a}$$

$$[e_2] = \frac{1}{\sqrt{4}} [1 \quad -1 \quad 1 \quad -1]^t \tag{21b}$$

$$[e_3] = \frac{1}{\sqrt{4}} [1 \quad -1 \quad -1 \quad 1]^t \tag{21c}$$

$$[e_4] = \frac{1}{\sqrt{4}} [1 \quad 1 \quad -1 \quad -1]^t \tag{21d}$$

(c) The eigen-networks are determined by exciting the device with different eigenvectors.

The application of $[e_1]$ results in an open circuit along plane A-A' and B-B', Fig. 11-1a, and the corresponding eigen-network is as shown in Fig. 11-1b.

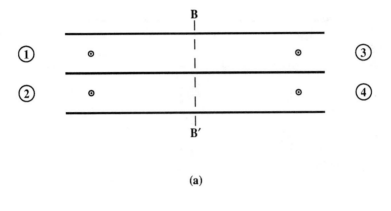

(a)

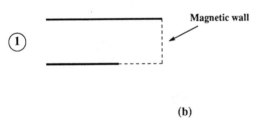

(b)

Figure 11-1: Excitation by $[e_1]$

318

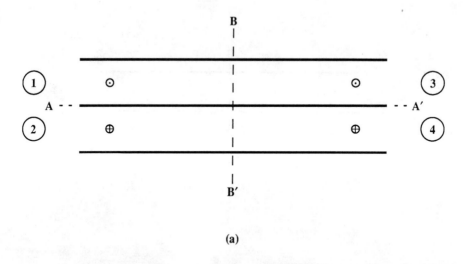

(a)

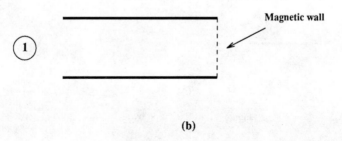

(b)

Figure 11-2: Excitation by [e_2]

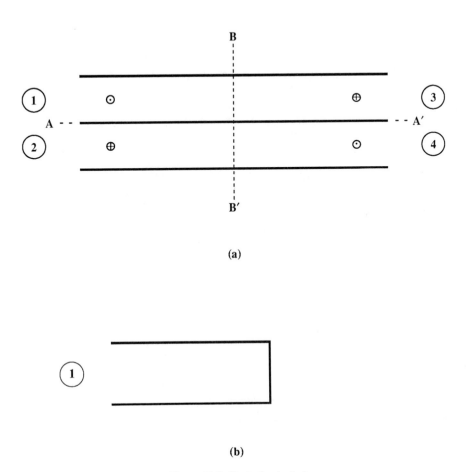

(a)

(b)

Figure 11-3: Excitation by [e₃]

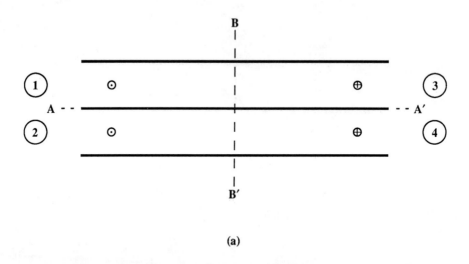

(a)

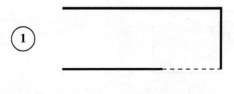

(b)

Figure 11-4: Excitation by [e₄]

The application of $[e_2]$, Fig. 11-2a requires an open circuit along plane B-B′ and a short circuit along A-A′, and the corresponding eigen-network is shown in Fig. 11-2b.

The eigen-networks for $[e_3]$ and $[e_4]$ are shown in Fig. 11-3 and Fig. 11-4, respectively.

(d) The scattering matrix of the coupler is obtained by diagonalization of the general scattering matrix for a symmetrical device with ports which are indistinguishable.

$$[S] = \begin{bmatrix} s_{11} & s_{12} & s_{13} & s_{14} \\ s_{12} & s_{11} & s_{14} & s_{13} \\ s_{13} & s_{14} & s_{11} & s_{12} \\ s_{14} & s_{13} & s_{12} & s_{11} \end{bmatrix} \tag{22}$$

Then

$$[D_\lambda] = [E]^t [S] [E]$$

$$= \begin{bmatrix} s_{11}+s_{12}+s_{13}+s_{14} & 0 & 0 & 0 \\ 0 & s_{11}-s_{12}+s_{13}-s_{14} & 0 & 0 \\ 0 & 0 & s_{11}-s_{12}-s_{13}+s_{14} & 0 \\ 0 & 0 & 0 & s_{11}+s_{12}-s_{13}-a_{14} \end{bmatrix} \tag{23a}$$

where

$$[E] = [\ [e_1] \quad [e_2] \quad [e_3] \quad [e_4] \ ]^t \tag{23b}$$

Since one also has

$$[D_\lambda] = \text{diag} [\ \lambda_1 \quad \lambda_2 \quad \lambda_3 \quad \lambda_4 \ ]^t \tag{23c}$$

therefore,

$$\lambda_1 = s_{11} + s_{12} + s_{13} + s_{14} \tag{24a}$$

$$\lambda_2 = s_{11} - s_{12} + s_{13} - s_{14} \tag{24b}$$

$$\lambda_3 = s_{11} - s_{12} - s_{13} + s_{14} \tag{24c}$$

$$\lambda_4 = s_{11} + s_{12} - s_{13} - s_{14} \tag{24d}$$

12. Problems

1. A symmetrical directional coupler is as shown in Fig. 12-1. The coupling holes are along plane P_1 and these holes are symmetrical with respect to plane P_2. The guide is filled with isotropic dielectric. Determine the scattering matrix.

2. If the directional coupler as described in Problem 1 is lossless, show that the scattering matrix has the following form.

$$[S] = \begin{bmatrix} 0 & \beta & 0 & \delta \\ \beta & 0 & \delta & 0 \\ 0 & \delta & 0 & \beta \\ \delta & 0 & \beta & 0 \end{bmatrix}$$

3. Determine the symmetry operators $[G_1]$ and $[G_2]$ corresponding to the planes of symmetry P_1 and P_2, respectively.

   Show that the linear combination of these symmetry operators will also commute with $[S]$.

4. Show that the matrix of eigenvectors of the above lossless directional coupler is

$$[E] = \frac{1}{2} \begin{bmatrix} 1 & 1 & 1 & 1 \\ 1 & -1 & 1 & -1 \\ 1 & 1 & -1 & -1 \\ 1 & -1 & -1 & 1 \end{bmatrix}$$

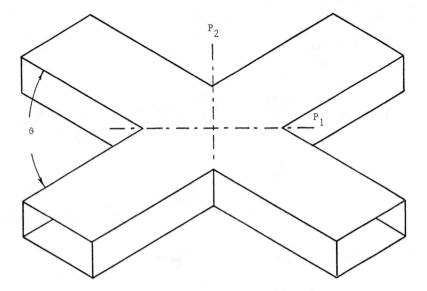

Figure 12-1: A symmetrical directional coupler

CHAPTER VII Impedance and Mode Transformers

1. Quarter-Wave Transformer

A microwave system usually involves several different types of waveguides, each with its own characteristic impedance and mode of propagation. One of the problems in microwave engineering is the design of the interface between different guides to optimize the overall system efficiency.

Quarter-wave transformers are widely used as the matching devices between guides with different characteristics - these can be guides filled with different dielectrics or guides with different physical dimensions or both.

The scattering matrix of a general two-port device, Fig. 1-1a, is given by

$$[S] = \begin{bmatrix} s_{11} & s_{12} \\ s_{21} & s_{22} \end{bmatrix} \tag{1a}$$

and

$$[b] = [S][a] \tag{1b}$$

$$b_1 = s_{11}a_1 + s_{12}a_2 \tag{1c}$$

$$b_2 = s_{21}a_1 + s_{22}a_2 \tag{1d}$$

where

$$[a] = [\, a_1 \quad a_2 \,]^t \qquad [b] = [\, b_1 \quad b_2 \,]^t \tag{1e}$$

When the output port is connected to a load z_L, Fig. 1-1b, then

$$s_L = \frac{b_L}{a_L} = \frac{a_2'}{b_2'} \qquad \text{or} \qquad a_2' = s_L b_2' \tag{2}$$

The primed quantities represent the parameters when the two-port device is terminated by a load. Then (1) becomes

$$[b'] = [S][a'] \tag{3a}$$

$$b_1' = s_{11}a_1' + s_{12}a_2' \tag{3b}$$

$$b_2' = s_{21}a_1' + s_{22}a_2' \tag{3c}$$

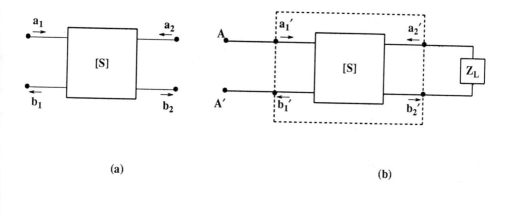

(a)

(b)

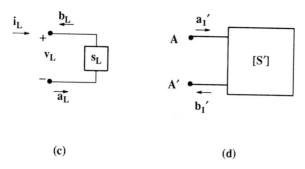

(c)

(d)

Figure 1-1: A two-port device

The substitution of (2) into (3c) yields

$$b_2' = s_{21}a_1' + s_{22}s_L b_2' = \frac{s_{21}a_1'}{1 - s_{22}s_L} \tag{4a}$$

and (3b) becomes

$$b_1' = s_{11}a_1' + s_{12}s_L b_2' = s_{11}a_1' + s_{12}s_L \frac{s_{21}a_1'}{1 - s_{22}s_L} \tag{4b}$$

If the two-port device, Fig. 1-1a, is a matched-terminated uniform waveguide of length L, then it is symmetrical and reciprocal; consequently [Section (III.2.1)]

$$s_{11} = s_{22} = 0$$

$$s_{12} = s_{21} = e^{-j\beta L} = e^{-j\theta} \qquad \theta \equiv \beta L \tag{IV.4.2-4}$$

Then its scattering matrix is given by [Eq. (1a)]

$$[S] = \begin{bmatrix} s_{11} & s_{12} \\ s_{12} & s_{11} \end{bmatrix} = \begin{bmatrix} 0 & e^{-j\theta} \\ e^{-j\theta} & 0 \end{bmatrix} \tag{5}$$

Equation (4b) then reduces to

$$b_1' = s_L e^{-j2\theta} a_1'$$

and

$$s_{11}' = \frac{b_1'}{a_1'} = s_L e^{-j2\theta} \tag{6}$$

where $\theta = \beta L$ and L is the length of the waveguide. s_{11}' is the reflection coefficient at the input of a uniform waveguide when its output is terminated by an impedance Z_L whose reflection coefficient is s_L, Fig. 1-1d.

Since the reflection coefficient is also given by (Fig. 1-2)

$$\rho = \frac{Z_{in} - Z_o}{Z_{in} + Z_o} = s_{11}' = \frac{\dfrac{Z_{in}}{Z_o} - 1}{\dfrac{Z_{in}}{Z_o} + 1} \tag{7}$$

where Z_o is the characteristic impedance of the guide connected to the input, therefore

$$\frac{Z_{in}}{Z_o} = \frac{1 + s_{11}'}{1 - s_{11}'} \tag{8}$$

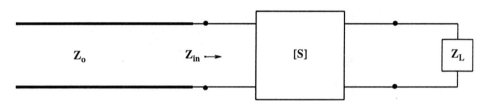

Figure 1-2: A terminated two-port network

Z_o can also be the characteristic impedance of the guide that made up the two-port device. Then s_{11}' is the reflection coefficient in the guide which is a continuation of the two-port device when the latter is terminated in Z_L. The input impedance can be expressed as [Eq. (6) and (8)]

$$Z_{in} = Z_o \frac{1 + s_L e^{-j2\theta}}{1 - s_L e^{-j2\theta}} \tag{9}$$

The voltage v_L and the current i_L at the load is related to the incident and reflected parameters by [Eq. (III.3-6)] (with appropriate change of notation)

$$\frac{v_L}{\sqrt{Z_o}} = a_L + b_L \tag{10a}$$

$$\sqrt{Z_o}\, i_L = a_L - b_L \tag{10b}$$

where the characteristic impedance of the guide is used as the normalizing resistance. The normalized load impedance is given by the ratio of (10a) to (10b).

$$\frac{Z_L}{Z_o} = \frac{a_L + b_L}{a_L - b_L} = \frac{1 + s_L}{1 - s_L} \tag{11a}$$

s_L can be expressed in terms of the load impedance.

$$s_L = \frac{Z_L - Z_o}{Z_L + Z_o} \tag{11b}$$

The substitution of (11b) into (9) yields

$$\frac{Z_{in}}{Z_o} = \frac{(Z_L + Z_o) + (Z_L - Z_o)\, e^{-j2\theta}}{(Z_L + Z_o) - (Z_L - Z_o)\, e^{-j2\theta}}$$

$$= \frac{Z_L (1 + e^{-j2\theta}) + Z_o (1 - e^{-j2\theta})}{Z_L (1 - e^{-j2\theta}) + Z_o (1 + e^{-j2\theta})}$$

$$= \frac{Z_L + jZ_o \tan\theta}{Z_o + jZ_L \tan\theta} \tag{12}$$

When the length of the guide is chosen to be $\lambda_g/4$,

$$\theta = \beta L = \frac{2\pi}{\lambda_g} \frac{\lambda_g}{4} = \frac{\pi}{2}$$

then (12) becomes

$$\frac{Z_{in}}{Z_o} = \frac{Z_o}{Z_L} \qquad \text{or} \qquad Z_o = \sqrt{Z_{in} Z_L} \qquad (13)$$

The relation between the characteristic, input, and load impedances of a lossless guide whose length is $\lambda_g/4$ is given by (13). In other words, a given load Z_L can be transformed into Z_{in} by a quarter-wavelength transformer whose characteristic impedance is chosen according to (13).

In the application of matching load to a lossless guide with characteristic impedance Z_{o1} = real, then $Z_{in} = Z_{o1}$ = real. Consequently, (13) is feasible only if $Z_L = R_L$ = real, since the quarter-wavelength transformer is assumed to be lossless in the analysis.

Figure 1-3 illustrates the matching of a load R_L to a guide Z_{o1} by a quarter-wavelength transformer whose characteristic impedance is Z_{o2}. The reflection coefficient in the main guide (guide 1, Fig. 1-3) is

$$\rho = \frac{Z_2 - Z_{o1}}{Z_2 + Z_{o1}}$$

with $Z_2 = Z_{in}$, given by (12), $Z_o = Z_{o2}$, and $Z_L = R_L$. Then

$$\rho \equiv \frac{Z_{o2}R_L + jZ_{o2}^2\tan\theta - Z_{o1}Z_{o2} - jZ_{o1}R_L \tan\theta}{Z_{o2}R_L + jZ_{o2}^2\tan\theta + Z_{o1}Z_{o2} + jZ_{o1}R_L\tan\theta}$$

$$= \frac{Z_{o2}(R_L - Z_{o1}) + j(Z_{o2}^2 - Z_{o1}R_L)\tan\theta}{Z_{o2}(R_L + Z_{o1}) + j(Z_{o2}^2 + Z_{o1}R_L)\tan\theta} \qquad (14)$$

When the operating frequency lies in the region adjacent to $L = \lambda_g/4$, (14) can be approximated by using (13),

$$\rho \approx \frac{R_L - Z_{o1}}{R_L + Z_{o1} + j2\sqrt{Z_{o1}R_L}\tan\theta} \qquad (15)$$

and the magnitude of ρ is

$$|\rho| = \frac{R_L - Z_{o1}}{\sqrt{(R_L + Z_{o1})^2 + (2\sqrt{Z_{o1}R_L}\tan\theta)^2}}$$

$$= \frac{R_L - Z_{o1}}{\sqrt{(R_L + Z_{o1})^2 + 4Z_{o1}R_L(\sec^2\theta - 1)}}$$

$$\approx \frac{R_L - Z_{o1}}{\sqrt{(R_L - Z_{o1})^2 + 4Z_{o1}R_L\sec^2\theta}}$$

$$|\rho| = \cfrac{1}{\sqrt{1 + \left[\cfrac{2\sqrt{Z_{o1}R_L} \ \sec \theta}{R_L - Z_{o1}}\right]^2}} \tag{16}$$

The magnitude of the reflection coefficient is periodic with frequency, since θ is periodic; $|\rho|$ increases rapidly as θ is moved away from $\pi/2$. The bandwidth of the device is determined by the maximum acceptable value of $|\rho|$. With $|\rho_{max}|$ specified, one can estimate the corresponding angle θ_{max} or f_M. The bandwidth is then given by

$$\delta f = 2 \, (f_o - f_M) \tag{17}$$

where f_o is the center frequency, for which $\theta = \lambda_g/4$.

It should be noted that in the above analysis it was assumed that the characteristic impedances Z_{o1} and Z_{o2} were independent of frequency. This is a good approximation for transmission lines, but the wave impedances of waveguides vary with frequency, thus making the analysis considerably more complex.

Further, reactive fields are excited at the junctions of discontinuities. These junction effects usually can be represented by shunt susceptance at each junction. The addition of such susceptances will alter the performance of the above analysis, which did not take them into account.

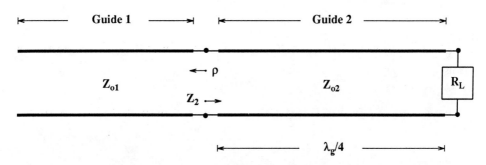

Figure 1-3: A quarter-wave transformer

2. Small-Reflection Theory

The bandwidth of a quarter-wave transformer can be improved by using a number of them in cascade. Approximation techniques for analyzing a device with multiple reflecting segments will now be developed.

Two reflecting planes, a-a' and b-b', separating regions with different characteristic impedances are shown in Fig. 2-1. The distance between these planes is L or its electrical length is $\theta = \beta L$. The reflection coefficients and transmission coefficients at each plane are as follows.

$$\rho_{a1} = \frac{Z_{o2} - Z_{o1}}{Z_{o2} + Z_{o1}} \tag{1a}$$

$$\rho_{a2} = \frac{Z_{o1} - Z_{o2}}{Z_{o1} + Z_{o2}} = -\rho_{a1} \tag{1b}$$

$$\tau_{a2} = 1 + \rho_{a1} = \frac{2Z_{o2}}{Z_{o1} + Z_{o2}} \tag{1c}$$

$$\tau_{a1} = 1 + \rho_{a2} = \frac{2Z_{o1}}{Z_{o1} + Z_{o2}} \tag{1d}$$

$$\rho_b = \frac{Z_{o3} - Z_{o2}}{Z_{o3} + Z_{o2}} \tag{1e}$$

When a wave of amplitude M is incident on surface a-a', it will be partially reflected and partially transmitted. The reflected portion is $\rho_{a1}M$, and the transmitted portion is $\tau_{a2}M$.

When the transmitted wave $\tau_{a2}M$ reaches the surface b-b', a portion of it will be reflected and its amplitude is $\rho_b\tau_{a2}Me^{-j\theta}$. Region 3 is assumed to extend to infinity.

As this second reflected wave reaches surface a-a', it will generate two waves:

(a) A transmitted wave into region 1 with an amplitude

$$\tau_{a1}\rho_b\tau_{a2}Me^{-j2\theta} = \tau_{a1}\tau_{a2}\rho_bMe^{-j2\theta}$$

(b) A reflected wave with an amplitude

$$\rho_{a2}\rho_b\tau_{a2}Me^{-j2\theta}$$

This process will continue indefinitely, and the sequence of reflections and transmissions is depicted in Fig. 2-1.

The total wave reflected back into region 1 is given by the sum of individual partial reflections.

332

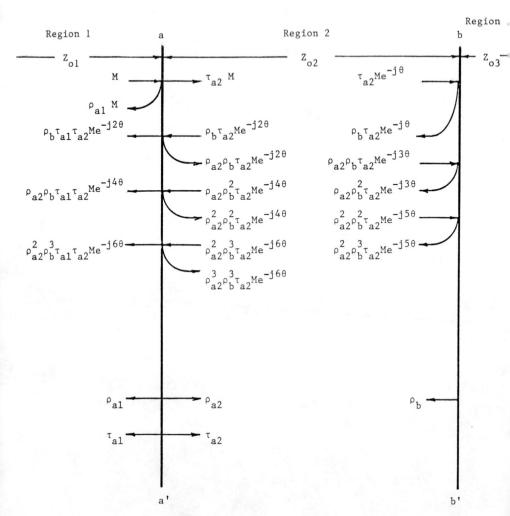

Figure 2-1: Regions of discontinuities

$$\rho M = \left[\rho_{a1} + \rho_b \tau_{a1} \tau_{a2} e^{-j2\theta} + \rho_{a2} \rho_b^2 \tau_{a1} \tau_{a2} e^{-j4\theta} \right.$$

$$\left. + \rho_{a2}^2 \rho_b^3 \tau_{a1} \tau_{a2} e^{-j6\theta} + \cdots \right] M$$

$$= \left[\rho_{a1} + \tau_{a1} \tau_{a2} \rho_b e^{-j2\theta} \left(1 + \rho_{a2} \rho_b e^{-j2\theta} \right. \right.$$

$$\left. \left. + (\rho_{a2} \rho_b)^2 e^{-j4\theta} + \cdots \right) \right] M$$

or

$$\rho = \rho_{a1} + \tau_{a1} \tau_{a2} \rho_b e^{-j2\theta} \sum_{m=0}^{\infty} \left[\rho_{a2} \rho_b e^{-j2\theta} \right]^m \tag{2}$$

But

$$\sum_{m=0}^{\infty} x^m = \frac{1}{1 - x} \tag{3}$$

Therefore, (2) becomes

$$\rho = \rho_{a1} + \frac{\tau_{a1} \tau_{a2} \rho_b e^{-j2\theta}}{1 - \rho_{a2} \rho_b e^{-j2\theta}} \tag{4}$$

Using the relations in (1) to eliminate τ_{a1}, τ_{a2}, and ρ_{a2}, one gets

$$\rho = \rho_{a1} + \frac{(1 - \rho_{a1})(1 + \rho_{a1}) \rho_b e^{-j2\theta}}{1 + \rho_{a1} \rho_b e^{-j2\theta}} = \frac{\rho_{a1} + \rho_b e^{-j2\theta}}{1 + \rho_{a1} \rho_b e^{-j2\theta}} \tag{5}$$

For small reflections, $\rho_{a1} < 0.1$ and $\rho_b < 0.1$, a first-order approximation of (5) may be obtained by neglecting the product terms, since

$$\rho_{a1} \rho_b < 0.01$$

Then (5) becomes

$$\rho \approx \rho_{a1} + \rho_b e^{-j2\theta} \tag{6}$$

For small reflections, the first-order approximation is given by the sum of each individual reflection with appropriate phase delay. The interactions between separate discontinuities are neglected.

In terms of the scattering matrix in Fig. 2-2, (6) can be expressed as

$$s_{11}' \approx s_{11} + s_L e^{-j2\theta} \tag{7}$$

$$[S] = \begin{bmatrix} s_{11} & s_{12} \\ s_{21} & s_{22} \end{bmatrix}$$

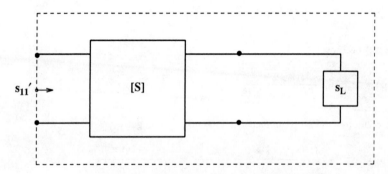

Figure 2-2: A one-port representation for Figure 2-1

3. Multistep Impedance Transformer

(a) A number of quarter-wave transformers are connected in tandem as shown in Fig. 3-1. The quarter-wave transformers are made of lossless guides with characteristic resistance R_k, $k = 0, 1, 2, \ldots, m$. The reflection coefficient at the kth junction is

$$\rho_k = \frac{R_{k+1} - R_k}{R_{k+1} + R_k} \qquad k = 0, 1, 2, \ldots, m-1. \tag{1a}$$

and for the last section,

$$\rho_m = \frac{R_L - R_m}{R_L + R_m} \tag{1b}$$

Each section has the same length, equal to one-quarter of the wavelength at the center frequency, f_o.

The total reflection coefficient is given approximately by the small-reflection theory [Eq. (2-6)].

$$\rho = \rho_0 + \rho_1 e^{-j2\theta} + \rho_2 e^{-j4\theta} + \cdots + \rho_m e^{-j2m\theta} \tag{2a}$$

$$\text{for } \rho_k < 0.1 \qquad k = 0, 1, 2, \ldots, m \tag{2b}$$

and $\theta = \beta L$.

(b) It is convenient to arrange the transformers such that they are symmetrical with respect to the mid-plane. That is,

$$\rho_0 = \rho_m, \quad \rho_1 = \rho_{m-1}, \quad \rho_2 = \rho_{m-2}, \quad \cdots$$

Then (2) becomes,

$$\rho = e^{-jm\theta} \left[\rho_0 \left(e^{jm\theta} + e^{-jm\theta} \right) + \rho_1 \left(e^{j(m-2)\theta} + e^{-j(m-2)\theta} \right) + \cdots \right]$$

$$= 2e^{-jm\theta} \left[\rho_0 \cos m\theta + \rho_1 \cos (m-2)\theta + \cdots + \frac{1}{2}\rho_{\frac{m}{2}} \right] \qquad m = \text{even} \tag{3a}$$

$$= 2e^{-jm\theta} \left[\rho_0 \cos m\theta + \rho_1 \cos (m-2)\theta + \cdots + \rho_{\frac{m-1}{2}} \cos \theta \right] \qquad m = \text{odd} \tag{3b}$$

The resultant reflection coefficient is thus expressed in the form of a trigonometric series, with the parameter $\theta = \beta L$, which is frequency dependent. By Fourier theory, one could obtain certain ρ as a function of frequency by proper choice of the coefficients ρ_k, $k = 0, 1, 2, \ldots, m/2$. Two most popular preferences are the maximally flat and equal-ripple pass-band transformers. These will be discussed in the following sections.

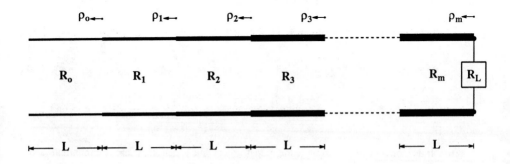

$$L = \lambda/4$$

Figure 3-1: Quarter-wave transformers in tandem

4. Maximally Flat Transformer

(a) The scattering parameters of a multistep transformer, Fig. 4-1, are found to be [Eq. (3.2)]

$$S = s_0 + s_1 e^{-j2\theta} + s_2 e^{-j4\theta} + \cdots + s_m e^{-j2m\theta} \qquad \theta = \beta L \qquad (1)$$

where reflection coefficients ρ_i are replaced by scattering coefficients s_i.

For a symmetrical m-step transformer, (1) can be expressed as [Eq. (3.3)]

$$S = 2e^{-jm\theta} \left[s_0 \cos m\theta + s_1 \cos (m-2)\theta + \cdots + s_k \cos (m-2k)\theta \right] \qquad (2a)$$

$$k = \frac{m}{2} \quad \text{and} \quad s_k = \frac{1}{2} s_{\frac{m}{2}} \qquad \text{for} \quad m = \text{even} \qquad (2b)$$

$$k = \frac{m-1}{2} \qquad \text{for} \quad m = \text{odd} \qquad (2c)$$

Equation (2) can be expressed in terms of $\cos \theta$ by the following trigonometric identities (Section 4.1).

$$\cos 2\theta = 2 \cos^2 \theta - 1 \qquad (3a)$$

$$\cos 3\theta = 4 \cos^3 \theta - 3 \cos \theta \qquad (3b)$$

$$\cos 4\theta = 8 \cos^4 \theta - 8 \cos^2 \theta + 1 \qquad (3c)$$

Therefore, (2a) can be written in the form of a power series in $\cos \theta$.

$$S = A_0 + A_1 \cos \theta + A_2 \cos^2 \theta + \cdots + A_m \cos^m \theta \qquad (4)$$

where the A_j's are some new constants made up of s_k's.

(b) The ideal $|S|$-function for a multistep transformer would have S remaining flat zero over the entire band as specified. This requires that the slope of such a response curve should be zero; that is, the first derivative of S with respect to its parameter should vanish.

If the curvature of the response is also made equal to zero, the flatness of the curve will improve, i.e., the second derivative of S should also vanish.

By the same argument, the higher order of the rate of change of $|S|$ should also be required to be zero to further improve the flatness, i.e., this requires

$$\frac{d^k |S(p)|}{dp^k} = 0$$

for $k = 1, 2, \ldots, m-1$, and $p = \cos \theta$ is the parameter of S.

By setting the derivatives of $|S|$ with respect to its parameters equal to zero, the result is

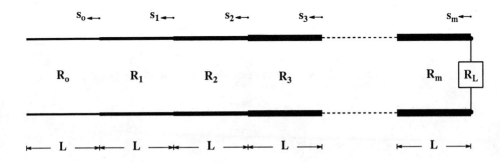

$$L = \lambda/4$$

Figure 4-1: A multi-step transformer

$$A_1 = A_2 = A_3 = \cdots = A_{m-1} = 0 \tag{5}$$

Requiring the mth derivative to be zero will force $A_m = 0$ and will create $S = A_o = $ constant, which is an all-pass transformer. This is not an acceptable approximation.

The parameter of S, (4), is $\cos \theta$ and the first derivative is

$$\frac{dS}{d(\cos \theta)} = A_1 + 2A_2 \cos \theta + 3A_3 \cos^2 \theta + \cdots + mA_m \cos^{m-1} \theta \tag{6}$$

At the center frequency, f_o, $L = \lambda_o/4$, $\theta = \theta_o = \beta_o L = \pi/2$, and $\cos \theta_o = 0$. Therefore,

$$\left[\frac{dS}{d(\cos \theta)} \right]_{\theta=\theta_o} = 0 = A_1 \tag{7}$$

The second derivative is

$$\left[\frac{d^2S}{d(\cos \theta)^2} \right]_{\theta=\theta_o} = 0 = 2A_2 + 6A_3 \cos \theta_o + \cdots + m(m-1)A_m \cos^{m-2} \theta_o$$

or

$$A_2 = 0 \tag{8}$$

The continuation of this process to the $(m-1)$th derivative will result in

$$A_1 = A_2 = A_3 = \cdots = A_{m-1} = 0 \tag{9}$$

With the result of (9), (4) becomes

$$S = A_o + A_m \cos^m \theta \tag{10}$$

The scattering parameter should be zero at the center frequency, f_o ($\theta = \theta_o$). Therefore,

$$S(\theta = \theta_o) = 0 = A_o \tag{11}$$

The final form of the scattering parameter is

$$S = A_m \cos^m \theta = A_m \left[\frac{1}{2} (e^{j\theta} + e^{-j\theta}) \right]^m$$

$$= \frac{A_m}{2^m} e^{jm\theta} [1 + e^{-j2\theta}]^m \tag{12a}$$

$$|S| = \frac{A_m}{2^m} \left| [1 + e^{-j2\theta}]^m \right| \tag{12b}$$

The constant A_m can be determined by imposing the condition that when $\theta = 0$ or $L = 0$, the scattering parameter is

$$S(\theta = 0) = \frac{Z_L - Z_0}{Z_L + Z_0} \tag{13}$$

From (12), one has

$$S(\theta = 0) = \frac{A_m}{2^m} 2^m = A_m \tag{14}$$

Equating (13) with (14) yields

$$A_m = \frac{Z_L - Z_0}{Z_L + Z_0} \tag{15}$$

Hence

$$S = \frac{1}{2^m} \frac{Z_L - Z_0}{Z_L + Z_0} [1 + e^{-j2\theta}]^m \tag{16}$$

To make (16) have the same form as (1), (16) is expanded by the binomial theorem.

$$S = \frac{1}{2^m} \frac{Z_L - Z_0}{Z_L + Z_0} \sum_{k=0}^{m} C_k^m e^{-j2k\theta} \tag{17a}$$

$$C_k^m = \frac{m!}{(m-k)! \, k!} = \text{binomial coefficients} \tag{17b}$$

A direct comparison of (17a) with (1) shows that

$$s_k = \frac{C_k^m}{2^m} \frac{Z_L - Z_0}{Z_L + Z_0} \tag{18}$$

If the scattering parameters of a multistep transformer are arranged to satisfy (18), then one would have a maximally flat transformer. Since the coefficients of the polynomial are proportional to the binomial coefficients, this type of transformer is also known as a binomial transformer.

(c) A further approximation may be used to simplify the impedance ratio. The series expansion of a logarithmic function is

$$\log x = 2 \left[\frac{x-1}{x+1} + \frac{(x-1)^3}{3(x+1)^3} + \frac{(x-1)^5}{5(x+1)^5} + \cdots \right] \qquad x > 0 \tag{19}$$

For $x = Z_L/Z_0$, then one has

$$\log \frac{Z_L}{Z_o} = \left[\frac{Z_L - Z_o}{Z_L + Z_o} + \frac{(Z_L - Z_o)^3}{3(Z_L + Z_o)^3} + \cdots \right]$$

$$\approx 2 \frac{Z_L - Z_o}{Z_L + Z_o} \quad \text{for} \quad (Z_L - Z_o)^2 \ll (Z_L + Z_o)^2 \tag{20}$$

Then (18) becomes

$$2s_k = \frac{C_k^m}{2^m} \log \frac{Z_L}{Z_o} \tag{21}$$

The scattering parameter is also given by

$$s_k = \frac{Z_{k+1} - Z_k}{Z_{k+1} + Z_k} \approx \frac{1}{2} \log \frac{Z_{k+1}}{Z_k} \tag{22}$$

The combination of (21) and (22) gives

$$\log \frac{Z_{k+1}}{Z_k} = \frac{C_k^m}{2^m} \log \frac{Z_L}{Z_o} \tag{23}$$

This equation provides a solution in terms of the logarithm of impedances.

(d) As an example consider a two-step transformer, m = 2, Fig. 4-2. Equation (23) yields:
For k = 0:

$$\log \frac{Z_1}{Z_o} = \frac{C_0^2}{2^2} \log \frac{Z_L}{Z_o}$$

$$C_0^2 = \frac{2!}{(2 - 0)! \, 0!} = 1$$

$$\log \frac{Z_1}{Z_o} = \frac{1}{4} \log \frac{Z_L}{Z_o}$$

$$\frac{Z_1}{Z_o} = \left[\frac{Z_L}{Z_o} \right]^{\frac{1}{4}}$$

$$Z_1 = Z_L^{\frac{1}{4}} Z_o^{\frac{3}{4}} \tag{24}$$

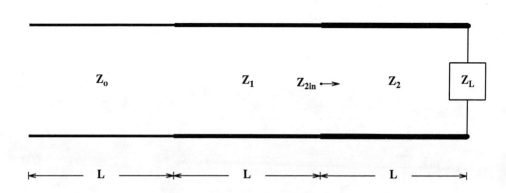

$$L = \lambda/4$$

Figure 4-2: A two-step transformer

For k = 1:

$$\log \frac{Z_2}{Z_1} = \frac{C_1^2}{2^2} \log \frac{Z_L}{Z_o}$$

$$C_1^2 = \frac{2!}{(2-1)!\,1!} = 2$$

$$\log \frac{Z_2}{Z_1} = \frac{1}{2} \log \frac{Z_L}{Z_o}$$

or

$$Z_2 = Z_1 \sqrt{\frac{Z_L}{Z_o}} = \left[Z_L^{\frac{1}{4}} Z_o^{\frac{3}{4}} \right] \sqrt{\frac{Z_L}{Z_o}}$$

$$= Z_L^{\frac{3}{4}} Z_o^{\frac{1}{4}} \tag{25}$$

The exact solution of the problem (Fig. 4-2) is [Eq. (1-13)]

$$Z_2 = \sqrt{Z_{2in} Z_L} \quad \text{or} \quad Z_{2in} = \frac{Z_2^2}{Z_L} \tag{26}$$

$$Z_1 = \sqrt{Z_o Z_{2in}} \quad \text{or} \quad Z_{2in} = \frac{Z_1^2}{Z_o} \tag{27}$$

The combination of (26) and (27) yields

$$\frac{Z_o}{Z_L} = \left[\frac{Z_1}{Z_2} \right]^2 \qquad \text{exact solution} \tag{28}$$

From the approximate solution, (24) and (25), one has

$$\frac{Z_1}{Z_2} = \frac{Z_L^{\frac{1}{4}} Z_o^{\frac{3}{4}}}{Z_L^{\frac{3}{4}} Z_o^{\frac{1}{4}}} = \sqrt{\frac{Z_o}{Z_L}} \qquad \text{approximate solution} \tag{29}$$

The approximate solution satisfies the condition of the exact solution (28).

(e) The bandwidth of a maximally flat transformer can be evaluated for a maximum tolerable scattering parameter S_M. The corresponding angle θ_M is given by (12), (15), and (20).

$$S = A_m \cos^m \theta = \frac{Z_L - Z_0}{Z_L + Z_0} \cos^m \theta$$

$$= \frac{1}{2} \log \frac{Z_L}{Z_0} \cos^m \theta$$

The maximum tolerable scattering parameter S_M is defined by the relation

$$S_M = \frac{1}{2} \log \frac{Z_L}{Z_0} \cos^m \theta_M$$

The corresponding maximum tolerable electrical length θ_M is

$$\theta_M = \cos^{-1} \left[\frac{2S_M}{\log \dfrac{Z_L}{Z_0}} \right]^{\frac{1}{m}} \tag{30}$$

and

$$\theta = \beta L = \frac{2\pi f}{v_0} \frac{\lambda_g}{4} = \frac{\pi f}{\dfrac{2v_0}{\lambda_g}}$$

Since $\lambda_g = v_0/f_0$, hence

$$\frac{f}{f_0} = \frac{2\theta}{\pi}$$

Similarly,

$$\frac{f_m}{f_0} = \frac{2\theta_M}{\pi}$$

where f_0 is the center frequency and f_m is the frequency corresponding to $\theta = \theta_M$.

The bandwidth is

$$\Delta f = 2 (f_0 - f_m)$$

The bandwidth normalized with respect to the center frequency is

$$\frac{\Delta f}{f_o} = \frac{2(f_o - f_m)}{f_o} = 2 - \frac{2f_m}{f_o} = 2 - \frac{4\theta_M}{\pi}$$

$$= 2 - \frac{4}{\pi} \cos^{-1} \left[\frac{2S_M}{\log \dfrac{Z_L}{Z_o}} \right]^{\frac{1}{m}} \tag{31}$$

The normalized bandwidth of a multistep transformer is given by (31) as a function of the maximum tolerable scattering parameter S_M.

4.1. Example: Expansion of cos mθ

Show that $\cos m\theta$ can be expressed in terms of powers of $\cos \theta$.

Solution:

By the use of the following trigonometric identities

$$\cos (a + b) = \cos a \cos b - \sin a \sin b \tag{1}$$

$$\cos (a - b) = \cos a \cos b + \sin a \sin b \tag{2}$$

one obtains

$$\cos (k + 1) \theta = \cos k\theta \cos \theta - \sin k\theta \sin \theta \tag{3}$$

$$\cos (k - 1) \theta = \cos k\theta \cos \theta + \sin k\theta \sin \theta \tag{4}$$

The sum of (3) and (4) yields

$$\cos (k + 1)\theta + \cos (k - 1)\theta = 2 \cos k\theta \cos \theta$$

or

$$\cos (k + 1)\theta = 2 \cos k\theta \cos \theta - \cos (k - 1)\theta \tag{5}$$

Let $m = k + 1$; then

$$\cos m\theta = 2 \cos (m - 1)\theta \cos \theta - \cos (m - 2)\theta \tag{6}$$

One can employ Eq. (6) to evaluate $\cos m\theta$ for any value of m provided $\cos (m - 1)\theta$ and $\cos (m - 2)\theta$ are known.

Let $x = \cos \theta$; then

$m = 0 \qquad \cos m\theta = \cos 0 = 1$

$m = 1 \qquad \cos \theta = x$

$m = 2 \qquad \cos 2\theta = 2 \cos^2 \theta - 1 = 2x^2 - 1$

$m = 3 \qquad \cos 3\theta = 2 \cos 2\theta \cos \theta - \cos \theta = 2(2x^2 - 1)x - x = 4x^3 - 3x$

$m = 4 \qquad \cos 4\theta = 2 \cos 3\theta \cos\theta - \cos 2\theta = 8x^4 - 8x^2 + 1$

$m = 5 \qquad \cos 5\theta = 16x^5 - 20x^3 + 5x$

5. Chebycheff Transformer

In the maximally flat transformer, the matching deteriorates as the frequency departs from the center frequency and the performance is worse at the edge of the pass-band. An alternative design is to allow the scattering parameter S to vary between $\pm S_M$ in an oscillatory sense over the pass-band. Such a design is also known as equal-ripple response, and it provides an increase in bandwidth over the maximally flat design.

The equal-ripple characteristics are provided by making use of the Chebycheff polynomials. The scattering parameter of an m-step transformer is found to be [Eq. (4.1)]

$$S = s_0 + s_1 e^{-j2\theta} + s_2 e^{-j4\theta} + \cdots + s_m e^{-j2m\theta}$$

$$= e^{-jm\theta} \left[s_0 e^{jm\theta} + s_1 e^{j(m-2)\theta} + s_2 e^{-j(m-4)\theta} + \cdots + s_m e^{-jm\theta} \right] \tag{1}$$

where $\theta = \beta L$ is the electrical length of each section. For a symmetrical m-step transformer, i.e.,

$$s_0 = s_m, \quad s_1 = s_{m-1}, \quad \cdots \tag{2}$$

(1) becomes [Eq. (3.3)]

$$S = 2e^{-jm\theta} \left[s_0 \cos m\theta + s_1 \cos (m-2)\theta + \cdots + s_k \cos (m-2k)\theta \right] \tag{3a}$$

where

$$k = \frac{m}{2} \quad \text{and} \quad s_k = \frac{\frac{s_m}{2}}{2} \quad \text{for} \quad m = \text{even} \tag{3b}$$

$$k = \frac{m-1}{2} \quad \text{for} \quad m = \text{odd} \tag{3c}$$

A Chebycheff polynomial is defined as (Section 5.1)

$$C_m(x) = \cos (m \cos^{-1} x) \quad \text{for} \quad |x| \le 1 \tag{4a}$$

$$= \cosh (m \cosh^{-1} x) \quad \text{for} \quad |x| \ge 1 \tag{4b}$$

$$x \equiv \cos \theta \quad \text{or} \quad \theta = \cos^{-1} x \tag{4c}$$

Hence, (3) can be expressed in terms of Chebycheff polynomials.

$$S = 2e^{-jm\theta} \left[A_0 C_m(\cos \theta) + A_1 C_{m-2}(\cos \theta) + \cdots + A_k C_{m-2k}(\cos \theta) \right] \tag{5}$$

where the coefficients A_i are to be determined.

The recursion relation, (5.1-4f), shows that the higher-order Chebycheff polynomials contain all lower-order polynomials. Further, in the present problem, either odd- or even-order terms are involved; therefore, only the highest-order term need be retained.

$$S = e^{-jm\theta}AC_m(\cos\theta) \tag{6}$$

The variable of the Chebycheff polynomial should be normalized, since the available values, curves, or tables, of Chebycheff polynomials are based upon normalized variables.

Let the specification be such that over the bandwidth the maximum $|S|$ is not to exceed S_M. The corresponding bandwidth is

$$\theta_M \leq \theta \leq \pi - \theta_M \tag{7}$$

and the mid-band is at $\theta = \pi/2$.

It is convenient to use $\cos\theta_M$ as the normalizing factor.

$$S = e^{-jm\theta}AC_m\left[\frac{\cos\theta}{\cos\theta_M}\right] = e^{-jm\theta}AC_m(\cos\theta\sec\theta_M) \tag{8}$$

The coefficient A can be determined by imposing the condition at $\theta = 0$:

$$S(\theta = 0) = \frac{Z_L - Z_o}{Z_L - Z_o} = \left[e^{-jm\theta}AC_m(\cos\theta\sec\theta_M)\right]_{\theta=0}$$

or

$$A = \frac{Z_L - Z_o}{(Z_L + Z_o)\,C_m(\sec\theta_M)} \tag{9}$$

Consequently, (8) becomes

$$S = e^{-jm\theta}\frac{Z_L - Z_o}{Z_L + Z_o}\frac{C_m(\cos\theta\sec\theta_M)}{C_m(\sec\theta_M)} \tag{10}$$

The maximum value of $C_m(\cos\theta\sec\theta_M)$ is unity in the pass-band; hence

$$S_M = \max|S| = \frac{Z_L - Z_o}{(Z_L + Z_o)\,C_m(\sec\theta_M)} \tag{11}$$

The scattering parameter may be expressed in terms of S_M by substituting (11) into (10).

$$S = S_M e^{-jm\theta}C_m(\cos\theta\sec\theta_M) \tag{12}$$

The angle θ_M at the band edge is related to S_M by (11).

$$C_m(\sec\theta_M) = \frac{Z_L - Z_o}{(Z_L + Z_o)\,S_M} \tag{13}$$

But from (4b) one has

$$C_m(\sec \theta_M) = \cosh [m \cosh^{-1} (\sec \theta_M)] \tag{14}$$

Then

$$\cosh [m \cosh^{-1} (\sec \theta)] = \frac{Z_L - Z_o}{(Z_L + Z_o) S_M}$$

$$\sec \theta = \cosh \left[\frac{1}{m} \cosh^{-1} \frac{Z_L - Z_o}{(Z_L + Z_o) S_M} \right] \tag{15}$$

The design procedure is to equate (3a) with (12).

$$S = 2e^{-jm\theta} [s_o \cos m\theta + s_1 \cos (m-2)\theta + \cdots + s_k \cos (m-2k)\theta]$$

$$= S_M e^{-jm\theta} C_m(\cos \theta \sec \theta_M)$$

or

$$2 [s_o \cos m\theta + s_1 \cos (m-2)\theta + \cdots + s_k \cos (m-2k)\theta] = S_M C_m(\cos \theta \sec \theta_M) \tag{16a}$$

$$k = \frac{m}{2} \quad \text{and} \quad s_k = \frac{\dfrac{s_m}{2}}{2} \quad \text{for} \quad m = \text{even} \tag{16b}$$

$$k = \frac{m-1}{2} \quad \text{for} \quad m = \text{odd} \tag{16c}$$

Next, expand C_m for the given number m, and the value of s_i is determined by identifying the coefficients of $\cos p\theta$ terms on both sides of (16a).

5.1. Chebycheff Polynomials

(a) Chebycheff polynomials of order n are defined as

$$C_n(x) = \cos(n \cos^{-1} x) \qquad \text{for} \qquad -1 \le x \le 1 \tag{1a}$$

$$= \cosh(n \cosh^{-1} x) \qquad \text{for} \qquad |x| \ge 1 \tag{1b}$$

Let $x = \cos\theta$; then

$$C_n(\cos\theta) = \cos[n \cos^{-1}(\cos\theta)] = \cos n\theta \tag{2}$$

The trigonometric function $\cos n\theta$ can be expressed as [Eq. (4.1-6)]

$$\cos n\theta = 2 \cos(n-1)\theta \cos\theta - \cos(n-2)\theta \tag{3}$$

Then, Chebycheff polynomials are given by

$$C_0(x) = 1 \tag{4a}$$

$$C_1(x) = \cos\theta = x \tag{4b}$$

$$C_2(x) = \cos 2\theta = 2x^2 - 1 \tag{4c}$$

$$C_3(x) = \cos 3\theta = 4x^3 - 3x \tag{4d}$$

$$C_4(x) = \cos 4\theta = 8x^4 - 8x^2 + 1 \tag{4e}$$

Equation (3) also provides the recursion relation.

$$C_n(x) \equiv 2xC_{n-1}(x) - C_{n-2}(x) \tag{4f}$$

(b) In the region $|x| \le 1$, $x = \cos\theta$, one has

$$C_n(\cos\theta) = \cos n\theta \le 1 \tag{5}$$

$C_n(x)$ is real and has a magnitude no greater than unity.

(c) When $|x| > 1$, then θ becomes imaginary. The polynomials of (4) are still valid.

(d) For odd (even) values of n, $C_n(x)$ is an odd (even) function of x. When the value of x is large, only the highest-degree term of the polynomials need be retained.

(e) In the region $-1 \leq x \leq 1$, the complex behavior of $C_n(x)$ can be obtained graphically (Fig. 5.1-1).

(e.1) Plot the $\cos \theta$ vs. θ curve, Fig. 5.1-1. The $C_1(x)$ vs. x curve is a straight line of unit slope, Fig. 5.1-2a, since $x = \cos \theta$, and from (4b),

$$C_1(x) \; = \; C_1(\cos \theta) \; = \; \cos \theta$$

(e.2) Plot $\cos 2\theta$ vs. θ on the same graph of $\cos \theta$ vs. θ, Fig. 5.1-1. For each value of $x = \cos \theta$, one can find the corresponding value of $\cos 2\theta$. Therefore, one can construct the curve (Fig. 5.1-2b)

$$C_2(x) \; = \; \cos 2\theta \;\; \text{vs.} \;\; x$$

(e.3) The above procedure can be repeated for values of $n = 3, 4, \ldots$.

(f) Some properties of Chebycheff polynomials are:

(f.1) $C_n(x)$ is always equal to $+1$ at $x = +1$.

(f.2) For n = even, $C_n(x)$ is always equal to $+1$ at $x = -1$.

For n = odd, $C_n = -1$ at $x = -1$.

(f.3) $C_n(x)$ has n nulls in the interval $-1 < x < 1$, for n either even or odd.

(f.4) For n = odd, $C_n(x) = 0$ at $x = 0$.

For n = even, $C_n(x) = +1$ at $x = 0$ if $(n/4)$ = integer; and $C_n(x) = -1$ at $x = 0$ if $(n/4) \neq$ integer.

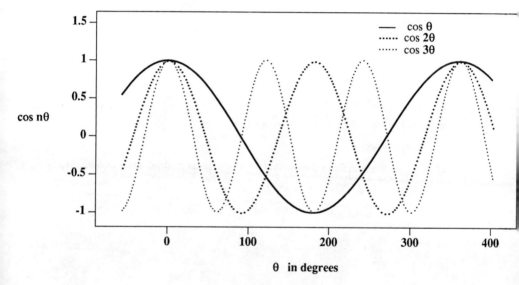

Figure 5.1-1: cos nθ vs θ

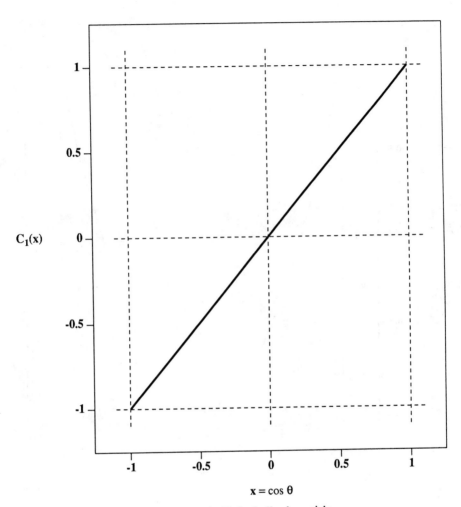

Figure 5.1-2(a): Chebycheff polynomials

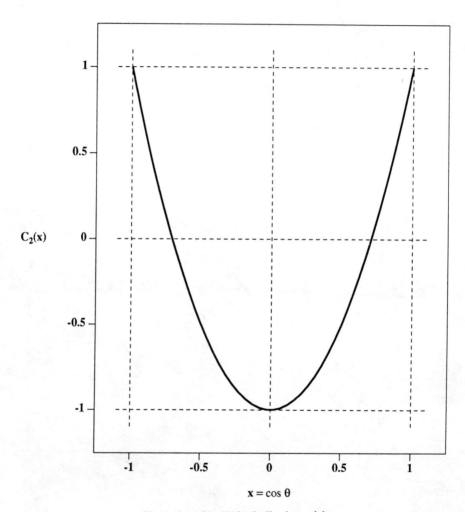

$$x = \cos \theta$$

Figure 5.1-2(b): Chebycheff polynomials

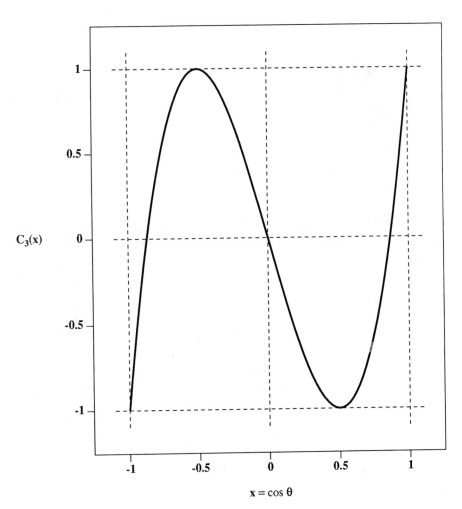

Figure 5.1-2(c): Chebycheff polynomials

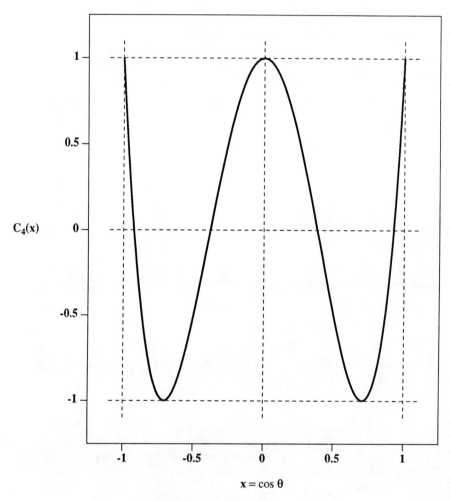

Figure 5.1-2(d): Chebycheff polynomials

5.2. Example: Chebycheff Transformer

Design a 2-section Chebycheff transformer (Fig. 5.2-1) with

$$S_M = 0.05 \qquad Z_o = 1 \text{ ohm} \qquad Z_L = 2 \text{ ohms}$$

and determine the bandwidth of the transformer.

Solution:

Equating (5-12) with (5-3a) and setting m = 2,

$$S_M C_2(\cos \theta \sec \theta_M) = 2s_o \cos 2\theta + s_1 \tag{1}$$

By Eq. (5-13),

$$C_m(\sec \theta_M) = \frac{Z_L - Z_o}{S_M (Z_L + Z_o)}$$

$$C_2(\sec \theta_M) = \frac{2 - 1}{0.05 (2 + 1)} = 6.667 \tag{2}$$

From Eq. (5.1-4c),

$$C_2(\sec \theta_M) = 2 (\sec \theta_M)^2 - 1 = 6.667$$

$$\sec \theta_M = \sqrt{\frac{1}{2} (6.667 + 1)} = \pm 1.958 \tag{3}$$

Similarly,

$$C_2(\cos \theta \sec \theta_M) = 2 \cos^2 \theta \sec^2 \theta_M - 1 = 2 \frac{1}{2} (\cos 2\theta + 1) (1.958)^2 - 1$$

$$= 3.834 \cos 2\theta + 2.834 \tag{4}$$

The substitution of (4) into (1) yields

$$S_M (3.834 \cos 2\theta + 2.834) = 2s_o \cos 2\theta + s_1$$

$$(3.834 S_M - 2s_o) \cos 2\theta + (2.834 S_M - s_1) = 0 \tag{5}$$

For (5) to be valid, terms within each pair of parentheses must vanish independently. Thus,

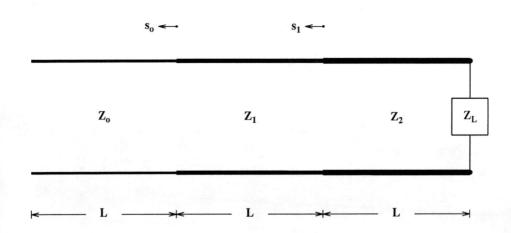

$$L = \lambda/4$$

Figure 5.2-1: A two-section Chebycheff transformer

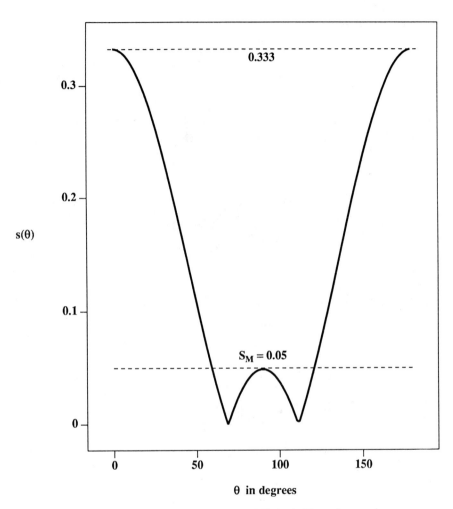

Figure 5.2-2: Frequency response of Chebycheff transformer

$$s_o = \frac{1}{2} S_M \times 3.834 = \frac{1}{2} \times 0.05 \times 3.834 = 0.0958 \qquad (6a)$$

$$s_1 = 0.05 \times 2.834 = 0.142 \qquad (6b)$$

At $\theta = 0$, one has, from (5-12),

$$| S(\theta = 0) | = | S_M e^{-j2\theta} C_2(\cos \theta \sec \theta_M) |_{\theta=0} = S_M [2 (\sec \theta_M \cos \theta)^2 - 1]_{\theta=0}$$

$$= 0.05 [(2 \times 3.834) - 1] = 0.333$$

The bandwidth normalized with respect to the center frequency is given by (4-31) (Fig. 5.2-2).

$$BW = \frac{\Delta f}{f_o} = \frac{\Delta \theta}{\pi/2} = \frac{2}{\pi} 2 (\frac{\pi}{2} - \theta_M)$$

$$= \frac{4}{\pi} (\frac{\pi}{2} - 59.29^o) \frac{\pi}{180^o} = 0.682 \text{ rad}$$

6. Perturbation in a Cavity

To perturb means to disturb or to change slightly. Perturbation methods are useful in evaluating the changes in some parameters caused by a small variation of some other parameters in the problem. In such a case, two problems are involved: the unperturbed problem with an already known solution, and the perturbed problem which is slightly different from the unperturbed one.

The average stored electric and magnetic energies in a resonant cavity are equal. When a small perturbation is introduced into the cavity (Fig. 6-1) it will, in general, change one type of energy more than the other. Consequently, the resonance frequency will shift to equalize the energies.

An unperturbed, air-filled (μ_0, ε_0) cavity has a resonance frequency ω_0 and volume V; the electromagnetic fields are $\vec{E}_0$ and $\vec{H}_0$.

Maxwell's equations for the unperturbed cavity at resonance frequency are

$$\nabla \times \vec{E}_0 = -j\omega_0\mu_0\vec{H}_0 \tag{1a}$$

$$\nabla \times \vec{H}_0 = j\omega_0\varepsilon_0\vec{E}_0 \tag{1b}$$

The perturbation is in the form of a volume ΔV (characterized by μ and ε) introduced into the cavity. The field of the perturbed cavity is denoted by $\vec{E}$ and $\vec{H}$; the corresponding resonance frequency is ω. Maxwell's equations for the perturbed cavity at resonance are

$$\nabla \times \vec{E} = -j\omega\mu_0\vec{H} \qquad \text{exterior to } \Delta V \tag{2a}$$

$$= -j\omega\mu\vec{H} \qquad \text{interior to } \Delta V \tag{2b}$$

$$\nabla \times \vec{H} = j\omega\varepsilon_0\vec{E} \qquad \text{exterior to } \Delta V \tag{2c}$$

$$= j\omega\varepsilon\vec{E} \qquad \text{interior to } \Delta V \tag{2d}$$

As indicated at the beginning, the perturbation will generally change the balance of the stored energies in the system. The stored energies of the system can be obtained by forming the following products.

$$\Omega_1 \equiv \nabla \cdot (\vec{E}_0^* \times \vec{H}) = \vec{H} \cdot \nabla \times \vec{E}_0^* - \vec{E}_0^* \cdot \nabla \times \vec{H} \tag{3a}$$

$$\Omega_2 \equiv \nabla \cdot (\vec{E} \times \vec{H}_0^*) = \vec{H}_0^* \cdot \nabla \times \vec{E} - \vec{E} \cdot \nabla \times \vec{H}_0^* \tag{3b}$$

Let

$$\Omega = \Omega_1 + \Omega_2 \tag{3c}$$

and integrate Ω over the entire volume of the cavity.

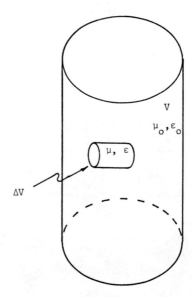

Figure 6-1: Perturbation in a cavity

$$\int_V \Omega \, dv = \int_{V-\Delta V} \Omega \, dv + \int_{\Delta V} \Omega \, dv = \int_{V-\Delta V} \Omega_e \, dv + \int_{\Delta V} \Omega_i \, dv \tag{4a}$$

$$\Omega_e = \Omega \text{ exterior to } \Delta V \tag{4b}$$

$$\Omega_i = \Omega \text{ interior to } \Delta V \tag{4c}$$

Then

$$\Omega_{1e} = \Omega_1 \text{ exterior to } \Delta V = \vec{H} \cdot [\text{Eq. 1a}]^* - \vec{E}_o^* \cdot [\text{Eq. 2c}]$$

$$= \vec{H} \cdot (-j\omega_o\mu_o\vec{H}_o)^* - \vec{E}_o^* \cdot (j\omega\varepsilon_o\vec{E})$$

$$= j\omega_o\mu_o\vec{H} \cdot \vec{H}_o^* - j\omega\varepsilon_o\vec{E} \cdot \vec{E}_o^* \tag{5a}$$

$$\Omega_{1i} = \Omega_1 \text{ interior to } \Delta V = \vec{H} \cdot [\text{Eq. 1a}]^* - \vec{E}_o^* \cdot [\text{Eq. 2d}]$$

$$= j\omega_o\mu_o\vec{H} \cdot \vec{H}_o^* - \vec{E}_o^* \cdot (j\omega\varepsilon\vec{E}) \tag{5b}$$

and

$$\Omega_{2e} = \Omega_2 \text{ exterior to } \Delta V = \vec{H}_o^* \cdot [\text{Eq. 2a}] - \vec{E} \cdot [\text{Eq. 1b}]^*$$

$$= -j\omega\mu_o\vec{H} \cdot \vec{H}_o^* + j\omega_o\varepsilon_o\vec{E} \cdot \vec{E}_o^* \tag{5c}$$

$$\Omega_{2i} = \Omega_2 \text{ interior to } \Delta V = \vec{H}_o^* \cdot [\text{Eq. 2b}] - \vec{E} \cdot [\text{Eq. 1b}]^*$$

$$= -j\omega\mu\vec{H} \cdot \vec{H}_o^* + j\omega_o\varepsilon_o\vec{E} \cdot \vec{E}_o^* \tag{5d}$$

Then

$$\Omega_e = \Omega_{1e} + \Omega_{2e} = j(\omega_o - \omega)\mu_o\vec{H} \cdot \vec{H}_o^* + j(\omega_o - \omega)\varepsilon_o\vec{E} \cdot \vec{E}_o^* \tag{6a}$$

The combination of (5b) and (5d) gives

$$\Omega_i \equiv \Omega_{1i} + \Omega_{2i}$$

$$= j[\,(\omega_0 - \omega)\mu_0 \,+\, \omega(\mu_0 - \mu)\,]\,\vec{H} \cdot \vec{H}_0^*$$

$$+\; j[\,(\omega_0 - \omega)\varepsilon_0 \,+\, \omega(\varepsilon_0 - \varepsilon)\,]\,\vec{E} \cdot \vec{E}_0^* \tag{6b}$$

where

$$\omega_0\mu_0 \,-\, \omega\mu \;=\; \omega_0\mu_0 \,-\, \omega\mu \,+\, \omega\mu_0 \,-\, \omega\mu_0$$

$$= (\omega_0 - \omega)\mu_0 \,+\, \omega(\mu_0 - \mu) \tag{6c}$$

$$\omega_0\varepsilon_0 \,-\, \omega\varepsilon \;=\; (\omega_0 - \omega)\varepsilon_0 \,+\, \omega(\varepsilon_0 - \varepsilon) \tag{6d}$$

The substitution of (6) into (4a) yields

$$\int\limits_{V-\Delta V} \Omega_e\, dv \;=\; \int\limits_{V-\Delta V} \left[-j(\omega - \omega_0)\mu_0\vec{H} \cdot \vec{H}_0^* \,-\, j(\omega - \omega_0)\varepsilon_0\vec{E} \cdot \vec{E}_0^* \right] dv \tag{7a}$$

$$\int\limits_{\Delta V} \Omega_i\, dv \;=\; \int\limits_{\Delta V} \left[\, j[\,(\mu_0 - \mu)\omega \,-\, (\omega - \omega_0)\mu_0\,]\,\vec{H} \cdot \vec{H}_0^* \right.$$

$$\left. +\; j[\,(\varepsilon_0 - \varepsilon)\omega \,-\, (\omega - \omega_0)\varepsilon_0\,]\,\vec{E} \cdot \vec{E}_0^* \,\right] dv \tag{7b}$$

The addition of (7a) and (7b) gives

$$\int\limits_V \Omega\, dv \;=\; \int\limits_V -j[\,(\omega - \omega_0)\,\mu_0\vec{H} \cdot \vec{H}_0^* \,+\, (\omega - \omega_0)\varepsilon_0\vec{E} \cdot \vec{E}_0^*\,]\, dv$$

$$-\; \int\limits_{\Delta V} j\omega[\,(\mu - \mu_0)\vec{H} \cdot \vec{H}_0^* \,+\, (\varepsilon - \varepsilon_0)\vec{E} \cdot \vec{E}_0^*\,]\, dv \tag{8}$$

But

$$\int\limits_V \Omega\, dv \;=\; \int\limits_S [\,\vec{E} \times \vec{H}_0^* \,+\, \vec{E}_0^* \times \vec{H}\,] \cdot \hat{n}\, da \;=\; 0 \tag{9}$$

since the tangential components of the $\vec{E}$-field are zero on the conducting surface of the cavity. Therefore, (8) becomes

$$0 = j(\omega - \omega_o) \int_V [\mu_o \vec{H} \cdot \vec{H}_o^* + \varepsilon_o \vec{E} \cdot \vec{E}_o^*] \, dv$$

$$+ j\omega \int_{\Delta V} [(\mu - \mu_o)\vec{H} \cdot \vec{H}_o^* + (\varepsilon - \varepsilon_o)\vec{E} \cdot \vec{E}_o^*] \, dv$$

or

$$\frac{\omega - \omega_o}{\omega} = \frac{- \int_{\Delta V} [(\mu - \mu_o) \vec{H} \cdot \vec{H}_o^* + (\varepsilon - \varepsilon_o) \vec{E} \cdot \vec{E}_o^*] \, dv}{\int_V [\mu_o \vec{H} \cdot \vec{H}_o^* + \varepsilon_o \vec{E} \cdot \vec{E}_o^*] \, dv} \tag{10}$$

Equation (10) is an exact expression and no approximation has been introduced. However, if the volume of perturbation ΔV is small such that the perturbation outside the volume ΔV is small, then

$$\vec{E} \approx \vec{E}_o \quad \text{and} \quad \vec{H} \approx \vec{H}_o \qquad \text{exterior to } \Delta V \tag{11}$$

Then the denominator of (10) becomes

$$\int_V [\mu_o \vec{H} \cdot \vec{H}_o^* + \varepsilon_o \vec{E} \cdot \vec{E}_o^*] \, dv \approx \int_V [\mu_o \vec{H}_o \cdot \vec{H}_o^* + \varepsilon_o \vec{E}_o \cdot \vec{E}_o^*] \, dv$$

$$\approx 4 (<w_m> + <w_e>) = 4w_o \tag{12}$$

where w_o is the average stored energy in the volume V. Then, (10) becomes

$$\frac{\omega - \omega_o}{\omega} = \frac{-1}{4w_o} \int_{\Delta V} [(\mu - \mu_o) \vec{H} \cdot \vec{H}_o^* + (\varepsilon - \varepsilon_o) \vec{E} \cdot \vec{E}_o^*] \, dv \tag{13}$$

When the perturbing element is introduced into a region of negligible $\vec{H}$-field, (13) becomes

$$\frac{\omega - \omega_o}{\omega} = \frac{-1}{4w_o} \int_{\Delta V} (\varepsilon - \varepsilon_o) \vec{E} \cdot \vec{E}_o^* \, dv \tag{14}$$

If, on the other hand, the element is introduced into a region of negligible $\vec{E}$-field, (13) becomes

$$\frac{\omega - \omega_o}{\omega} = \frac{-1}{4w_o} \int_{\Delta V} (\mu - \mu_o) \vec{H} \cdot \vec{H}_o^* \, dv \tag{15}$$

In order to make the evaluation of (14) and (15) easier, one should choose a simple boundary contour for the perturbing element. The main boundary should be either parallel to the $\vec{E}$-field,

$$\vec{E}_o = \vec{E} \tag{16a}$$

or perpendicular to the $\vec{E}$-field,

$$\vec{D}_o = \vec{D} \tag{16b}$$

7. Perturbation in Waveguides

An air-filled waveguide is perturbed by a sample with a cross-sectional area ΔS. Both the guide and the perturbing element are uniform in the z-direction (Fig. 7-1).

The unperturbed waveguide is characterized by μ_o and ε_o and a propagation constant γ_u. The unperturbed fields are $\vec{E}_o$ and $\vec{H}_o$.

$$\vec{E}_o(x, y, z) = \vec{e}_o(x, y) \, e^{j(\omega t - \gamma_u z)} \equiv \vec{e}_o e^{j\theta_u} \tag{1a}$$

$$\vec{H}_o(x, y, z) = \vec{h}_o(x, y) \, e^{j(\omega t - \gamma_u z)} = \vec{h}_o e^{j\theta_u} \tag{1b}$$

$$\theta_u = \omega t - \gamma_u z \tag{1c}$$

The subscript u stands for unperturbed.

Maxwell's equations for the unperturbed fields are:

$$\nabla \times \vec{E}_o = -j\omega\mu_o\vec{H}_o \tag{2a}$$

$$\nabla \times \vec{H}_o = j\omega\varepsilon_o\vec{E}_o \tag{2b}$$

By the vector identity

$$\nabla \times (\vec{A}\phi) = \phi\nabla \times \vec{A} + \nabla\phi \times \vec{A} \tag{3}$$

one has

$$\nabla \times \vec{E}_o = \nabla \times (\vec{e}_o e^{j\theta_u}) = (\nabla \times \vec{e}_o - j\gamma_u\hat{z} \times \vec{e}_o)\, e^{j\theta_u} \tag{4a}$$

$$\nabla \times \vec{H}_o = (\nabla \times \vec{h}_o - j\gamma_u\hat{z} \times \vec{h}_o)\, e^{j\theta_u} \tag{4b}$$

Then (2) becomes

$$(\nabla \times \vec{e}_o - j\gamma_u\hat{z} \times \vec{e}_o)\, e^{j\theta_u} = -j\omega\mu_o\vec{h}_o e^{j\theta_u}$$

or

$$\nabla \times \vec{e}_o = j\gamma_u\hat{z} \times \vec{e}_o - j\omega\mu_o\vec{h}_o \tag{5a}$$

$$(\nabla \times \vec{h}_o - j\gamma_u\hat{z} \times \vec{h}_o)\, e^{j\theta_u} = j\omega\varepsilon_o\vec{e}_o e^{j\theta_u}$$

or

$$\nabla \times \vec{h}_o = j\gamma_u\hat{z} \times \vec{h}_o + j\omega\varepsilon_o\vec{e}_o \tag{5b}$$

The perturbing element is characterized by μ and ε and its propagating constant γ. The perturbed fields are $\vec{E}$ and $\vec{H}$. The sbuscript p stands for perturbed.

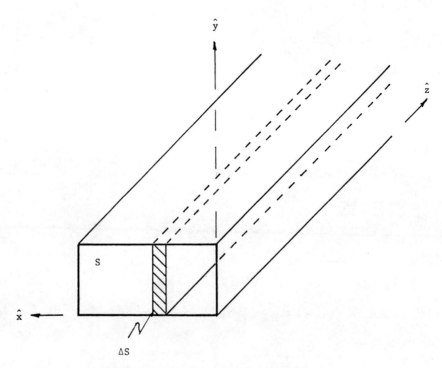

Figure 7-1: A perturbed waveguide

$$\vec{E}(x, y, z) = \vec{e}(x, y)\, e^{j(\omega t - \gamma_p z)} = \vec{e}(x, y)\, e^{j\theta_p} \tag{6a}$$

$$\vec{H}(x, y, z) = \vec{h}(x, y)\, e^{j\theta_p} \tag{6b}$$

$$\theta_p = \omega t - \gamma_p z \tag{6c}$$

The corresponding Maxwell's equations are

$$\nabla \times \vec{E} = -j\omega\mu_o\vec{H} \qquad \text{exterior to } \Delta S \tag{7a}$$

$$= -j\omega\mu\vec{H} \qquad \text{interior to } \Delta S \tag{7b}$$

$$\nabla \times \vec{H} = j\omega\varepsilon_o\vec{E} \qquad \text{exterior to } \Delta S \tag{7c}$$

$$= j\omega\varepsilon\vec{E} \qquad \text{interior to } \Delta S \tag{7d}$$

The application of (3) to (7) yields

$$\nabla \times \vec{E} = (\nabla \times \vec{e} - j\gamma_p\hat{z} \times \vec{e})\, e^{j\theta_p}$$

$$= -j\omega\mu_o\vec{h}e^{j\theta_p} \qquad \text{exterior to } \Delta S$$

$$= -j\omega\mu\vec{h}e^{j\theta_p} \qquad \text{interior to } \Delta S$$

or

$$\nabla \times \vec{e} = j\gamma_p\hat{z} \times \vec{e} - j\omega\mu_o\vec{h} \qquad \text{exterior to } \Delta S \tag{8a}$$

$$\equiv j\gamma_p\hat{z} \times \vec{e} - j\omega\mu\vec{h} \qquad \text{interior to } \Delta S \tag{8b}$$

$$\nabla \times \vec{H} = (\nabla \times \vec{h} - j\gamma_p\hat{z} \times \vec{h})\, e^{j\theta_p}$$

$$= j\omega\varepsilon_o\vec{e}e^{j\theta_p} \qquad \text{exterior to } \Delta S$$

$$= j\omega\varepsilon\vec{e}e^{j\theta_p} \qquad \text{interior to } \Delta S$$

or

$$\nabla \times \vec{h} = j\gamma_p\hat{z} \times \vec{h} + j\omega\varepsilon_o\vec{e} \qquad \text{exterior to } \Delta S \tag{8c}$$

$$= j\gamma_p\hat{z} \times \vec{h} + j\omega\varepsilon\vec{e} \qquad \text{interior to } \Delta S \tag{8d}$$

Apply similar perturbation techniques as in the previous section and form the following products.

$$\Omega_1 = \nabla \cdot \vec{e}_o^* \times \vec{h} = \vec{h} \cdot \nabla \times \vec{e}_o^* - \vec{e}_o^* \cdot \nabla \times \vec{h} \tag{9a}$$

$$\Omega_2 = \nabla \cdot \vec{e} \times \vec{h}_o^* = \vec{h}_o^* \cdot \nabla \times \vec{e} - \vec{e} \cdot \nabla \times \vec{h}_o^* \tag{9b}$$

Let

$$\Omega = \Omega_1 + \Omega_2 \tag{9c}$$

and integrate Ω over the volume V of the waveguide.

$$\int_V \Omega \, dv = \int_{V-\Delta V} \Omega_e \, dv + \int_{\Delta V} \Omega_i \, dv \tag{10a}$$

where

$$\Omega_e = \Omega \text{ exterior to } \Delta V \qquad \Omega_i = \Omega \text{ interior to } \Delta V \tag{10b}$$

$\Omega_{1e} = \Omega_1$ exterior to ΔV

$$= \vec{h} \cdot [\text{Eq. 5a}]^* - \vec{e}_o^* \cdot [\text{Eq. 8c}]$$

$$= -j\gamma_u \hat{z} \cdot (\vec{e}_o^* \times \vec{h}) + j\omega\mu_o \vec{h} \cdot \vec{h}_o^* + j\gamma_p \hat{z} \cdot (\vec{e}_o^* \times \vec{h}) - j\omega\varepsilon_o \vec{e} \cdot \vec{e}_o^*$$

$$= j(\gamma_p - \gamma_u)\hat{z} \cdot (\vec{e}_o^* \times \vec{h}) + j\omega\mu_o \vec{h} \cdot \vec{h}_o^* - j\omega\varepsilon_o \vec{e} \cdot \vec{e}_o^* \tag{11a}$$

and

$\Omega_{2e} = \Omega_2$ exterior to ΔV

$$= \vec{h}_o^* \cdot [\text{Eq. 8a}] - \vec{e} \cdot [\text{Eq. 5b}]^*$$

$$= j\gamma_p \hat{z} \cdot (\vec{e} \times \vec{h}_o^*) - j\omega\mu_o \vec{h} \cdot \vec{h}_o^* + j\gamma_u \hat{z} \cdot (\vec{h}_o^* \times \vec{e}) + j\omega\varepsilon_o \vec{e} \cdot \vec{e}_o^*$$

$$= j(\gamma_p - \gamma_u)\hat{z} \cdot (\vec{e} \times \vec{h}_o^*) - j\omega\mu_o \vec{h} \cdot \vec{h}_o^* + j\omega\varepsilon_o \vec{e} \cdot \vec{e}_o^* \tag{11b}$$

The addition of (11a) and (11b) yields

$$\Omega_e = \Omega_{1e} + \Omega_{2e} = j(\gamma_p - \gamma_u)\hat{z} \cdot (\vec{e}_o^* \times \vec{h} + \vec{e} \times \vec{h}_o^*) \tag{12}$$

Similarly,

$$\Omega_{1i} = \Omega_1 \text{ interior to } \Delta V$$

$$= \vec{h} \cdot [\text{Eq. 5a}]^* - \vec{e}_o^* \cdot [\text{Eq. 8d}]$$

$$= -j\gamma_u \hat{z} \cdot (\vec{e}_o^* \times \vec{h}) + j\omega\mu_o \vec{h} \cdot \vec{h}_o^* - j\gamma_p \hat{z} \cdot (\vec{h} \times \vec{e}_o^*) - j\omega\varepsilon\vec{e} \cdot \vec{e}_o^*$$

$$= j(\gamma_p - \gamma_u)\hat{z} \cdot (\vec{e}_o^* \times \vec{h}) + j\omega\mu_o \vec{h} \cdot \vec{h}_o^* - j\omega\varepsilon\vec{e} \cdot \vec{e}_o^* \tag{13a}$$

and

$$\Omega_{2i} = \Omega_2 \text{ interior to } \Delta V$$

$$= \vec{h}_o^* \cdot [\text{Eq. 8b}] - \vec{e} \cdot [\text{Eq. 5b}]^*$$

$$= j\gamma_p \hat{z} \cdot (\vec{e} \times \vec{h}_o^*) - j\omega\mu \vec{h} \cdot \vec{h}_o^* + j\gamma_u \hat{z}(\vec{h}_o^* \times \vec{e}) + j\omega\varepsilon_o \vec{e} \cdot \vec{e}_o^*$$

$$= j(\gamma_p - \gamma_u)\hat{z} \cdot (\vec{e} \times \vec{h}_o^*) - j\omega\mu \vec{h} \cdot \vec{h}_o^* + j\omega\varepsilon_o \vec{e} \cdot \vec{e}_o^* \tag{13b}$$

The sum of (13a) and (13b) gives

$$\Omega_i = \Omega_{1i} + \Omega_{2i}$$

$$= j(\gamma_p - \gamma_u)\hat{z} \cdot (\vec{e}_o^* \times \vec{h} + \vec{e} \times \vec{h}_o^*) + j\omega(\mu_o - \mu)\vec{h} \cdot \vec{h}_o^* + j\omega(\varepsilon_o - \varepsilon)\vec{e} \cdot \vec{e}_o^* \tag{14}$$

The substitution of (12) and (14) into (10a) yields

$$\int_V \Omega \, dv = \int_{V-\Delta V} \Omega_e \, dv + \int_{\Delta V} \Omega_i \, dv$$

$$= \int_{V-\Delta V} j(\gamma_p - \gamma_u)\hat{z} \cdot (\vec{e}_o^* \times \vec{h} + \vec{e} \times \vec{h}_o^*) \, dv$$

$$+ \int_{\Delta V} [j(\gamma_p - \gamma_u)\hat{z} \cdot (\vec{e}_o^* \times \vec{h} + \vec{e} \times \vec{h}_o^*)$$

$$+ j\omega(\mu_o - \mu)\vec{h} \cdot \vec{h}_o^* + j\omega(\varepsilon_o - \varepsilon)\vec{e} \cdot \vec{e}_o^*] \, dv$$

$$= j(\gamma_p - \gamma_u)\int_V \hat{z} \cdot (\vec{e}_o^* \times \vec{h} + \vec{e} \times \vec{h}_o^*) \, dv$$

$$+ \int_{\Delta V} [j\omega(\mu_o - \mu)\vec{h} \cdot \vec{h}_o^* + j\omega(\varepsilon_o - \varepsilon)\vec{e} \cdot \vec{e}_o^*] \, dv \tag{15}$$

But

$$\int_V \Omega \, dv = \int_V [\nabla \cdot \vec{e}_o^* \times \vec{h} + \nabla \cdot \vec{e} \times \vec{h}_o^*] \, dv$$

$$= \int_S (\vec{e}_o^* \times \vec{h} + \vec{e} \times \vec{h}_o^*) \cdot \hat{n} \, da \tag{16}$$

The surface S of the surface integral comprises the walls of the guide and the cross-sectional surface at the input and the output ports. The integral vanishes on the walls of the guide since the tangential components of the $\vec{E}$-field are zero on the conducting surfaces. For a lossless waveguide, the power entering at the input must equal that emerging at the output port. Thus, the surface integrals over the cross-sectional surfaces sum up to zero. Therefore, (15) becomes

$$(\gamma_p - \gamma_u) \int_V [\, \hat{z} \cdot (\vec{e}_o^* \times \vec{h}) + \hat{z} \cdot (\vec{e} \times \vec{h}_o^*) \,] \, dv$$

$$= \omega \int_{\Delta V} [\, (\mu - \mu_o) \vec{h} \cdot \vec{h}_o^* + (\varepsilon - \varepsilon_o) \vec{e} \cdot \vec{e}_o^* \,] \, dv \tag{17}$$

The volume integrals can be replaced by surface integrals over the cross-sectional areas since the guide is uniform.

$$(\gamma_p - \gamma_u) \int_S (\vec{e}_o^* \times \vec{h} + \vec{e} \times \vec{h}_o^*) \, da$$

$$= \omega \int_{\Delta S} [\, (\mu - \mu_o) \vec{h} \cdot \vec{h}_o^* + (\varepsilon - \varepsilon_o) \vec{e} \cdot \vec{e}_o^* \,] \, da$$

or

$$\gamma_p - \gamma_u = \frac{\omega \int_{\Delta S} [\, (\mu - \mu_o) \vec{h} \cdot \vec{h}_o^* + (\varepsilon - \varepsilon_o) \vec{e} \cdot \vec{e}_o^* \,] \, da}{\int_S (\vec{e}_o^* \times \vec{h} + \vec{e} \times \vec{h}_o^*) \, da} \tag{18}$$

Since

$$\vec{e}_o^* \times \vec{h} = \vec{e} \times \vec{h}_o^*$$

hence

$$\int_S (\vec{e}_o^* \times \vec{h} + \vec{e} \times \vec{h}_o^*) \, da = 2 \int_S \vec{e}_o^* \times \vec{h} \, da = 4P \tag{19}$$

Therefore,

$$\gamma_p - \gamma_u = \frac{\omega}{4P} \int_{\Delta S} [\, (\mu - \mu_o) \vec{h} \cdot \vec{h}_o^* + (\varepsilon - \varepsilon_o) \vec{e} \cdot \vec{e}_o^* \,] \, da \tag{20}$$

If the perturbation is small, i.e.,

$$\Delta S \ll S \tag{21}$$

then

$$\vec{E} \approx \vec{E}_o \quad \text{and} \quad \vec{H} \approx \vec{H}_o \qquad \text{exterior to } \Delta V \tag{22}$$

Hence (20) becomes

$$\gamma_p - \gamma_u \approx \frac{\omega}{4P_o} \int_{\Delta S} [(\mu - \mu_o) \vec{h} \cdot \vec{h}_o^* + (\varepsilon - \varepsilon_o) \vec{e} \cdot \vec{e}_o^*] \, da \tag{23}$$

$$4P_o = \int_S \vec{e}_o^* \times \vec{h}_o \cdot \hat{z} \, da \tag{24}$$

If the perturbing material is lossy, then ε (or μ) is complex, as is γ_p; then

$$\varepsilon = \varepsilon_o \left[\varepsilon' - \frac{j\sigma}{\omega\varepsilon_o} \right] = \varepsilon_o (\varepsilon' - j\varepsilon'') \tag{25a}$$

$$\gamma_p = \beta_p - j\alpha_p \tag{25b}$$

For a nonmagnetic dielectric, $\mu = \mu_o$, and (23) becomes

$$(\beta_p - j\alpha_p) - \beta_u = \omega\varepsilon_o [(\varepsilon' - j\varepsilon'') - 1] \frac{1}{4P_o} \int_{\Delta S} \vec{e} \cdot \vec{e}_o^* \, da \tag{26}$$

Equating the real and the imaginary parts separately,

$$\beta_p - \beta_u = \frac{\omega\varepsilon_o(\varepsilon' - 1)}{4P_o} \int_{\Delta S} \vec{e} \cdot \vec{e}_o^* \, da \tag{27a}$$

$$\alpha_p = \frac{\omega\varepsilon_o\varepsilon''}{4P_o} \int_{\Delta S} \vec{e} \cdot \vec{e}_o^* \, da \tag{27b}$$

The insertion of a nonmagnetic dielectric into a uniform waveguide will produce a phase shift as well as an attenuation between the incoming and the outgoing waves.

8. Dielectric Phase Shifter

A rectangular waveguide can be made to be a simple phase shifter by inserting a thin dielectric sheet parallel to the electric field. For the dominant mode, TE_{10}, the arrangement is shown in Fig. 8-1. The lossless dielectric sheet is characterized by μ_o and ε, its thickness δ, height h, and length L, and its location $x = x_d$.

The exact propagation constant of this composite arrangement is very difficult to evaluate for the case when the height h of the dielectric sheet is different from that of the guide, b.

A fairly accurate solution may be obtained by perturbation techniques (Section 7). The change in phase constant is given by (7-27a).

$$\beta_p - \beta_u = \frac{\omega\varepsilon_o(\varepsilon' - 1)}{4P_o} \int_{\Delta A} \vec{e} \cdot \vec{e}_o^* \, da \tag{1}$$

where

$$\varepsilon = \varepsilon_o (\varepsilon' - j\varepsilon'')$$

β_p = phase constant of the guide with the dielectric sheet $\equiv \dfrac{2\pi}{\lambda_g}$

λ_g = guide wavelength with dielectric sheet

β_u = phase constant of guide without the dielectric sheet $\equiv \dfrac{2\pi}{\lambda_{go}}$

λ_{go} = guide wavelength without dielectric sheet

$4P_o = \displaystyle\int_A \vec{e}_o^* \times \vec{h}_o \cdot \hat{z} \, da$

A = cross-sectional area of waveguide = ab

ΔA = cross-sectional area of dielectric sheet = $h\delta$

$\vec{e}_o = \vec{e}_o(x, y) = \vec{E}$-field of the unperturbed guide

$\vec{h}_o = \vec{h}_o(x ,y) = \vec{H}$-field of the unperturbed guide

$\vec{e} = \vec{e}(x ,y) = \vec{E}$-field of the perturbed guide

It is to be noted that (1) is valid for small perturbations with negligible disturbance exterior to the dielectric sheet.

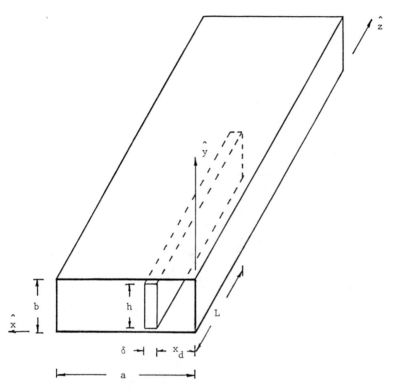

Figure 8-1: A phase shifter

For the TE_{10} mode, the $\vec{e}_o$-field is given by

$$\vec{e}_o(x,y) = \hat{y}E_M \sin \frac{\pi x}{a}$$ (2)

The boundary condition requires

$$\vec{e} = \vec{e}_o \qquad \text{at} \qquad x = x_d$$ (3)

Then

$$4P_o = \int_A \vec{e}_o^* \times \vec{h}_o \cdot \hat{z}\, da = \frac{1}{\eta_{TE}} \int_A |\vec{e}_t|^2\, da$$

$$= \frac{1}{\eta_{TE}} \int_0^b \int_0^a \left| E_M \sin \frac{\pi x}{a} \right|^2 dx\, dy$$

$$= \frac{E_M^2}{\eta_{TE}}\, ab = \frac{AE_M^2}{\eta_{TE}}$$ (4)

where the wave impedance for the TE mode is

$$\eta_{TE} = \sqrt{\frac{\mu_o}{\varepsilon_o}}\, \frac{\lambda_{go}}{\lambda}$$ (5)

and for small ΔA,

$$\int_{\Delta A} \vec{e} \cdot \vec{e}_o^*\, da = \int_{x_d-\frac{\delta}{2}}^{x_d+\frac{\delta}{2}} \int_0^h \left| E_M \sin \frac{\pi x}{a} \right|^2 dx\, dy$$

$$\approx E_M^2 \sin^2 \frac{\pi x_d}{a} \int_{x_d-\frac{\delta}{2}}^{x_d+\frac{\delta}{2}} dx \int_0^h dy$$

$$= E_M^2 \sin^2 \frac{\pi x_d}{a}\, h\delta$$

$$= E_M^2 \sin^2 \frac{\pi x_d}{a}\, \Delta A$$ (6)

where the following approximation is used.

$$\sin \frac{\pi \left(x_d \pm \frac{\delta}{2}\right)}{a} \approx \sin \frac{\pi x_d}{a}$$

The substitution of (4) to (6) into (1) yields

$$\beta_p - \beta_u = \omega \varepsilon_o (\varepsilon' - 1) \frac{E_M^2 \sin^2 \frac{\pi x_d}{a} \Delta A}{\dfrac{E_M^2 A}{\eta_{TE}}}$$

$$= \omega \varepsilon_o (\varepsilon' - 1) \frac{\Delta A}{A} \sin^2 \frac{\pi x_d}{a} \frac{\lambda_{go}}{\lambda} \sqrt{\frac{u_o}{\varepsilon_o}}$$

$$= (\varepsilon' - 1) \omega \sqrt{\mu_o \varepsilon_o} \frac{\lambda_{go}}{\lambda} \frac{\Delta A}{A} \sin^2 \frac{\pi x_d}{a}$$

$$= 2\pi (\varepsilon' - 1) \frac{\lambda_{go}}{\lambda^2} \frac{\Delta A}{A} \sin^2 \frac{\pi x_d}{a} \tag{7}$$

since $\omega \sqrt{\mu_o \varepsilon_o} = 2\pi/\lambda$, where λ is the wavelength in free space.

$$\left[\frac{1}{\lambda_g} - \frac{1}{\lambda_{go}} \right] = (\varepsilon' - 1) \frac{\lambda_{go}}{\lambda^2} \frac{\Delta A}{A} \sin^2 \frac{\pi x_d}{a} \tag{8}$$

This approximation shows the dependence of the composite guide wavelength λ_g upon the relative dielectric constant of the dielectric sheet, the relative cross-sectional area $\Delta A/A$, and the location of the sheet $x_d = a/2$. The effect will be minimal at $x = 0$.

In accordance with the small-perturbation theory, (8) is valid when all the following conditions are satisfied.

$$\frac{\delta}{a} \ll 1 \tag{9a}$$

$$\varepsilon \approx 1 \tag{9b}$$

$$\lambda_g \approx \lambda_{go}$$

$$\frac{h}{\delta} \gg 1 \tag{9d}$$

A practical realization of this device is a thin sheet of dielectric with tapered ends to minimize reflection. The sheet is made movable in the x-direction.

The total phase shift is, from (7) and (8),

$$(\beta_p - \beta_u) L = 2\pi L \left[\frac{1}{\lambda_g} - \frac{1}{\lambda_{go}} \right] \tag{10}$$

The phase shift is proportional to λ_{go}/λ^2 as indicated by (7), and λ_{go}/λ is a function of frequency.

9. Strip Attenuator

An attenuator can be constructed in a similar way to the phase shifter discussed in the previous section if the lossless dielectric sheet is replaced by a thin sheet of lossy material. The attenuation coefficient of such a structure is approximately given by (7-27b).

$$\alpha_p = \frac{\omega \varepsilon_o \varepsilon''}{4P_o} \int_{\Delta A} \vec{e} \cdot \vec{e}_o^* \, da \tag{1}$$

$$\varepsilon = \varepsilon_o (\varepsilon' - j\varepsilon'') = \varepsilon_o \left[\varepsilon' - \frac{j\sigma}{\omega \varepsilon_o} \right]$$

= complex permittivity of the lossy sheet

A = cross-sectional area of the unperturbed guide

ΔA = cross-sectional area of the lossy sheet

$\vec{e}_o, \vec{h}_o$ = unperturbed fields

$\vec{e}$ = perturbed field

For the arrangement shown in Fig. 8-1, and for the TE_{10} mode, the unperturbed $\vec{E}$-field is

$$\vec{e}_o(x, y) = \hat{y} E_M \sin \frac{\pi x}{a} \tag{2}$$

Then Eqs. (8-4) and (8-6) give

$$4P_o = \frac{A E_M^2}{\eta_{TE}} \tag{3a}$$

$$\int_{\Delta A} \vec{e} \cdot \vec{e}_o^* \, da = E_M^2 \sin^2 \frac{\pi x_d}{a} \Delta A \tag{3b}$$

$$\eta_{TE} = \sqrt{\frac{\mu_o}{\varepsilon_o}} \frac{\lambda_{go}}{\lambda} \tag{3c}$$

and (1) becomes

$$\alpha_p = \omega\varepsilon_o\varepsilon'' \; \frac{E_M^2 \sin^2 \dfrac{\pi x_d}{a} \, \Delta A}{\dfrac{E_M^2 A}{\eta_{TE}}}$$

$$= \omega\varepsilon_o \frac{\sigma}{\omega\varepsilon_o} \frac{\Delta A}{A} \sqrt{\frac{\mu_o}{\varepsilon_o}} \frac{\lambda_{go}}{\lambda} \sin^2 \frac{\pi x_d}{a}$$

$$= \sigma \frac{h\delta}{ab} \sqrt{\frac{\mu_o}{\varepsilon_o}} \frac{\lambda_{go}}{\lambda} \sin^2 \frac{\pi x_d}{a}$$

$$= \frac{h}{R_s} \frac{1}{ab} \sqrt{\frac{\mu_o}{\varepsilon_o}} \frac{\lambda_{go}}{\lambda} \sin^2 \frac{\pi x_d}{a} \tag{4a}$$

$$R_s = \frac{1}{\sigma\delta} = \text{surface resistance} \tag{4b}$$

or

$$\sigma = \frac{1}{R_s\delta} \tag{4c}$$

for δ is less than the skin depth of the lossy sheet.

The attenuation coefficient is inversely proportional to the surface resistance R_s. The smallest R_s for which the perturbation approximation is not violated is typically about 50 ohms per square.

10. Polarization of Plane Waves

(a) A complex wave distribution can be considered as the sum of several simple plane waves with different magnitudes, phases, and directions of propagation. The behavior of the complex wave can be analyzed by considering the behavior of each individual wave.

The orientations of the field vectors in these waves are called the polarizations of the waves. The discussion is focused primarily on sinusoidal waves of identical frequency.

A wave with arbitrary orientations of field vectors, arbitrary amplitudes, and random phases is known as an unpolarized wave.

A wave with its $\vec{E}$-field vector always lying in a given direction is known as a linearly polarized wave. A linearly polarized wave is the result of the superposition of waves with $\vec{E}$-vectors in the same direction or in different directions but of exactly the same phase.

The combination of two uniform plane waves at the same frequency but having different amplitudes, phases, and orientations of field vectors results in an elliptically polarized wave. This may be visualized by considering two traveling waves as given by

$$\vec{E} = \vec{E}_1 + \vec{E}_2 \tag{1a}$$

$$\vec{E}_1 = \hat{x}E_x e^{j(\omega t - \beta z)} \tag{1b}$$

$$\vec{E}_2 = \hat{y}E_y e^{j(\omega t - \beta z + \phi)} \tag{1c}$$

where ϕ is the relative phase angle. The analysis to follow will be valid at any location z. To simplify the notation, the reference point is chosen at $z = 0$. Then

$$\vec{E}_1 = \hat{x}E_x e^{j\omega t} \tag{2a}$$

$$\vec{E}_2 = \hat{y}E_y e^{j(\omega t + \phi)} \tag{2b}$$

The real field is taken to be

$$e_x = \text{Re}\,[E_1] = E_x \cos \omega t \tag{3a}$$

$$e_y = \text{Re}\,[E_2] = E_y \cos (\omega t + \phi) \tag{3b}$$

These are the parametric equations for e_x and e_y in the xy-plane. The variation of e_x and e_y in the xy-plane can be determined by eliminating the time variable t.

$$\left[\frac{e_x}{E_x}\right]^2 + \left[\frac{e_y}{E_y}\right]^2 = \cos \omega t + \cos^2 (\omega t + \phi)$$

$$= \cos^2 \omega t + (\cos \omega t \cos \phi - \sin \omega t \sin \phi)^2$$

$$= \cos^2 \omega t \, (1 + \cos^2 \phi) + (1 - \cos^2 \omega t) \sin^2 \phi - 2 \cos \omega t \sin \omega t \cos \phi \sin \phi$$

$$= \sin^2 \phi + 2 \cos^2 \omega t \cos^2 \phi - 2 \cos \omega t \sin \omega t \cos \phi \sin \phi$$

$$= \sin^2 \phi + 2 \cos \omega t \cos \phi \, (\cos \omega t \cos \phi - \sin \omega t \sin \phi)$$

$$= \sin^2 \phi + 2 \cos \omega t \cos \phi \cos (\omega t + \phi)$$

$$= \sin^2 \phi + 2 \frac{e_x}{E_x} \frac{e_y}{E_y} \cos \phi$$

$$\left[\frac{e_x}{E_x} \right]^2 + \left[\frac{e_y}{E_y} \right]^2 - \frac{2 e_x e_y}{E_x E_y} \cos \phi = \sin^2 \phi \qquad (4)$$

This is a general equation for an ellipse. The equation

$$A x^2 + B x y + C y^2 + D x + E y + F = 0$$

is an equation for an ellipse if

$$B^2 - 4AC < 0$$

For Eq. (4), one has

$$B^2 - 4AC = (-2 \cos \phi)^2 - 4 = 4 \, (\cos^2 \phi - 1) < 0$$

Therefore, (4) is an equation for ellipse, if $\phi \neq 0$.

(b) For the case $\phi = \pm \, \pi/2$, Eq. (4) becomes

$$\left[\frac{e_x}{E_x} \right]^2 + \left[\frac{e_y}{E_y} \right]^2 = 1 \qquad (5)$$

This is an equation for an ellipse for unequal E_x and E_y. When $E_x = E_y = E_o$, then

$$e_x^2 + e_y^2 = E_o^2 \qquad (6)$$

This represents a circle with a radius $r = E_o$.

(c) For $\phi = -\pi/2$, Eq. (3) becomes

$$e_x = E_x \cos \omega t \qquad (7a)$$

$$e_y = E_y \sin \omega t \qquad (7b)$$

At t = 0 (Fig. 10-1a),

$$\vec{E}_1 = \hat{x}e_x = \hat{x}E_x$$

$$\vec{E}_2 = \hat{y}e_y = 0$$

and

$$\vec{E} = \vec{E}_1 + \vec{E}_2 = \vec{E}_1$$

At $\omega t = \pi/2$ (Fig. 10-1b),

$$\vec{E}_1 = 0 \qquad \text{and} \qquad \vec{E}_2 = \hat{y}E_y$$

$$\vec{E} = \vec{E}_1 + \vec{E}_2 = \vec{E}_2$$

The $\vec{E}$-vector rotates in the counterclockwise direction as the wave propagates in the positive z-direction. This is known as a right-hand polarized wave.

(d) For $\phi = \pi/2$, Eq. (3) becomes

$$e_x = E_x \cos \omega t \tag{8a}$$

$$e_y = -E_y \sin \omega t \tag{8b}$$

At $\omega t = 0$ (Fig. 10-2a),

$$\vec{E}_1 = \hat{x}e_x = \hat{x}E_x$$

$$\vec{E}_2 = \hat{y}e_y = 0$$

At $\omega t = \pi/2$ (Fig. 10-2b),

$$\vec{E}_1 = 0$$

$$\vec{E}_2 = -\hat{y}E_y$$

The $\vec{E}$-vector rotates in the clockwise direction as the wave propagates in the positive z-direction. This is known as a left-hand polarized wave.

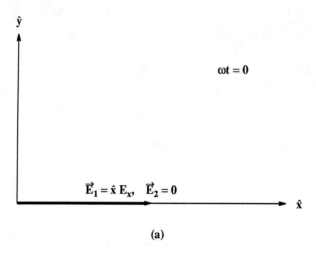

(a)

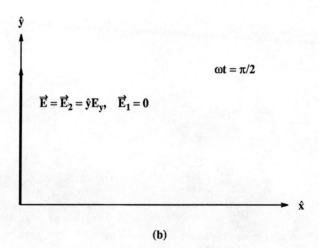

(b)

Figure 10-1: A right-hand polarization

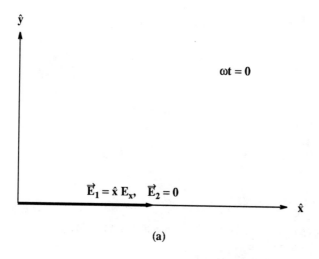

(a)

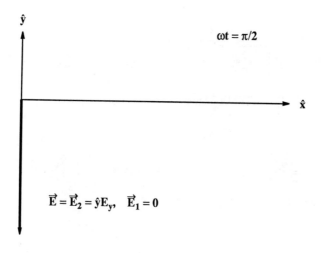

(b)

Figure 10-2: A left-hand polarization

(e) Equation (1) may be expressed in parametric form.

$$[E^+] = \begin{bmatrix} \alpha \\ -j\alpha \end{bmatrix} \qquad \text{right-hand polarization} \tag{9a}$$

$$[E^-] = \begin{bmatrix} \alpha \\ j\alpha \end{bmatrix} \qquad \text{left-hand polarization} \tag{9b}$$

where

$$\alpha = E_o e^{j\theta} \tag{9c}$$

$$\theta = \omega t - \beta z + \Omega \tag{9d}$$

for a wave propagating along the positive z-direction, with arbitrary phase angle Ω. Then (1) can be written as

$$[\vec{E}] = [\ \hat{x} \quad \hat{y}\ ]\ [E] \tag{10}$$

The right-hand circularly polarized wave is given by

$$\vec{E}^+ = [\ \hat{x} \quad \hat{y}\ ]\ [E^+] = \hat{x}\alpha - \hat{y}j\alpha$$

$$\text{Re}\ [\vec{E}^+] = \text{Re}\left[\ \hat{x}E_o e^{j\theta} - \hat{y}jE_o e^{j\theta}\ \right] = \hat{x}E_o\cos\theta + \hat{y}E_o\sin\theta \tag{11}$$

Equation (11) represents a right-hand polarized wave.

The left-hand circularly polarized wave is given by

$$\vec{E}^- = [\ \hat{x} \quad \hat{y}\ ]\ [E^-] = \hat{x}\alpha + \hat{y}j\alpha$$

$$\text{Re}\ [\vec{E}^-] = \text{Re}\ [\ \hat{x}E_o e^{j\theta} + j\hat{y}E_o e^{j\theta}\ ] = \hat{x}E_o\cos\theta - \hat{y}E_o\sin\theta \tag{12}$$

Equation (12) represents a left-hand polarized wave.

11. Quarter-Wave Plate

A linearly polarized wave can be converted into a circularly polarized wave by decomposing the incoming wave into mutually perpendicular components of equal magnitude but with a relative phase difference of $90°$. A quarter-wave plate is one type of circular wave polarizer.

A quarter-wave plate is composed of a circular waveguide with a thin dielectric sheet oriented at $45°$ with respect to the x- and y-axes (Fig. 11-1). An incoming wave polarized linearly along the x-axis will have equal components parallel to and normal to the dielectric sheet. These components will propagate with different phase constants, β_p for the parallel component and β_n for the normal component along the waveguide.

The component normal to the dielectric sheet remains essentially unperturbed and $\beta_n \approx \beta_u$, where β_u is the phase constant for the unperturbed waveguide. The phase constant $\beta_{parallel}$ for the component parallel to the dielectric sheet can be analyzed in a manner similar to that presented in Section 7, and $\beta_p > \beta_o$, i.e., β_p is increased. In this case, $\beta_{parallel} = \beta_p =$ the phase constant of the perturbed guide.

The output wave will be circularly polarized provided

$$(\beta_p - \beta_n) L = \pm 90°$$

where L is the length of the dielectric sheet.

(a) A circular waveguide is operating in its dominant TE_{11} mode and contains a dielectric sheet of finite length located along its diameter. The dielectric sheet is arranged so that the degenerate TE_{11} modes, polarized parallel to and perpendicular to the sheet, are matched.

Further, the length, thickness, and dielectric constant of the dielectric sheet are chosen such that the mode polarized along the sheet is delayed $90°$ more than the mode polarized perpendicularly to the sheet. Since a differential phase of $90°$ is created between these two modes, this type of polarizer is known as a quarter-wave plate. This structure is essentially anisotropic in the transverse plane.

In analyzing a quarter-wave plate, it is considered as a four-port device even though physically there is only one input port and one output port. This is because of the decomposition of the field with the E-field parallel to and perpendicular to the dielectric sheet.

The field with the $\vec{E}$-vector perpendicular to the dielectric sheet enters at port 1 and that with the $\vec{E}$-vector parallel to the sheet enters port 2 (Fig. 11-2). A randomly polarized wave can be considered as the vector sum of waves entering at port 1 and port 2.

Port 3 is the output port of port 1 and port 4 is the output port of port 2. Ports 1 and 2 form a right-handed system with the direction of propagation z.

Let the input wave to port 1 be represented by a_1. It will experience a phase delay of ϕ_1 as it propagates through the device. The output at port 3 is then given by

$$b_3 = a_1 e^{-j\phi_1} \tag{1}$$

Similarly, let the input to port 2 be a_2, and its phase delay through the device is $\phi_1 + \pi/2$. The output at port 4 is

$$b_4 = a_2 e^{-j(\phi_1 + \frac{\pi}{2})} \tag{2}$$

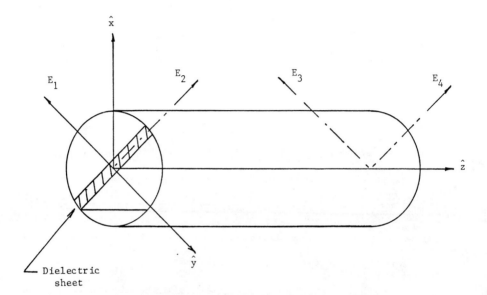

Figure 11-1: Quarter-wave plate

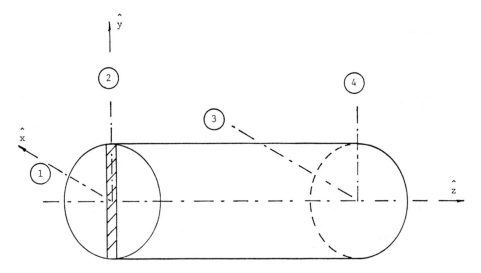

Figure 11-2: A four-port representation of a quarter-wave plate

Since the structure is physically indistinguishable between its ends, reciprocity properties exist, i.e.,

$$b_1 = a_3 e^{-j\phi_1} \tag{3}$$

$$b_2 = a_4 e^{-j(\phi_1 + \frac{\pi}{2})} \tag{4}$$

The scattering matrix for the quarter-wave plate can be written from the relation

$$[b] = [S][a] \tag{5}$$

or

$$\left[S_{\frac{\pi}{2}} \right] = e^{-j\phi_1} \begin{bmatrix} 0 & 0 & 1 & 0 \\ 0 & 0 & 0 & -j \\ 1 & 0 & 0 & 0 \\ 0 & -j & 0 & 0 \end{bmatrix} \tag{6}$$

(b) Let the input wave be

$$a_1 = a_2 = a e^{j\phi_o} \tag{7}$$

where ϕ_o is an arbitrary phase angle. This is the case for a linearly polarized wave whose plane of polarization makes an angle of 45° with respect to the dielectric sheet, or 45° with respect to the x-axis (Fig. 11-3). The output is then

$$[b] = e^{-j\phi_1} \begin{bmatrix} 0 & 0 & 1 & 0 \\ 0 & 0 & 0 & -j \\ 1 & 0 & 0 & 0 \\ 0 & -j & 0 & 0 \end{bmatrix} e^{j\phi_o} \begin{bmatrix} a \\ a \\ 0 \\ 0 \end{bmatrix}$$

$$= e^{-j(\phi_1 - \phi_o)} \begin{bmatrix} 0 \\ 0 \\ a \\ -ja \end{bmatrix} \tag{8}$$

Equation (8) represents a right-hand circularly polarized wave, (10-9a). Thus, a linearly polarized wave oriented at 45° with respect to the dielectric sheet is transformed into a right-hand circularly polarized wave.

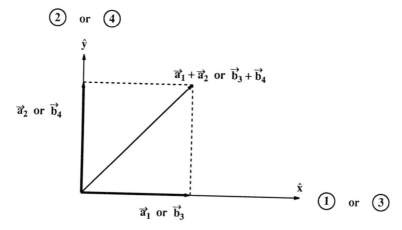

Figure 11-3: A right-hand circular polarization

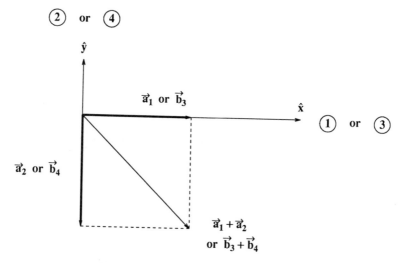

Figure 11-4: A left-hand circular polarization

(c) Consider the input wave

$$a_1 = -a_2 = ae^{j\phi_o} \tag{9}$$

This is a linearly polarized wave oriented at −45° with respect to the x-axis (Fig. 11-4). The output is

$$[b] = e^{-j\phi_1} \begin{bmatrix} 0 & 0 & 1 & 0 \\ 0 & 0 & 0 & -j \\ 1 & 0 & 0 & 0 \\ 0 & -j & 0 & 0 \end{bmatrix} e^{j\phi_o} \begin{bmatrix} a \\ -a \\ 0 \\ 0 \end{bmatrix}$$

$$= e^{-j(\phi_1 - \phi_o)} \begin{bmatrix} 0 \\ 0 \\ a \\ ja \end{bmatrix} \tag{10}$$

The output is a left-hand circularly polarized wave. Therefore, a linearly polarized input wave oriented at −45° with respect to the x-axis is transformed into a left-hand circularly polarized wave, (10-9b).

(d) Consider the input wave to be a right-hand circularly polarized wave

$$a_1 = ja_2 = ae^{j\phi_o} \tag{11}$$

and the output is

$$[b] = e^{-j\phi_1} \begin{bmatrix} 0 & 0 & 1 & 0 \\ 0 & 0 & 0 & -j \\ 1 & 0 & 0 & 0 \\ 0 & -j & 0 & 0 \end{bmatrix} e^{j\phi_o} \begin{bmatrix} a \\ -ja \\ 0 \\ 0 \end{bmatrix}$$

$$= e^{-j(\phi_1 - \phi_o)} \begin{bmatrix} 0 \\ 0 \\ a \\ -a \end{bmatrix} \tag{12}$$

A right-hand circularly polarized wave is transformed into a linearly polarized wave with its plane of polarization oriented at −45° with respect to the x-axis (Fig. 11-4).

(e) Consider a left-hand circularly polarized input wave,

$$a_1 = -ja_2 = ae^{j\phi_o} \tag{13}$$

$$[b] = e^{-j\phi_1} \begin{bmatrix} 0 & 0 & 1 & 0 \\ 0 & 0 & 0 & -1 \\ 1 & 0 & 0 & 0 \\ 0 & -j & 0 & 0 \end{bmatrix} e^{j\phi_o} \begin{bmatrix} a \\ ja \\ 0 \\ 0 \end{bmatrix}$$

$$= e^{-j(\phi_1 - \phi_o)} \begin{bmatrix} 0 \\ 0 \\ a \\ a \end{bmatrix} \tag{14}$$

A left-hand circularly polarized input wave is transformed into a linearly polarized wave with its plane of polarization oriented at 45° with respect to the x-axis (Fig. 11-3).

12. Problems

1. A p-hole directional coupler is shown in Fig. 12-1. Let the voltage coupling coefficient at the kth hole be c_k with a phase angle

$$\theta_k = (k - 1) \theta + \theta_h$$

where θ_h is the phase delay due to the coupling hole. For a coupler with identical holes, θ_h is simply an added constant phase angle and may be neglected.

   (a) Show that for a unit incident wave at port 1, the total coupled wave at port 2 is given by

$$C_r = \sum_{m=1}^{p} c_m e^{-j2(m - 1)\theta}$$

   (b) Show that the total coupled wave at port 4 is

$$C_f = e^{j(p - 1)\theta} \sum_{m=1}^{p} c_m$$

2. For a p-hole directional coupler which is symmetrical with respect to plane A–A′, Fig. 12-1, show that the outputs at port 2 and port 4 are given, respectively, as follows:

$$C_r' = 2 [c_1 \cos (p - 1)\theta + c_2 \cos (p - 3)\theta + \cdots + c_{p/2} \cos \theta] \qquad \text{for} \quad p = \text{even}$$

$$C_f' = e^{-j\frac{p-1}{2}\theta} \sum_{m=1}^{p} c_m$$

3. Design a Chebycheff 5-hole, 20-db (coupling = 20 db = $-20 \log |C_f'|$) directional coupler with a bandwidth $BW = f_U - f_L$ specified by

$$m = \frac{\lambda_U}{\lambda_L} = 2$$

where λ_U and λ_L are the wavelengths at the upper and the lower cutoff frequency, respectively. At the center frequency, $C_r = 0$, i.e., $d = \lambda/4$.

Ans. $c_1 = 8 \times 10^{-3}$, $c_2 = 24 \times 10^{-3}$, $c_3 = 33 \times 10^{-3}$.

4. One of the techniques in network synthesis is to rearrange the normalized admittance matrix as follows:

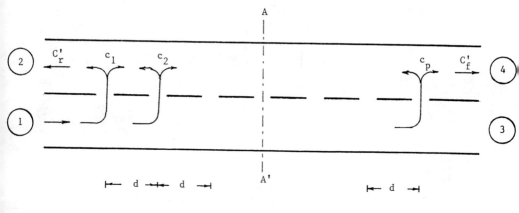

$$\vartheta = \beta d$$

Figure 12-1: A p-port directional coupler

$$[\hat{y}] = \begin{bmatrix} y_{11} & y_{12} & \cdots & y_{1m} \\ y_{21} & y_{22} & \cdots & y_{2m} \\ \cdots & \cdots & \cdots & \cdots \\ y_{m1} & y_{m2} & \cdots & y_{mm} \end{bmatrix}$$

$$= \begin{bmatrix} y_{11} & 0 & \cdots & 0 \\ 0 & 0 & \cdots & 0 \\ \cdots & \cdots & \cdots & \cdots \\ 0 & 0 & \cdots & 0 \end{bmatrix} + \begin{bmatrix} 0 & 0 & \cdots & 0 \\ 0 & y_{22} & \cdots & 0 \\ \cdots & \cdots & \cdots & \cdots \\ 0 & 0 & \cdots & 0 \end{bmatrix} + \cdots$$

$$+ \begin{bmatrix} 0 & 0 & \cdots & 0 \\ 0 & 0 & \cdots & 0 \\ \cdots & \cdots & \cdots & \cdots \\ 0 & 0 & \cdots & y_{mm} \end{bmatrix}$$

$$+ \begin{bmatrix} 0 & y_{21} & \cdots & 0 \\ y_{12} & 0 & \cdots & 0 \\ \cdots & \cdots & \cdots & \cdots \\ 0 & 0 & \cdots & 0 \end{bmatrix} + \cdots + \begin{bmatrix} 0 & 0 & \cdots & y_{1m} \\ 0 & 0 & \cdots & 0 \\ \cdots & \cdots & \cdots & \cdots \\ y_{m1} & 0 & \cdots & 0 \end{bmatrix} + \cdots$$

$$+ \begin{bmatrix} 0 & 0 & 0 & \cdots & 0 \\ 0 & 0 & y_{23} & \cdots & 0 \\ 0 & y_{32} & 0 & \cdots & 0 \\ \cdots & \cdots & \cdots & \cdots & \cdots \\ 0 & 0 & 0 & \cdots & 0 \end{bmatrix} + \cdots$$

$$= \sum_{i=1}^{p} \hat{Y}_{ii} + \sum_{i=1}^{p} \sum_{k=1}^{p} \hat{Y}_{ik}$$

This final expression can be realized by interpreting $\hat{Y}_{ii}$ as the normalized self-admittance at the ith port, and $\hat{Y}_{ik}$ as the normalized admittance between the ith and the kth port.

Show that a four-port structure with a scattering matrix

$$[S] = \frac{-j}{\sqrt{2}} \begin{bmatrix} 0 & 0 & 1 & 1 \\ 0 & 0 & 1 & -1 \\ 1 & 1 & 0 & 0 \\ 1 & -1 & 0 & 0 \end{bmatrix}$$

can be realized as Fig. 12-2, where

$\hat{Y}_{\lambda/4}$ = normalized admittance of a $\lambda/4$ line

$\hat{Y}_{3\lambda/4}$ = normalized admittance of a $3\lambda/4$ line

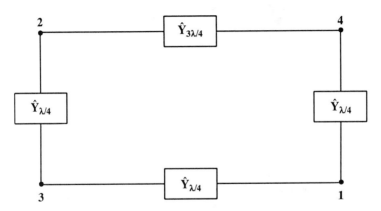

Figure 12-2: A realization of a normalized admittance

CHAPTER VIII   Ferrite Devices

1. Propagation in Ferrite

Ferrites are of great interest to microwave engineering. The basic principles and characteristics of ferrite will be investigated in this chapter. More information is available in many excellent textbooks and some of these are listed in the bibliography [D].

The resistivity of some ferrites is as high as 10^{10} ohm-meter as compared to 10^{-7} ohm-meter for iron. This high resistivity reduces the eddy current losses in ferrites to negligible values even at microwave frequencies. They also show spontaneous magnetization below a certain temperature known as the Curie temperature.

Ferrite materials have nonreciprocal electrical properties, i.e., the transmission characteristics are not the same for propagation in different directions. The operation of a ferrite device can be easily analyzed once the basic behavior of microwave propagation in an unbounded ferrite medium is understood.

Einstein's special theory of relativity shows that Newtonian mechanics is accurate only for velocities small compared to that of light. Dirac solved Schrodinger's wave equation taking relativity into consideration and found that significant modification is necessary for low-velocity phenomena. Uhlenbeck and Goudsmit (Naturwiss., vol. 13, page 593, 1925) introduced the hypothesis that the electron spins about an axis and has an angular momentum, i.e., an angular momentum in addition to its orbital motion in an atom. This intrinsic property of electrons is known as "spin." This postulate supplies the consistency between theoretical predictions and experimental observations. The spin angular momentum of an electron, $\vec{L}_{spin}$, is defined by

$$\vec{L}_{spin} \equiv \vec{s} \frac{h}{2\pi} \qquad h = \text{Planck's constant} \tag{1a}$$

$$|\vec{s}| = \sqrt{n_s(n_s + 1)} \tag{1b}$$

$$n_s = \frac{1}{2} = \text{spin quantum number} \tag{1c}$$

$$\vec{s} = \hat{s}s = \pm\vec{s} \qquad \text{parallel or antiparallel} \tag{1d}$$

For the purpose of the present analysis, the "spin" may be taken as an actual mechanical spin. Up to now, there has not been any explanation for the existence of "spin" except that it provides an agreement between theory and experiments.

Since the electron describes an orbit about the nucleus, it has an orbital angular momentum, $\vec{L}_{orb}$, which can be obtained from classical theory [Eq. (1.2-5)]. By virtue of the orbital motion, the electron has an orbital magnetic moment, $\vec{m}_{orb}$ [Eq. (1.1-3)].

The spin magnetic moment, $\vec{m}_{spin}$, is difficult if not impossible to calculate since there is insufficient information about the shape of an electron and the distribution of its charge. The spin magnetic moment is defined, for the purpose of obtaining agreement between theory and experimental results, as

$$\vec{m}_{spin} = 2\frac{e}{2m_o}\vec{L}_{spin} = 2\vec{s}\frac{he}{4\pi m_o} \qquad m_o = \text{mass of the electron} \qquad (2)$$

The macroscopic theory of microwave ferrite devices is based on the equation of motion of the magnetization vector. This equation can be derived by considering ferrite to be made of spinning electrons.

(a) Precession

A spinning electron is characterized by a magnetic dipole moment $\vec{m}$ and angular momentum $\vec{L}$. The subscript "spin" has been omitted because only the spinning electron is considered in this analysis. Since the electron carries a negative charge, $\vec{L}$ and $\vec{m}$ are pointed in opposite directions [Eq. (1.1-14), Fig. 1-1] and are related by the gyromagnetic ratio γ_e.

$$\vec{m} = \gamma_e\vec{L} \qquad (3a)$$

$$\gamma_e \equiv -\left| \frac{e}{m_o} \right| \qquad (3b)$$

where e is the charge and m_o is the mass of the electron. The gyromagnetic ratio is established by quantum theory, and the classical determination of γ_e is in error by a factor of 2. Classical models are used in this analysis because their simplicity allows one to gain insight into the problem without being bogged down in mathematics.

If an electron is located in a uniform static magnetic field $\vec{B}_o = \hat{z}B_o$, the spinning electron (Fig. 1-2) will experience a torque given by

$$\vec{\tau} = \vec{m} \times \vec{B}_o = \gamma_e\vec{L} \times \vec{B}_o \qquad (4)$$

by virtue of (3a).

From the theory of mechanics, the torque acting on a body is equal to the rate of change of its angular momentum, (1.2-8),

$$\frac{d\vec{L}}{dt} = \vec{\tau} \qquad (5a)$$

Combining (4) and (5a) gives

$$\frac{d\vec{L}}{dt} = \gamma_e\vec{L} \times \vec{B}_o \qquad (5b)$$

Equation (5) can be expressed in terms of magnetic dipole moment. Differentiation of (3a) with respect to time yields

$$\frac{d\vec{m}}{dt} = \gamma_e\frac{d\vec{L}}{dt} \qquad (6)$$

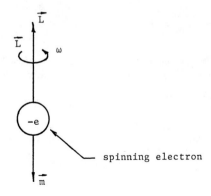

Figure 1-1: A spinning electron.

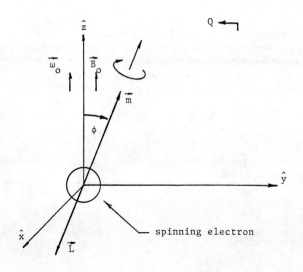

Figure 1-2: A spinning electron in uniform B-field

Then (5a) becomes

$$\vec{\tau} = \frac{d\vec{L}}{dt} = \frac{1}{\gamma_e} \frac{d\vec{m}}{dt} \tag{7}$$

Equating (7) to (4) gives

$$\frac{d\vec{m}}{dt} = \gamma_e \vec{m} \times \vec{B}_o \tag{8}$$

for a single magnetic dipole moment, $\vec{m}$. If the medium is described by a density of N dipole moments per unit volume, then

$$\frac{d\vec{M}}{dt} = \gamma_e \vec{M} \times \vec{B}_o \tag{9a}$$

where

$$\vec{M} = N\vec{m} \tag{9b}$$

This is the equation of motion for the magnetic moment vector.

The physical meaning of (9a) can be obtained from Fig. 1-2. Consider the instant when $\vec{m}$ is in the plane of paper, the yz-plane. The sectional view at Q-Q', Fig. 1-2, is shown in Fig. 1-3. It is to be noted that neither $\vec{m}$ nor $\vec{L}$ is in the plane of the paper, the xz-plane, in Fig. 1-3: $\vec{m}$ is pointing out of the xz-plane and $\vec{L}$ is pointing into the xz-plane.

For an increment of time Δt, (5a) becomes

$$\Delta\vec{L} = \vec{\tau} \Delta t \tag{10}$$

The change of angular momentum is in the same direction as the torque. The resultant angular momentum, $\vec{L} + \Delta\vec{L}$, is tilted away from the yz-plane. The exaggerated figure, Fig. 1-3, seems to indicate that the angular momentum $\vec{L}$ and, consequently, $\vec{m}$ change their magnitudes. The angular momentum at the end of the time interval Δt is the vector sum $\vec{L} + \Delta\vec{L}$. Since $\Delta\vec{L}$ is perpendicular to $\vec{L}$ and is assumed to be very small as compared to $\vec{L}$, the resultant angular momentum has the same magnitude as the original one but has a different direction. Hence, the tip of the angular momentum vector swings around in a horizontal circle as time increases (Fig. 1-4).

(b) Damping of Precession

If there were no damping, the magnetic dipole moment could precess forever and one would never be able to magnetize the ferrite material, i.e., $\vec{m}$ would never be aligned with $\vec{B}_o$. Landau and Lifshitz (Physik Zeitschrift Sowjetunion, vol. 8, page 153, 1935) proposed a damping term to be added to the equation of motion, (9a).

$$\text{damping} = \frac{\alpha}{M} \vec{M} \times (\vec{M} \times \vec{B}_o) \tag{11}$$

where α is a parameter which determines the amplitude of damping. This represents a change of magnetic moment or equivalently a change of angular momentum. The effect of (11) can be determined as follows.

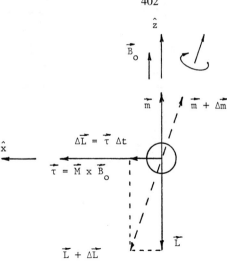

Figure 1-3: A sectional view across plane Q-Q', Figure 1-2.

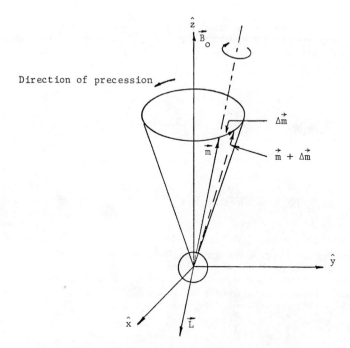

Figure 1-4: Precession of a spinning electron.

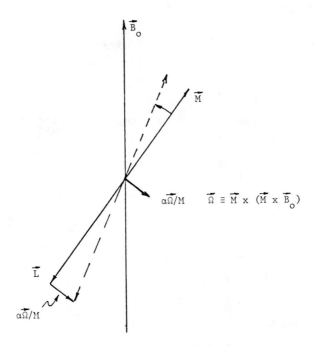

Figure 1-5: Precession with damping.

The relative positions of the magnetic dipole moment $\vec{M}$ and the magnetic field intensity $\vec{B}_o$ are shown in Fig. 1-5. The addition of the damping term, (11), to the equation of motion tends to reduce the angle between the magnetic dipole moment and the $\vec{B}_o$-field. Consequently, the moment $\vec{M}$ will precess at a smaller angle. The damping will also reduce the torque $\vec{\tau} = \vec{m} \times \vec{B}_o$. The effects will be compounded and thus bring $\vec{M}$ into alignment with the $\vec{B}_o$-field.

With the damping term added, (9a) becomes

$$\frac{d\vec{M}}{dt} = \gamma_e \vec{M} \times \vec{B}_o + \frac{\gamma_e \alpha}{M} \vec{M} \times (\vec{M} \times \vec{B}_o) \tag{12}$$

The factor of γ_e in the last term of (12) is added to simplify the final result. The term $\gamma_e \vec{M} \times \vec{B}_o$ is solved from (12).

$$\gamma_e \vec{M} \times \vec{B}_o = \frac{-\alpha}{M} \vec{M} \times \gamma_e (\vec{M} \times \vec{B}_o) + \frac{d\vec{M}}{dt} \tag{13}$$

The substitution of (13) into the last term in (12) yields

$$\frac{d\vec{M}}{dt} = \gamma_e \vec{M} \times \vec{B}_o + \frac{\alpha}{M} \vec{M} \times \left[\frac{-\alpha}{M} \vec{M} \times \gamma_e (\vec{M} \times \vec{B}_o) + \frac{d\vec{M}}{dt} \right]$$

$$= \gamma_e \vec{M} \times \vec{B}_o - \frac{\alpha^2}{M^2} \vec{M} \times \vec{M} \times \gamma_e (\vec{M} \times \vec{B}_o) + \frac{\alpha}{M} \vec{M} \times \frac{d\vec{M}}{dt} \tag{14}$$

For small values of the attenuation constant α, which is the case for most ferrites, the second term on the right-hand side is negligible. Then one has

$$\frac{d\vec{M}}{dt} = \gamma_e \vec{M} \times \vec{B}_o + \frac{\alpha}{M} \vec{M} \times \frac{d\vec{M}}{dt} \tag{15}$$

This aproximation was suggested by T. A. Gilbert, Armour Research Foundation, Report No. 11, Jan. 25, 1955.

1.1. Magnetic Moment and Angular Momentum of Atomic Models

The magnetic moment and angular momentum for two classical models of electronic motions in an atom will be derived in this section. The purpose is to gain some insight into these properties without highly complicated mathematics. The results provide the essential features found in quantum theory; however, the numerical values are different.

(a) Orbital Model (Bohr Model)

The orbital model represents the atom by a stationary nucleus and a circularly orbiting electron. The orbit has a radius R and its center is located at the nucleus (Fig. 1.1-1). The nucleus has a charge of +e and the electron has a charge of −e. The electron has a constant angular velocity of ω radians per second.

The orbiting electron gives rise to a magnetic dipole moment $\vec{m}$ directed normally into the paper.

$$\vec{m} = i\vec{A} \tag{1}$$

where i is the current resulting from the moving electron and $\vec{A}$ is the area enclosed by the current. The subscript "orb" is omitted for the orbital magnetic moment (and later for the orbital angular momentum), as only the orbital case is considered in this subsection. The associated current is

$$i = -ef = -e\frac{\omega}{2\pi} \tag{2}$$

where f is the frequency of the orbiting motion and $\omega = 2\pi f$ is the angular frequency. Therefore,

$$\vec{m} = -e\frac{\omega}{2\pi}\pi R^2\hat{m} = -\frac{1}{2}e\omega R^2\hat{m} \qquad \hat{m} \equiv \frac{\vec{m}}{|\vec{m}|} \tag{3}$$

This magnetic dipole moment is known as the orbital magnetic moment since it is the result of the orbital motion of the electron. The direction of the magnetic dipole moment is given by the right-hand rule. In accordance with the direction of movement, this direction is normally outward from the paper for a positive charge. Because the charge of the electron is negative, its direction is normally into the paper.

The angular momentum of an orbiting particle with a mass m_0 is given by Eq. (1.2-5),

$$\vec{L} = \vec{r} \times \vec{p} \tag{4}$$

where $\vec{r}$ is the position vector of the particle and $\vec{p}$ is the linear momentum given by Eq. (1.2-2).

$$\vec{p} = m_0\vec{v} \tag{5}$$

where $\vec{v}$ is the linear velocity of the particle and m_0 is the mass of the particle. The orbital angular momentum of the Bohr model is therefore

$$\vec{L} = \vec{R} \times m_0\vec{v} \tag{6}$$

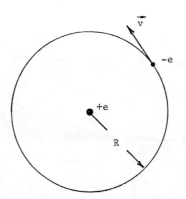

Figure 1.1-1: Orbital model of an electron.

The direction of the angular momentum is pointed normally outward from the paper. Thus, the magnetic dipole moment $\vec{m}$ of an arbitrary negatively charged particle and its angular momentum are oppositely directed. The ratio of their magnitudes is

$$\frac{m}{L} = \frac{e\omega R^2}{2m_0 R v} = \frac{e}{2m_0} \tag{7a}$$

$$v = R\omega \tag{7b}$$

The magnetic dipole moment and the corresponding angular momentum of an orbiting charged particle are linearly related.

(b) Spherical Charged-Cloud Model

In the spherical charged-cloud model, an atom is visualized as a positive nucleus of charge +e with some charge −e moving around the nucleus. The nucleus has a diameter of the order of 10^{-15} m while the radius of the electron orbit is of the order of 10^{-10} m. It is therefore permissible to consider the nucleus as a point charge and represent the orbiting charges by the total negative charge −e distributed uniformly throughout a sphere of radius R, where $R \approx 10^{-10}$ m (Fig. 1.1-2). The negative charge cloud rotates with a constant angular frequency ω about an axis passing through its center. This model may also be used to approximate the "spin" of an electron, in which case R would be the equivalent radius of an electron.

The magnetic moment of the rotating charge cloud can be determined from the contribution of the charges in the cylindrical shell between r and r + dr. The height of the cylinder is

$$h = 2\sqrt{R^2 - r^2} \tag{8}$$

The associated current of this shell is the amount of charge passing through the cross section (h dr) per second.

$$di = \rho(r\omega)h\,dr \tag{9}$$

The charge density ρ is given by

$$\rho = \frac{-e}{\frac{4\pi}{3} R^3} \tag{10}$$

The contribution to the magnetic dipole moment, dm (without the subscript), is

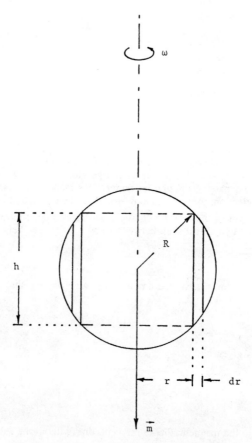

Figure 1.1-2: Charged-cloud model of an electron.

$$dm = \pi r^2 \, di = \pi r^2 \, \frac{-e}{\dfrac{4\pi R^3}{3}} \, \omega r h \, dr$$

$$= \frac{-3}{4} \frac{e}{R^3} \, \omega h r^3 \, dr$$

$$= \frac{-3}{2} \frac{e}{R^3} \, \omega \, \sqrt{R^2 - r^2} \, r^3 \, dr \tag{11}$$

The total magnetic moment is then

$$m = \frac{-3e}{2R^3} \, \omega \int_0^R \sqrt{R^2 - r^2} \, r^3 \, dr$$

$$= \frac{-3e\omega}{2R^3} \left[\frac{1}{5} (R^2 - r^2)^{\frac{5}{2}} - \frac{R^2}{3} (R^2 - r^2)^{\frac{3}{2}} \right]_0^R$$

$$= \frac{-3e\omega}{2R^3} \left[-\frac{R^5}{5} + \frac{R^5}{3} \right]$$

$$= \frac{-1}{5} \, e\omega R^2$$

or

$$\vec{m} = \frac{-1}{5} \, eR^2\vec{\omega} \tag{12}$$

The magnetic dipole moment for the electron cloud model and that for the Bohr model differ by a constant. This is expected since the concepts of modeling are quite different in the two cases.

The angular momentum with the moving charge within the cylindrical shell is

$$dL = |\vec{r} \times d(m_o\vec{v})| = |\vec{r} \times \vec{v}\, dm_o|$$

$$= \frac{m_o}{\dfrac{4\pi R^3}{3}} (2\pi r h\, dr)(r\omega)\, r$$

$$= \frac{3}{2} \frac{m_o}{R^3} h\omega r^3\, dr$$

$$= \frac{3}{2} \frac{m_o\omega}{R^3}\, 2\sqrt{R^2 - r^2}\, r^3\, dr$$

where m_o is the total mass of the charge cloud, and dm_o is the mass of the cylindrical shell.

$$L = \frac{3m_o\omega}{R^3} \int_o^R \sqrt{R^2 - r^2}\, r^3\, dr = \frac{2}{5} m_o\omega R^2 \quad \text{or} \quad \omega R^2 = \frac{5L}{2m_o} \tag{13}$$

and

$$\vec{m} = \frac{-e}{2m_o} \vec{L} \tag{14}$$

1.2. Angular Momentum

(a) Figure 1.2-1 shows a force $\vec{F}$ acting on a particle situated at point P, which is located by a position vector $\vec{r}$.

The torque $\vec{\tau}$ acting on the particle with respect to the origin is defined by

$$\vec{\tau} = \vec{r} \times \vec{F} \tag{1}$$

The torque is acting about an axis located at the origin, and this axis is normal to the plane containing the vectors $\vec{r}$ and $\vec{F}$. Its direction is given by (1) in accordance with the right-hand rule. Torque is also known as the moment of force.

The linear momentum $\vec{p}$ of a moving particle is defined as the product of its mass m_0 and its velocity $\vec{v}$.

$$\vec{p} = m_0\vec{v} \tag{2}$$

The rate of change of momentum of a body is proportional to the resultant force acting on the body and it points in the direction of that force.

$$\vec{F} = \frac{d\vec{p}}{dt} \tag{3}$$

For a single particle of constant mass m_0, one has

$$\vec{F} = \frac{d\vec{p}}{dt} = \frac{d(m_0\vec{v})}{dt} = m_0\frac{d\vec{v}}{dt} = m_0\vec{a} \tag{4}$$

(b) The angular momentum $\vec{L}$ of a particle with respect to the origin is defined to be

$$\vec{L} = \vec{r} \times \vec{p} \tag{5}$$

Angular momentum is also known as moment of linear moment.

The relation between torque and angular momentum can be derived as follows.

$$\vec{\tau} = \vec{r} \times \vec{F} = \vec{r} \times \frac{d\vec{p}}{dt} \tag{6}$$

The differentiation of (5) with respect to time is

$$\frac{d\vec{L}}{dt} = \frac{d(\vec{r} \times \vec{p})}{dt} = \frac{d\vec{r}}{dt} \times \vec{p} + \vec{r} \times \frac{d\vec{p}}{dt} \tag{7}$$

Since $\vec{v} = \frac{d\vec{r}}{dt}$ and

$$\frac{d\vec{r}}{dt} \times \vec{p} = \vec{v} \times (m_0\vec{v}) = 0$$

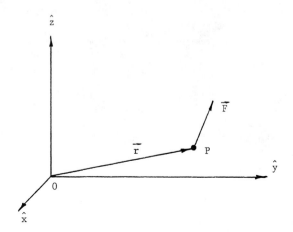

Figure 1.2-1: Torque on a particle.

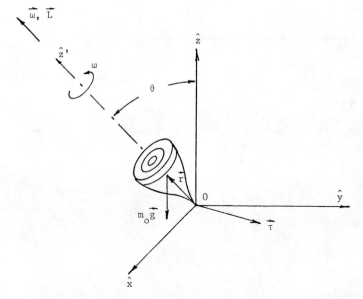

Figure 1.2-2: A spinning top.

hence

$$\frac{d\vec{L}}{dt} = \vec{r} \times \frac{d\vec{p}}{dt} = \vec{r} \times \vec{F} = \vec{\tau} \tag{8}$$

The time rate of change of the angular momentum of a particle is equal to the torque acting on it.

(c) A top spinning about its axis of symmetry is shown in Fig. 1.2-2. The sharp point of the top is supported at the origin of an inertial reference frame. Experimental experience shows that the axis of the rapidly spinning top moves around the vertical axis and sweeps out a cone. This motion is known as precession.

Let the top be rotating at an angular velocity $\vec{\omega}$ about its own axis. The angular momentum of the top is $\vec{L}$. The vector $\vec{\omega}$ always points along the fixed axis of rotation of a spinning body, but the vector $\vec{L}$, in general, does not [see part (e)]. For a rigid body with symmetry about the axis of rotation, both $\vec{\omega}$ and $\vec{L}$ are directed along the axis of symmetry if this axis remains fixed.

In this analysis, $\vec{\omega}$ and $\vec{L}$ will be assumed to be parallel, which implies that the rate of precession ω_p is very slow in comparison to $\vec{\omega}$, and the movement of the z′-axis (Fig. 1.2-2) is extremely slow. Both $\vec{\omega}$ and $\vec{L}$ make an angle θ with the vertical axis.

The upward force at the pivot 0 has a moment arm of zero value and produces no torque about the origin.

The weight $m_o \vec{g}$ exerts a torque about point 0, and it is given by

$$\vec{\tau} = \vec{r} \times \vec{F} = \vec{r} \times m_o \vec{g} \tag{9}$$

where $\vec{r}$ is the position vector to the center of mass of the top. This torque is perpendicular to the plane containing the vectors $\vec{r}$ and $m_o \vec{g}$ in accordance with the right-hand rule.

This torque $\vec{\tau}$, due to gravity, changes the angular momentum of the spinning top according to (8).

$$\vec{\tau} = \frac{d\vec{L}}{dt} = \lim_{\Delta t \to 0} \frac{\Delta \vec{L}}{\Delta t}$$

or

$$\Delta \vec{L} = \vec{\tau} \Delta t \tag{10}$$

Since $\vec{\tau}$ is perpendicular to $\vec{L}$ (Fig. 1.2-2), so is $\Delta \vec{L}$.

The resultant angular momentum is given by the vector sum of the original $\vec{L}$ and the incremental $\Delta \vec{L}$ as shown in Fig. 1.2-3. One sees that the movement of the precessing $\vec{L}$ traces out a cone.

(d) The angular speed of precession ω_p is given by (Fig. 1.2-3)

$$\omega_p = \lim_{\Delta t \to 0} \frac{\Delta \phi}{\Delta t} \tag{11}$$

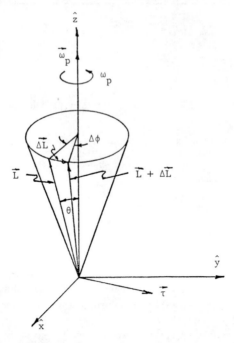

Figure 1.2-3: The resultant angular momentum of a spinning top.

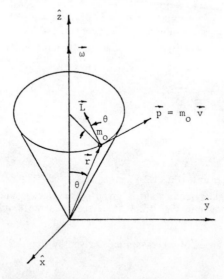

Figure 1.2-4: The angular momentum of an orbiting particle.

For small incremental $\Delta\vec{L}$, $\Delta L \ll L$, one has

$$\tan \Delta\phi = \frac{\Delta L}{L \sin \theta} \tag{12}$$

Equation (12) can be approximated for small angles as

$$\tan \Delta\phi \approx \Delta\phi = \frac{\Delta L}{L \sin \theta} = \frac{\tau \Delta t}{L \sin \theta}$$

or

$$\frac{\Delta\phi}{\Delta t} = \frac{\tau}{L \sin \theta} \tag{13}$$

But from (9)

$$\tau = r m_o g \sin \theta \tag{14}$$

Therefore,

$$\frac{\Delta\phi}{\Delta t} = \frac{r m_o g}{L} \tag{15}$$

and

$$\omega_p = \lim_{\Delta t \to 0} \frac{\Delta\phi}{\Delta t} = \frac{r m_o g}{L} \tag{16}$$

From (13), one also has

$$\omega_p = \lim_{\Delta t \to 0} \frac{\Delta\phi}{\Delta t} = \lim_{\Delta t \to 0} \frac{\tau}{L \sin \theta}$$

or

$$\tau = \omega_p L \sin \theta \tag{17}$$

Equation (17) can be written in vector form.

$$\vec{\tau} = \vec{\omega}_p \times \vec{L} \tag{18}$$

This is the general vector expression relating the precessional angular velocity $\vec{\omega}_p$, the angular momentum $\vec{L}$, and the torque $\vec{\tau}$.

(e) A single particle of mass m_o moving at a linear velocity $\vec{v}$ along a circular path about the z-axis is shown in Fig. 1.2-4. Its angular velocity $\vec{\omega}$ is directed in the positive z-direction. The corresponding angular momentum about the origin is given by (5).

$$\vec{L} = \vec{r} \times \vec{p} \tag{19}$$

$\vec{L}$ is a vector prependicular to the plane containing the vectors $\vec{r}$ and $\vec{p}$, and is shown in Fig. 1.2-4. It is obvious in this case that $\vec{\omega}$ and $\vec{L}$ are not coincident with each other.

(f) Consider two diametrically located particles of identical mass and moving such that their relative position remains unchanged, (Fig. 1.2-5). A cross-sectional view in the plane passing through both particles and the origin is shown in Fig. 1.2-6. The resultant angular momentum of these symmetrically arranged particles is the sum of the momentum of each particle and is seen to be pointed in the same direction as $\vec{\omega}$.

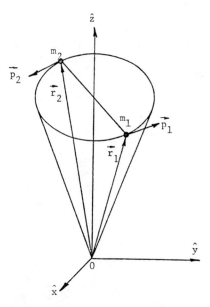

Figure 1.2-5: The orbiting of a symmetrically positioned particles.

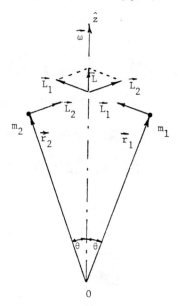

Figure 1.2-6: The angular momentum of symmetrically arranged particles in
Figure 1.2-5: Two diametrically locted identical particles

2. Permeability Tensor

(a) An infinite unbounded ferrite material is immersed in a uniform steady magnetic field $\vec{H}_o$. In the absence of the RF field, the ferrite is in the saturated state, i.e., all the magnetic dipole moments are aligned with the DC field. Under the saturated condition, one has

$$\vec{H}_o = \hat{z}H_o \tag{1a}$$

$$\vec{M}_o = \hat{z}M_o \tag{1b}$$

When an RF field of frequency ω is superimposed onto the DC field, then one has

$$\vec{H} = \vec{H}_o + \vec{H}'e^{j\omega t} \tag{2a}$$

$$\vec{M} = \vec{M}_o + \vec{M}'e^{j\omega t} \tag{2b}$$

where the primed items belong to the RF field with the appropriate time factor.

The equation of motion of the magnetization vector, (1-9), is

$$\frac{d\vec{M}}{dt} = \gamma_e \mu_o \vec{M} \times \vec{H} \tag{3}$$

The substitution of (2) into (3) yields

$$\frac{d}{dt} (\vec{M}_o + \vec{M}'e^{j\omega t}) = \gamma_e \mu_o (\vec{M}_o + \vec{M}'e^{j\omega t}) \times (\vec{H}_o + \vec{H}'e^{j\omega t})$$

$$j\omega\vec{M}'e^{j\omega t} = \gamma_e \mu_o (\hat{z}M_o + \vec{M}'e^{j\omega t}) \times (\hat{z}H_o + \vec{H}'e^{j\omega t})$$

$$= \gamma_e \mu_o (\hat{z}M_o \times \vec{H}'e^{j\omega t} + \vec{M}' \times \vec{H}'e^{j2\omega t} + \vec{M}' \times \hat{z}H_o e^{j\omega t}) \tag{4}$$

If the RF quantities are small compared with the DC quantities, the product of two RF items can be neglected and (4) reduces to

$$j\omega\vec{M}'e^{j\omega t} = \gamma_e \mu_o (\hat{z}M_o \times \vec{H}' + \vec{M}' \times \hat{z}H_o) e^{j\omega t}$$

or

$$j\omega\vec{M}' = \gamma_e \mu_o (\hat{z}M_o \times \vec{H}' + \vec{M}' \times \hat{z}H_o) \tag{5}$$

Equation (5) can be decomposed into its components.

$$j\omega M_x' = \gamma_e \mu_o (-M_o H_y' + M_y' H_o) \tag{6a}$$

$$j\omega M_y' = \gamma_e \mu_o (M_o H_x' - M_x' H_o) \tag{6b}$$

$$j\omega M_z' = 0 \tag{6c}$$

Equation (6) can be arranged in matrix form.

$$\begin{bmatrix} j\omega & -\gamma_e\mu_o H_o \\ \gamma_e\mu_o H_o & j\omega \end{bmatrix} \begin{bmatrix} M_x' \\ M_y' \end{bmatrix} = \begin{bmatrix} -\gamma_e\mu_o M_o H_y' \\ \gamma_e\mu_o M_o H_x' \end{bmatrix} \tag{7a}$$

$$\Delta = (-\gamma_e\mu_o H_o)^2 - \omega^2 = \omega_o^2 - \omega^2 \tag{7b}$$

$$\omega_o \equiv -\gamma_e\mu_o H_o \tag{7c}$$

where ω_o is the gyromagnetic resonance frequency; its physical meaning will be evident later [Eq. (9)]. When $\omega = \omega_o$, the susceptance becomes infinitely large.

It is convenient to introduce

$$\omega_m = -\gamma_e\mu_o M_o \tag{7d}$$

and ω_m is a frequency which depends upon the saturation magnetization of the material; thus, it is a property of the material. Equation (7a) then becomes

$$\begin{bmatrix} j\omega & \omega_o \\ -\omega_o & j\omega \end{bmatrix} \begin{bmatrix} M_x' \\ M_y' \end{bmatrix} = \begin{bmatrix} \omega_m H_y' \\ -\omega_m H_x' \end{bmatrix} \tag{8}$$

The components of $\vec{M}'$ can now be solved from (8).

$$M_x' = \frac{1}{\Delta} \begin{bmatrix} \omega_m H_y' & \omega_o \\ -\omega_m H_x' & j\omega \end{bmatrix}$$

$$= \frac{1}{\Delta} (\omega_o\omega_m H_x' + j\omega\omega_m H_y')$$

$$= \chi_{xx} H_x' + \chi_{xy} H_y' \tag{9a}$$

$$M_y' = \frac{1}{\Delta} \begin{bmatrix} j\omega & \omega_m H_y' \\ -\omega_o & -\omega_m H_x' \end{bmatrix}$$

$$= \frac{1}{\Delta} (-j\omega\omega_m H_x' + \omega_o\omega_m H_y')$$

$$= \chi_{yx} H_x' + \chi_{yy} H_y' \tag{9b}$$

where

$$\chi_{xx} = \frac{\omega_o \omega_m}{\Delta} = \chi_{yy} \equiv \overline{\chi} \tag{9c}$$

$$\chi_{xy} = \frac{j\omega \omega_m}{\Delta} = -\chi_{yx} \equiv -jK \tag{9d}$$

Therefore,

$$M_x' = \overline{\chi} H_x' - jK H_y' \tag{10a}$$

$$M_y' = jK H_x' + \overline{\chi} H_y' \tag{10b}$$

$$M_z' = 0 \tag{10c}$$

The matrix form of (10) is

$$\begin{bmatrix} M_x' \\ M_y' \\ M_z' \end{bmatrix} = \begin{bmatrix} \overline{\chi} & -jK & 0 \\ jK & \overline{\chi} & 0 \\ 0 & 0 & 0 \end{bmatrix} \begin{bmatrix} H_x' \\ H_y' \\ H_z' \end{bmatrix} \tag{11a}$$

or

$$[M'] = [\chi] [H'] \tag{11b}$$

where the susceptance matrix is given by

$$[\chi] = \begin{bmatrix} \overline{\chi} & -jK & 0 \\ jK & \overline{\chi} & 0 \\ 0 & 0 & 0 \end{bmatrix} \tag{11c}$$

The magnetic field intensity is given by

$$\vec{B}' = \mu_o (\vec{H}' + \vec{M}') \tag{12}$$

$$= (\mu_o [u] + \mu_o [\chi]) [H'] = \mu_o ([u] + [\chi]) [H']$$

$$\equiv [\mu] [H'] \tag{13}$$

The permeability tensor is

$$[\mu] = \mu_o [u] + \mu_o [\chi]$$

$$= \begin{bmatrix} \mu_o(1 + \overline{\chi}) & -jK\mu_o & 0 \\ jK\mu_o & \mu_o(1 + \overline{\chi}) & 0 \\ 0 & 0 & \mu_o \end{bmatrix}$$

$$= \begin{bmatrix} \mu & -jk & 0 \\ jk & \mu & 0 \\ 0 & 0 & \mu_o \end{bmatrix} \tag{14a}$$

where

$$k \equiv \mu_o K \quad \text{and} \quad \mu \equiv \mu_o (1 + \overline{\chi}) \tag{14b}$$

(b) Practical ferrite medium has finite losses, and this will significantly affect the properties of the material. The equation of motion for the medium with damping is given by (1-15).

$$\frac{d\vec{M}}{dt} = \gamma_e \mu_o \vec{M} \times \vec{H} + \frac{\alpha}{M} \vec{M} \times \frac{d\vec{M}}{dt} \tag{15}$$

where the total $\vec{H}$-field is considered in the present analysis. This equation is similar to the one without damping except for the addition of the term containing damping factor α. The substitution of (2) into the last term of (15) produces

$$\frac{\alpha}{M} \vec{M} \times \frac{d\vec{M}}{dt} = \frac{\alpha}{M} (\vec{M}_o + \vec{M}'e^{j\omega t}) \times \frac{d}{dt} (\vec{M}_o + \vec{M}'e^{j\omega t})$$

$$= \frac{\alpha}{M} (\hat{z}M_o + \vec{M}'e^{j\omega t}) \times j\omega\vec{M}'e^{j\omega t}$$

$$\approx \frac{j\omega\alpha M_o}{M} (\hat{z} \times \vec{M}') e^{j\omega t} \tag{16}$$

since $\vec{M}' \times \vec{M}' = 0$. Let

$$M'e^{j\omega t} = M_r' + jM_i'$$

Then

$$M = | \vec{M}_o + \vec{M}'e^{j\omega t} | \approx \sqrt{(M_o + M_r')^2 + M_i'^2} \approx M_o \tag{17}$$

This is the small-signal approximation, and it is assumed that the products of the RF terms are negligible. Thus

$$\frac{\alpha}{M} \vec{M} \times \frac{d\vec{M}}{dt} \approx \frac{j\omega\alpha M_o}{M_o} (\hat{y}M_x' - \hat{x}M_y') e^{j\omega t} \tag{18}$$

The substitution of (2) and (18) into (15) yields the results of both (5) and (6), for the corresponding case with damping.

$$j\omega\vec{M}' = \gamma_e\mu_o (\hat{z}M_o \times \vec{H}' + \vec{M}' \times \hat{z}H_o) + j\omega\alpha (\hat{z} \times \vec{M}') \tag{19}$$

Decomposing (19) into its components yields

$$j\omega M_x' = \gamma_e\mu_o (-M_oH_y' + M_y'H_o) - j\omega\alpha M_y'$$

$$= -\gamma_e\mu_oM_oH_y' + (\gamma_e\mu_oH_o - j\omega\alpha) M_y'$$

$$= \omega_mH_y' - (\omega_o + j\omega\alpha) M_y' \tag{20a}$$

$$j\omega M_y' = \gamma_e\mu_o (M_oH_x' - M_x'H_o) + j\omega\alpha M_x'$$

$$= \gamma_e\mu_oM_oH_x' - (\gamma_e\mu_oH_o - j\omega\alpha) M_x'$$

$$= -\omega_mH_x' + (\omega_o + j\omega\alpha) M_x' \tag{20b}$$

$$j\omega M_z' = 0 \tag{20c}$$

Equation (20) can be arranged in matrix form.

$$\begin{bmatrix} j\omega & \omega_o + j\omega\alpha \\ -(\omega_o + j\omega\alpha) & j\omega \end{bmatrix} \begin{bmatrix} M_x' \\ M_y' \end{bmatrix} = \begin{bmatrix} \omega_mH_y' \\ -\omega_mH_x' \end{bmatrix} \tag{21}$$

A comparison of (21) and (8) indicates that the damping term can be introduced by replacing ω_o by $(\omega_o + j\omega\alpha)$ in the lossless components.

M_x' and M_y' can be determined from (21)

$$M_x' = \frac{1}{\Delta} \begin{bmatrix} \omega_m H_y' & \omega_o + j\omega\alpha \\ -\omega_m H_x' & j\omega \end{bmatrix}$$

$$= \frac{1}{\Delta} [\, \omega_m (\, \omega_o + j\omega\alpha \,) H_x' + j\omega\omega_m H_y' \,]$$

$$= \chi_{xx} H_x' + \chi_{xy} H_y' \tag{22a}$$

$$M_y' = \frac{1}{\Delta} \begin{bmatrix} j\omega & \omega_m H_y' \\ -(\omega_o + j\omega\alpha) & -\omega_m H_x' \end{bmatrix}$$

$$= \frac{1}{\Delta} [\, \omega_m (\, \omega_o + j\omega\alpha \,) H_y' - j\omega\omega_m H_x' \,]$$

$$= \chi_{yy} H_y' + \chi_{yx} H_x' \tag{22b}$$

where

$$\Delta = (\, \omega_o + j\omega\alpha \,)^2 - \omega^2 \tag{22c}$$

$$\chi_{xx} = \frac{\omega_m}{\Delta} (\, \omega_o + j\omega\alpha \,) = \chi_{yy} = \overline{\chi} \tag{22d}$$

$$\chi_{xy} = \frac{j\omega\omega_m}{\Delta} = -\chi_{yx} = -jK \tag{22e}$$

The magnetization vector can be expressed in terms of the susceptance tensor.

$$[M'] = [\chi] [H'] \tag{23a}$$

where

$$[\chi] = \begin{bmatrix} \chi_{xx} & \chi_{xy} & \chi_{xz} \\ \chi_{yx} & \chi_{yy} & \chi_{yz} \\ \chi_{zx} & \chi_{zy} & \chi_{zz} \end{bmatrix} = \begin{bmatrix} \overline{\chi} & -jK & 0 \\ jK & \overline{\chi} & 0 \\ 0 & 0 & 0 \end{bmatrix} \tag{23b}$$

The elements of the susceptance tensors for practical ferrite materials with finite losses are complex quantities. The real and imaginary parts of $\overline{\chi}$ and K are found to be (Section 2.1)

$$\overline{\chi} = \chi' - j\chi'' \tag{24a}$$

$$\chi' = \frac{\omega_m\omega_o}{D} [\omega_o^2 - (1 - \alpha^2) \omega^2] \tag{24b}$$

$$\chi'' = \frac{\omega\omega_m\alpha}{D} [\omega_o^2 + (1 + \alpha^2) \omega^2] \tag{24c}$$

$$D = [\omega_o^2 - (1 + \alpha^2) \omega^2]^2 + 4\omega^2\omega_o^2\alpha^2 \tag{24d}$$

and

$$K = K' - jK'' \tag{25a}$$

$$K' = \frac{-\omega\omega_m}{D} [\omega_o^2 - (1 + \alpha^2) \omega^2] \tag{25b}$$

$$K'' = \frac{-1}{D} 2\omega^2\omega_m\omega_o\alpha \tag{25c}$$

The frequency dependence of the real and imaginary parts of $\overline{\chi}$ can be evaluated from (24).

At $\omega = 0$, the real part χ' is equal to the static value.

$$\chi'(\omega=0) = \frac{\omega_m}{\omega_o} = \frac{M_o}{H_o} \tag{26}$$

The real part vanishes as the frequency ω approaches infinity. It is obvious from (24b) that

$$\chi' = \text{positive} \qquad \text{when} \quad \omega < \frac{\omega_o}{\sqrt{1 - \alpha^2}} \tag{27a}$$

$$= \text{negative} \qquad \text{when} \quad \omega > \frac{\omega_o}{\sqrt{1 - \alpha^2}} \tag{27b}$$

$$= 0 \qquad \text{when} \quad \omega = \frac{\omega_o}{\sqrt{1 - \alpha^2}} \tag{27c}$$

For small value of damping parameter α, the value of χ' can have a maximum value when the denominator D is minimum. This behavior can be more conveniently investigated by introducing the following variables. At $\omega_r \approx \omega_0$ and $\Delta\omega \ll \omega_0$, let

$$\omega_r^2 \equiv (1 + \alpha^2)\omega^2 \tag{28a}$$

$$\Delta\omega \equiv \omega_0 - \omega_r \tag{28b}$$

$$\omega_0^2 - \omega_r^2 = (\omega_0 + \omega_r)(\omega_0 - \omega_r) \approx 2\omega_0\Delta\omega \tag{28c}$$

Then

$$D = (2\omega_0\Delta\omega)^2 + k \qquad k \equiv 4\omega^2\omega_0^2\alpha^2 \approx \frac{4\omega_0^4\alpha^2}{1 + \alpha^2}$$

$$\approx 4\omega_0^4 D_0 \tag{29a}$$

$$D_0 \equiv \frac{D}{4\omega_0^4} = \frac{1}{4}[1 - (1 + \alpha^2)\Omega^2]^2 + (\Omega\alpha)^2$$

$$= \Delta\Omega^2 + \frac{\alpha^2}{1 + \alpha^2} \tag{29b}$$

$$\Delta\Omega \equiv \frac{\Delta\omega}{\omega_0} \qquad\qquad \Omega \equiv \frac{\omega}{\omega_0} \tag{29c}$$

$$\chi' = \frac{\omega_m\omega_0}{D}\left[\omega_0^2 - \frac{1 - \alpha^2}{1 + \alpha^2}\omega_r^2\right]$$

$$= \frac{\omega_m\omega_0}{D}\left[\omega_0^2 - (1 - \alpha^2)(1 - \alpha^2 + \alpha^4 - \alpha^6 + \cdots)\omega_r^2\right] \tag{29d}$$

$$\approx \frac{2\Omega_m}{4D_0}\left[\Delta\Omega + \frac{\alpha^2}{1 + \alpha^2}\right] \tag{29e}$$

$$\Omega_m \equiv \frac{\omega_m}{\omega_0} \tag{29f}$$

The frequencies at which the extreme values of χ' exist can be determined by defferentiating (29b) with respect to $\Delta\omega$ and set the resultant expression equal to zero.

$$\frac{d\,\chi'}{d(\Delta\omega)} = 2\omega_m\omega_o^2 \left[\frac{1}{D} - \frac{\Delta\omega}{D^2} \frac{dD}{d(\Delta\omega)} \right] = 0$$

$$D - \Delta\omega [2(2\omega_o\Delta\omega)2\omega_o] = 0$$

$$4(\omega_o\Delta\omega)^2 = k = 4\omega_o^4\alpha^2$$

$$\Delta\omega = \frac{\pm\alpha\omega_o}{\sqrt{1+\alpha^2}} \approx \frac{\pm\alpha\omega_r}{\sqrt{1+\alpha^2}} = \pm\,\alpha\omega \equiv \Delta\omega_\pm \tag{30}$$

From (28)

$$\Delta\omega = \omega_o - \omega_r = \omega_o - \sqrt{1+\alpha^2}\omega = \pm\,\alpha\omega$$

or

$$\omega = \frac{\omega_o}{\sqrt{1+\alpha^2} \pm \alpha} = \omega_o \left[\sqrt{1+\alpha^2} \mp \alpha \right] \equiv \omega_\pm \tag{31}$$

where ω_+ is the frequency corresponding to $\Delta\omega_+ \equiv +\alpha\omega$, and ω_- corresponds to $\Delta\omega_- \equiv -\alpha\omega$.
The substitution of $\omega = \omega_+$ into (29) yields

$$D(\omega{=}\omega_+) = (2\omega_o\alpha\omega_+)^2 + k$$

$$= 8\omega_o^2\alpha^2\omega_+^2$$

$$= 8\omega_o^4\alpha^2 \left[1 - 2\alpha\sqrt{1+\alpha^2} + 2\alpha^2 \right] \tag{32a}$$

$$\chi'_{max} = \chi'(\omega{=}\omega_+) = \frac{2\omega_m\omega_o^2}{D} \alpha\omega_+$$

$$= \frac{2\alpha\omega_m\omega_o^3 \left[\sqrt{1+\alpha^2} - \alpha \right]}{8\omega_o^4\alpha^2 \left[1 - 2\alpha\sqrt{1+\alpha^2} + 2\alpha^2 \right]}$$

$$\approx \frac{\omega_m}{4\omega_o\alpha} \frac{1-\alpha}{1-2\alpha} \tag{32b}$$

At $\Delta\omega = -\alpha\omega$ and $\omega = \omega_-$, then one has

$$D = 8\omega_0^2\alpha^2\omega_-^2$$

$$= 8\omega_0^4\alpha^2 \left[1 + 2\alpha\sqrt{1 + \alpha^2} + 2\alpha^2 \right] \tag{33a}$$

$$\chi'_{min} = \chi'(\omega=\omega_-) = \frac{2\omega_m\omega_0^2}{D} [-\alpha\omega_-]$$

$$= \frac{-2\omega_0^3\omega_m\alpha \left[\sqrt{1 + \alpha^2} + \alpha \right]}{8\omega_0^4\alpha^2 \left[1 + 2\alpha\sqrt{1 + \alpha^2} + 2\alpha^2 \right]}$$

$$\approx \frac{-\omega_m}{4\omega_0\alpha} \frac{1 + \alpha}{1 + 2\alpha} \tag{33b}$$

The frequency response of the real part of $\overline{\chi}$, χ', is sketched in Fig. 2-1.

The imaginary part of $\overline{\chi}$ is zero at $\omega = 0$ as well at ω approaches infinity. Introducing the notation of (28), then, χ'' has the form

$$\chi'' = \frac{\omega\omega_m\alpha (\omega_0^2 + \omega_r^2)}{(2\omega_0\Delta\omega)^2 + k} \tag{34a}$$

$$\approx \frac{1}{4D_0} \frac{2\Omega_m\alpha}{\sqrt{1 + \alpha^2}} \tag{34b}$$

where D_0 is given by (29b). Equation (34) exhibits a maximum when $\Delta\omega = 0$, i.e., at $\omega_0 = \omega_r$,

$$\chi''_{max} = \chi''(\omega_0=\omega_r) = \frac{\omega_0\omega_m\alpha}{k} (\omega_0^2 + \omega_0^2)$$

$$= \frac{\omega_m}{2\omega_0} \frac{1 + \alpha^2}{\alpha} \tag{35}$$

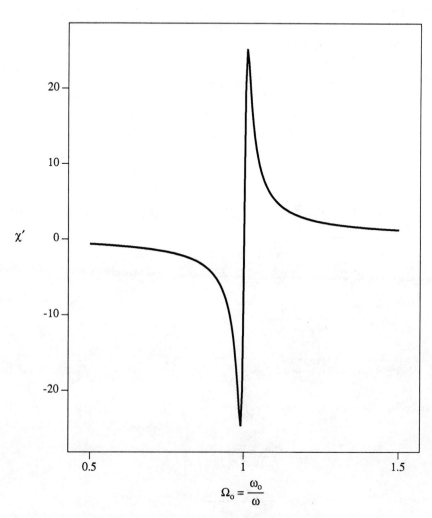

Figure 2-1: χ' vs Ω_O, ($\Omega_m = 1.0$ and $\alpha = 0.01$)

The frequency response of the imaginary part of $\overline{\chi}$, χ'', is sketched in Fig. 2-2. The width of the bell curve at magnitude equal to one-half that of the maximum occurs when the denominator of (34) is equal to 2k, i.e.,

$$(2\omega_o\Delta\omega)^2 = k \qquad \text{or} \qquad \Delta\omega = \frac{\alpha\omega_o}{\sqrt{1 + \alpha^2}} \tag{36}$$

The frequency response of the real and imaginary parts of K can be investigated similarly.

The real part K' vanishes at $\omega = 0$ and at $\omega = \infty$. With the notation of (28), K' has the form

$$K' = \frac{-\omega\omega_m}{D} [\omega_o^2 - \omega_r^2] \tag{37a}$$

$$\approx \frac{-1}{4D_o} \frac{2\Omega_m\Delta\Omega}{\sqrt{1 + \alpha^2}} \qquad \omega_r \approx \omega_o \tag{37b}$$

Hence,

$$K' > 0 \qquad \text{for} \qquad \omega_r^2 > \omega_o^2 \tag{38a}$$

$$K' = 0 \qquad \text{for} \qquad \omega_r^2 = \omega_o^2 \tag{38b}$$

$$K' < 0 \qquad \text{for} \qquad \omega_r^2 < \omega_o^2 \tag{38c}$$

With the exception of the minus sign, Eq. (37a) has a similar format to (29b); therefore the frequency response of K' has the inverted shape of Fig. 2-1. This is shown in Fig. 2-3.

The imaginary part K", (25c), vanishes at $\omega = 0$ and at $\omega = \infty$. With the notation of (28), K" becomes

$$K'' \approx \frac{-2\omega^2\omega_m\omega_o\alpha}{(2\omega_o\Delta\omega)^2 + k} \tag{39a}$$

$$= \frac{-1}{4D_o} \frac{2\alpha\Omega_m}{\sqrt{1 + \alpha^2}} \tag{39b}$$

Equation (39) has the same form as (34), and hence, the frequency response of K" has the inverted appearance of Fig. 2-2. This is shown in Fig. 2-4.

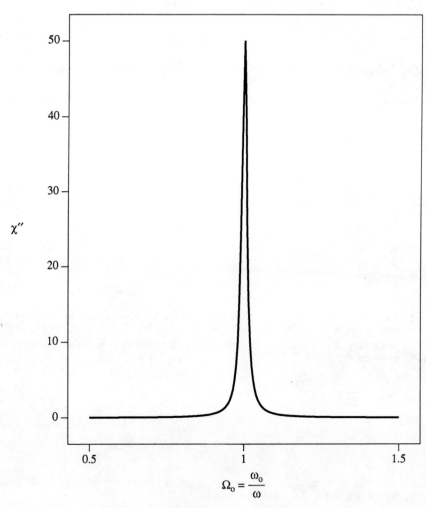

Figure 2-2: χ'' vs Ω_o, ($\Omega_m = 1.0$ and $\alpha = 0.01$)

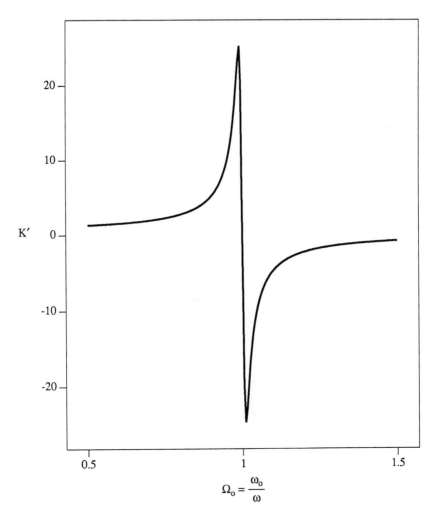

Figure 2-3: K′ vs Ω_O, ($\Omega_m = 1.0$ and $\alpha = 0.01$)

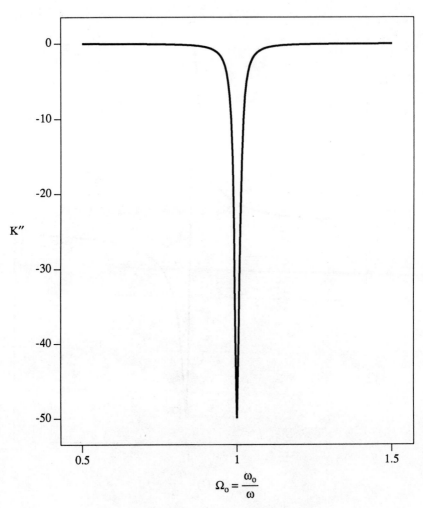

Figure 2-4: K″ vs Ω_O, (Ω_m = 1.0 and α = 0.01)

2.1. Components of Susceptance Elements

Equation (2-22d) can be expanded as follows.

$$\bar{\chi} = \frac{\omega_m (\omega_0 + j\omega\alpha)}{(\omega_0 + j\omega\alpha)^2 - \omega^2}$$

$$= \frac{\omega_m (\omega_0 + j\omega\alpha)}{\omega_0^2 - (1 + \alpha^2) \omega^2 + j2\omega\omega_0\alpha}$$

$$= \frac{\omega_m (\omega_0 + j\omega\alpha)}{\omega'' + j2\omega\omega_0\alpha} \tag{1a}$$

$$\omega'' = \omega_0^2 - (1 + \alpha^2) \omega^2 \tag{1b}$$

Multiplying and dividing (1a) by the complex conjugate of its denominator yields

$$\bar{\chi} = \frac{1}{D} [\omega_m\omega_0(\omega'' + 2\omega^2\alpha^2) + j\omega\omega_m\alpha(\omega'' - 2\omega_0^2)] \equiv \chi' - j\chi'' \tag{2a}$$

$$D = \omega''^2 + 4\omega^2\omega_0^2\alpha^2 \tag{2b}$$

$$\chi' = \frac{\omega_m\omega_0}{D} [\omega_0^2 - (1 + \alpha) \omega^2 + 2\omega^2\alpha^2]$$

$$= \frac{\omega_m\omega_0}{D} [(\omega_0^2 - \omega^2) + \omega^2\alpha^2] \tag{2c}$$

and

$$\chi'' = \frac{-\omega\omega_m\alpha}{D} [\omega_0^2 - (1 + \alpha^2) \omega^2 - 2\omega_0^2]$$

$$= \frac{\omega\omega_m\alpha}{D} [\omega_0^2 + (1 + \alpha^2) \omega^2] \tag{2d}$$

Equation (2-22e) can be expanded as

$$K = \frac{-\omega\omega_m}{(\omega_o + j\omega\alpha)^2 - \omega^2}$$

$$= \frac{-\omega\omega_m}{\omega_o^2 + j2\omega\omega_o\alpha - \omega^2\alpha^2 - \omega^2}$$

$$= \frac{-\omega\omega_m}{\omega'' + j2\omega\omega_o\alpha} \tag{3}$$

where ω'' is defined by (1b). The rationalization of (3) yields

$$K = \frac{-\omega\omega_m}{D} [\omega'' - j2\omega\omega_o\alpha] \equiv K' - jK'' \tag{4a}$$

$$K' = \frac{-\omega\omega_m\omega''}{D} = \frac{-\omega\omega_m}{D} [\omega_o^2 - \omega^2(1 + \alpha^2)] \tag{4b}$$

$$K'' = \frac{-1}{D} 2\omega^2\omega_m\omega_o\alpha \tag{4c}$$

3. Scalar Susceptibility

(a) The RF magnetization vector $\vec{M}'$ and the RF magnetic field $\vec{H}'$ are related by the susceptance tensor, (2-23).

$$[M'] = [\chi] [H'] \tag{1a}$$

$$[\chi] = \begin{bmatrix} \overline{\chi} & -jK & 0 \\ jK & \overline{\chi} & 0 \\ 0 & 0 & 0 \end{bmatrix} \tag{1b}$$

$$[M'] = [M_x' \quad M_y' \quad M_z']^t \tag{1c}$$

$$[H'] = [H_x' \quad H_y' \quad H_z']^t \tag{1d}$$

The elements $\overline{\chi}$ and K are defined by (2-22). Equation (1a) becomes a scalar relation if the following condition is satisfied:

$$[\chi] [H'] = \lambda^\chi [H'] \tag{2}$$

where λ^χ is a scalar. Equation (2) is the eigenvalue equation of $[\chi]$. In order to have a scalar $[\chi]$, the RF field $\vec{H}'$ should be adjusted to be the eigenvector of $[\chi]$. When the RF field is adjusted to satisfy (2), the system is said to be operating in its normal mode. The eigenvalue λ^χ can be determined from the characteristic equation.

$$\det \left| \; [\chi] - \lambda^\chi [u] \; \right| = \begin{bmatrix} \overline{\chi} - \lambda^\chi & -jK & 0 \\ jK & \overline{\chi} - \lambda^\chi & 0 \\ 0 & 0 & \lambda^\chi \end{bmatrix}$$

$$= -\lambda^\chi [(\overline{\chi} - \lambda^\chi)^2 - K^2] = 0 \tag{3}$$

The eigenvalues are, with the superscript removed,

$$\lambda_1 = 0 \tag{4a}$$

$$\lambda_2 = \overline{\chi} + K \equiv \overline{\chi}_- \tag{4b}$$

$$\lambda_3 = \overline{\chi} - K \equiv \overline{\chi}_+ \tag{4c}$$

where $\overline{\chi}_\pm$ are the susceptibilities for the positive and negative circularly polarized modes, respectively. This will be obvious later [Eqs. (25) and (27)]. There is one normal mode corresponding to each eigenvalue.

The real and imaginary parts of $\overline{\chi}_\pm$ can be obtained by substituting (2-24) and (2-25) into (4).

$$\overline{\chi}_+ = (\chi' - j\chi'') - (K' - jK'') = \chi_+' - j\chi_+'' \tag{5a}$$

$$\chi_+' = \chi' - K'$$

$$= \frac{\omega_m}{D} \left[\omega_o [\omega_o^2 - (1 - \alpha^2)\omega^2] + \omega[\omega_o^2 - (1 + \alpha^2)\omega^2] \right]$$

$$= \frac{\omega_m}{D} \left[(\omega_o^2 - \omega^2)(\omega_o + \omega) + \alpha^2\omega^2(\omega_o - \omega) \right]$$

$$= \frac{\omega_m}{D} (\omega_o - \omega) \left[(\omega_o + \omega)^2 + \alpha^2\omega^2 \right] \tag{5b}$$

$$\chi_+'' = \chi'' - K''$$

$$= \frac{\omega\omega_m\alpha}{D} \left[\omega_o^2 + (1 + \alpha^2)\omega^2 + 2\omega\omega_o \right]$$

$$= \frac{\omega\omega_m\alpha}{D} \left[(\omega_o + \omega)^2 + \alpha^2\omega^2 \right] \tag{5c}$$

The components of $\overline{\chi}_-$ can be obtained similarly.

$$\overline{\chi}_- = \overline{\chi} + K = \chi_-' - j\chi_-'' \tag{6a}$$

$$\chi_-' = \chi' + K' = \frac{\omega_m}{D} (\omega_o + \omega) \left[(\omega_o - \omega)^2 + \alpha^2\omega^2 \right] \tag{6b}$$

$$\chi_-'' = \chi'' + K'' = \frac{\omega\omega_m\alpha}{D} \left[(\omega_o - \omega)^2 + \alpha^2\omega^2 \right] \tag{6c}$$

The frequency response of $\overline{\chi}_+$ can be estimated as follows. The values of the real part of $\overline{\chi}_+$ at $\omega = 0$ and at $\omega = \infty$ are

$$\chi'_+(\omega=0) = \frac{\omega_m}{\omega_o} \tag{7a}$$

$$\chi'_+(\omega=\infty) = 0 \tag{7b}$$

At $\omega_r \approx \omega_0$, one has [Eq. (2-28)]

$$\omega_0 - \omega = (\omega_r + \Delta\omega) - a\omega_r \qquad\qquad a \equiv \frac{1}{\sqrt{1 + \alpha^2}}$$

$$= (1 - a)\omega_r + \Delta\omega$$

$$\approx (1 - a)\omega_0 + \Delta\omega$$

$$= \omega_0[1 - a + \Delta\Omega] \qquad\qquad \Delta\Omega \equiv \frac{\Delta\omega}{\omega_0} \qquad\qquad (8a)$$

$$(\omega_0 + \omega)^2 + \alpha^2\omega^2 = (\omega_0^2 + 2\omega\omega_0 + \omega^2) + \alpha^2\omega^2$$

$$= \omega_0^2 + 2a\omega_r\omega_0 + \omega_r^2$$

$$\approx 2\omega_0^2(1 + a) \qquad\qquad (8b)$$

Then (5b) becomes [Eqs.(2-29e) and (2-37a)]

$$\chi'_+ = \frac{2\Omega_m}{4D_o}\left[(1 + a)\Delta\Omega + \frac{\alpha^2}{1 + \alpha_2}\right] \qquad\qquad (9)$$

$$D_o = \Delta\Omega^2 + \frac{\alpha^2}{1 + \alpha^2} \qquad\qquad (2\text{-}29b)$$

Equation (5b) indicates that

$$\chi'_+ = \text{positive} \qquad \text{when} \qquad \omega_0 > \omega \qquad\qquad (10a)$$
$$= 0 \qquad\qquad \text{when} \qquad \omega_0 = \omega \qquad\qquad (10b)$$
$$= \text{negative} \qquad \text{when} \qquad \omega_0 < \omega \qquad\qquad (10c)$$

To determine the location of the maximum value of χ'_+, take the derivative of χ'_+ with respect to $\Delta\omega$ and set the resultant expression to zero. The frequency response of χ'_+ is sketched in Fig. 3-1.

The imaginary part of $\overline{\chi}_+$, χ''_+, vanishes at $\omega = 0$ and at $\omega = \infty$. At $\omega_r \approx \omega_0$, χ''_+ can be expressed as [Eqs. (2-34b) and (2-39b)]

$$\chi''_+ \approx \frac{2\Omega_m\alpha}{4D_o} \frac{\sqrt{1 + \alpha^2} + \alpha}{1 + \alpha^2} \qquad\qquad (11)$$

χ''_+ has the same format as χ'' [Eq. (2-34)]; hence its frequency response, Fig. 3-2, is similar to Fig. 2-2.

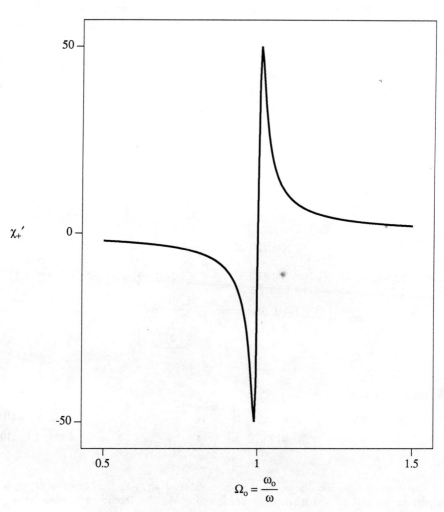

Figure 3-1: χ_+' vs Ω_0, ($\Omega_m = 1.0$ and $\alpha = 0.01$)

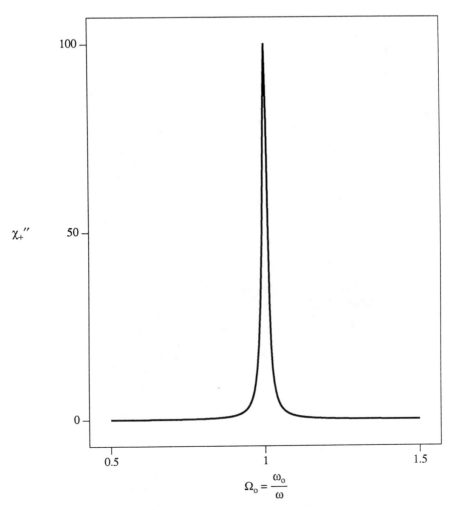

Figure 3-2: χ_+'' vs Ω_o, ($\Omega_m = 1.0$ and $\alpha = 0.01$)

The real part of $\overline{\chi}_-$ vanishes at $\omega = \infty$ and has a static value at $\omega = 0$.

$$\chi'_-(\omega{=}0) = \frac{\omega_m}{\omega_o} \tag{12}$$

For $\omega_r \approx \omega_o$, χ'_- becomes

$$\chi'_- \approx \frac{2\Omega_m}{4D_o} \left[\Delta\Omega \left(1 - \frac{1}{\sqrt{1 + \alpha^2}} \right) + \frac{\alpha^2}{1 + \alpha^2} \right\} \tag{13}$$

By setting the derivative of χ'_- with respect to $\Delta\Omega$ equal to zero, one can determine the location of the maximum value of χ'_-.

The frequency response of χ'_- is shown in Fig. 3-3. Figure 3-3a shows the resonance effect at $\Delta\Omega = 0$. It is to be noted that χ'_- in Fig. 3-3a does not approach its static value as the frequency approaches zero, since Eq. (13) is valid only for small values of $\Delta\Omega$.

The imaginary part χ''_- vanishes at $\omega = 0$ and at $\omega = \infty$. At $\omega_r \approx \omega_o$, χ''_- has the form

$$\chi''_- \approx \frac{2\Omega_m\alpha}{4D_o} \frac{\sqrt{1 + \alpha^2} - \alpha}{1 + \alpha^2} \tag{15}$$

The frequency response of χ''_- is shown in Fig. 3-4.

(b) The eigenvectors are determined by the substitution of each eigenvalue into (2). Expanding (2), with the use of (1b), one gets,

$$\overline{\chi}H_x' - jKH_y' = \lambda^x H_x' \quad \text{or} \quad -jKH_y' = (\lambda^x - \overline{\chi}) H_x' \tag{16a}$$

$$jKH_x' + \overline{\chi}H_y' = \lambda^x H_y' \quad \text{or} \quad jKH_x' = (\lambda^x - \overline{\chi}) H_y' \tag{16b}$$

$$0 = \lambda^x H_z' \tag{16c}$$

For $\lambda^x = \lambda_1 = 0$, (16) gives

$$-jKH_y' = -\overline{\chi}H_x' \tag{17a}$$

$$jKH_x' = -\overline{\chi}H_y' \tag{17b}$$

$$0 = 0\, H_z' \tag{17c}$$

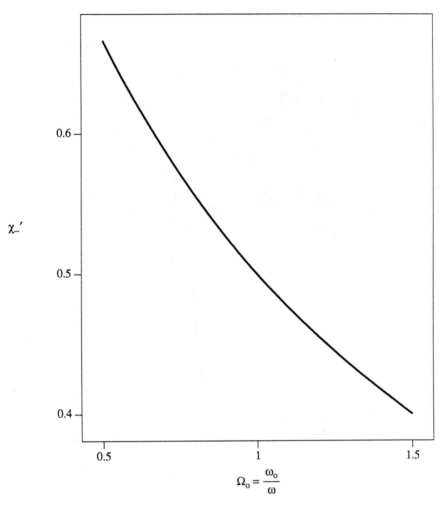

Figure 3-3: χ_-' vs Ω_O, $(\Omega_m = 1.0$ and $\alpha = 0.01)$

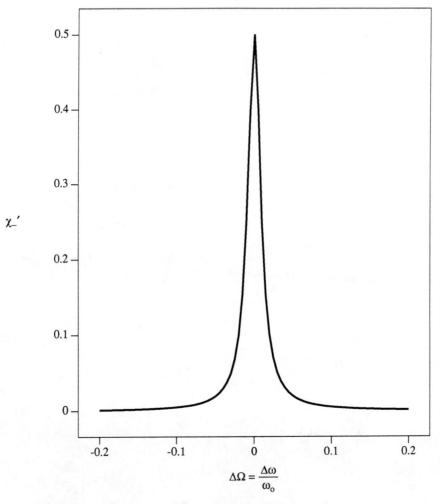

Figure 3-3a: χ_-' vs $\Delta\Omega$, ($\Omega_m = 1.0$ and $\alpha = 0.01$)

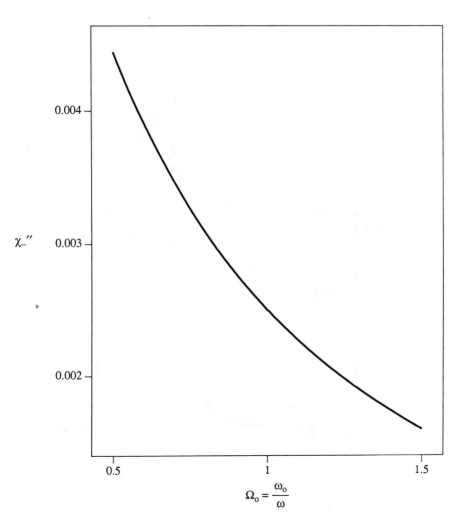

Figure 3-4: χ_-'' vs Ω_o, ($\Omega_m = 1.0$ and $\alpha = 0.01$)

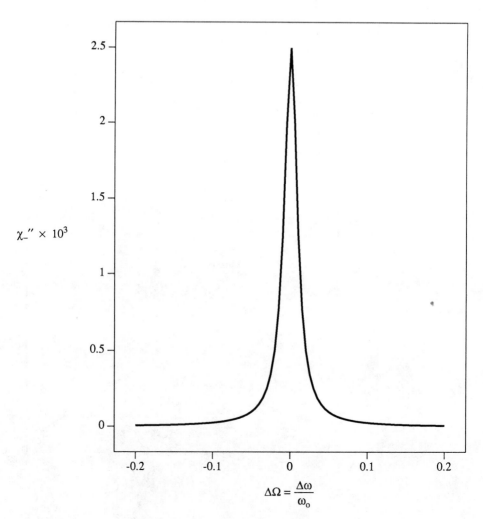

Figure 3-4(a): χ_-'' vs $\Delta\Omega$, ($\Omega_m = 1.0$ and $\alpha = 0.01$)

Any arbitrary value of H_z' will satisfy (17c); therefore,

$$H_z' = h_o \tag{18}$$

where h_o is any constant. From (17b), one has

$$H_y' = \frac{-jK}{\overline{\chi}} H_x' \tag{19}$$

The substitution of (19) into (17a) yields

$$[(-jK)^2 + \overline{\chi}^2] H_x' = 0 \quad \text{or} \quad H_x' = 0 \tag{20}$$

Consequently, (19) gives

$$H_y' = 0 \tag{21}$$

The normalized eigenvector for $\lambda^{\chi} = \lambda_1 = 0$ is therefore

$$[H_1] = [0 \quad 0 \quad 1]^t \tag{22}$$

The eigenvalue

$$\lambda^{\chi} = \lambda_1 = 0$$

requires

$$H_x' = 0 = H_y'$$

Consequently,

$$M_x' = 0 = M_y' \quad \text{since} \quad [M'] = [\chi] [H']$$

This type of field will not interact with the polarizing $\vec{H}_o$-field, (2-15), and is of little special interest.

For $\lambda^{\chi} = \lambda_2 = \overline{\chi} + K$, (16) becomes

$$-jKH_y' = (\overline{\chi} + K - \overline{\chi}) H_x' = KH_x' \tag{23a}$$

$$jKH_x' = KH_y' \tag{23b}$$

$$0 = (\overline{\chi} + K) H_z' \tag{23c}$$

Thus,

$$H_y' = jH_x' \tag{24a}$$

$$H_z' = 0 \tag{24b}$$

and the corresponding normalized eigenvector is

$$[H_2] = H_x' [\, 1 \quad j \quad 0 \,]^t = \frac{1}{\sqrt{2}} [\, 1 \quad j \quad 0 \,]^t = [H_-] \tag{25}$$

This is a negative or left-hand circularly polarized traveling wave in the positive z-direction [Section VII.10(d)]; hence it is represented by $[H_-]$.

Similarly, for $\lambda^x = \lambda_3 = \overline{\chi} - K$, (16) becomes

$$-jKH_y' = -KH_x' \tag{26a}$$

$$jKH_x' = -KH_y' \tag{26b}$$

$$H_z' = 0 \tag{26c}$$

The corresponding normalized eigenvector is

$$[H_3] = [H_+] = \frac{1}{\sqrt{2}} [\, 1 \quad -j \quad 0 \,]^t \tag{27}$$

This is a positive or right-hand circularly polarized traveling wave in the positive z-direction [Section VII.10(c)].

These three normalized eigenvectors form an orthonormal set.

When ferrite is operating in its normal modes, it behaves like a conventional magnetic material with a scalar susceptibility. These normal modes are the positive and negative circularly polarized waves. These two normal modes have different susceptibilities as given by (4).

(c) The corresponding scalar permeability can be similarly obtained [Eq. (2-12)].

$$\vec{B}' = \mu_o (\vec{H}' + \vec{M}') = \mu_o ([u] + [\chi]) \vec{H}' = [\mu] \vec{H}' \tag{28a}$$

$$[\mu] = \mu_o ([u] + [\chi]) \tag{28b}$$

For scalar susceptibility, $[\chi] = \lambda^\chi$ [Eqs. (2) and (2-14)]; then (28b) becomes

$$\mu_i = \mu_o (1 + \lambda_i^\chi) \tag{29}$$

Therefore, (4) and (29) give

$$\mu_1 = \mu_o (1 + \lambda_1) = \mu_o \tag{30a}$$

$$\mu_2 = \mu_- = \mu_o (1 + \overline{\chi}_-) = \mu_o (1 + \overline{\chi} + K) = \mu + k \tag{30b}$$

$$\mu_3 = \mu_+ = \mu - k \tag{30c}$$

$$\mu = \mu_o (1 + \overline{\chi}) \tag{30d}$$

$$k = \mu_o K \tag{30e}$$

4. Faraday Rotation

(a) A linearly polarized wave is propagating in an infinite ferrite medium magnetized in the direction of propagation. Let the RF $\vec{H}$-field be represented by

$$\vec{H}' = \vec{h}\, e^{j(\omega t - \gamma_z z)} = (\hat{x} h_x + \hat{y} h_y)\, e^{j(\omega t - \gamma_z z)} \tag{1}$$

where h_x and h_y are field amplitudes which are constants for uniform plane waves. Maxwell's equations for ferrite medium are

$$\nabla \times \vec{E}' = -j\omega\vec{B}' = -j\omega\,[\mu]\,\vec{H}' \tag{2a}$$

$$\nabla \times \vec{H}' = j\omega\varepsilon\vec{E}' \tag{2b}$$

Then

$$\nabla \times \nabla \times \vec{H}' = j\omega\varepsilon\nabla \times \vec{E}' = \omega^2\varepsilon\,[\mu]\,\vec{H}' \tag{3a}$$

The permeability tensor is given by (2-14).

$$[\mu] = \begin{bmatrix} \mu & -jk & 0 \\ jk & \mu & 0 \\ 0 & 0 & \mu_o \end{bmatrix} \tag{3b}$$

The substitution of (1) into the left-hand side of (3a) gives

$$\nabla \times \nabla \times \vec{H}' = \nabla \times \nabla \times \left[\vec{h}\, e^{j(\omega t - \gamma_z z)} \right]$$

$$= \nabla \times \left[\hat{z} \times \vec{h}\, \frac{\delta}{\delta z} \left(e^{j(\omega t - \gamma_z z)} \right) \right]$$

$$= \hat{z} \times (\hat{z} \times \vec{h})\, \frac{\delta^2}{\delta z^2} \left(e^{j(\omega t - \gamma_z z)} \right)$$

$$= \vec{h}\gamma_z^2 e^{j(\omega t - \gamma_z z)} = \gamma_z^2\vec{H}' \tag{4}$$

Hence, (3a) becomes

$$\gamma_z^2\vec{H}' = \omega^2\varepsilon\,[\mu]\,\vec{H}' \tag{5}$$

This is the eigenvalue equation for the $[\mu]$-matrix. The eigenvalues γ can be determined from the characteristic equation.

$$\det \left| \omega^2 \varepsilon [\mu] - \gamma_z^2 [u] \right| = 0$$

$$\begin{bmatrix} \omega^2\varepsilon\mu - \gamma_z^2 & -j\omega^2\varepsilon k \\ j\omega^2\varepsilon k & \omega^2\varepsilon\mu - \gamma_z^2 \end{bmatrix} = 0$$

or

$$(\omega^2\varepsilon\mu - \gamma_z^2)^2 - (\omega^2\varepsilon k)^2 = 0$$

$$\omega^2\varepsilon\mu - \gamma_z^2 = \pm\omega^2\varepsilon k$$

$$\gamma_z^2 = \omega^2\mu\varepsilon \mp \omega^2\varepsilon k$$

$$\gamma_z = \pm\omega\sqrt{\varepsilon(\mu \mp k)} \tag{6}$$

Equation (6) has four roots.

$$\gamma_1 = \omega\sqrt{\varepsilon(\mu + k)} = \gamma_{F-} \tag{7a}$$

$$\gamma_2 = \omega\sqrt{\varepsilon(\mu - k)} = \gamma_{F+} \tag{7b}$$

$$\gamma_3 = -\omega\sqrt{\varepsilon(\mu - k)} = \gamma_{B-} \tag{7c}$$

$$\gamma_4 = -\omega\sqrt{\varepsilon(\mu + k)} = \gamma_{B+} \tag{7d}$$

where the subscripts F and B refer to the forward and the backward traveling wave, respectively; the + and − refer to right-hand and left-hand circular polarization with respect to the direction of propagation. The validity of the physical meaning of these notations is verified in Section 4.2. It is to be noted that the right-hand circularly polarized wave with respect to the negative z-axis (the direction of propagation of a $-\hat{z}$ traveling wave) is left-hand polarized with respect to the positive $\hat{z}$-axis; see Section 4.1.

(b) To obtain the behavior of a linearly polarized plane wave in a ferrite magnetized in the direction of propagation, the z-axis, it is sufficient to consider just one component of the $\vec{H}'$-field as given by (1). The other component can be treated similarly. Consider the x-component only.

$$\vec{H}' = \hat{x}h_x e^{j\Omega} \tag{8a}$$

$$\Omega = \omega t - \beta z \tag{8b}$$

where $\gamma_z = \beta$ for the lossless case. Equation (8a) can be rearranged as

$$\vec{H}' = \left[\hat{x} + \frac{1}{2} (j\hat{y} - j\hat{y}) \right] h_x e^{j\Omega}$$

$$= \frac{1}{2} [(\hat{x} - j\hat{y}) + (\hat{x} + j\hat{y})] h_x e^{j\Omega}$$

$$= \vec{H}_+ + \vec{H}_- \tag{9a}$$

where

$$\Omega \equiv \omega t - \beta z$$

$$\vec{H}_+ \equiv \frac{1}{2} (\hat{x} - j\hat{y}) h_x e^{j\Omega}$$

$$= \frac{h_x}{2} \left[\hat{x} e^{j\Omega} + \hat{y} e^{j(\Omega - \frac{\pi}{2})} \right]$$

$$= \text{right-hand circularly polarized wave at a given location z} \tag{9b}$$

and

$$\vec{H}_- \equiv \frac{h_x}{2} [\hat{x} e^{j\Omega} + \hat{y} e^{j(\Omega + \frac{\pi}{2})}]$$

$$= \text{left-hand circularly polarized wave at a given location z} \tag{9c}$$

A linearly polarized wave can be decomposed into a positive (right-hand) and a negative (left-hand) circularly polarized wave. Since these waves have different propagation constants, the symbols in (9a) should be modified as follows.

$$\vec{H}' = \frac{h_x}{2} \left[\left[\hat{x} e^{j\Omega_+} + \hat{y} e^{j(\Omega_+ - \frac{\pi}{2})} \right] + \left[\hat{x} e^{j\Omega_-} + \hat{y} e^{j(\Omega_- + \frac{\pi}{2})} \right] \right]$$

$$= \frac{h_x}{2} \left[\hat{x} \left[e^{j\Omega_+} + e^{j\Omega_-} \right] + \hat{y} \left[e^{j(\Omega_+ - \frac{\pi}{2})} + e^{j(\Omega_- + \frac{\pi}{2})} \right] \right] \tag{10a}$$

$$\Omega_+ = \omega t - \beta_+ z \tag{10b}$$

$$\Omega_- = \omega t - \beta_- z \tag{10c}$$

$\beta_\pm$ are the phase constants for the right-hand and the left-hand polarized wave, respectively. Terms within the brackets can be combined as follows.

$$e^{j\Omega_+} + e^{j\Omega_-} = e^{j(\omega t - \beta_1 z)} [e^{-j\beta_2 z} + e^{j\beta_2 z}]$$

$$= 2 \cos \beta_2 z \; e^{j(\omega t - \beta_1 z)} \tag{11a}$$

$$\beta_1 = \frac{\beta_+ + \beta_-}{2} \qquad \text{and} \qquad \beta_2 = \frac{\beta_+ - \beta_-}{2} \tag{11b}$$

or

$$\beta_+ = \beta_1 + \beta_2 \qquad \text{and} \qquad \beta_- = \beta_1 - \beta_2 \tag{11c}$$

and

$$e^{j(\Omega_+ - \frac{\pi}{2})} + e^{j(\Omega_- + \frac{\pi}{2})} = e^{j(\omega t - \beta_1 z)} \left[e^{-j(\beta_2 z + \frac{\pi}{2})} + e^{j(\beta_2 z + \frac{\pi}{2})} \right]$$

$$= -2 \sin \beta_2 z \; e^{j(\omega t - \beta_1 z)} \tag{11d}$$

The substitution of (11) into (10a) yields

$$\vec{H}' = h_x [\hat{x} \cos \beta_2 z - \hat{y} \sin \beta_2 z] \; e^{j(\omega t - \beta_1 z)} \tag{12}$$

This is a linearly polarized wave, and the plane of polarization makes an angle θ with the x-axis. The angle θ is given by

$$\theta = \tan^{-1} \frac{-\sin \beta_2 z}{\cos \beta_2 z} = -\tan^{-1} [\tan \beta_2 z] = -\beta_2 z$$

$$= \frac{-1}{2} (\beta_+ - \beta_-) z \tag{13}$$

The plane of polarization rotates as a function of the distance of propagation z. A linearly polarized plane wave polarized in the x-direction [Eq. (8)] is equivalent to a wave with components in both the x-direction and the y-direction [Eq. (12)] and propagates with an effective phase constant

$$\beta = \beta_1 = \frac{1}{2} (\beta_+ + \beta_-)$$

and its plane of polarization undergoes a rotation of θ given by (13) as it progresses along the z-axis.

The above analysis can be repeated for a wave traveling in the negative z-direction. Consider the linearly polarized wave

$$\vec{H}'' = \hat{x} h_x e^{j\phi} \tag{14a}$$

$$\phi = \omega t + \beta z \tag{14b}$$

which is propagating in a ferrite material magnetized in the positive z-direction. This wave can be expressed as [Eq. (4.1-5)]

$$\vec{H}'' = h_x (\hat{x} \cos \beta_2 z - \hat{y} \sin \beta_2 z) e^{j(\omega t + \beta_1 z)} \tag{15}$$

where $\beta_{1,2}$ are defined by (11). This is a linearly polarized wave with plane of polarization inclined at an angle θ with respect to the x-axis.

$$\theta = - (\beta_+ - \beta_-) \frac{z}{2} \tag{16}$$

For a given distance of travel, $z = z_1$, the plane of polarization rotates by the same amount with respect to the x-axis for waves traveling in either the positive or the negative z-direction. The phase delays are also identical for both waves.

In other words, a wave traveling a distance z_1 in one direction will have its plane of polarization rotated through an angle θ_1 with respect to the x-axis. When this wave is reflected and returned back to its starting point, it is rotated again by θ_1. The total angle of rotation is $2\theta_1$ with respect to the incident wave, and the wave is not rotated back to its original orientation. This rotation of the plane of polarization is known as Faraday rotation, and it is nonreciprocal.

(c) For the general case with damping, the susceptance tensor is given by (2-22) and (2-23).

$$[\chi] = \begin{bmatrix} \overline{\chi} & -jK & 0 \\ jK & \overline{\chi} & 0 \\ 0 & 0 & 0 \end{bmatrix} \tag{17a}$$

$$\overline{\chi} = \frac{\omega_m}{\Delta} (\omega_o + j\omega\alpha) = \chi' - j\chi'' \tag{17b}$$

$$K = \frac{-\omega\omega_m}{\Delta} = K' - jK'' \tag{17c}$$

$$\Delta = (\omega_o + j\omega\alpha)^2 - \omega^2 \tag{17d}$$

The permeability tensor is given by (2-14).

$$[\mu] = \mu_o [u] + \mu_o [\chi]$$

$$= \begin{bmatrix} \mu & -jk & 0 \\ jk & \mu & 0 \\ 0 & 0 & \mu_o \end{bmatrix} \tag{18a}$$

$$\mu = \mu_o (1 + \overline{\chi}) = \mu' - j\mu'' \tag{18b}$$

$$k = \mu_o K = \mu_o (K' - jK'') = k' - jk'' \tag{18c}$$

Then the propagation constant, (6), becomes

$$\gamma_z \equiv \gamma_\pm = \pm j\omega \sqrt{\varepsilon (\mu \mp k)}$$

$$= \pm j\omega \sqrt{\varepsilon [(\mu' - j\mu'') \mp (k' - jk'')]}$$

$$= \pm j\omega \sqrt{\mu_o \varepsilon} \sqrt{(1 + j\chi'_\pm) - j\chi''_\pm} \equiv \gamma_\pm \tag{19}$$

where

$$\overline{\chi}_+ = \chi'_+ - j\chi''_+ \qquad \chi'_+ = \chi' - K' \qquad \chi'' = \chi'' - K'' \tag{3-5}$$

$$\overline{\chi}_- = \chi'_- - j\chi''_- \qquad \chi'_- = \chi' + K' \qquad \chi''_- = \chi'' + K'' \tag{3-6}$$

The propagation constant is complex in general. To evaluate the real and the imaginary parts of $\gamma_\pm$, let

$$\gamma_\pm = \alpha_\pm + j\beta_\pm \tag{20a}$$

$$A_\pm = 1 + \chi'_\pm \tag{20b}$$

$$B_\pm = \chi''_\pm \tag{20c}$$

$$C = \omega \sqrt{\mu_o \varepsilon} \equiv \beta_o \tag{20d}$$

Then (19) becomes, neglecting the subscripts $\pm$ for the time being,

$$\alpha + j\beta = \pm jC \sqrt{A - jB}$$

or

$$\alpha^2 - \beta^2 + j2\alpha\beta = -C^2 (A - jB) \tag{21}$$

This can be decomposed into two equations.

$$\alpha^2 - \beta^2 = -C^2 A \tag{22a}$$

$$2\alpha\beta = BC^2 \tag{22b}$$

α can be eliminated from (22a) by using (22b)

$$\left[\frac{BC^2}{2\beta} \right]^2 - \beta^2 = -AC^2$$

or

$$\beta^4 - (AC^2) \beta^2 - \left[\frac{BC^2}{2} \right]^2 = 0 \tag{23}$$

This is a quadratic equation and the solution is

$$\beta^2 = \frac{1}{2} \left[AC^2 \pm \sqrt{(AC^2)^2 + 4 \left[\frac{BC^2}{2} \right]^2} \right]$$

$$= \frac{C^2}{2} \left[A + \sqrt{A^2 + B^2} \right] \tag{24}$$

Only the plus sign before the radical is retained since β is a positive real number by assumption.

$$\beta = \frac{C}{\sqrt{2}} \sqrt{A + \sqrt{A^2 + B^2}} \tag{25}$$

α can be obtained from (22b).

$$\alpha = \frac{BC^2}{2\beta} = \frac{BC^2}{2\dfrac{C}{\sqrt{2}}\sqrt{A + \sqrt{A^2 + B^2}}}$$

$$= \frac{CB}{\sqrt{2}}\frac{\sqrt{\sqrt{A^2 + B^2} - A}}{\sqrt{(A^2 + B^2) - A^2}}$$

$$= \frac{C}{\sqrt{2}}\sqrt{\sqrt{A^2 + B^2} - A} \tag{26}$$

The substitution of (20) into (25) and (26) yields

$$\beta_{\pm} = \omega\sqrt{\mu_0\varepsilon/2}\left[\sqrt{(1+\chi'_{\pm})^2 + (\chi''_{\pm})^2} + (1+\chi'_{\pm})\right]^{\frac{1}{2}} \tag{27a}$$

$$\alpha_{\pm} = \omega\sqrt{\mu_0\varepsilon/2}\left[\sqrt{(1+\chi'_{\pm})^2 + (\chi''_{\pm})^2} - (1+\chi'_{\pm})\right]^{\frac{1}{2}} \tag{27b}$$

Typical sketches of $\beta_{\pm}$ and $\alpha_{\pm}$ are shown in Figs. 4-1 to 4-4. The precessional frequency $\omega_0 = -\gamma_e\mu_0H_0$ is used as the parameter for the plot. Note that α_- is always very small and α_+ is large in the vicinity of the resonant frequency, $\omega_0 \approx \omega$.

For frequency ω_0 considerably above the operating frequency ω, both α_+ and α_- become small. β_+ and β_- have approximately the same value. Consequently, the rate of Faraday rotation will be small.

At low values of ω_0, i.e., H_0 is small, the rate of rotation becomes greater, especially where β_+ goes through a minimum, which occurs when μ'_+ is negative.

It is to be noted that $\beta_+ < \beta_-$ when $\omega_0 < \omega$. Hence, the direction of Faraday rotation is different in the regions above and below the resonant frequency.

The permeability for the forward traveling waves with positive (right-hand) and negative (left-hand) circular polarization, respectively, can be obtained from (6).

$$\mu_{F_+} = \mu - k \tag{28a}$$

$$\mu_{F_-} = \mu + k \tag{28b}$$

where μ and k are defined by (3-30d) and (3-30e).

$$\mu = \mu_0(1 + \bar{\chi}) = \mu_0(1 + \chi' - j\chi'') \tag{29a}$$

$$k = \mu_0 K = \mu_0(K' - jK'') \tag{29b}$$

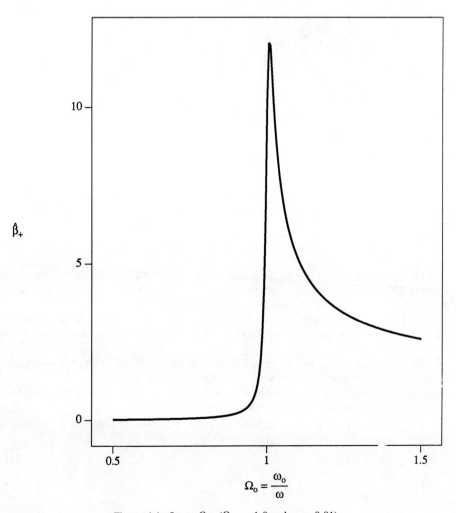

Figure 4-1: β_+ vs Ω_O, ($\Omega_m = 1.0$ and $\alpha = 0.01$)

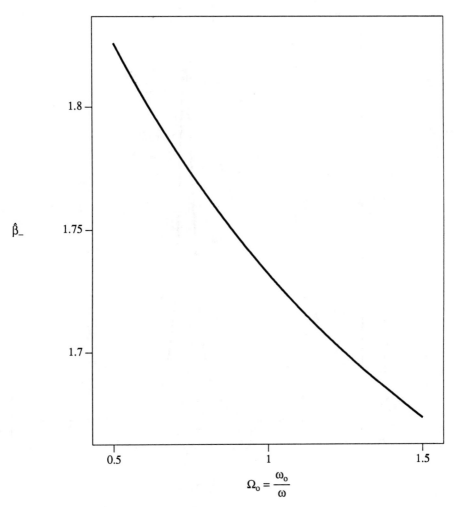

Figure 4-2: $\beta_- = \dfrac{\beta_-}{\beta_o}$ vs Ω_o, ($\Omega_m = 1.0$ and $\alpha = 0.01$)

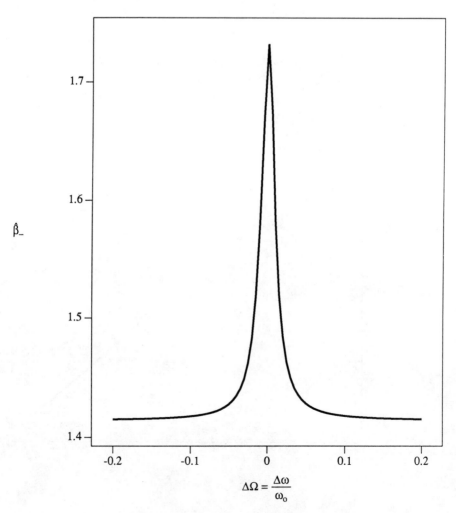

Figure 4-2(a): β_- vs $\Delta\Omega$, (Ω_m = 1.0 and α = 0.01)

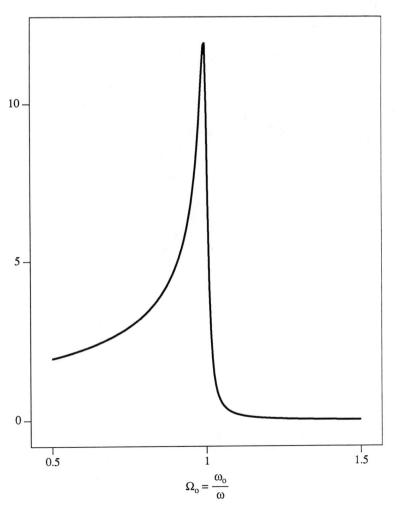

Figure 4-3: $\hat{\alpha}_+ = \dfrac{\alpha_+}{\beta_O}$ vs Ω_O, ($\Omega_m = 1.0$ and $\alpha = 0.01$)

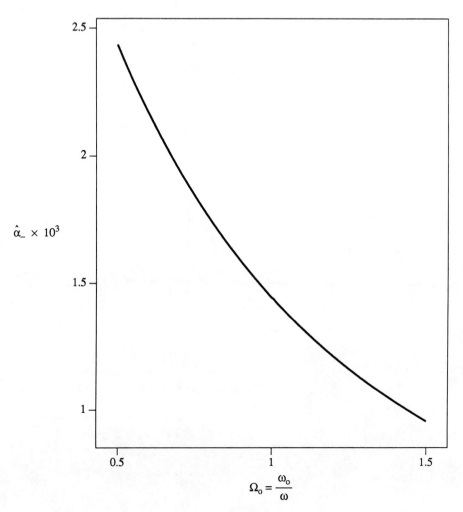

Figure 4-4: $\hat{\alpha}_- = \dfrac{\alpha_-}{\beta_O}$ vs Ω_O, ($\Omega_m = 1.0$ and $\alpha = 0.01$)

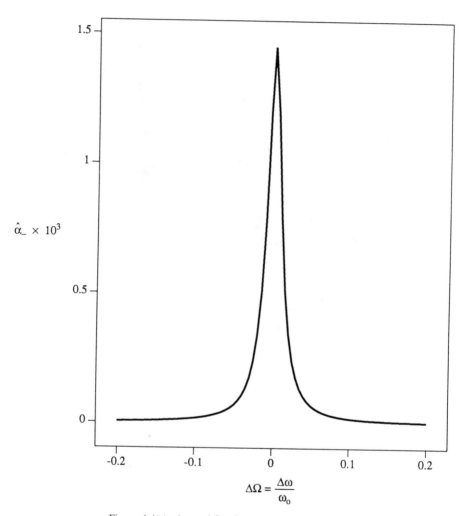

Figure 4-4(a): $\hat{\alpha}_-$ vs $\Delta\Omega$, ($\Omega_m = 1.0$ and $\alpha = 0.01$)

Then

$$\mu_{F_+} = \mu_o [(1 + \chi' - K') - j (\chi'' - K'')] = \mu_{F_+}' - j\mu_{F_+}'' \tag{30a}$$

$$\mu_{F_-} = \mu_o [(1 + \chi' + K') - j (\chi'' + K'')] = \mu_{F_-}' - j\mu_{F_-}'' \tag{30b}$$

Typical plots of the normalized $\hat{\mu}_\pm' \equiv \mu_{F_\pm}'/\mu_o$ and $\hat{\mu}_\pm'' \equiv \mu_{F_\pm}''/\mu_o$ are shown in Figs. 4-5 to 4-8.

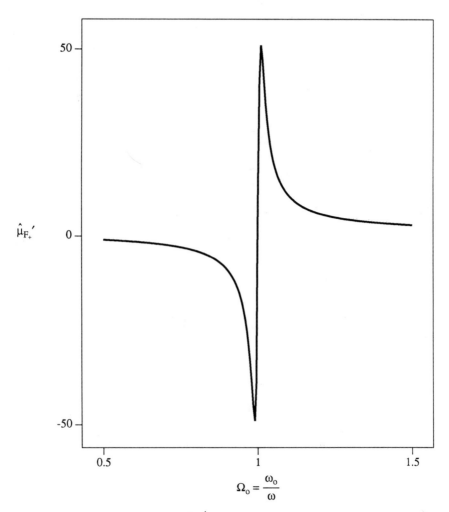

Figure 4-5: $\hat{\mu}_{F_+}' = \dfrac{\mu_{F-}'}{\mu_O}$ vs Ω_O, ($\Omega_m = 1.0$ and $\alpha = 0.01$)

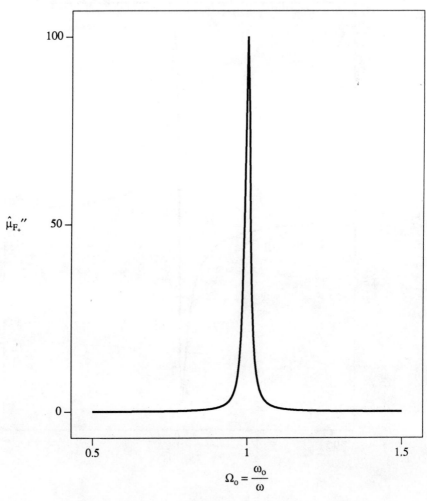

Figure 4-6: $\hat{\mu}_{F_+}'' = \dfrac{\mu_{F_+}''}{\mu_O}$ vs Ω_O, ($\Omega_m = 1.0$ and $\alpha = 0.01$)

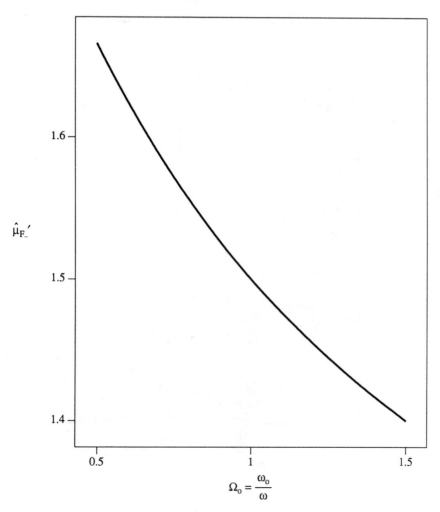

Figure 4-7: $\hat{\mu}_{F-}{}' = \dfrac{\mu_{F-}{}'}{\mu_O}$ vs Ω_O, ($\Omega_m = 1.0$ and $\alpha = 0.01$)

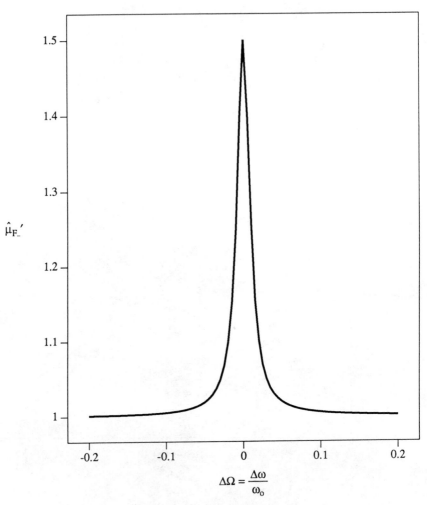

Figure 4-7(a): $\hat{\mu}_{F_-}'$ vs $\Delta\Omega$, ($\Omega_m = 1.0$ and $\alpha = 0.01$)

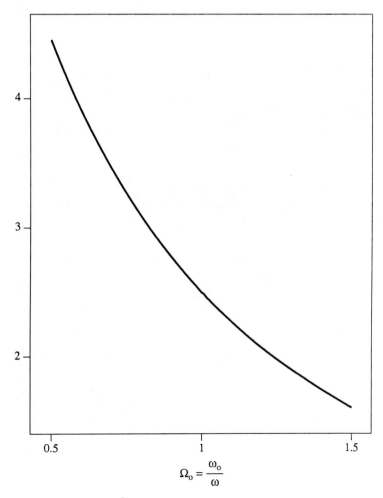

$\hat{\mu}_{F_-}{}'' \times 10^3$

Figure 4-8: $\hat{\mu}_{F_-}{}'' = \dfrac{\mu_{F_-}{}''}{\mu_O}$ vs Ω_o, ($\Omega_m = 1.0$ and $\alpha = 0.01$)

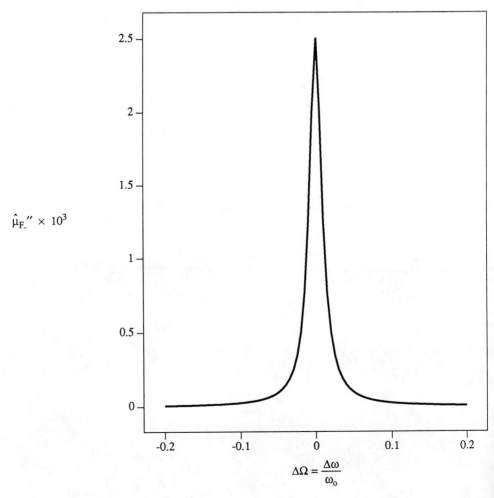

$\hat{\mu}_{F_-}'' \times 10^3$

Figure 4-8(a): $\hat{\mu}_{F_-}'' =$ vs $\Delta\Omega$, ($\Omega_m = 1.0$ and $\alpha = 0.01$)

4.1. Traveling Wave in the Negative $\hat{z}$-Direction

A traveling wave in the negative z-direction in a ferrite medium which is magnetized in the positive z-direction can be analyzed in a manner similar to that for a positive traveling wave. Let the linearly polarized wave be

$$\vec{H}'' = \hat{x}h_x e^{j\phi} \tag{1a}$$

$$\phi = \omega t + \beta z \tag{1b}$$

A right-hand circularly polarized wave with respect to the negative z-axis, the direction of propagation, is

$$\vec{H}_+ = \hat{x}e^{j\phi} + \hat{y}e^{j(\phi + \frac{\pi}{2})} = (\hat{x} + j\hat{y}) e^{j\phi_+} \tag{2a}$$

$$\phi_+ = \omega t + \beta_+ z \tag{2b}$$

where ϕ_+ is the phase constant for the right-hand polarized wave.

The left-hand circularly polarized wave is

$$\vec{H}_- = \hat{x}e^{j\phi} + \hat{y}e^{j(\phi - \frac{\pi}{2})} = (\hat{x} - j\hat{y}) e^{j\phi_-} \tag{2c}$$

$$\phi_- = \omega t + \beta_- z \tag{2d}$$

and ϕ_- is the phase constant for the left-hand polarized wave.

Since a linearly polarized wave can be decomposed into a right-hand and a left-hand polarized wave [Section (4b)], therefore, (1) can be rearranged as [Eq. (4-9)]

$$\begin{aligned}
\vec{H}'' &= \frac{1}{2} [(\hat{x} + j\hat{y}) e^{j\phi_+} + (\hat{x} - j\hat{y}) e^{j\phi_-}] h_x \\
&= \frac{h_x}{2} \left[\hat{x}e^{j\phi_+} + \hat{y}e^{j(\phi_+ + \frac{\pi}{2})} + \hat{x}e^{j\phi_-} + \hat{y}e^{j(\phi_- - \frac{\pi}{2})} \right] \\
&= \frac{h_x}{2} \left[\hat{x} (e^{j\phi_+} + e^{j\phi_-}) + \hat{y} \left[e^{j(\phi_+ + \frac{\pi}{2})} + e^{j(\phi_- - \frac{\pi}{2})} \right] \right]
\end{aligned} \tag{3}$$

The terms within parentheses can be manipulated to the following form.

$$\begin{aligned}
e^{j\phi_+} + e^{j\phi_-} &= e^{j(\omega t + \beta_+ z)} + e^{j(\omega t + \beta_- z)} \\
&= e^{j(\omega t + \beta_1 z)} (e^{j\beta_2 z} + e^{-j\beta_2 z}) \\
&= 2 \cos \beta_2 z \, e^{j(\omega t + \beta_1 z)}
\end{aligned} \tag{4a}$$

and

$$e^{j(\phi_+ + \frac{\pi}{2})} + e^{j(\phi_- - \frac{\pi}{2})} = e^{j(\omega t + \beta_1 z)} \left[e^{j(\beta_2 z + \frac{\pi}{2})} + e^{-j(\beta_2 z + \frac{\pi}{2})} \right]$$

$$= -2 \sin \beta_2 z \, e^{j(\omega t + \beta_1 z)} \tag{4b}$$

$$\beta_1 = \frac{1}{2} (\beta_+ + \beta_-) \tag{4c}$$

$$\beta_2 = \frac{1}{2} (\beta_+ - \beta_-) \tag{4d}$$

The substitution of (4) into (3) yields

$$\vec{H}'' = h_x (\hat{x} \cos \beta_2 z - \hat{y} \sin \beta_2 z) \, e^{j(\omega t + \beta_1 z)} \tag{5}$$

This is a linearly polarized wave traveling in the negative z-direction. The plane of polarization makes an angle θ with respect to the x-axis.

$$\theta = -\frac{z}{2} (\beta_+ - \beta_-) \tag{6}$$

This angle increases linearly with the distance of travel. The phase constant of the traveling wave is

$$\beta = \beta_1 = \frac{1}{2} (\beta_+ + \beta_-)$$

For a given length of travel, $z = L_1$, the plane of polarization rotates by the same amount with respect to the x-axis for traveling waves in either the positive or the negative z-direction. The phase delays are also identical for both waves.

4.2. Verification of Equation (4-7)

The meaning of the notation can be verified by substituting γ_i, $i = 1, 2, 3, 4$, for γ_z, one at a time, into (4-5). Equation (4-5) can be written as

$$(\omega^2 \varepsilon [\mu] - [u]\gamma_z^2) \vec{H}' = 0$$

or

$$\begin{bmatrix} \omega^2\varepsilon\mu - \gamma_z^2 & -j\omega^2\varepsilon k \\ j\omega^2\varepsilon k & \omega^2\varepsilon\mu - \gamma_z^2 \end{bmatrix} \begin{bmatrix} h_x \\ h_y \end{bmatrix} = 0 \tag{1}$$

Then

$$\omega^2\varepsilon\mu - \gamma_1^2 = \omega^2\varepsilon\mu - [\,\omega\sqrt{\varepsilon\,(\mu + k)}\,]^2 = -\omega^2\varepsilon k \tag{2a}$$

$$\omega^2\varepsilon\mu - \gamma_2^2 = \omega^2\varepsilon k \tag{2b}$$

$$\omega^2\varepsilon\mu - \gamma_3^2 = \omega^2\varepsilon k \tag{2c}$$

$$\omega^2\varepsilon\mu - \gamma_4^2 = -\omega^2\varepsilon k \tag{2d}$$

For $\gamma = \gamma_1$, (1) and (2a) yield

$$\omega^2\varepsilon k \begin{bmatrix} -1 & -j \\ j & -1 \end{bmatrix} \begin{bmatrix} h_x \\ h_y \end{bmatrix} = 0$$

or

$$h_y = jh_x \tag{3}$$

With (3), Eq. (4-1) gives

$$\vec{H}' = h_x (\hat{x} + j\hat{y}) e^{j\Omega_1} \tag{4a}$$

$$\Omega_1 = \omega t - |\gamma_1|z \tag{4b}$$

Equation (4) represents a left-hand polarized traveling wave in the positive z-direction, (4-9c).
Equations (1) and (2b) give

$$\omega^2\varepsilon k \begin{bmatrix} 1 & -j \\ j & 1 \end{bmatrix} \begin{bmatrix} h_x \\ h_y \end{bmatrix} = 0$$

or

$$h_y = -jh_x \tag{5}$$

Then (4-1) yields

$$\vec{H}' = h_x (\hat{x} - j\hat{y}) e^{j\Omega_2} \tag{6a}$$

$$\Omega_2 = \omega t - |\gamma_2| z \tag{6b}$$

Equation (6) is the expression of a right-hand polarized traveling wave in the positive z-direction, (4-9b).

Equations (1) and (2c) produce the same relation as (5). Then (4-1) gives

$$\vec{H}' = h_x (\hat{x} + j\hat{y}) e^{j\Omega_3} \tag{7a}$$

$$\Omega_3 = \omega t + |\gamma_3| z \tag{7b}$$

Equation (7) represents a left-hand polarized traveling wave in the negative z-direction.

Equations (1) and (2d) produce the same result as (3). Then (4-1) gives

$$\vec{H}' = h_x (\hat{x} - j\hat{y}) e^{j\Omega_4} \tag{8a}$$

$$\Omega_4 = \omega t + |\gamma_4| z \tag{8b}$$

Equation (8) represents a right-hand polarized traveling wave in the negative z-direction.

4.3 Frequency Response of the Propagation Constants

The frequency response of $\gamma_\pm$ can be investigated by a similar procedure to that employed in Section 2.

$$\gamma_\pm = \alpha_\pm + j\beta_\pm \tag{4-20a}$$

$$\beta_\pm = C\left[\sqrt{(1+\chi'_\pm)^2 + (\chi''_\pm)^2} + (1+\chi'_\pm)\right]^{\frac{1}{2}} \tag{4-27a}$$

$$\alpha_\pm = C\left[\sqrt{(1+\chi'_\pm)^2 + (\chi''_\pm)^2} - (1+\chi'_\pm)\right]^{\frac{1}{2}} \tag{4-27b}$$

where $C \equiv \omega\sqrt{\mu_o\varepsilon}$. The following properties of the variables χ were found in Section 3.

$$\chi'_+(\omega=0) = \frac{\omega_m}{\omega_o} \equiv \Omega_m \qquad \chi'_+(\omega\to\infty) = \frac{1}{\omega} \tag{2-7}$$

$$\chi''_+(\omega=0) = 0 \qquad \chi''_+(\omega\to\infty) = \frac{1}{\omega}$$

$$\chi'_-(\omega=0) = \Omega_m \qquad \chi'_-(\omega\to\infty) = \frac{1}{\omega} \tag{2-12}$$

$$\chi''_-(\omega=0) = 0 \qquad \chi''_-(\omega\to\infty) = \frac{1}{\omega}$$

Hence

$$\beta_\pm(\omega=0) = C\sqrt{2(1 + \Omega_m)} \tag{1a}$$

$$\beta_\pm(\omega\to\infty) = C\left[\sqrt{\left[1+\frac{1}{\omega}\right]^2 + \frac{1}{\omega^2}} + \left[1 + \frac{1}{\omega}\right]\right]^{\frac{1}{2}} \tag{1b}$$

$$\alpha_\pm(\omega=0) = 0 \tag{1c}$$

$$\alpha_\pm(\omega\to\infty) = C\left[\sqrt{\left[1+\frac{1}{\omega}\right]^2 + \frac{1}{\omega^2}} - \left[1 + \frac{1}{\omega}\right]\right]^{\frac{1}{2}} \tag{1d}$$

The values of $\alpha_\pm$ are generally very small in comparison with those of $\beta_\pm$. This is the result of the contribution from the difference of two approximately equal quantities.

At frequencies $\omega_r \approx \omega_o$ and $\Delta\omega \ll \omega_o$ [Eq. 2-28] the real and the imaginary parts of $\bar{\chi}_\pm$ have the following approximate forms [Eqs. (3-8) and (3-11)].

$$N = (1 - a) + \Delta\Omega \qquad\qquad a = \frac{1}{\sqrt{1 + \alpha^2}} \tag{3a}$$

$$D = 4\omega_0^4 \Delta\Omega^2 + k \qquad\qquad k = \frac{4\omega_0^4 \alpha^2}{1 + \alpha^2} \tag{3b}$$

$$\chi'_+ = \frac{2}{D}\,\omega_0^4 \Omega_m \left[(1 - a^2) + (1 + a)\Delta\Omega \right] \tag{3c}$$

$$\chi''_+ = \frac{2}{D}\,\omega_0^4 \Omega_m \alpha(1 + a) \tag{3d}$$

and

$$1 + \chi'_+ = \frac{1}{D} \left[D + 2\omega_0^4 \Omega_m \left[(1 - a^2) + (1 + a)\Delta\Omega \right] \right]$$

$$= \frac{4\omega_0^4}{D} \left[\Delta\Omega^2 + \frac{\Omega_m}{2}(1 + a)\Delta\Omega + \alpha_1 \left[1 + \frac{\Omega_m}{2} \right] \right]$$

$$= \frac{4\omega_0^4}{D} \left[\Delta\Omega^2 + b_1\Delta\Omega + b_2 \right] \tag{4a}$$

$$b_1 \equiv (1 + a)\frac{\Omega_m}{2} \approx \left[1 - \frac{\alpha^2}{4} \right]\Omega_m \tag{4b}$$

$$b_2 \equiv \alpha_1 \left[1 + \frac{\Omega_m}{2} \right] \approx \alpha^2 \left[1 + \frac{\Omega_m}{2} \right] \tag{4c}$$

$$\alpha_1 \equiv \frac{\alpha^2}{1 + \alpha^2} \tag{4d}$$

Terms containing a factor of α with power larger than two have been omitted, since $\alpha^2 \ll 1$.

$$(1 + \chi'_+)^2 \approx \frac{16\omega_0^8}{D^2} \left[(b_1^2 + 2b_2)\Delta\Omega^2 + 2b_1b_2\Delta\Omega + b_2^2 \right]$$

$$= \frac{16\omega_0^8}{D^2} \left[b_2^2 + p_1\Delta\Omega + p_2\Delta\Omega^2 \right] \tag{5a}$$

$$p_1 \equiv 2b_1b_2 \approx 2\alpha^2\Omega_m \left[1 + \frac{\Omega_m}{2} \right] \tag{5b}$$

$$p_2 \equiv b_1^2 + 2b_2 \approx \Omega_m^2 + \alpha^2 \left[2 + \Omega_m - \frac{\Omega_m^2}{2} \right] \tag{5c}$$

Terms of $\Delta\Omega$ with power of three and higher are neglected ($\Delta\omega \ll \omega_o$).

$$M \equiv (1 + \chi'_+)^2 + (\chi''_+)^2$$

$$= \frac{16\omega_o^8}{D^2} \left[b_2^2 + p_1\Delta\Omega + p_2\Delta\Omega^2 \right] + \frac{4\omega_o^8}{D^2} \Omega_m^2\alpha^2(1 + a)^2$$

$$= \frac{16\omega_o^8}{D^2} \left[p_o + p_1\Delta\Omega + p_2\Delta\Omega^2 \right] \tag{6a}$$

$$p_o \equiv b_2^2 + \frac{\Omega_m^2}{4} \alpha^2(1 + a)^2 \approx \alpha^2\Omega_m^2 \tag{6b}$$

$$M^{\frac{1}{2}} = \frac{4}{D} \omega_o^4\sqrt{p_o} \left[1 + \frac{p_1}{p_o}\Delta\Omega + \frac{p_2}{p_o}\Delta\Omega^2 \right]^{\frac{1}{2}}$$

$$\approx \frac{4}{D} \omega_o^4\sqrt{p_o} \left[1 + \frac{p_1}{2p_o}\Delta\Omega + \frac{p_2}{2p_o}\Delta\Omega^2 \right] \tag{7}$$

Then

$$\left[\frac{\beta_+}{C} \right]^2 = M^{\frac{1}{2}} + (1 + \chi'_+)$$

$$= \frac{4}{D} \omega_o^4\sqrt{p_o} \left[1 + \frac{p_1}{2p_o}\Delta\Omega + \frac{p_2}{2p_o}\Delta\Omega^2 \right] + \frac{4}{D} \omega_o^4 \left[\Delta\Omega^2 + b_1\Delta\Omega + b_2 \right]$$

$$= \frac{4}{D} \omega_o^4 \left[q_o + q_1\Delta\Omega + q_2\Delta\Omega^2 \right] \tag{8a}$$

$$q_o = b_2 + \sqrt{p_o} \approx \alpha\Omega_m + \alpha^2(1 + 0.50\Omega_m) \tag{8b}$$

$$q_1 = b_1 + \frac{p_1}{2\sqrt{p_o}} \approx \Omega_m + \alpha(1 + 0.50\Omega_m) - 0.25\alpha^2\Omega_m \tag{8c}$$

$$q_2 = 1 + \frac{p_2}{2\sqrt{p_o}} \approx 1 + \frac{\Omega_m^2 + \alpha^2(2 + \Omega_m - 0.50\Omega_m^2)}{2\alpha\Omega_m} \qquad (8d)$$

Hence

$$\beta_+ = 2\omega_o^2 C \sqrt{\frac{N_1}{D}} \qquad (9a)$$

$$N_1 = q_o + q_1\Delta\Omega + q_2\Delta\Omega^2 \qquad (9b)$$

To find the location of the maximum value of β_+, differentiate β_+ with respect to $\Delta\Omega$ and then set the resultant expression equal to zero.

$$\frac{d\beta_+}{d(\Delta\Omega)} = 0 = \frac{\omega_o C}{\sqrt{\frac{N_1}{D}}} \left[\frac{1}{D} \frac{dN_1}{d(\Delta\Omega)} - \frac{N_1}{D^2} \frac{dD}{d(\Delta\Omega)} \right]$$

or

$$D\frac{dN_1}{d(\Delta\Omega)} - N_1\frac{dD}{d(\Delta\Omega)} = 0 \qquad (10a)$$

$$4\omega_o^4(\Delta\Omega^2 + \alpha_1)(q_1 + 2q_2\Delta\Omega) - (q_o + q_1\Delta\Omega + q_2\Delta\Omega^2)8\omega_o^4\Delta\Omega = 0$$

$$-q_1\Delta\Omega^2 + 2(\alpha_1 q_2 - q_o)\Delta\Omega + q_1\alpha_1 = 0 \qquad (10b)$$

The substitution of (8b) to (8c) into (10b) yields

$$-[\Omega_m + \alpha(1 + 0.50\Omega_m) - 0.25\alpha^2\Omega_m]\Delta\Omega^2 - \Omega_m(\alpha + \alpha^2)\Delta\Omega$$

$$+ [\Omega_m + \alpha(1 + 0.50\Omega_m) - 0.25\alpha^2\Omega_m]\frac{\alpha^2}{1 + \alpha^2} = 0$$

Neglecting the terms with a factor of α^2, one has

$$[\Omega_m + \alpha(1 + 0.50\Omega_m)]\Delta\Omega^2 - \alpha\Omega_m\Delta\Omega = 0 \qquad (11a)$$

and the roots of (11a) are

$$\Delta\Omega_1 = 0 \quad \text{and} \quad \Delta\Omega_2 = \frac{-\alpha\Omega_m}{\Omega_m + \alpha(1 + 0.50\Omega_m)} \qquad (11b)$$

By direct substitution of (11) into (9), one finds that

$$\beta_+(\Delta\Omega_2) < \beta_+(\Delta\Omega_1) \qquad (12a)$$

and consequently

$$\beta_{+_{max}} = \beta_+(\Delta\Omega_1) \approx C\sqrt{\frac{q_o}{\alpha^2}} \tag{12b}$$

The attenuation constant α_+ is given by (4-27b).

$$\left[\frac{\alpha_+}{C}\right]^2 = M^{\frac{1}{2}} - (1 + \chi'_+)$$

$$= \frac{4\omega_o^4}{D}\left[s_o + s_1\Delta\Omega + s_2\Delta\Omega^2\right] \tag{13a}$$

$$s_o \equiv \sqrt{p_o} - b_2 \approx \alpha\Omega_m - \alpha^2(0.50\Omega_m + 1) \tag{13b}$$

$$s_1 \equiv \frac{p_1}{2\sqrt{p_o}} - b_1 = -\Omega_m + \alpha(0.50\Omega_m + 1) + 0.25\alpha^2\Omega_m \tag{13c}$$

$$s_2 = \frac{p_2}{2\sqrt{p_o}} - 1$$

$$= \frac{1}{2\alpha\Omega_m}[\Omega_m^2 - 2\alpha\Omega_m + \alpha^2(2 + \Omega_m - 0.50\Omega_m^2)] \tag{13d}$$

Thus

$$\alpha_+ = C\left[M^{\frac{1}{2}} - (1 + \chi'_+)\right]^{\frac{1}{2}} = 2\omega_o^2C\sqrt{\frac{N_2}{D}} \tag{14a}$$

$$N_2 \equiv s_o + s_1\Delta\Omega + s_2\Delta\Omega^2 \tag{14b}$$

Equating the derivative of α_+ with respect to $\Delta\Omega$ equal to zero yields

$$D\frac{dN_2}{d(\Delta\Omega)} - N_2\frac{dD}{d(\Delta\Omega)} = 0 \tag{15}$$

$$(\Delta\Omega^2 + \alpha_1)(s_1 + 2s_2\Delta\Omega) - (s_o + s_1\Delta\Omega + s_2\Delta\Omega^2)2\Delta\Omega = 0$$

$$-s_1\Delta\Omega^2 + 2(\alpha_1s_2 - s_o)\Delta\Omega + \alpha_1s_1 = 0$$

With the values of s_i given by (13), one has

$$-[-\Omega_m + \alpha(1 + 0.50\Omega_m) + 0.25\alpha^2\Omega_m]\Delta\Omega^2 - 0.50\alpha(1 - \alpha)\Omega_m\Delta\Omega$$

$$+ \alpha^2[-\Omega_m + \alpha(1 + 0.50\Omega_m) + 0.24\alpha^2\Omega_m] = 0$$

Neglecting terms with a factor of α with power of 2 and higher, the preceding expression becomes

$$\Delta\Omega\left\{ \left[-\Omega_m - \alpha(1 + 0.50\Omega_m) \right] \Delta\Omega - 0.50\alpha\Omega_m \right\} = 0$$

and the roots for the expression are

$$\Delta\Omega_3 = 0 \tag{16a}$$

$$\Delta\Omega_4 = \frac{-\alpha\Omega_m}{\Omega_m - \alpha(1 + 0.50\Omega_m)} \tag{16b}$$

It can be easily verified, by direct substitution, that the maximum value of α_+ is given by

$$\alpha_{+_{max}} = \alpha_+(\Delta\Omega_4) \tag{17}$$

The analysis of the frequency response of the propagation constant for the negative circularly polarized mode is summarized as follow.

$$\chi'_- \approx = \frac{(1 + a)}{D}\omega_0^4\Omega_m\left[\Delta\Omega^2 + 2(1 - a) \right] \tag{3-13}$$

$$\chi''_- \approx \frac{\alpha}{D}\omega_0^4\Omega_m\left[\Delta\Omega^2 + 2(1 - a) \right] \tag{3-15}$$

$$1 + \chi'_- \approx \frac{4\omega_0^4}{D}\left[b_3\Delta\Omega^2 + b_4 \right] \tag{18a}$$

$$b_3 = 1 + \frac{\Omega_m}{2} - \frac{\alpha^2\Omega_m}{8} \tag{18b}$$

$$b_4 = \alpha^2\left[1 + \frac{\Omega_m}{2} \right] \tag{18c}$$

$$(1 + \chi'_-)^2 = \frac{16\omega_0^8}{D^2}\left[b_3^2\Delta\Omega^4 + 2b_3b_4\Delta\Omega^2 + b_4^2\right] \tag{19}$$

$$M_- \equiv (1 + \chi'_-)^2 + (\chi''_-)^2 = \frac{16\omega_0^8}{D^2}\left[p_5\Delta\Omega^4 + p_4\Delta\Omega^2 + p_3\right] \tag{20a}$$

$$p_3 \equiv b_4^2 + \frac{\Omega_m^2\alpha^2}{4}(1 - a)^2 \approx \alpha^4\left[1 + \frac{\Omega_m}{2}\right]^2 \tag{20b}$$

$$p_4 = 2b_3b_4 + \frac{1 - a}{4}\Omega_m^2\alpha^2 \approx 2\alpha^2(1 + 0.50\Omega_m)^2 \tag{20c}$$

$$p_5 = b_3^2 + \frac{\Omega_m^2\alpha^2}{16} \approx (1 + 0.50\Omega_m)^2 - 0.25\alpha^2\Omega_m(1 + 0.25\Omega_m) \tag{20d}$$

$$M_-^{\frac{1}{2}} \approx \frac{4\omega_0^4}{D}\sqrt{p_3}\left[1 + \frac{p_4}{2p_3}\Delta\Omega^2 + \frac{p_5}{2p^3}\Delta\Omega^4\right] \tag{21}$$

$$\left[\frac{\beta_-}{C}\right]^2 = M_-^{\frac{1}{2}} + (1 + \chi'_-) \approx \frac{4\omega_0^4}{D}[q_3\Delta\Omega^2 + q_4] \tag{22a}$$

$$q_3 \equiv b_3 + \frac{p_4}{2\sqrt{p_3}} \approx 2 + \Omega_m - \frac{\alpha^2\Omega_m}{8} \tag{22b}$$

$$q_4 \equiv b_4 + \sqrt{p_3} \approx 2\alpha^2\left[1 + \frac{\Omega_m}{2}\right] \tag{22c}$$

Then

$$\beta_- = 2\omega_0^2C\sqrt{\frac{N_3}{D}} \tag{23a}$$

$$N_3 \equiv q_3\Delta\Omega^2 + q_4 \tag{23b}$$

The location of the maximum value of β_- is determined by the condition

$$\frac{d\beta_-}{d(\Delta\Omega)} = 0 \quad \text{or} \quad D\frac{dN_3}{d(\Delta\Omega)} - N_3\frac{dD}{d(\Delta\Omega)} = 0$$

which results in

$$(2\alpha_1q_3 - 2q_{4)}\Delta\Omega = 0$$

This implies that the maximum value of β_- occurs at $\Delta\Omega = 0$ since

$$\alpha_1 q_3 - q_4 \neq 0$$

The value of the maximum β_- is

$$\beta_{-_{max}} = \beta_-(\Delta\Omega{=}0) = 2\omega_o^2 C \sqrt{\frac{q_4}{\alpha_1}} = C\sqrt{(1 + \alpha^2)(2 + \Omega_m)} \tag{24}$$

The corresponding attenuation constant, α_-, is obtained as follows.

$$\left[\frac{\alpha_-}{C}\right]^2 = M_-^{\frac{1}{2}} - (1 + \chi'_-) \tag{25}$$

Because of the difference terms in (25), higher order of approximations are needed in the analysis of α_-. Including terms with a factor of α^4, the parameters (18), (20), and (22) are approximated as follows.

$$1 + a \approx 2 - \frac{\alpha^2}{2} + \frac{3\alpha^4}{8} \tag{26a}$$

$$1 - a \approx \frac{\alpha^2}{2} - \frac{3\alpha^4}{8} \tag{26b}$$

$$1 - a^2 \approx \alpha^2 - \alpha^4 \tag{26c}$$

$$\alpha_1 \approx \alpha^2(1 - \alpha^2) \tag{26d}$$

$$b_3 = 1 + (1 + a)\frac{\Omega_m}{4} \approx 1 + \frac{\Omega_m}{2} - \alpha^2\frac{\Omega_m}{8} + \alpha^4\frac{3\Omega_m}{32} \tag{27a}$$

$$b_4 = \alpha_1 + (1 - a^2)\frac{\Omega_m}{2} \approx (\alpha^2 - \alpha^4)\left[1 + \frac{\Omega_m}{2}\right] \tag{27b}$$

$$p_3 \approx \alpha^4\left[1 + \frac{\Omega_m}{2}\right]^2 \tag{28a}$$

$$p_4 \approx 2\alpha^2\left[1 + \frac{\Omega_m}{2}\right]^2 - \alpha^4\left[2 + \frac{5\Omega_m}{4} + \frac{\Omega_m^2}{2}\right] \tag{28b}$$

$$p_5 \approx \left[1 + \frac{\Omega_m}{2} \right]^2 - \frac{\alpha^2 \Omega_m}{4} \left[1 + \frac{\Omega_m}{4} \right] + \frac{\alpha^4 \Omega_m}{16} \left[3 + \frac{7\Omega_m}{4} \right] \quad (28c)$$

$$\left[\frac{\alpha_-}{C} \right]^2 \approx \frac{4\omega_0^4}{D} \left[s_3 + s_4 \Delta\Omega^2 \right] \quad (29a)$$

$$s_3 = \sqrt{p_3} - b_4 \approx \alpha^4 \left[1 + \frac{\Omega_m}{2} \right] \quad (29b)$$

$$s_4 = \frac{p_4}{2\sqrt{p_3}} - b_3 \approx -\alpha^2 \left[1 + \frac{(3/16)\Omega_m^2}{1 + 0.50\Omega_m} \right] - \alpha^4 \frac{3\Omega_m}{32} \quad (29c)$$

Hence

$$\alpha_- = 2\omega_0^2 C \sqrt{\frac{N_4}{D}} \quad (30a)$$

$$N_4 = s_3 + s_4 \Delta\Omega^2 \quad (30b)$$

The maximum value of α_- is determined by the condition

$$\frac{d\alpha_-}{d(\Delta\Omega)} = 0 \quad \text{or} \quad D\frac{dN_4}{d(\Delta\Omega)} - N_4\frac{dD}{d(\Delta\Omega)} = 0 \quad (31)$$

which yields

$$(\alpha_1 s_4 - s_3)\Delta\Omega = 0$$

Since

$$\alpha_1 s_4 - s_3 \neq 0$$

therefore

$$\Delta\Omega = 0$$

The maximum value of α_- is given by

$$\alpha_{-max} = \alpha_-(\Delta\Omega=0) = 2\omega_0^2 C \sqrt{\frac{s_3}{k}} \approx C\sqrt{\alpha^2(1 + \alpha^2)(1 + 0.50\Omega_m)} \quad (32)$$

5. Isolator

An ideal isolator is a device which permits transmission without attenuation in one direction and a very large attenuation in the opposite direction. An isolator which makes use of the property of Faraday rotation is composed of a rectangular-to-circular guide transition, a circular guide with a ferrite rod at its center, and a second circular-to-rectangular transition, Fig. 5-1. A resistance card is placed in the transition section to minimize waves polarized in the broad side of the rectangular guide. The nonreciprocal ferrite section provides a 45° Faraday rotation. Port 2 is physically rotated 45° from the y-axis as shown.

When an incident wave with its $\vec{E}$-field polarized along the x-axis enters port 1, it will be transmitted through the transition section without loss onto the ferrite section. The plane of polarization will be rotated as it passes through the section with ferrite. At the exit of the ferrite section, the wave has rotated 45°. Since the $\vec{E}$-field is perpendicular to the resistance card of the output transition section, it will be transmitted without loss.

When a wave with its $\vec{E}$-field oriented normal to the resistance card enters port 2, it will pass through the ferrite section. The ferrite section will cause a rotation of 45° further away from the x-axis. This rotated wave has its $\vec{E}$-field tangential to the resistance card of port 1; hence it is absorbed and on output appears at port 1.

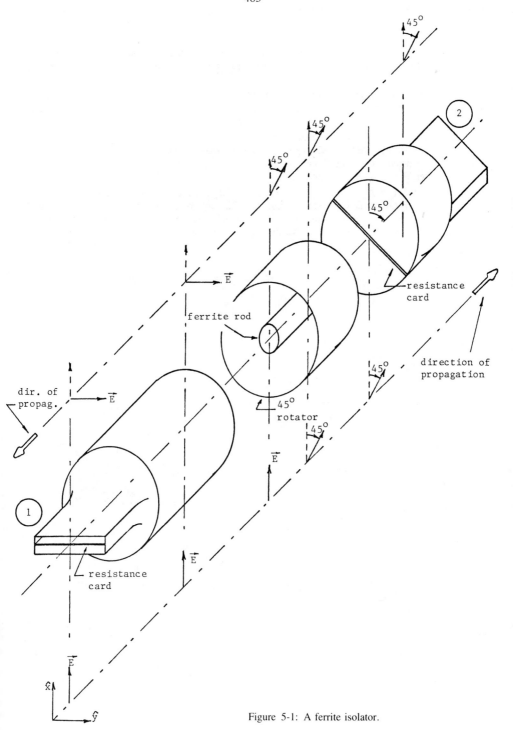

Figure 5-1: A ferrite isolator.

6. Gyrator

A gyrator is a two-port device characterized by a difference in phase shift of 180° for the transmission from port 1 to port 2 relative to the phase shift for the transmission from port 2 to port 1.

A Faraday rotation gyrator consists of a 90° rectangular twist, a transition from rectangular to circular guide, a ferrite 90° rotator, and a second circular-to-rectangular guide transition, Fig. 6-1.

A wave with its E-field polarized in the positive x-direction enters port 1. The plane of polarization will be rotated by 90° when this wave passes through the waveguide twist. The ferrite section provides an additional 90° rotation. Hence, a total of 180° rotation will have occurred when the wave reaches port 2.

When a wave with its E-field oriented in the negative x-direction enters port 2, it will undergo a 90° rotation as it passes through the Faraday rotator. The waveguide twist will rotate the wave back to its original plane of polarization at the output of port 1. As a result, there is no phase rotation when the wave propagates from port 2 to port 1.

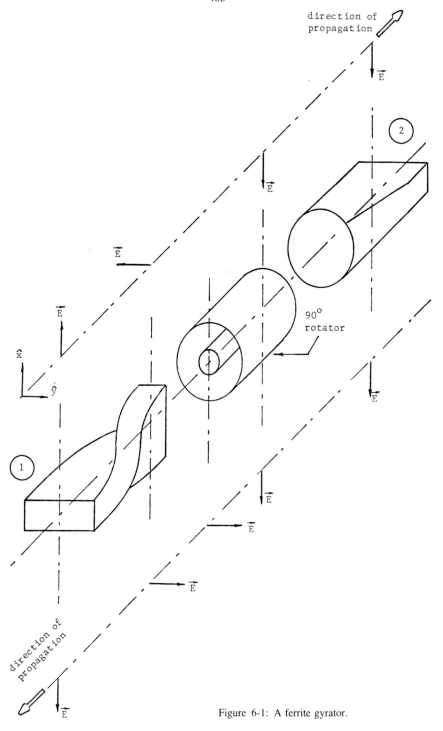

Figure 6-1: A ferrite gyrator.

7. Polarization of Guided Wave

(a) The magnetic field of the TE_{10} mode in a rectangular waveguide is given by [Sections II.4 and II.12]

$$H_x = \pm H^\pm \frac{\pi}{a} \sin \frac{\pi x}{a} e^{j(\omega t \mp \beta z)} \tag{1a}$$

$$H_z = -jH^\pm \frac{k_{c10}^2}{\beta} \cos \frac{\pi x}{a} e^{j(\omega t \mp \beta z)}$$

$$= H^\pm \frac{k_{c10}^2}{\beta} \cos \frac{\pi x}{a} e^{j(\omega t \mp \beta z - \frac{\pi}{2})} \tag{1b}$$

$$k_{cmn}^2 = k_o^2 - \beta^2 \tag{1c}$$

$$k_o^2 = \omega^2 \mu_o \varepsilon_o \tag{1d}$$

$$k_{cmn}^2 = (m\pi/a)^2 + (n\pi/b)^2 \qquad k_{c10} = \frac{\pi}{a} \tag{1e}$$

The superscript $\pm$ indicates waves traveling in the $\pm z$-direction, and a and b are the dimensions of the guide in the x- and y-direction, respectively.

Since H_x and H_z have a phase difference of $90°$, circular polarization occurs when

$$|H_x| = |H_z| \tag{2a}$$

That is,

$$\pm \frac{\beta}{k_{c10}} \sin \frac{\pi x}{a} = \cos \frac{\pi x}{a} \tag{2b}$$

But

$$k_{c10}^2 = (\pi/a)^2 = (2\pi/\lambda_c)^2 \qquad \lambda_c = 2a \tag{3a}$$

$$\beta = \frac{2\pi}{\lambda_g} \tag{3b}$$

Therefore,

$$\frac{\beta}{k_{c10}} = \frac{\lambda_c}{\lambda_g} \tag{3c}$$

The use of (3) in (2b) gives

$$\sin \frac{\pi x}{a} = \pm \frac{\lambda_g}{\lambda_c} \cos \frac{\pi x}{a} \tag{4}$$

This relation can be satisfied by an appropriate choice of x, such that

$$\tan \frac{\pi x'}{a} = + \frac{\lambda_g}{\lambda_c} \tag{5a}$$

and

$$\tan \frac{\pi x''}{a} = - \frac{\lambda_g}{\lambda_c} \tag{5b}$$

Equation (5b) can be rearranged as

$$\tan \frac{\pi x''}{a} = - \frac{\lambda_g}{\lambda_c} = -\tan \frac{\pi x'}{a} = \tan \left[\pi - \frac{\pi x'}{a} \right]$$

$$= \tan (a - x') \frac{\pi}{a} \tag{6a}$$

Therefore,

$$x'' = a - x' \tag{6b}$$

That is, (4) can be satisfied at x' and x" (Fig. 7-1), with x' given by (5a),

$$x' = \frac{a}{\pi} \tan^{-1} \frac{\lambda_g}{\lambda_c} \tag{7}$$

(b) For a traveling wave in the positive z-direction, the components of the H-field are given by [Eq. (1)]

$$H_z = H^+ k_{c10} \frac{\lambda_g}{\lambda_c} \cos \frac{\pi x}{a} e^{j\Omega} \tag{8a}$$

$$H_x = H^+ k_{c10} \sin \frac{\pi x}{a} e^{j(\Omega + \frac{\pi}{2})} \tag{8b}$$

$$\Omega = \omega t - \beta z - \frac{\pi}{2} \tag{8c}$$

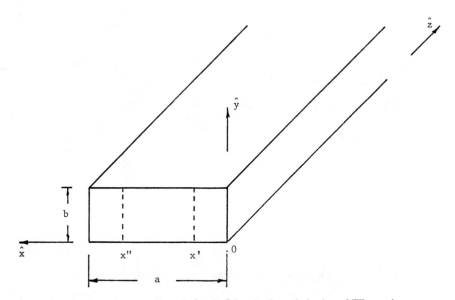

Figure 7-1: The locations of χ' and χ'' for circular polarization of TE_{01} mode.

direction of
propagation

Figure 7-2: The left-hand and the right-hand polarization for traveling wave in the positive z-direction.

At $x = x'$, (5a) and (3) yield

$$\sin \frac{\pi x'}{a} = \frac{\lambda_g}{\lambda_c} \cos \frac{\pi x'}{a} \tag{9}$$

The substitution of (9) into (8) yields

$$H_z = A_0^+ e^{j\Omega} \tag{10a}$$

$$H_x = H^+ k_{c10} \left[\frac{\lambda_g}{\lambda_c} \cos \frac{\pi x'}{a} \right] = A_0^+ e^{j(\Omega + \frac{\pi}{2})} \tag{10b}$$

$$A_0^+ = H^+ k_{c10} \frac{\lambda_g}{\lambda_c} \cos \frac{\pi x'}{a} \tag{10c}$$

The total H-field is then

$$\vec{H}(x') = \hat{z} H_z + \hat{x} H_x = A_0^+ \left[\hat{z} e^{j\Omega} + \hat{x} e^{j(\Omega + \frac{\pi}{2})} \right] \tag{11}$$

The physical field is given by the real part of (11)

$$\text{Re} [\vec{H}(x')] = A_0^+ [\; \hat{z} \cos (\omega t - \beta z) - \hat{x} \sin (\omega t - \beta z) \;] \tag{12}$$

Equation (12) represents a left-hand circularly polarized wave with respect to the y-axis, at x = x' (Fig. 7-2).

(c) At $x = x'' = a - x'$, (6a) gives

$$\sin \frac{\pi x''}{a} = - \frac{\lambda_g}{\lambda_c} \cos \frac{\pi x''}{a} \tag{13}$$

Equation (8) then becomes

$$H_x = -A_0^+ e^{j(\Omega + \frac{\pi}{2})} \tag{14a}$$

$$H_z = A_0^+ e^{j\Omega} \tag{14b}$$

and the total field is

$$\vec{H}(x'') = A_0^+ \left[\hat{z} e^{j\Omega} - \hat{x} e^{j(\Omega + \frac{\pi}{2})} \right] \tag{15a}$$

and

$$\text{Re} \, [\vec{H}(x'')] \; = \; A_o^+ \, [\; \hat{z} \cos \, (\omega t - \beta z) \; + \; \hat{x} \sin \, (\omega t - \beta z) \;] \tag{15b}$$

This is a right-hand circularly polarized wave with respect to the y-axis, at x = x'' (Fig. 7-2).

(d) For the traveling wave in the negative z-direction, (1) becomes

$$H_x \; = \; -H^- k_c \, \sin \frac{\pi x}{a} \; e^{j(\phi + \frac{\pi}{2})} \tag{16a}$$

$$H_z \; = \; H^- k_c \, \frac{\lambda_g}{\lambda_c} \, \cos \frac{\pi x}{a} \; e^{j\phi} \tag{16b}$$

$$\phi \; = \; \omega t \; + \; \beta z \; - \; \frac{\pi}{2} \tag{16c}$$

At x = x', Eq. (4) gives

$$\sin \frac{\pi x'}{a} \; = \; \frac{\lambda_g}{\lambda_c} \, \cos \frac{\pi x'}{a}$$

Then

$$H_x \; = \; -A_o^- e^{j(\phi + \frac{\pi}{2})} \tag{17a}$$

$$H_z \; = \; A_o^- e^{j\phi} \tag{17b}$$

$$A_o^- \; = \; H^- k_{c10} \, \frac{\lambda_g}{\lambda_c} \, \cos \frac{\pi x'}{a} \tag{17c}$$

The total field is

$$\vec{H}(x') \; = \; A_o^- \left[\hat{z} e^{j\phi} \; - \; \hat{x} e^{j(\phi + \frac{\pi}{2})} \right] \tag{18a}$$

$$\text{Re} \, [\vec{H}(x')] \; = \; A_o^- \, [\; \hat{z} \cos \, (\omega t + \beta z) \; - \; \hat{x} \sin \, (\omega t + \beta z) \;] \tag{18b}$$

This is a right-hand circularly polarized wave with respect to the y-axis , at x = x' (Fig. 7-3).

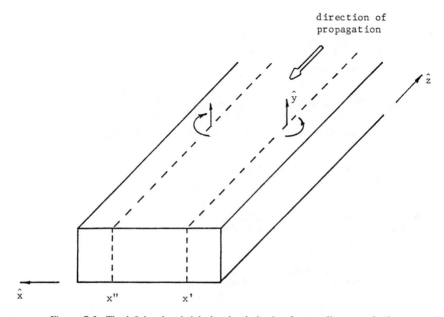

Figure 7-3: The left-hand and right-hand polarization for traveling wave in the negative z-direction.

At $x = x''$,

$$\sin \frac{\pi x''}{a} = \frac{-\lambda_g}{\lambda_c} \cos \frac{\pi x''}{a} \tag{19}$$

and (16) becomes

$$H_x = A_o^- e^{j(\phi + \frac{\pi}{2})} \tag{20a}$$

$$H_z = A_o^- e^{j\phi} \tag{20b}$$

$$\vec{H}(x'') = A_o^- \left[\hat{z} e^{j\phi} + \hat{x} e^{j(\phi + \frac{\pi}{2})} \right] \tag{21a}$$

$$\mathrm{Re}\,[\vec{H}(x'')] = A_o^- [\,\hat{z}\cos(\omega t + \beta z) - \hat{x}\sin(\omega t + \beta z)\,] \tag{21b}$$

This is a left-hand circularly polarized wave with respect to the y-axis, at $x = x''$, (Fig. 7-3).

(e) In summary, to each direction of propagation at a given x = constant plane, there corresponds a definite circular polarization:

Positive z-traveling wave:

at $x = x'$, left-hand circular polarization
at $x = x''$, right-hand circular polarization

Negative z-traveling wave:

at $x = x'$, right-hand circular polarization
at $x = x''$, left-hand circular polarization

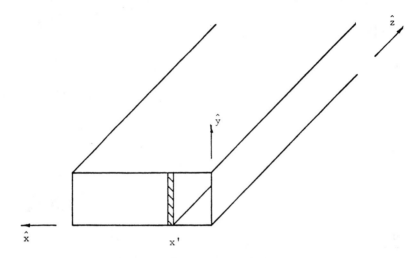

Figure 8-1: A resonance isolator.

8. Resonance Isolator

A resonance isolator makes use of the property that the attenuation constant is always very small for left-hand circular polarization (Section 4c) whereas for right-hand circular polarization, it is very large in the vicinity of the resonance, $\omega_0 \approx \omega$.

In the TE_{10} mode of a rectangular waveguide, it was shown in the previous section that along some x = constant planes, traveling waves in opposite directions have opposite circular polarization with respect to an axis transverse to the direction of propagation.

Consider a rectangular waveguide with a thin ferrite sheet placed along the plane where the RF H-field is circularly polarized, i.e., at $x = x'$ or $x = x''$ (Section 7). The ferrite is magnetized by a DC field along the y-axis (Fig. 8-1).

Since the sense of the circular polarization depends upon the direction of propagation, waves propagating in one direction corresponding to left-hand circular polarization, will suffer little attenuation. On the other hand, waves whose direction of propagation corresponds to right-hand circular polarization will be greatly attenuated, and the device thus serves as an isolator.

9. Problems

1. A circulator is a device which has the characteristic that the input to the kth port appears as the output from the (k + 1)th port. For a circulator with p ports, the input to the pth port appears as the output of the first port.

    Construct a three-port circulator by means of gyrators.

2. Verify that the arrangement of Fig. 9-1 is a circulator.

3. Determine whether the four-port device shown in Fig. 9-2 is a phase shifter, an isolator, or a circulator.

4. Show that the scattering matrix of a lossless, matched, nonreciprocal three-port microwave device is given by

$$[S] = \begin{bmatrix} 0 & 0 & s_{13} \\ s_{21} & 0 & 0 \\ 0 & s_{23} & 0 \end{bmatrix}$$

Hence, such a junction is a perfect three-port circulator.

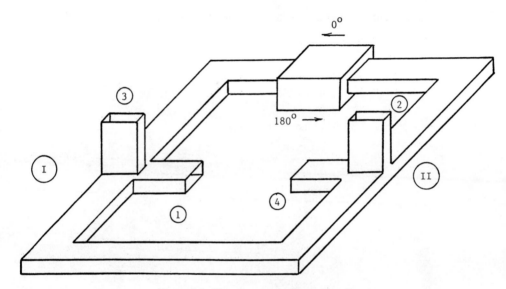

Figure 9-1: The arrangement for problem 2.

497

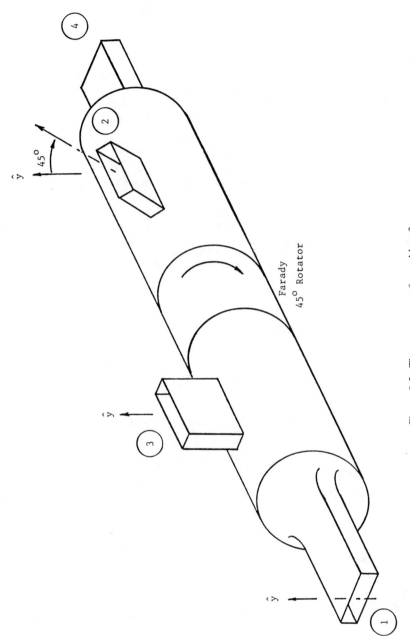

Figure 9-2: The arrangement for problem 3.

CHAPTER IX   Review on Resonators

1. Q Factor

Cavity resonators find applications in filter networks, wavemeters, and tuned circuits in oscillators. Microwave cavities rely on the field distribution in lines or guides bounded by electric or magnetic walls instead of the lumped LC elements used at the lower frequencies.

The losses in a circuit are specified by a quality factor Q which is defined as

$$Q = \omega_o \frac{\text{time average energy stored in the circuit}}{\text{power dissipated per unit time}} \tag{1}$$

where ω_o is the angular frequency at which Q is defined, usually chosen to be the resonance frequency.

In a simple series RLC-circuit (Fig. 1-1), the resistive element r_e represents the equivalent resistance of the losses present in the nonideal capacitor and the nonideal inductor.

The total energy stored in the circuit is

$$w_s = \frac{Li^2}{2} = \frac{Cv^2}{2} \tag{2}$$

and the dissipated power is

$$w_d = \frac{r_e i^2}{2} \tag{3}$$

The unloaded quality factor, Q_u, for the series RLC-circuit is therefore given by

$$Q_u = \frac{\omega_o \dfrac{Li^2}{2}}{\dfrac{r_e i^2}{2}} = \frac{\omega_o L}{r_e} \qquad \text{for series RLC-circuit} \tag{4}$$

This is defined as the unloaded Q_u since there is no load connected to the circuit and r_e is the equivalent resistance of the L and C elements.

When the resonator is coupled to the external source and load (Fig. 1-2), the dissipated power in the circuit becomes

$$w_d = \frac{i^2}{2} \left[r_e + \frac{R_g}{n_g^2} + \frac{R_L}{n_L^2} \right] = \frac{R_t i^2}{2} \tag{5a}$$

$$R_t = r_e + \frac{R_g}{n_g^2} + \frac{R_L}{n_L^2} = r_e + R_g' + R_L' \tag{5b}$$

where R_g' and R_L' are the generator and the load resistance reflected into the series RLC-circuit, respectively.

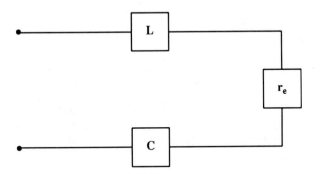

Figure 1-1: A series RLC circuit.

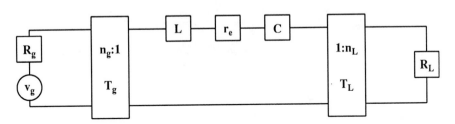

Figure 1-2: A resonator coupled to a source and a load.

The loaded quality factor Q_L is then given by

$$Q_L = \frac{\omega_o L}{R_t} \tag{6}$$

Equation (6) can be rearranged to give

$$\frac{1}{Q_L} = \frac{R_t}{\omega_o L} = \frac{r_e + R_g' + R_L'}{\omega_o L}$$

$$= \frac{1}{Q_u} + \frac{1}{Q_{eG}} + \frac{1}{Q_{eL}} \tag{7}$$

where Q_{eG} and Q_{eL} are the external Q factor of the generator and the load, respectively.

An alternative expression for the Q factor for a series resonator is given by the use of the slope parameters. The reactance slope parameter X' is defined as

$$X' = \frac{\omega_o}{2} \left[\frac{dX}{d\omega} \right]_{\omega=\omega_o} \tag{8}$$

where the total impedance of the resonance circuit, Fig. 1-2, is given by

$$Z = R_t + jX = R_t + j \left[\omega L - \frac{1}{\omega C} \right] \tag{9}$$

The reactance slope parameter is

$$X' = \frac{\omega_o}{2} \frac{d}{d\omega} \left[\omega L - \frac{1}{\omega C} \right]_{\omega=\omega_o}$$

$$= \frac{\omega_o}{2} \left[L + \frac{1}{\omega^2 C} \right]_{\omega=\omega_o}$$

$$= \omega_o L \tag{10}$$

since $\omega_o = \frac{1}{\sqrt{LC}}$. The loaded Q_L is given by, from (6) and (10),

$$Q_L = \frac{X'}{R_t} \tag{11}$$

The unloaded Q_u is

$$Q_u = \frac{X'}{r_e} = \frac{\omega_o L}{r_e} \tag{12}$$

When a resonator is represented by a parallel circuit, the shunt admittance is

$$Y = G_t + jB = G_t + j\left[\omega C - \frac{1}{\omega L}\right] \tag{13}$$

The susceptance slope parameter is defined by

$$B' = \frac{\omega_o}{2}\left[\frac{dB}{d\omega}\right]_{\omega=\omega_o} = \omega_o C \tag{14}$$

and the corresponding loaded Q factor for a parallel RLC-circuit is

$$Q_L = \frac{B'}{G_t} = \omega C R_t \tag{15}$$

2. Waveguide Resonant Circuits

Waveguides and transmission lines when terminated with an open circuit or a short circuit exhibit similar immitance slope parameters near discrete frequencies to those of tuned resonant circuits at lower frequencies. Waveguides arranged in this manner can be used as resonators at microwave frequencies.

The input impedance of a tuned series RLC-circuit is given by

$$Z = R + j \left[\omega L - \frac{1}{\omega C} \right]$$

$$= R + j\sqrt{\frac{L}{C}} \left[\omega\sqrt{LC} - \frac{1}{\omega\sqrt{LC}} \right]$$

$$= R + j\sqrt{\frac{L}{C}} \left[\frac{\omega}{\omega_0} - \frac{\omega_0}{\omega} \right] \qquad \omega_0^2 \equiv \frac{1}{LC} \tag{1}$$

The reactance is positive when $\omega > \omega_0$, and it is negative when $\omega < \omega_0$. The reactive term can be rearranged to be more informative by the following identities.

$$\sqrt{\frac{L}{C}} = [\omega_0\sqrt{LC}] \sqrt{\frac{L}{C}} = \omega_0 L \tag{2a}$$

$$\sqrt{\frac{L}{C}} = \frac{\sqrt{\frac{L}{C}}}{\omega_0\sqrt{LC}} = \frac{1}{\omega_0 C} \tag{2b}$$

Then (1) becomes, with the use of (2a),

$$Z = R + j\omega_0 L_e \qquad \text{for} \qquad \omega > \omega_0 \tag{3a}$$

$$L_e = L \left[\frac{\omega}{\omega_0} - \frac{\omega_0}{\omega} \right] \tag{3b}$$

and with (2b), one has

$$Z = R + \frac{1}{j\omega_0 C_e} \qquad \text{for} \qquad \omega < \omega_0 \tag{3c}$$

$$C_e = \frac{C}{\dfrac{\omega}{\omega_0} - \dfrac{\omega_0}{\omega}} \tag{3d}$$

The input impedance of a guide terminated by an open circuit (Fig. 2-1) is given by

$$Z_{in} = Z_0 \coth (\alpha + j\beta)d \qquad (4)$$

The hyperbolic function can be expanded as follows [H.B. Dwight, Tables of Integrals and Other Mathematical Data, Fourth Edition, Eq. (655.4)].

$$\coth (\alpha + j\beta)d = \frac{\sinh 2\alpha d - j \sin 2\beta d}{\cosh 2\alpha d - \cos 2\beta d} \qquad (5)$$

For a low-loss guide, the value of αd is very small, and (5) can be approximated as

$$\coth (\alpha + j\beta) d \approx \frac{2\alpha d - j \sin 2\beta d}{1 - \cos 2\beta d} \qquad (6)$$

In the vicinity of the resonance frequency ω_0, $d = \lambda/4$, and

$$\beta = \beta_0 + \Delta\beta \qquad (7)$$

where

$$\beta = \frac{\omega}{v} \qquad (8a)$$

$$\beta_0 = \frac{\omega_0}{v} \qquad (8b)$$

$$\Delta\beta = \beta - \beta_0 = \frac{\omega - \omega_0}{v} = \frac{\Delta\omega}{v} \qquad (8c)$$

$$\Delta\omega = \omega - \omega_0 \qquad (8d)$$

$$\frac{\Delta\beta}{\beta_0} = \frac{\Delta\omega}{\omega_0} \qquad (8e)$$

where v is the phase velocity of the guide. Then

$$\sin 2\beta d = \sin (2\beta_0 d + 2\Delta\beta d) = \sin \left[\pi + 2\Delta\beta d \frac{\frac{\pi}{2}}{\beta_0 d} \right]$$

$$= -\sin \pi\frac{\Delta\beta}{\beta_0} \approx -\frac{\Delta\beta}{\beta_0} \pi \qquad \text{for small } \frac{\Delta\beta}{\beta_0} \qquad (9a)$$

504

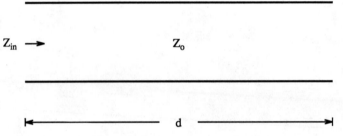

Figure 2-1: A waveguide terminated by an open circuit.

$$\beta_o d = \frac{\pi}{2} \qquad \text{for} \qquad d = \frac{\lambda}{4} \tag{9b}$$

$$\cos 2\beta d = \cos \left[\pi + \frac{\Delta\beta}{\beta_o} \pi \right] = \cos \pi \cos \frac{\Delta\beta}{\beta_o} \pi \approx -1 \tag{9c}$$

The substitution of (9) into (6) yields

$$\coth (\alpha + j\beta)d \approx \frac{2\alpha d - j \dfrac{-\Delta\beta \, \pi}{\beta_o}}{1 - (-1)} = \alpha d + j \frac{\Delta\beta\pi}{2\beta_o}$$

$$= \alpha d + j \frac{\pi}{2} \frac{\Delta\omega}{\omega_o} \tag{10}$$

Hence, (4) becomes

$$Z_{in} \approx Z_o \left[\alpha d + j \frac{\pi}{2} \frac{\Delta\omega}{\omega_o} \right] \qquad \text{for low-loss case} \tag{11}$$

Equation (11) can be arranged to have the same form as (1) by introducing (8d) into the terms within brackets in (1), i.e.,

$$\frac{\omega}{\omega_o} - \frac{\omega_o}{\omega} = \frac{\omega_o + \Delta\omega}{\omega_o} - \frac{\omega_o}{\omega_o + \Delta\omega} = \left[1 + \frac{\Delta\omega}{\omega_o} \right] - \frac{1}{1 + \dfrac{\Delta\omega}{\omega_o}}$$

$$\approx \left[1 + \frac{\Delta\omega}{\omega_o} \right] - \left[1 - \frac{\Delta\omega}{\omega_o} \right] = \frac{2\Delta\omega}{\omega_o} \tag{12}$$

for $\Delta\omega \ll \omega_o$. With (12), Eq. (11) takes the form

$$Z_{in} \approx Z_o \alpha d + j Z_o \frac{\pi}{4} \left[\frac{\omega}{\omega_o} - \frac{\omega_o}{\omega} \right] = R_{in} + j X_{in} \tag{13a}$$

where

$$R_{in} = Z_o \alpha d \tag{13b}$$

$$X_{in} = Z_o \frac{\pi}{4} \left[\frac{\omega}{\omega_o} - \frac{\omega_o}{\omega} \right] = \omega_o L_{in} \left[\frac{\omega}{\omega_o} - \frac{\omega_o}{\omega} \right] \tag{13c}$$

$$\omega_o L_{in} = Z_o \frac{\pi}{4} \qquad\qquad C_{in} = \frac{1}{\omega_o^2 L_{in}} \qquad\qquad (13d)$$

The input impedance of an open-circuit waveguide has the identical form as that of a series RLC-circuit in the vicinity of resonance. The duality between an open-circuited waveguide and a series resonance circuit is thus established (Fig. 2-2).

The duality between a shunt resonant circuit and a short-circuited waveguide can be obtained similarly.

The input admittance of a shunt resonant circuit (Fig. 2-3), is given by

$$Y_{in} = G + j \left[\omega C - \frac{1}{\omega L} \right] = G + j \sqrt{\frac{L}{C}} \left[\frac{\omega}{\omega_o} - \frac{\omega_o}{\omega} \right]$$

$$= G + j\omega_o C \left[\frac{\omega}{\omega_o} - \frac{\omega_o}{\omega} \right] \qquad \text{for} \qquad \omega > \omega_o \qquad (14a)$$

$$= G + j\frac{1}{\omega_o L} \left[\frac{\omega}{\omega_o} - \frac{\omega_o}{\omega} \right] \qquad \text{for} \qquad \omega < \omega_o \qquad (14b)$$

The input admittance of a short-circuited guide (Fig. 2-4) is, from (10) and (12),

$$Y_{in} = Y_o \coth (\alpha + j\beta)d$$

$$\approx Y_o \left[\alpha d + j \frac{\pi}{4} \left[\frac{\omega}{\omega_o} - \frac{\omega_o}{\omega} \right] \right] \qquad \text{for low--loss case} \qquad (15a)$$

$$\equiv G_{in} + jB_{in}$$

$$G_{in} = Y_o \alpha d \qquad \text{and} \qquad B_{in} = \frac{\pi}{4} \left[\frac{\omega}{\omega_o} - \frac{\omega_o}{\omega} \right] \qquad (15b)$$

$$B_{in} = Y_o \frac{\pi}{4} \left[\frac{\omega}{\omega_o} - \frac{\omega_o}{\omega} \right] = \omega_o C_{in} \left[\frac{\omega}{\omega_o} - \frac{\omega_o}{\omega} \right] \qquad (15c)$$

$$\omega_o C_{in} = Y_o \frac{\pi}{4} \qquad\qquad (15d)$$

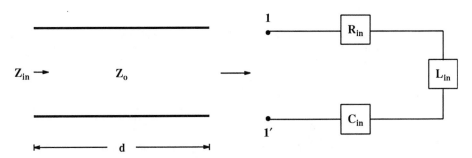

Figure 2-2: A equivalent RLC circuit for a waveguide terminated by open circuit.

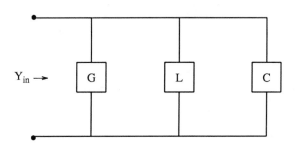

Figure 2-3: A parallel RLC circuit.

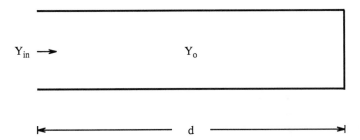

Figure 2-4: A waveguide terminated by a short circuit.

3. Rectangular Cavity Resonators

(a) A rectangular cavity resonator is obtained by terminating a section of rectangular waveguide by conducting planes at both ends. When a guide is terminated by a load which is characterized by a reflection coefficient s_L (Fig. 3-1), the input scattering coefficient is given by Eq. (VII.1-4b).

$$s_{11}' = \frac{b_1'}{a_1'} = s_{11} + \frac{s_{21}^2 s_L}{1 - s_{22} s_L} \qquad\qquad s_{12} = s_{21} \qquad (1)$$

If the guide is terminated by short circuits at both input and output ports, the boundary conditions are

$$s_{11}' = -1 \qquad \text{and} \qquad s_L = -1 \qquad (2)$$

Then (1) becomes

$$-1 = s_{11} - \frac{s_{21}^2}{1 + s_{22}} \qquad (3)$$

The scattering matrix for a lossless guide is given by [Eq. (IV.4.2-5)]

$$[S] = \begin{bmatrix} 0 & e^{-j\theta} \\ e^{-j\theta} & 0 \end{bmatrix} = \begin{bmatrix} s_{11} & s_{21} \\ s_{21} & s_{22} \end{bmatrix} \qquad (4a)$$

$$\theta = \beta_g d = \frac{2\pi d}{\lambda_g} \qquad (4b)$$

where d is the length of the guide and λ_g is the wavelength of the guided wave. Hence, (3) reduces to

$$1 = s_{21}^2 = e^{-j2\theta} \qquad (5)$$

The solution of (5) is

$$e^{-j2\theta} = 1 = e^{\pm 2p\pi} \qquad p = \text{integer}$$

$$\theta = \beta_g d = p\pi \qquad p = 1, 2, \ldots \qquad (6)$$

For the waveguide shown in Fig. 3-2, the phase coefficient for different guided modes is given by

$$[S] = \begin{bmatrix} s_{11} & s_{21} \\ s_{21} & s_{22} \end{bmatrix}$$

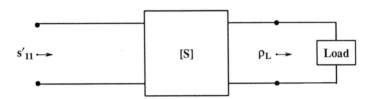

Figure 3-1: A terminated two-port device.

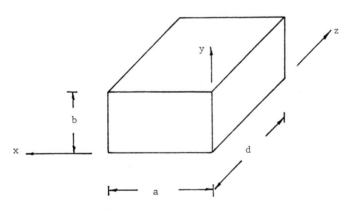

Figure 3-2: A rectangular cavity.

$$\beta_{gmn} = \sqrt{\omega^2\mu\varepsilon - \left[\frac{m\pi}{a}\right]^2 - \left[\frac{n\pi}{b}\right]^2} \qquad (7a)$$

$$= \frac{2\pi}{\lambda_{gmn}} \qquad (7b)$$

The corresponding wavelength is

$$\left[\frac{1}{\lambda_{gmn}}\right]^2 = \left[\frac{1}{\lambda_o}\right]^2 - \left[\frac{m}{2a}\right]^2 - \left[\frac{n}{2b}\right]^2 \qquad (8)$$

From (6), one has

$$\beta_{gmn}d = p\pi \qquad \text{or} \qquad \frac{1}{\lambda_{gmn}} = \frac{p}{2d} \qquad (9)$$

The combination of (8) and (9) yields

$$\left[\frac{1}{\lambda_o}\right]^2 = \left[\frac{p}{2d}\right]^2 + \left[\frac{m}{2a}\right]^2 + \left[\frac{n}{2b}\right]^2 \qquad (10)$$

For a given waveguide and a given set of integers, m, n, and p, both TE and TM modes resonate at the same frequency. Such modes with different field patterns but identical resonant frequencies are known as degenerate modes.

The concept of defining a resonant mode by the standing-wave pattern for an incident and reflected waveguide mode is not unique. This is because the mode identification is dependent on the choice of the axis of propagation for the respective mode. A TE_{101} mode for a wave propagating in the positive z-direction becomes a TM_{110} mode if the positive y-axis is taken to be the direction of propagation.

(b) The fields of TE_{mn} modes are given by (II.4-4) and (II.4-5),

$$\vec{h}_t = \frac{\mp j\gamma_z}{\gamma^2} \nabla_t h_z \qquad (11a)$$

$$\vec{e}_t = \frac{\mp \omega\mu}{\gamma_z} \hat{z} \times \vec{h}_t \qquad (11b)$$

For a wave traveling in the positive z-direction [Eq. (II.12-5a)],

$$H_z^+ = H_{mn}^+ \cos\frac{m\pi x}{a} \cos\frac{n\pi y}{b} e^{-j\gamma_z z} \qquad \gamma_z = \alpha + j\beta \qquad (12)$$

and the associated TE_{10} mode is

$$H_z^+ \;=\; H^+ \cos \frac{\pi x}{a}\, e^{-j\gamma_z z} \qquad H^+ \equiv H_{10}^+ \tag{13a}$$

The transverse components of fields are

$$\vec{H}_t^+ \;=\; \frac{-j\gamma_z}{\gamma^2}\left[\hat{x}\frac{\delta}{\delta x} + \hat{y}\frac{\delta}{\delta y} \right]\left[H^+ \cos \frac{\pi x}{a}\, e^{-j\gamma_z z} \right]$$

$$=\; \hat{x}\, \frac{j\gamma_z}{\gamma}\, H^+ \sin \frac{\pi x}{a}\, e^{-j\gamma_z z} \tag{13b}$$

and

$$\vec{E}_t^+ \;=\; \frac{-\omega\mu}{\gamma_z}\, \hat{z}\times\left[\hat{x}\, \frac{j\gamma_z}{\gamma} \sin \frac{\pi x}{a}\, e^{-j\gamma_z z} \right]$$

$$=\; -\hat{y}\, \frac{j\omega\mu}{\gamma}\, H^+ \sin \frac{\pi x}{a}\, e^{-j\gamma_z z} \tag{13c}$$

since $\gamma = \gamma_x = \dfrac{\pi}{x}$ for the TE_{10} mode.

For a TE_{10} wave traveling in the negative z-direction, one has

$$H_z^- \;=\; H^- \cos \frac{\pi x}{a}\, e^{j\gamma_z z} \qquad H^- \equiv H_{10}^- \tag{14a}$$

$$\vec{H}_t^- \;=\; \frac{j\gamma_z}{\gamma^2}\, \hat{x}\frac{\delta}{\delta x}\left[H^- \cos \frac{\pi x}{a}\, e^{j\gamma z} \right]$$

$$=\; \hat{x}\, \frac{j\gamma_z}{\gamma}\, H^- \sin \frac{\pi x}{a}\, e^{j\gamma_z z} \tag{14b}$$

and

$$\vec{E}_t^- \;=\; \frac{\omega\mu}{\gamma_z}\, \hat{z}\times\left[-\hat{x}\, \frac{j\gamma_z}{\gamma}\, H^- \sin \frac{\pi x}{a}\, e^{j\gamma_z z} \right]$$

$$=\; -\hat{y}\, \frac{j\omega\mu}{\gamma}\, H^- \sin \frac{\pi x}{a}\, e^{j\gamma_z z} \tag{14c}$$

The total field is equal to the sum of the negative and positive traveling waves.

$$H_z \;=\; H_z^- + H_z^+ \;=\; \left(H^- e^{j\gamma_z z} + H^+ e^{-j\gamma_z z} \right)\cos \frac{\pi x}{a} \tag{15a}$$

$$H_x = \frac{j\gamma_z}{\gamma} (H^- e^{j\gamma_z z} + H^+ e^{-j\gamma_z z}) \sin \frac{\pi x}{a} \qquad (15b)$$

$$E_y = \frac{-j\omega\mu}{\gamma} (H^- e^{j\gamma_z z} + H^+ e^{-j\gamma_z z}) \sin \frac{\pi x}{a} \qquad (15c)$$

The coefficients, H^+ and H^-, can be evaluated by invoking the boundary conditions. At $z = 0$, one has

$$E_y(z = 0) = 0, \quad \text{hence} \quad H^+ = -H^- \equiv -\frac{H_o}{2}$$

$$E_y = \frac{-j\omega\mu}{\gamma} \frac{H_o}{2} (e^{j\gamma_z z} - e^{-j\gamma_z z}) \sin \frac{\pi x}{a}$$

$$= \frac{\omega\mu}{\gamma} H_o \sin \beta z \sin \frac{\pi x}{a} \qquad (16a)$$

$$H_x = j \frac{\beta}{\gamma} H_o \cos \beta z \sin \frac{\pi x}{a} \qquad (16b)$$

$$H_z = j H_o \sin \beta z \cos \frac{\pi x}{a} \qquad (16c)$$

for a lossless guide, $\gamma_z = \beta$.

One also has at $z = d$, $E_y(z = d) = 0$; therefore,

$$\sin \beta d = 0 \qquad \text{or} \qquad \beta d = p\pi$$

$$\beta_p = \frac{p\pi}{d} \qquad p = 1, 2, \ldots \qquad (16d)$$

Consequently, the total fields are

$$E_y = H_o \frac{\omega\mu}{\gamma} \sin \frac{p\pi z}{d} \sin \frac{\pi x}{a} \qquad (17a)$$

$$H_x = j H_o \frac{\beta}{\gamma} \cos \frac{p\pi z}{d} \sin \frac{\pi x}{a} \qquad (17b)$$

$$H_z = j H_o \sin \frac{p\pi z}{d} \cos \frac{\pi x}{a} \qquad (17c)$$

4. Scattering Matrix of a Lossless Resonator

Figure 4-1a represents a lossless resonator by a series LC-circuit. The corresponding scattering matrix can be obtained by means of eigen-networks. The in-phase and opposite-phase eigen-networks are shown in Fig. 4-1b and Fig. 4-1c.

The input impedances of these eigen-networks are

$$Z_1 = Z_{oc} = \infty \tag{1a}$$

$$Z_2 = Z_{sc} = \frac{j\omega L}{2} + \frac{1}{j\omega 2C} = j\omega_0 L\delta \tag{1b}$$

$$\delta = \frac{1}{2}\left[\frac{\omega}{\omega_0} - \frac{\omega_0}{\omega}\right] \tag{1c}$$

$$\omega_0 = \frac{1}{\sqrt{LC}} \tag{1d}$$

The eigenvalues of the scattering matrix are (Section III.9)

λ_n^s = reflection coefficient at nth port

$$= \frac{Z_n - Z_0}{Z_n + Z_0} \tag{2a}$$

where Z_n is the terminating impedance of the nth port and Z_0 is the characteristic impedance. Therefore, with the simplified notation $\lambda_n \equiv \lambda_n^s$, one has

$$\lambda_1 = \frac{Z_1 - Z_0}{Z_1 + Z_0} = 1 \tag{2b}$$

$$\lambda_2 = \frac{j\omega_0 L\delta - R_0}{j\omega_0 L\delta + R_0} = \frac{jQ_L \delta - 1}{jQ_L \delta + 1} \tag{2c}$$

where

$$Q_L = \frac{\omega_0 L}{R_0} = \text{loaded Q factor} \tag{2d}$$

The scattering elements for symmetrical and reciprocal devices are related to the eigen-values by (III.9-4),

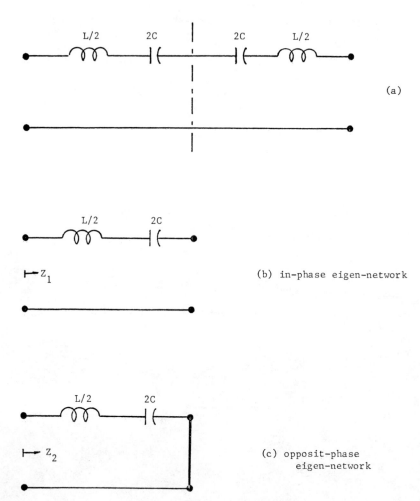

Figure 4-1: A lossless resonator and its eigen-networks.

$$s_{11} = \frac{\lambda_1 + \lambda_2}{2} \tag{3a}$$

$$s_{21} = \frac{\lambda_1 - \lambda_2}{2} \tag{3b}$$

Thus

$$s_{11} = \frac{1}{2} \left[1 + \frac{jQ_L\delta - 1}{jQ_L\delta + 1} \right] = \frac{jQ_L\delta}{1 + jQ_L\delta} \tag{3c}$$

$$s_{21} = \frac{1}{1 + jQ_L\delta} \tag{3d}$$

The physical meaning of δ can be obtained from the expression (1c) by replacing ω by

$$\omega = \omega_o + \Delta\omega \tag{4}$$

where $\Delta\omega$ is the small deviation from resonance. Then

$$2\delta = \frac{\omega_o + \Delta\omega}{\omega_o} - \frac{\omega_o}{\omega_o + \Delta\omega}$$

$$\approx 2 \frac{\Delta\omega}{\omega_o} = \text{normalized bandwidth} \tag{5}$$

by virtue of (2-12). Hence, $2\delta\omega_o$ is approximately twice the frequency deviation.

The square of the magnitude of s_{21} is

$$|s_{21}|^2 = \frac{1}{1 + (Q_L\delta)^2} \tag{6}$$

The bandwidth of the device is defined to be the maximum frequency deviation when $|s_{21}|^2$ is reduced to one-half of its value at resonance or when $Q_L\delta = 1$ (Fig. 4-2).

$$Q_L = \frac{1}{\delta} = \frac{1}{\dfrac{\Delta\omega}{\omega}} = \frac{\omega_o}{\omega - \omega_o} \qquad \text{or} \qquad \frac{1}{Q_L} = \frac{\Delta\omega}{\omega} \tag{7a}$$

and the bandwidth is

$$BW = 2\Delta\omega = 2(\omega - \omega_o) \tag{7b}$$

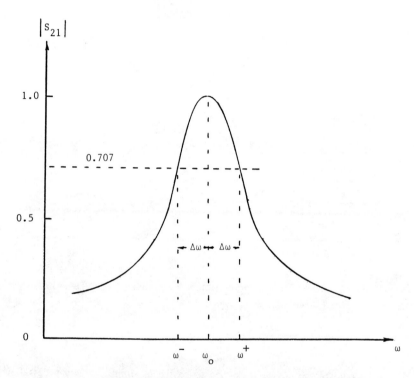

Figure 4-2: The frequency response of $|S_{21}|$.

5. Resonator with Damping

Figure 5-1a shows a resonator with losses. The corresponding eigen-networks are shown in (b) and (c) of Fig. 5-1. Following a similar procedure as for the lossless case, the following results are obtained.

(a) The input impedances of the eigen-networks are

$$Z_1 = \infty \tag{1a}$$

$$Z_2 = \frac{r_e}{2} + j\frac{1}{2}\left[\omega L - \frac{1}{\omega C}\right] = \frac{r_e}{2} + j\omega_o L\delta$$

$$= \frac{r_e}{2}(1 + jQ_u 2\delta) \tag{1b}$$

where

$$Q_u = \frac{\omega_o L}{r_e} \tag{1c}$$

$$2\delta = \frac{\omega}{\omega_o} - \frac{\omega_o}{\omega} \tag{1d}$$

(b) The eigenvalues of the scattering matrix are

$$\lambda_1 = \frac{Z_1 - R_o}{Z_1 + R_o} = 1 \tag{2a}$$

$$\lambda_2 = \frac{\frac{r_e}{2}(1 + j2Q_u\delta) - R_o}{\frac{r_e}{2}(1 + j2Q_u\delta) + R_o} = \frac{1 + j2Q_u\delta - \frac{2R_o}{r_e}}{1 + j2Q_u\delta + \frac{2R_o}{r_e}}$$

$$= \frac{1 - 2\alpha_e + j2Q_u\delta}{1 + 2\alpha_e + j2Q_u\delta} \tag{2b}$$

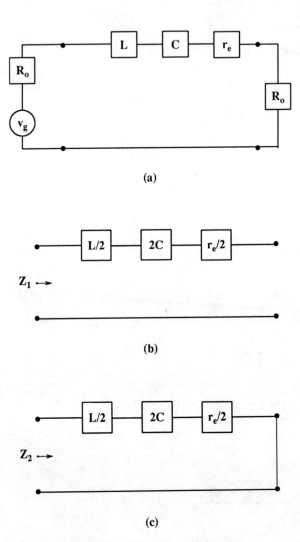

(a)

(b)

(c)

Figure 5-1: A resonator with losses.

where

$$\alpha_e = \frac{R_o}{r_e} = \frac{\dfrac{\omega_o L}{r_e}}{\dfrac{\omega_o L}{R_o}} = \frac{Q_u}{Q_e} \tag{2c}$$

$$Q_e = \frac{\omega_o L}{R_o} = \text{external } Q \tag{2d}$$

(c) The scattering elements are

$$s_{11} = \frac{\lambda_1 + \lambda_2}{2} = \frac{1}{2}\left[1 + \frac{1 - 2\alpha_e + j2Q_u\delta}{1 + 2\alpha_e + j2Q_u\delta}\right]$$

$$= \frac{1 + j2Q_u\delta}{1 + 2\alpha_e + j2Q_u\delta} \tag{3a}$$

$$s_{21} = \frac{\lambda_1 - \lambda_2}{2} = \frac{2\alpha_e}{1 + 2\alpha_e + j2Q_u\delta} = \frac{\dfrac{2\alpha_e}{1 + 2\alpha_e}}{1 + \dfrac{j2Q_u\delta}{1 + 2\alpha_e}}$$

$$= \frac{\dfrac{2\alpha_e}{1 + 2\alpha_e}}{1 + j2Q_L\delta} \tag{3b}$$

where

$$\frac{Q_u}{1 + 2\alpha_e} = \frac{Q_u}{1 + \dfrac{2Q_u}{Q_e}} = \frac{1}{\dfrac{1}{Q_u} + \dfrac{2}{Q_e}} = \frac{1}{Q_L} \tag{3c}$$

Q_L is the loaded Q-factor and $2/Q_e$ is the contribution from both the source and the load resistances.

6. Cavity with Shunt Elements

Shunt elements are introduced in cavities used for band-pass filters. Figure 6-1 shows the arrangement of two susceptances separated by a length of lossless line. The eigen-networks for such an arrangement are shown in Fig. 6-2.

(a) The input admittances of these eigen-networks are

$$Y_1 = j \left[B + Y_o \tan \frac{\theta}{2} \right] \tag{1a}$$

$$Y_2 = j \left[B - Y_o \cot \frac{\theta}{2} \right] \tag{1b}$$

where θ is the electrical length of the waveguide.

(b) The eigenvalues of the scattering matrix are

$$\lambda_1 = \frac{Y_o - Y_1}{Y_o + Y_1} = \frac{1 - y_1}{1 + y_1} \tag{2a}$$

$$\lambda_2 = \frac{Y_o - Y_2}{Y_o + Y_2} = \frac{1 - y_2}{1 + y_2} \tag{2b}$$

where

$$y_1 = \frac{Y_1}{Y_o} = j \left[\frac{B}{Y_o} + \tan \frac{\theta}{2} \right] \tag{2c}$$

$$y_2 = \frac{Y_2}{Y_o} = j \left[\frac{B}{Y_o} - \cot \frac{\theta}{2} \right] \tag{2d}$$

(c) The scattering elements are

$$s_{11} = \frac{\lambda_1 + \lambda_2}{2} = \frac{1}{2} \left[\frac{1 - y_1}{1 + y_1} + \frac{1 - y_2}{1 + y_2} \right] = \frac{1 - y_1 y_2}{D}$$

$$= \frac{1}{D} \left[2 + \frac{2B}{Y_o} \cot \theta - \frac{B^2}{Y_o^2} \right] \tag{3a}$$

and

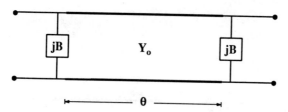

Figure 6-1: A waveguide terminated by a susceptance at each end.

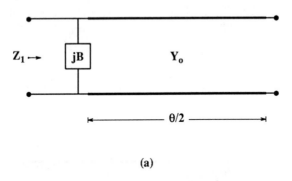

(a)

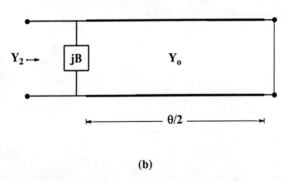

(b)

Figure 6-2: The eigen-networks for Figure 6-1.

$$1 - y_1 y_2 = 1 + \left[\frac{B^2}{Y_o^2} + \frac{B}{Y_o} \left[\tan \frac{\theta}{2} - \cot \frac{\theta}{2} \right] - 1 \right]$$

$$= -\frac{2B}{Y_o} \cot \theta + \frac{B^2}{Y_o^2} \tag{3b}$$

$$D = (1 + y_1)(1 + y_2) \tag{3c}$$

The denominator D can be further expanded as

$$D = 1 + y_1 y_2 + y_1 + y_2$$

$$= 1 - \left[\frac{B^2}{Y_o^2} + \frac{B}{Y_o} \left[\tan \frac{\theta}{2} - \cot \frac{\theta}{2} \right] - 1 \right] + j \left[\frac{2B}{Y_o} + \tan \frac{\theta}{2} - \cot \frac{\theta}{2} \right]$$

$$= 2 - \left[\frac{B^2}{Y_o^2} - \frac{2B}{Y_o} \cot \theta \right] + j \left[\frac{2B}{Y_o} - 2 \cot \theta \right] \tag{3d}$$

$$\tan \frac{\theta}{2} - \cot \frac{\theta}{2} = \frac{\sin \frac{\theta}{2}}{\cos \frac{\theta}{2}} - \frac{\cos \frac{\theta}{2}}{\sin \frac{\theta}{2}} = -2 \cot \theta \tag{3e}$$

$$s_{21} = \frac{\lambda_1 - \lambda_2}{2} = \frac{1}{2D} [(1 - y_1)(1 + y_2) - (1 + y_1)(1 - y_2)]$$

$$= \frac{y_2 - y_1}{D} = \frac{-j}{D} \left[\tan \frac{\theta}{2} + \cot \frac{\theta}{2} \right]$$

$$= \frac{j2}{D \sin \theta} \tag{3f}$$

(d) The square of the magnitude of the transmission coefficient is

$$|s_{21}|^2 = \frac{4}{| D \sin \theta |^2} \tag{4a}$$

and

$$|D|^2 = \left[2 + \frac{2B}{Y_0} \cot\theta - \frac{B^2}{Y_0^2} \right]^2 + \left[\frac{2B}{Y_0} - 2\cot\theta \right]^2$$

$$= 4 \left[1 + \frac{B}{Y_0} \cot\theta \right]^2 - 4\frac{B^2}{Y_0^2} \left[1 + \frac{B}{Y_0} \cot\theta \right] + \frac{B^4}{Y_0^4}$$

$$+ 4 \left[\frac{B^2}{Y_0^2} - \frac{2B}{Y_0} \cot\theta + \cot^2\theta \right]$$

$$= \left[4 + \frac{8B}{Y_0} \cot\theta + 4\frac{B^2}{Y_0^2} \cot^2\theta \right] - 4\frac{B^2}{Y_0^2} - 4\frac{B^3}{Y_0^3} \cot\theta$$

$$+ \frac{B^4}{Y_0^4} + 4\frac{B^2}{Y_0^2} - 8\frac{B}{Y_0} \cot\theta + 4\cot^2\theta$$

$$= 4 \left[\frac{B^2}{Y_0^2} + 1 \right] \cot^2\theta - 4\frac{B^3}{Y_0^3} \cot\theta + 4 + \frac{B^4}{Y_0^4}$$

$$= 4\cot^2\theta + \left[4\frac{B^2}{Y_0^2} \cot^2\theta - 4\frac{B^3}{Y_0^3} \cot\theta + \frac{B^4}{Y_0^4} \right] + 4$$

$$= 4(\cot^2\theta + 1) + \frac{B^2}{Y_0^2} \left[2\cot\theta - \frac{B}{Y_0} \right]^2$$

$$= 4 \left[\frac{\cos^2\theta}{\sin^2\theta} + 1 \right] + \frac{B^2}{Y_0^2} \left[2\cot\theta - \frac{B}{Y_0} \right]^2$$

$$= \frac{4}{\sin^2\theta} + \frac{B^2}{Y_0^2} \left[2\cot\theta - \frac{B}{Y_0} \right]^2 \tag{4b}$$

Thus

$$|s_{21}|^2 = \cfrac{\cfrac{4}{\sin^2 \theta}}{\cfrac{4}{\sin^2 \theta} + \cfrac{B^2}{Y_o^2} \left[2 \cot \theta - \cfrac{B}{Y_o} \right]^2}$$

$$= \cfrac{1}{1 + \cfrac{B^2}{4Y_o^2} \left[2 \cot \theta - \cfrac{B}{Y_o} \right]^2 \sin^2 \theta}$$

$$= \cfrac{1}{1 + \cfrac{B^2}{4Y_o^2} \left[2 \cos \theta - \cfrac{B}{Y_o} \sin \theta \right]^2} \tag{5}$$

(e) The resonance frequency of the resonator is defined by the condition $s_{21} = 1$, for which (5) yields

$$\left[\left[2 \cos \theta - \frac{B}{Y_o} \sin \theta \right]_{\omega = \omega_o} \right]^2 = 0 \qquad \text{or} \qquad \tan \theta_o = \frac{2Y_o}{B} \tag{6}$$

where $\theta_o = \beta_o d$ is the electrical length at resonance.

(f) It is convenient to express s_{21} in the standard form, (4-3d),

$$s_{21} = \frac{1}{1 + jQ_L \delta} \qquad \text{or} \qquad |s_{21}| = \frac{1}{\sqrt{1 + (Q_L \delta)^2}} \tag{7}$$

A comparison of (7) with (5) reveals that

$$Q_L \delta = \frac{B}{2Y_o} [2 \cos \theta - \frac{B}{Y_o} \sin \theta] \tag{8}$$

The trigonometric functions can be expanded in the vicinity of resonance θ_o.

$$\cos \theta = \cos (\theta_o + \Delta\theta) = \cos \theta_o \cos \Delta\theta - \sin \theta_o \sin \Delta\theta$$

$$\approx \cos \theta_o - \Delta\theta \sin \theta_o \tag{9a}$$

$$\sin \theta = \sin (\theta_o + \Delta\theta) \approx \sin \theta_o + \Delta\theta \cos \theta_o \tag{9b}$$

The values of $\sin \theta_o$ and $\cos \theta_o$ can be obtained from (6).

$$\tan \theta = \frac{\sqrt{1 - \cos^2 \theta}}{\cos \theta} = \sqrt{\cos^{-2} \theta - 1}$$

$$\cos \theta_o = \frac{1}{\sqrt{1 + \tan^2 \theta_o}} = \frac{1}{\sqrt{1 + (2Y_o/B)^2}} = \frac{B}{\sqrt{B^2 + 4Y_o^2}} \qquad (10a)$$

$$\tan \theta = \frac{\sin \theta}{\sqrt{1 - \sin^2 \theta}} = \frac{1}{\sqrt{\sin^{-2} \theta - 1}}$$

$$\sin \theta_o = \frac{\tan \theta_o}{\sqrt{1 + \tan^2 \theta_o}} = \frac{2Y_o}{\sqrt{B^2 + 4Y_o^2}} \qquad (10b)$$

The right-hand side of (8) then becomes

$$2 \cos \theta - \frac{B}{Y_o} \sin \theta \approx 2 (\cos \theta_o - \Delta\theta \sin \theta_o) - \frac{B}{Y_o} (\sin \theta_o + \Delta\theta \cos \theta_o)$$

$$= (2 - \frac{B}{Y_o} \Delta\theta) \cos \theta_o - (2\Delta\theta + \frac{B}{Y_o}) \sin \theta_o$$

$$= \frac{(2 - \frac{B}{Y_o} \Delta\theta) B - (2\Delta\theta + \frac{B}{Y_o}) 2Y_o}{\sqrt{B^2 + 4Y_o^2}}$$

$$= \frac{2B - \frac{B^2\Delta\theta}{Y_o} - 4Y_o\Delta\theta - 2B}{\sqrt{B^2 + 4Y_o^2}}$$

$$= \frac{\frac{-\Delta\theta}{Y_o} (B^2 + 4Y_o^2)}{\sqrt{B^2 + 4Y_o^2}} = - \Delta\theta \sqrt{\frac{B^2}{Y_o^2} + 4} \qquad (11)$$

The substitution of (11) into (8) gives

$$Q_L \delta = \frac{-B}{2Y_o} \Delta\theta \sqrt{\frac{B^2}{Y_o^2} + 4} \qquad (12)$$

The incremental electrical length $\Delta\theta$ is

$$\Delta\theta = \theta - \theta_o = \beta d - \beta_o d = (\omega - \omega_o) d\sqrt{\mu\varepsilon}$$

and

$$\frac{\Delta\theta}{\theta_o} = \frac{\omega - \omega_o}{\omega_o} \approx \frac{1}{2}\left[\frac{\omega}{\omega_o} - \frac{\omega_o}{\omega}\right] = \delta \approx \frac{\Delta\omega}{\omega_o} \tag{13}$$

because of (4-5).

$$\frac{\omega}{\omega_o} - \frac{\omega_o}{\omega} \approx \frac{\omega^2 - \omega_o^2}{\omega\omega_o} = \frac{(\omega + \omega_o)(\omega - \omega_o)}{\omega\omega_o} \approx \frac{2\omega\Delta\omega}{\omega\omega_o} = \frac{2\Delta\omega}{\omega_o}$$

Since $\theta_o = \dfrac{\pi}{2}$ at resonance, then

$$\Delta\theta = \Delta\theta \frac{\frac{\pi}{2}}{\theta_o} = \frac{\pi}{2}\delta \tag{14}$$

Hence, (12) becomes

$$Q_L\delta = \frac{-B}{2Y_o}\frac{\pi}{2}\delta\sqrt{\frac{B^2}{Y_o^2} + 4} = -\delta\frac{B\pi}{4Y_o}\sqrt{\frac{B^2}{Y_o^2} + 4}$$

and

$$Q_L = \frac{-\pi B}{4Y_o}\sqrt{\frac{B^2}{Y_o^2} + 4} \tag{15}$$

This is the equivalent loaded Q for a cavity with a shunt element. The parameter s_{21} for a cavity with a shunt element can thus be expressed as (7) with the corresponding Q_L given by (15).

7. Transformer-Coupled Resonator

(a) Figure 7-1 is the schematic diagram of a doubly terminated transformer-coupled series resonator. The eigen-networks are shown in Fig. 7-2.

(b) The input impedances of eigen-networks or eigenvalues of the impedance matrix are

$$Z_1 = j\omega L_1 \tag{1}$$

$$Z_2 = j\omega L_1 + Z_2' \tag{2a}$$

$Z_2' = $ reflected impedance from the secondary circuit [Eq.(7.2–2d)]

$$= \frac{(\omega M)^2}{\dfrac{j\omega L'}{2} + \dfrac{1}{j2\omega C}} = \frac{-j2 (\omega M)^2}{\omega_o L'} \frac{1}{\dfrac{\omega}{\omega_o} - \dfrac{\omega_o}{\omega}}$$

$$= \frac{-j(\omega M)^2}{\omega_o L' \delta} \tag{2b}$$

$$2\delta = \frac{\omega}{\omega_o} - \frac{\omega_o}{\omega} \tag{2c}$$

$$L' = 2L_2 + L \tag{2d}$$

(c) The impedance elements of the impedance matrix are

$$z_{11} = \frac{Z_1 + Z_2}{2} = j\omega L_1 - \frac{j(\omega M)^2}{2\omega_o L' \delta} \tag{3a}$$

$$z_{21} = \frac{Z_1 - Z_2}{2} = \frac{j(\omega M)^2}{2\omega_o L' \delta} \tag{3b}$$

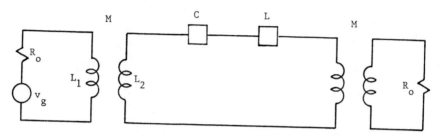

Figure 7-1: A terminated series resonator.

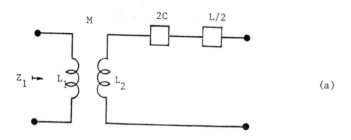

(a)

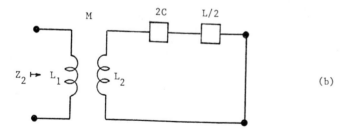

(b)

Figure 7-2: The eigen-networks for Figure 7-1.

7.1. Mutual Coupling

(a) Mutual induction is the phenomenon whereby two coils are arranged in such a way that some of the flux generated in one coil links the other coil and a change of current or flux in one coil induces a voltage in the other.

Figure 7.1-1 shows two ideal coils located in the vicinity of each other. The self-inductance of a coil in a linear system is defined by the relation

L_k = self-inductance of the kth coil = flux linkage of coil k per unit current i_k

$$= \frac{\Phi_k}{i_k} \tag{1}$$

where Φ_k is the total magnetic flux linking the kth coil and Φ_k is produced by the current i_k in the kth coil.

Similarly, the mutual inductance between the jth and the kth coils is defined as

$$M_{jk} = \text{mutual inductance between coils j and k} = \frac{\Phi_{jk}}{i_k} \tag{2}$$

where Φ_{jk} is the portion of flux produced by current i_k in the kth coil which links the jth coil.

The self- and mutual inductances of the arrangement shown in Fig. 7.1-1 are

$$L_1 = \frac{n_1 \Phi_1}{i_1} \tag{3a}$$

$$L_2 = \frac{n_2 \Phi_2}{i_2} \tag{3b}$$

$$M_{21} = \frac{n_2 \Phi_{21}}{i_1} \tag{3c}$$

$$M_{12} = \frac{n_1 \Phi_{12}}{i_2} \tag{3d}$$

where

$$\Phi_1 = \text{total flux generated by } i_1 \tag{3e}$$

$$\Phi_{21} = \text{portion of } \Phi_1 \text{ which links coil 2} \tag{3f}$$

$$\Phi_2 = \text{total flux generated by } i_2 \tag{3g}$$

$$\Phi_{12} = \text{portion of } \Phi_2 \text{ which links coil 1} \tag{3h}$$

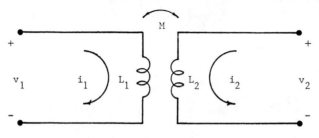

Figure 7.1-1: Two coupled coils.

(b) When the reluctance of the magnetic circuit is constant, the mutual flux Φ_{21} is linearly proportional to the total flux Φ_1, i.e.,

$$\Phi_{21} = k_{21}\Phi_1 \qquad \Phi_{12} = k_{12}\Phi_2 \tag{4}$$

where k_{21} and k_{12} are constants of proportionality and they have a maximum value of unity. Then (3c) and (3d) become

$$M_{21} = \frac{n_2 k_{21}\Phi_1}{i_1} = k_{21}\frac{n_2}{n_1}\frac{n_1\Phi_1}{i_1} = k_{21}\frac{n_2}{n_1}L_1 \tag{5a}$$

$$M_{12} = k_{12}\frac{n_1}{n_2}\frac{n_2\Phi_2}{i_2} = k_{12}\frac{n_1}{n_2}L_2 \tag{5b}$$

(c) To show that $M_{12} = M_{21}$, consider the case when the currents are initially zero in both coils. With the current in coil 2 maintained at zero level, $i_2 = 0$, the current in coil 1 is increased to its final value i_a. The energy required to accomplish this is given by

$$W_1 = \int_0^T i_1 L_1 \frac{di_1}{dt} dt = \int_0^{i_a} L_1 i_1 \, di_1$$

$$= \frac{1}{2} L_1 i_a^2 \tag{6}$$

Keeping $i_1 = i_a$ = constant, the current in coil 2 is now increased to its final value i_b. The energy associated with this process is

$$W_2 + W_{21} = \int_0^{i_b} L_2 i_2 \, di_2 + \int_0^{i_b} M_{21} i_a \, di_2$$

$$= \frac{1}{2} L_2 i_b^2 + M_{21} i_a i_b \tag{7}$$

The total energy required to establish i_a first and then i_b is the sum of (6) and (7).

$$W_a = W_1 + W_2 + W_{21} = \frac{1}{2}(L_1 i_a^2 + L_2 i_b^2) + M_{21} i_a i_b \tag{8}$$

Instead of establishing i_a in coil 1, keep $i_1 = 0$ and increase i_2 to i_b first. The required energy is

$$W_2 = \frac{1}{2} L_2 i_b^2 \tag{9}$$

Then let $i_2 = i_b$ = constant, and i_1 is increased to the value of i_a. The energy is

$$W_1 + W_{12} = \frac{1}{2} L_1 i_a^2 + M_{12} i_a i_b \tag{10}$$

The total energy of this process is the sum of (9) and (10).

$$W_b = W_1 + W_2 + W_{12} = \frac{1}{2} (L_1 i_a^2 + L_2 i_b^2) + M_{12} i_a i_b \tag{11}$$

Since the final states in both processes are identical, their energies must be equal also. Hence

$$M_{12} = M_{21} \equiv M \tag{12}$$

Then (5) yields

$$M_{12} M_{21} = M^2 = k_{12} k_{21} L_1 L_2 = k^2 L_1 L_2$$

$$M = k \sqrt{L_1 L_2} \tag{13}$$

Since M is reciprocal, consequently the constant of proportionality k is also reciprocal.

7.2. Transformer

(a) A transformer circuit is shown in Fig. 7.2-1. The mesh equations for this circuit are

$$(R_1 + j\omega L_1)\, i_1 + j\omega M i_2 = v_1 \tag{1a}$$

$$j\omega M i_1 + (Z_L + R_2 + j\omega L_2)\, i_2 = 0 \tag{1b}$$

i_2 can be eliminated from (1a) by using (1b).

$$i_2 = \frac{-j\omega M i_1}{Z_L + R_2 + j\omega L_2} \tag{2a}$$

$$\left[(R_1 + j\omega L_1) + \frac{(\omega M)^2}{Z_L + R_2 + j\omega L_2} \right] i_1 = v_1$$

or

$$\frac{v_1}{i_1} = Z_1 + Z_1' \tag{2b}$$

$$Z_1 = R_1 + j\omega L_1 \tag{2c}$$

$$Z_1' = \frac{(\omega M)^2}{Z_L + R_2 + j\omega L_2}$$

$$= \text{reflected impedance of the secondary circuit} \tag{2d}$$

(b) For a tightly coupled low-loss circuit, one has

$$k = 1 \tag{3a}$$

$$M = \sqrt{L_1 L_2} \tag{3b}$$

$$R_1 \ll j\omega L_1 \tag{3c}$$

$$R_2 \ll j\omega L_2 \tag{3d}$$

$$Z_L \ll R_2 + j\omega L_2 \tag{3e}$$

Then

$$Z_1 \approx j\omega L_1 \tag{4a}$$

$$Z_1' \approx \frac{\omega^2 L_1 L_2}{R_2 + j\omega L_2} \tag{4b}$$

and (2a) becomes,

$$\frac{i_1}{i_2} = \frac{Z_L + R_2 + j\omega L_2}{-j\omega M} \approx -\sqrt{\frac{L_2}{L_1}} = \frac{-n_2}{n_1} \tag{5a}$$

where (7.1-5)

$$\frac{L_1}{L_2} = \left[\frac{n_1}{n_2} \right]^2 \tag{5b}$$

was used to obtain the final form.

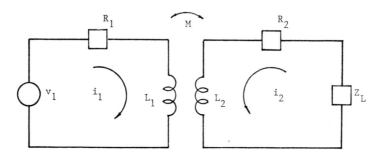

Figure 7.2-1: Two circuits coupled by a transformer.

8. Problems

1. An ideal Fabry-Perot resonator is represented by two infinite parallel plates with a spacing d, and the cross-sectional view is shown in Fig. 8-1. The fields in such a resonator are TEM standing waves and are given by:

$$E_x = E_0 \sin \beta_0 z \qquad \beta_0 = \omega \sqrt{\mu_0 \varepsilon_0}$$

$$H_y = jY_0 E_0 \cos \beta_0 z$$

where the region between plates is filled with vacuum.

If the plates have a surface resistance R_{sm} for the mth mode, show that the Q of the resonator is given by

$$Q = \frac{\omega \varepsilon_0 Z_0^2 d}{4 R_{sm}} = \frac{\pi k Z_0}{4 R_{sm}}$$

$$k \equiv \frac{2d}{\lambda_0}$$

2. A practical Fabry-Perot resonator can be constructed from two very large parallel plates which carry a square array of small circular holes. One end plate is illuminated by the field radiated from a source, and a receiver is located on the opposite side of the other plate (Fig. 8-2). Let the thin plate with holes be characterized by a normalized inductive susceptance, $-j\hat{B}_h$. If the electric field coupled into the resonator is E_0 at $z = 0$, show that the total field in the parallel plate region is

$$E = E_0 \frac{e^{j\beta_0 z} + \Gamma e^{-j\beta_0(2d - z)}}{1 - \Gamma^2 e^{-j2\beta_0 d}}$$

where $\Gamma = |\Gamma|e^{j\theta_r}$ is the reflection coefficient of the perforated plate.

3. For the resonator described in Problem 2, let $\hat{B}_h$ be made very large in order to gain high external Q. Determine the condition to yield the maximum E-field.

4. The fields of the TE_{011} mode in the rectangular cavity shown in Fig. 8-3 are given as follows.

$$E_x = E_0 \sin \frac{\pi y}{b} \sin \frac{\pi z}{c}$$

$$H_y = \frac{jbE_0}{\eta \sqrt{b^2 + c^2}} \sin \frac{\pi y}{b} \cos \frac{\pi z}{c}$$

$$H_z = \frac{-jcE_0}{\eta \sqrt{b^2 + c^2}} \cos \frac{\pi y}{b} \sin \frac{\pi z}{c}$$

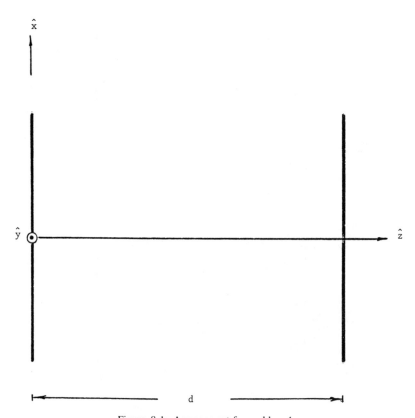

Figure 8-1: Arrangement for problem 1.

538

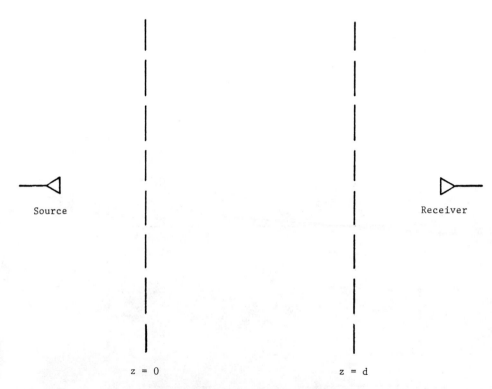

z = 0   z = d

Figure 8-2: Arrangement for problem 2.

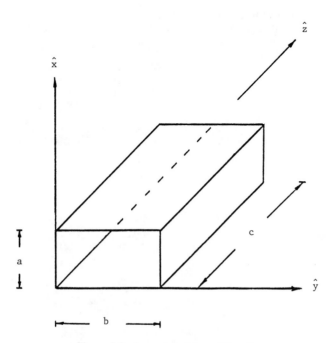

Figure 8-3: Arrangement for problem 3.

(a) Determine the Q of the cavity if the waveguide is made of a perfect conductor and is filled with a dielectric characterized by

$$\underline{\varepsilon} = \varepsilon' - j\varepsilon''$$

(b) Determine the Q of the cavity if the guide is filled with lossless dielectric and the waveguide is made of a good conductor characterized by a surface resistance R_s.

(c) What is the Q of the cavity if it is filled with a dielectric as in part (a) and the guide is made of the same metal as in (b)?

CHAPTER X   Review on System Functions

1. Linear System

The systems to be discussed are restricted to those having linear, lumped, and time-invariant elements. The description that is pertinent here is the system function of the complex-frequency variable.

The behavior of a system can be described by a set of linear simultaneous integro-differential equations with constant coefficients.

The nodal equations of an electrical system are

$$a_{11}v_1 + a_{12}v_2 + \cdots + a_{1n}v_n = i_1(t)$$

$$a_{21}v_1 + a_{22}v_2 + \cdots + a_{2n}v_n = i_2(t)$$

$$\cdots \cdots \cdots \qquad\qquad (1a)$$

$$a_{n1}v_1 + a_{n2}v_2 + \cdots + a_{nn}v_n = i_n(t)$$

where

v_j = voltage at the jth node

i_j = current source injected into the jth node

$$a_{jk} = G_{jk} + C_{jk}\frac{d}{dt} + \frac{1}{L_{jk}}\int dt$$

G_{jk} = self-conductance of a node pair for j = k,   (1b)

or mutual conductance between nodes j and k for k ≠ j

C_{jk} = self-capacitance for j = k, or mutual capacitance for j ≠ k

L_{jk} = self-inductance for j = k, or mutual inductance for j ≠ k

The response v_k can be obtained through the use of Laplace transformation. The Laplace transform of each of the equations in (1a) yields a set of algebraic equations which are functions of the complex transform variable, $s = \sigma + j\omega$.

$$b_{11}V_1 + b_{12}V_2 + \cdots + b_{1n}V_n = I_1 + I_{o1}$$

$$b_{21}V_1 + b_{22}V_2 + \cdots + b_{2n}V_n = I_2 + I_{o2}$$

$$\cdots\cdots\cdots \tag{2a}$$

$$b_{n1}V_1 + b_{n2}V_2 + \cdots + b_{nn}V_n = I_n + I_{on}$$

where

$$b_{jk} = G_{jk} + sC_{jk} + \frac{1}{sL_{jk}} \tag{2b}$$

$$I_j \equiv I_j(s) = \text{Laplace transform of } i_j(t) \tag{2c}$$

$$V_k \equiv V_k(s) = \text{Laplace transform of } v_k(t) \tag{2d}$$

$$I_{oj} \equiv I_{oj}(s) = \sum_{k=1}^{n} C_{jk}v_k(0^+) - \frac{i_{Lj}(0^+)}{s} = I_{oC} + I_{oL} \tag{2e}$$

I_{oC} = contributions from the initial values of voltages of the capacitances

I_{oL} = contributions from initial currents in inductances leaving the jth nodes

In matrix form,

$$[B]\,[V] = [\,I\,] + [\,I_o\,] \tag{3a}$$

$$[B] = \begin{bmatrix} b_{11} & b_{12} & \cdots & b_{1n} \\ \cdots & \cdots & \cdots & \cdots \\ b_{n1} & b_{n2} & \cdots & b_{nn} \end{bmatrix} \tag{3b}$$

$$[V] = [\,V_1 \quad V_2 \quad \cdots \quad V_n\,]^t \tag{3c}$$

$$[\,I\,] = [\,I_1 \quad I_2 \quad \cdots \quad I_n\,]^t \tag{3d}$$

$$[\,I_o\,] = [\,I_{o1} \quad I_{o2} \quad \cdots \quad I_{on}\,]^t \tag{3e}$$

The response can be determined by Cramer's rule.

$$V_k = \frac{\Delta_k}{\Delta} = \sum_{j=1}^{n} I_j \frac{\Delta_{jk}}{\Delta} + \sum_{j=1}^{n} I_{oj}(s) \frac{\Delta_{ojk}}{\Delta} \tag{4a}$$

$$\Delta = \text{determinant of } [B] \tag{4b}$$

$$\Delta_k = \text{the matrix } [B] \text{ with its kth column replaced by } [\,I\,] \tag{4c}$$

$$\Delta_{ok} = \text{the matrix } [B] \text{ with its kth column replaced by } [\,I_o\,] \tag{4d}$$

$$\Delta_{jk} = \text{the cofactor of element } I_j \text{ in the } \Delta_k \text{ matrix} \tag{4e}$$

$$\Delta_{ojk} = \text{the cofactor of element } I_{oj} \text{ in the } \Delta_{ok} \text{ matrix} \tag{4f}$$

When the excitation is zero, the first term on the right-hand side of (4a) vanishes. The resulting solution is called the free-response function. The free-response function is

$$V_k(s) \big|_{\text{free}} = \sum_{j=1}^{n} I_{oj}(s) \frac{\Delta_{ojk}}{\Delta} \tag{5}$$

Equation (5) can also be expressed in terms of partial fraction expansion.

$$V_k(s) \big|_{\text{free}} = \sum_{i=1}^{m_1} \frac{k_{1i}}{(s - s_1)^{m_1}} + \cdots + \sum_{i=1}^{m_n} \frac{k_{ni}}{(s - s_n)^{m_n}} \tag{6}$$

where roots of multiplicity m_i are assumed. A simple root has unit multiplicity.

The inverse Laplace transform yields the free response in the time domain.

$$v_k(t) \big|_{\text{free}} = \sum_{i=1}^{m_1} \frac{k_{1i} t^{i-1} e^{s_1 t}}{(i-1)!} + \cdots + \sum_{i=1}^{m_n} \frac{k_{ni} t^{i-1} e^{s_n t}}{(i-1)!} \tag{7}$$

The s_i's are the zeros of the determinant Δ and their values vary with the values of circuit elements and the configuration of the network. They are independent of the sources and the initial state of the network. Because the s_i's have the dimensions of frequency and describe the free or natural responses of the network, they are known as the natural frequencies of the network.

The values of the k_{ji}'s vary with the values of circuit elements, the configuration, and the initial state.

The nature of each term of (7) depends upon the values of s_i or its location in the complex-frequency plane, and also on the multiplicity m_j.

For a real natural frequency, $s_j = \sigma_j + j0$, the response has the form

$$\sum_{i=1}^{m_j} \frac{k_{ji}t^{i-1}e^{\sigma_j t}}{(i-1)!} \tag{8}$$

Natural frequencies lying on the imaginary axis are accompanied by a conjugate set of the same multiplicity. In addition, the respective coefficients in Eqs. (2) to (8) must also be conjugates. For $s_j = 0 \pm j\omega_j$, the response has the form

$$\sum_{i=1}^{m_j} \frac{2|k_{ji}| \ t^{i-1} \cos (\omega_j t + \phi_{ji})}{(i-1)!} \tag{9a}$$

$$k_{ji} = |k_{ji}| \ e^{j\phi_{ji}} \tag{9b}$$

When the natural frequency is complex, $s_j = \sigma_j + j\omega_j$, the contribution to the response is

$$\sum_{i=1}^{m_j} \frac{2|k_{ji}| \ t^{i-1}e^{\sigma_j t} \cos (\omega_j t + \phi_{ji})}{(i-1)!} \tag{10}$$

It is to be noted from (8) through (10) that the response will be finite as time grows indefinitely if the following conditions are satisfied.

(i) Natural frequencies lie in the left half-plane can have any multiplicity.

(ii) Natural frequencies lie on the $j\omega$-axis can have a multiplicity of unity only.

Any network which has such a bounded free response is known as a stable network. Hence, the above conditions are also known as stability conditions.

The impedance function

$$Z_{jk} = \frac{V_k}{I_j} = \frac{\Delta_{jk}}{\Delta} \tag{11}$$

is seen to be a ratio of two polynomials.

$$Z_{jk} = \frac{a_n s^n + a_{n-1}s^{n-1} + \cdots + a_o}{b_m s^m + b_{m-1}s^{m-1} + \cdots + b_o} \tag{12}$$

A rational function is a function which can be expressed as the quotient of two polynomial functions. Thus, the impedance function is a rational function.

1.1. Natural Frequencies

(a) Natural frequencies of a passive system are the values of s for which a response can exist without external excitation.

(b) The behavior of a system has the generic form

$$e^{s_k t} \tag{1}$$

In a lossy system, the natural behavior must be damped due to the dissipation of energy. Hence, the real part of the natural frequency must be negative.

$$\text{Re } [s_k] < 0 \tag{2}$$

In a lossless system, the natural frequencies are purely imaginary since there is no dissipation of energy.

$$\text{Re } [s_k] = 0 \tag{3}$$

The combination of both lossy and lossless cases yields the general property of natural frequencies.

$$\text{Re } [s_k] \leq 0 \tag{4}$$

Thus, the natural frequencies are located in the left half of the s-plane.

(c) An equivalent point of view is to calculate the natural frequencies as the roots of an equation which indicates that there exists a response without external excitation.

For the case of a one-terminal pair network (Fig. 1.1-1), Kirchhoff's voltage law yields

$$V(s) = I(s)Z(s) \tag{5a}$$

$$Z(s) = \text{impedance of the input loop} \tag{5b}$$

Without external excitation, one has Fig. 1.1-2, and (5a) becomes

$$0 = I(s)Z(s) \tag{5c}$$

Since $I(s) \neq 0$, then

$$Z(s) = 0 \tag{6}$$

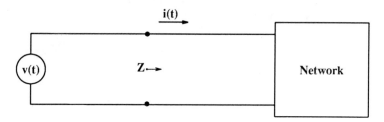

Figure 1.1-1: One-port network with external voltage source.

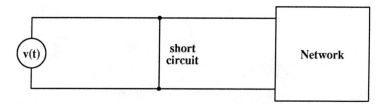

Figure 1.1-2: One-port network without external voltage source.

The natural frequencies of a one-terminal pair network are the roots of the expression, (6), for the sum of impedances in the input loop.

The dual case is shown in Fig. 1.1-3, and the relation is

$$I(s) = Y(s)V(s) \tag{7a}$$

$$Y(s) = \text{the sum of all the admittances between input nodes 1 and 1}' \tag{7b}$$

Without excitation, the arrangement becomes Fig. 1.1-4, and the expression is

$$Y(s) = 0 \tag{8}$$

The natural frequencies of a one-terminal pair network are the roots of the relations for the sum of all the admittances between the input nodes.

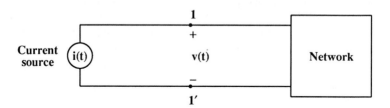

Figure 1.1-3: One-port network with external current source.

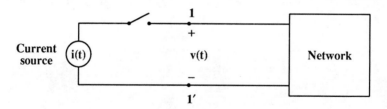

Figure 1.1-4: One-port network without external current source.

2. System Function

(a) The response of a system due to the excitation only is known as the forced response. This is given by the first term in (1-4a).

$$V_k(s) \Big|_{force} = \sum_{j=1}^{n} I_j(s) \frac{\Delta_{jk}}{\Delta} \tag{1}$$

The factor Δ_{jk}/Δ depends upon the system alone and is called the system function. The other factor, $I_j(s)$, is due to the excitation alone and is referred to as the excitation function.

The response of a system caused by the excitation function can be expressed as

$$\text{Response function} = \text{System function} \times \text{Excitation function} \tag{2a}$$

or

$$\text{System function} = \frac{\text{Response function}}{\text{Excitation function}} \tag{2b}$$

From (1),

$$\text{System function} \equiv N(s) = \frac{\Delta_{jk}}{\Delta} = \frac{a_n s^n + a_{n-1} s^{n-1} + \cdots + a_o}{b_m s^m + b_{m-1} s^{m-1} + \cdots + b_o} \equiv \frac{Q(s)}{D(s)} \tag{3}$$

where $Q(s)$ and $D(s)$ are polynomial functions in s with constant coefficients. These polynomials can be expressed in factor form.

$$N(s) = N_o \frac{(s - z_1)(s - z_2) \cdots (s - z_n)}{(s - p_1)(s - p_2) \cdots (s - p_m)} \tag{4}$$

where z_k's are zeros of $N(s)$, p_k's are poles of $N(s)$, and N_o is a constant factor.

A rational function is one which can be expressed as the quotient of two polynomial functions. The system function, $N(s)$, is a rational function.

(b) Since both a_j and b_j are real, poles and zeros of the system function, if complex, must occur in complex conjugate pairs.

(c) The determinant Δ appears in the denominator of (3). The zeros of Δ are the natural frequencies of the system; hence, the natural frequencies are also the poles of the system function.

The natural frequencies are identical with poles of the system function when the system function is the immittance function.

(d) There may be natural frequencies which are not poles of $N(s)$ and conversely.

(d.1)   If a branch is connected in series with the current source or a branch is connected in parallel with the voltage source, Fig. 2-1, the impedance of the branch will not appear in the system determinant. Consequently, neither branch will have contribution to the natural frequencies of the system.

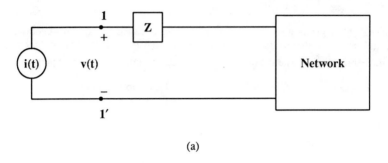

(a)

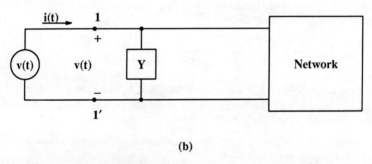

(b)

Figure 2-1: Arrangements with impedance of a brance which will not appear in the system function.

But, the system function v/i for Fig. 2-1a, or i/v for Fig. 2-1b, is a function of the branch immittance. Poles of the immittance function are also poles of the system function.

(d.2)   The desired system function can be either the ratio of two voltage functions or the ratio of two current functions. Figure 2-2 shows a system driven by a current source. The nodal analysis yields

$$V_k = I_1 \frac{\Delta_{1K}}{\Delta} \tag{5}$$

The current I_k is given by

$$I_k = V_k Y_k = \frac{\Delta_{1k}}{\Delta} I_1 Y_k \tag{6}$$

and the system function is

$$N(s) = \frac{I_k}{I_1} = \frac{\Delta_{1k}}{\Delta} Y_k \tag{7}$$

Hence, the poles of Y_k will appear as poles of $N(s)$ and are not natural frequencies of the system.

(d.3)   A cancellation of zeros of the cofactor Δ_{jk} and zeros of the determinant Δ may occur. In such cases, some of the natural frequencies are not poles of the system function.

(e) The stability condition requires that the poles of all system functions should lie in the left half-plane. If the poles lie on the imaginary $j\omega$-axis, they must be simple poles.

(f) When the poles of the system function all located within the interior of the left half-plane, the polynomial, $D(s)$, in (3) is known as a Hurwitz polynomial.

A polynomial $P(s)$ is said to be Hurwitz if the following conditions are satisfied.

(f.1) $P(s)$ is real when s is real.

(f.2) The roots of $P(s)$ have real parts which are zero or negative.

(g) In general, the same restriction for the poles as given in (f) cannot be said about the locations of zeros of $N(s)$, except when the system function is the driving-point function.

The driving-point function is defined to be the ratio of response to excitation at the same port. In such a case, the system function is either the driving-point impedance or the driving-point admittance function, i.e.,

$$N(s) = Z(s) = \frac{V_j}{I_j} = \frac{Q(s)}{D(s)} \tag{8a}$$

or

$$N(s) = Y(s) = \frac{D(s)}{Q(s)} \tag{8b}$$

The zeros of D(s) are the poles of Z(s) when the excitation is a current source. The locations of poles are specified by item (e).

When the excitation is a voltage source, the system function is given by (8b). The poles of Y(s) are the zeros of Q(s), which are also restricted by item (e). But the poles of (8b) are zeros of (8a) and vice versa. Hence, both poles and zeros of a driving-point system function are specified by the same restriction, item (e).

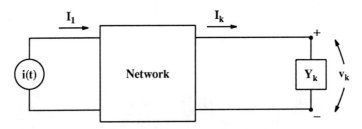

Figure 2-2: A system excited by a current source.

2.1 Properties of Network Function - One-Terminal Pair Network

(a) The immittance function of a one-terminal pair network is a rational function of s with real coefficients.

(b) As | s | approaches infinity, all capacitors become short circuits, all inductors become open circuits, and the resistors remain constant (real and positive). Therefore, as $s \to \infty$,

$$Z(s) \quad \text{or} \quad Y(s) \to Ks$$

$$\to K \tag{1}$$

$$\to \frac{K}{s}$$

where K is a real and positive constant.

(c) The zeros and poles of Z(s) and Y(s) are in the left half of the s-plane or on the $j\omega$-axis.

(d) Poles and zeros of Z(s) and Y(s) that lie on the $j\omega$-axis are simple; the residues at these poles are real and positive, and the derivatives at these zeros are real and positive (Section 3.2).

(e) For a passive network, the average power is given by

$$P = \frac{v^2 G}{2} = \frac{1}{2} v^2 \operatorname{Re} [\, Y(j\omega) \,] \tag{2a}$$

$$= \frac{i^2 R}{2} = \frac{1}{2} i^2 \operatorname{Re} [\, Z(j\omega) \,] \tag{2b}$$

Therefore, for $-\infty < \omega < \infty$,

$$\operatorname{Re} [\, Z(j\omega) \,] \geq 0 \tag{3a}$$

$$\operatorname{Re} [\, Y(j\omega) \,] \geq 0 \tag{3b}$$

A positive-real function is a function of s that has the five foregoing properties.

3. Properties of System Functions

(a) A system function $N(s)$ is a rational function of a complex variable, $s = \sigma + j\omega$, and can be expressed as the sum of a real and an imaginary component.

$$N(s) = N(\sigma + j\omega) = U(\sigma, \omega) + jV(\sigma, \omega) \tag{1}$$

where U and V are the real and the imaginary component of $N(s)$, respectively, and each is a function of two variables, σ and ω.

(b) When $N(s)$ is analytic, its real and imaginary parts are related by the Cauchy-Reimann condition.

$$\frac{\delta U}{\delta \sigma} = \frac{\delta V}{\delta \omega} \tag{2a}$$

$$\frac{\delta U}{\delta \omega} = -\frac{\delta V}{\delta \sigma} \tag{2b}$$

(c) The system function can also be divided into an even and an odd part. This can be achieved by separating the polynomials into even and odd polynomials.

$$N(s) = \frac{Q(s)}{D(s)} = \frac{q_1(s^2) + sq_2(s^2)}{d_1(s^2) + sd_2(s^2)} \tag{3}$$

where q_i and d_i are even polynomials. The following manipulation is then carried out.

$$N(s) = \left[\frac{q_1 + sq_2}{d_1 + sd_2}\right]\left[\frac{d_1 - sd_2}{d_1 - sd_2}\right]$$

$$= \frac{(q_1 d_1 - s^2 q_2 d_2) + s(q_2 d_1 - q_1 d_2)}{d_1^2 - s^2 d_2^2}$$

$$= N_e(s) + N_o(s) \tag{4a}$$

$$N_e(s) = \text{even function of } N(s)$$

$$= \frac{q_1 d_1 - s^2 q_2 d_2}{d_1^2 - s^2 d_2^2} \tag{4b}$$

$$N_o(s) = \text{odd function of } N(s)$$

$$= \frac{s(q_2 d_1 - q_1 d_2)}{d_1^2 - s^2 d_2^2} \tag{4c}$$

(d) Since the sinusoidal steady-state measurements of a network can be made simply and accurately, certain steady-state properties may be obtained by such a process. The special case when $s = j\omega$ is therefore of particular importance.

For the case $s = j\omega$, one has

$$N(j\omega) = U(0, \omega) + jV(0, \omega) \equiv U_0(\omega) + jV_0(\omega) \tag{5}$$

A comparison with (4a) leads to

$$N_e(j\omega) = \text{Re } [N(j\omega)] = U_0(\omega) \tag{6a}$$

$$N_0(j\omega) = j \{ \text{Im } [N(j\omega)] \} = jV_0(\omega) \tag{6b}$$

3.1. Definition of Postive-Real Function

A postive-real function has the following properties:

(a) It is a rational function with real coefficients.
(b) It behaves at $s = \infty$ as Ks, K, or K/s, with K being positive.
(c) It has no poles in the right half of the s-plane.
(d) Its poles and zeros on the $j\omega$-axis are simple and have real and positive residues; and the derivatives at these zeros are real and positive (Section 3.2).
(e) Re $[F(j\omega)] \geq 0$ for $-\infty < \omega < \infty$.

By the theory of functions of complex variables, an equivalent definition is:

A function $F(s)$ is positive-real when $F(\sigma)$ is real and Re $[F(s)]$ is greater than zero throughout the right half of the s-plane, i.e., $\sigma > 0$.

An alternative definition is:

A function is positive-real when

$$|\text{Arg } F(s)| \leq |\text{Arg } s|$$

throughout the right half of the s-plane, $\sigma > 0$.

The last two versions include functions which are not of rational form such as transmission lines, image immittances, and so on.

It is clear that any function $F(s)$ proposed for a synthesis as a driving-point immittance must be a positive-real function. In other words, this is a necessary and sufficient condition for a problem to be soluble.

3.2 Some Properties of Poles and Zeros Located on the Imaginary Axis.

(a) A pole of a positive-real function, $Z(s)$ or $Y(s)$, located on the $j\omega$-axis must be simple. If the pole has a multiplicity m, the natural behavior corresponding to this pole has the form

$$t^{m-1} \cos \omega_0 t \tag{1}$$

which increases with time, and this is impossible for a passive network.

(b) Consider the impedance $Z(s)$ with a simple pole at $s = j\omega_0$. It can be expressed in the partial fraction form.

$$Z(s) = \cdots + \frac{a + jb}{s - j\omega_0} + \cdots \tag{2}$$

where $A \equiv a + jb$ is the residue of $Z(s)$ at the pole $s = j\omega_0$. In a sinusoidal steady state, $s = j\omega$,

$$Z(j\omega) = \cdots + \frac{a + jb}{j(\omega - \omega_0)} + \cdots \equiv R(\omega) + jX(\omega) \tag{3a}$$

When $\omega = \omega_0$, the principal terms are

$$R(\omega) = \frac{b(\omega - \omega_0)}{(\omega - \omega_0)^2} = \text{Re}\,[Z(j\omega)] \tag{3b}$$

$$X(\omega) = \frac{-a(\omega - \omega_0)}{(\omega - \omega_0)^2} = \text{Im}\,[Z(j\omega)] \tag{3c}$$

Equation (3b) shows that $\text{Re}\,[Z(j\omega)]$ changes sign at $\omega = \omega_0$. This violates the property of a positive-real function which requires

$$\text{Re}\,[Z(j\omega)] \geq 0 \tag{4}$$

Thus, it is necessary that b vanish or that the residue A be real.

(c) The sign of the residue A can be investigated by considering dissipative circuit elements, Fig. 3.2-1. For uniform dissipation (Fig. 3.2-1), the effect of losses is simply the replacement of s by $(s + k)$. When $\omega = \omega_0$, one has

$$Z(j\omega) = \cdots + \frac{A}{(j\omega + k) - j\omega_0} + \cdots$$

$$= \ldots + A\,\frac{k - j(\omega + \omega_0)}{k^2 + (\omega - \omega_0)^2} + \ldots = R(\omega) + jX(\omega) \tag{5}$$

which requires $A > 0$, since $R(\omega) \geq 0$.

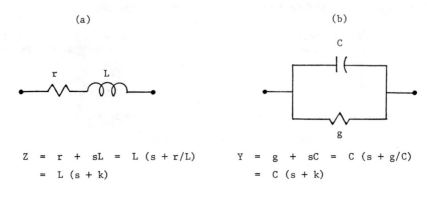

(a)

(b)

$$Z = r + sL = L (s + r/L)$$
$$= L (s + k)$$

$$Y = g + sC = C (s + g/C)$$
$$= C (s + k)$$

$$k := r/L := g/C := \text{uniform dissipation}$$

Figure 3.2-1: Circuits with dissipations.

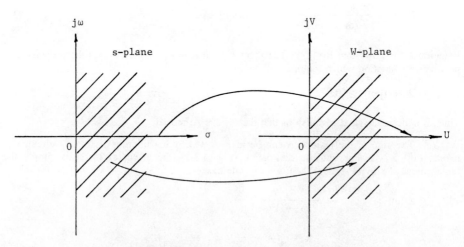

Figure 3.2-2: The mapping of a positive real function.

(d) The definition of a positive-real function can be interpreted as a mapping operation. Let a positive-real function $F(s)$ be

$$F(s) = U(s) + jV(s) \equiv W(s) \tag{6a}$$

$$s = \sigma + j\omega \tag{6b}$$

Then the positive-real function is said to map the real σ-axis in the s-plane into the real U-axis in the W-plane; and to map the right half of the s-plane into the right half of the W-plane, Fig. 3.2-2.

If $F_1(s)$ and $F_2(s)$ are positive-real, then the resultant function

$$F_3(s) = F_1[F_2(s)] \tag{7}$$

is also postive-real. This is because $F_2(s)$ maps the right half of the s-plane into the right half of the $F_2(s)$-plane, since $F_2(s)$ is positive-real. $F_1(s)$ maps the right half of F_2-plane into the right half of F_1-plane, since $F_1(s)$ is positive-real. The composite mapping therefore maps the right half of the s-plane into the right half of the F_3-plane and the real axis is preserved.

If $F(s)$ is positive-real, then $F(1/s)$ is also positive-real. Since

$$\frac{1}{s} = \frac{\sigma - j\omega}{\sigma^2 + \omega^2} \tag{8}$$

is a positive-real function, let

$$F_1(s) = F(s) \tag{9a}$$

$$F_2(s) = \frac{1}{s} \tag{9b}$$

and then (7) yields

$$F_3(s) = F\left[\frac{1}{s}\right] \tag{9c}$$

which is positive-real.

If $F(s)$ is positive-real, then its inverse is also positive-real. Let

$$F_1(s) = \frac{1}{F(s)} \tag{10a}$$

$$F_2(s) = s \tag{10b}$$

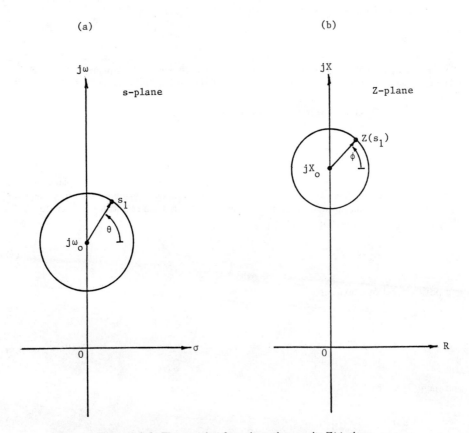

Figure 3.2-3: The mapping from the s-plane to the Z(s)-plane.

Then (7) gives

$$F_3(s) = \frac{1}{F(s)} \tag{10c}$$

which is positive-real.

In view of the fact that the reciprocal of a positive-real function is itself positive-real, it follows that a positive-real function can have no zeros in the right half of the s-plane.

Let a point on the $j\omega$-axis be mapped by a positive-real function $Z(s)$ into a point on the imaginary axis of the Z-plane (Fig. 3.2-3). If $j\omega_o$ is the point in question, then

$$Z(j\omega_o) = jX_o \tag{11}$$

where X_o is real, positive or negative or zero.

The function $Z(s_1)$, with s_1 a point in the neighborhood of $s = j\omega_o$, can be expanded by a Taylor expansion in the neigborhood of $j\omega_o$.

$$F(s_1) = F(j\omega_o) + F^{(1)}\big|_{j\omega_o} (s_1 - j\omega_o) + \cdots + F^{(k)}\big|_{j\omega_o} (s_1 - j\omega_o)^k + \cdots$$

$$Z(s_1) - jX_o = Z^{(m)}\big|_{j\omega_o} (s_1 - j\omega_o)^m + \cdots \tag{12}$$

where $F^{(m)}$ is the first nonvanishing derivative of $F(s)$. Let the following arguments be defined:

$$\phi = \text{Arg} [Z(s_1) - jX_o] \tag{13a}$$

$$\theta = \text{Arg} [s_1 - j\omega_o] \tag{13b}$$

$$\alpha = \text{Arg} [F^{(m)}\big|_{j\omega_o}] \tag{13c}$$

The relation between the arguments of (12) at $s = j\omega_o$ is

$$\phi = \alpha + m\theta \tag{14}$$

For $[F(s_1) - jX_o]$ to be positive-real, it is necessary that

$$\frac{-\pi}{2} \leq \phi \leq \frac{\pi}{2} \tag{15a}$$

when

$$\frac{-\pi}{2} \leq \theta \leq \frac{\pi}{2} \tag{15b}$$

Therefore, one concludes that

$$m = 1 \qquad \text{and} \qquad \alpha = 0 \tag{16}$$

This implies that the first nonvanishing derivative must be the first derivative and it must be real.

If any point on the $j\omega$-axis is mapped by a positive-real function $Z(s)$ onto the imaginary axis in the Z-plane, then at this point the derivative dZ/ds is real and positive.

4. Postive-Real Function

(a) The driving-point system function of a passive, dissipative system is a postive-real function. A function $F(s)$ is defined as a positive-real function if the following conditions are satisfied.

(a.1) $F(s)$ is real if s is real.

(a.2) Re $[F(s)] \geq 0$ if Re $[s] \geq 0$.

(b) The driving-point system function, $N_{d.p.}(s)$, is a rational function with real coefficients.

$$N_{d.p.}(s) = \frac{Q(s)}{D(s)} = N_o \frac{(s - z_1)(s - z_2) \cdots (s - z_n)}{(s - p_1)(s - p_2) \cdots (s - p_m)} \tag{1}$$

When the independent variable is real, $s = \sigma + j0$, then

$$N_{d.p.}(s = \sigma) \quad \text{is} \quad \text{real}$$

Condition (a.1) for a positive-real function is thus satisfied.

(c) The condition of stability requires that both poles and zeros of a driving-point system function should lie in the left half of the s-plane or on the $j\omega$-axis [Section 2-(g)], and any poles or zeros on the $j\omega$-axis should be simple.

(d) Since there are neither poles nor zeros in the right half of the s-plane, the driving-point system functions are analytic in the interior of the right half of the s-plane.

(e) Since the system is passive, the average power entering the system must be a positive quantity

$$P_{in} = \frac{1}{2} I^2 \text{ Re } [Z(j\omega)] \geq 0 \tag{2}$$

Since I^2 is non-negative, (2) implies that

$$\text{Re } [Z(j\omega)] \geq 0 \tag{3}$$

because $s = j\omega$ and the frequency is non-negative.

(f) In complex-function theory, the maximum modulus theorem states that if a function $F(s)$ is analytic within and on the boundary of a region, the maximum value of the magnitude of the function must appear on the boundary. This theorem may be extended in a region without zeros; then the minimum value of the magnitude of the function also occurs on the boundary. A three-dimensional plot is shown in Fig. 4-1.

(g) Similar arguments can be applied to the exponential function $e^{F(s)}$. When $F(s)$ is analytic in a region, the function $e^{F(s)}$ is not only analytic but also possesses no zeros in this region. Hence, both the maximum and the minimum of the amplitudes of exp $[F(s)]$ appear on the boundary. Since

$$| \exp [F(s)] | = e^{\text{Re } [F(s)]} \tag{4}$$

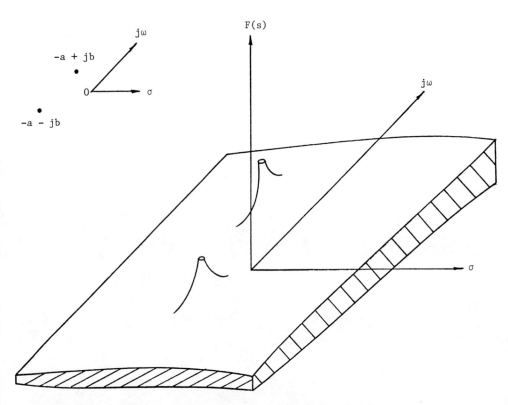

Figure 4-1: The three-dimensional representation of poles.

one therefore concludes that both the maximum and the minimum values of Re [F(s)] must appear on the boundary.

(h) For the present problem, the driving-point system function $N_{d.p.}$ is analytic within the right half of the s-plane. The boundary is the jω-axis, which extends to infinity.

If $N_{d.p.}(s)$ does not have poles on the jω-axis, then the minimum value of $N_{d.p.}(s)$ appears on the jω-axis. If $N_{d.p.}(s)$ is the impedance function Z(s), (3) implies that Re [Z(s)] in the right half-plane must also be non-negative, i.e.,

$$\text{Re } [Z(s)] = \text{Re } [N_{d.p.}(s)] \geq 0 \qquad \text{in RHP} \tag{5}$$

If $N_{d.p.}(s)$ has simple poles on the jω-axis, the same conclusion is valid except at points where the poles are located. The maximum modulus theorem can be modified to deal with the case where the contour is closed except for a definite number of points. Then the maximum modulus of the function on the contour is not exceeded at any points within the contour. The minimum values will appear at points away from the poles. The same conclusion as (5) is obtained.

(i) It follows from the above analysis that the driving-point system function of a passive system has the following properties.

(i.1) $N_{d.p.}(s)$ is real when s is real.

(i.2)   Re $[N_{d.p.}(s)]$ is non-negative when Re [s] is non-negative.

Thus the driving-point system function is a positive-real function by definition, item (a).

5. Properties of the Driving-Point System Function

(a) It was shown in a previous section, Section 3.1, that the driving-point system function is a positive-real function.

(b) The residues of the driving-point function at the poles on the $j\omega$-axis are real and positive. If the function has a pole at $j\omega_1$, the expansion of $N_{d.p.}(s)$ about $j\omega_1$ by Laurent series is

$$N_{d.p.}(s) = \frac{k_{-1}}{s_1 - j\omega_1} + k_o + k_1 (s - j\omega_1) + \cdots \tag{1}$$

The residue is the coefficient k_{-1}. The first term is dominant in the vicinity of $j\omega_1$; thus

$$N_{d.p.}(s)|_{\omega \approx \omega_1} \approx \frac{k_{-1}}{s - j\omega_1} = \frac{k_{-1}}{re^{j\phi}} \tag{2a}$$

$$s - j\omega_1 \equiv re^{j\phi} \tag{2b}$$

If k_{-1} is complex, it can be represented by

$$k_{-1} = |k_{-1}| e^{j\theta} \tag{2c}$$

Then

$$N_{d.p.}(s) = \frac{|k_{-1}|}{r} e^{j(\theta - \phi)} \tag{3}$$

and

$$\text{Re}[N_{d.p.}(s)] = \frac{|k_{-1}|}{r} \cos(\theta - \phi) \tag{4}$$

A point in the right half-plane corresponds to (Fig. 5-1)

$$\frac{-\pi}{2} \le \phi \le \frac{\pi}{2} \tag{5}$$

Since $\text{Re}[Z(s)] \ge 0$ in the right half-plane, one must therefore have

$$\cos(\theta - \phi) \ge 0 \quad \text{for} \quad \frac{-\pi}{2} \le \phi \le \frac{\pi}{2} \tag{6}$$

This implies that $\theta = 0$, and hence k_{-1} must be real [Eq. (2c)].

(c) The driving-point system function is generally given by

$$N_{d.p.}(s) = N_o \frac{s^n + a_{n-1}s^{n-1} + \cdots + a_o}{s^m + b_{m-1}s^{m-1} + \cdots + b_o} \tag{7}$$

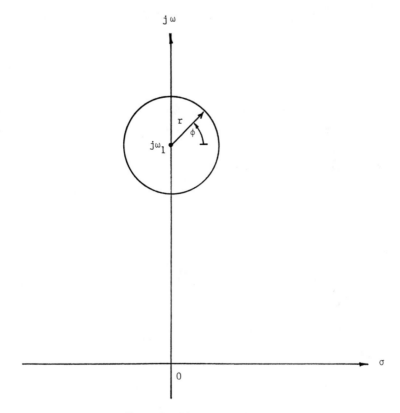

Figure 5-1: The $s = \sigma + j\omega$ plane.

At high frequencies,

$$N_{d.p.}(s) \rightarrow N_o s^{n-m} \tag{8}$$

Since the pole or zero at infinity must be simple, the possibilities are

$$N - m = 1$$

$$N - m = 0 \tag{9}$$

$$N - m = -1$$

(d) From (8) and properties of positive-real functions, the constant N_o must always be positive.

If $m - n = \pm 1$, either

$$N_{d.p.} = Z(s) \quad \text{or} \quad N_{d.p.} = Y(s)^{-1}$$

has a pole at infinity with residue equal to N_o or $1/N_o$ which must be postive-real.

If $m = 0$, N_o must also be positive, since it is equal to Re $[Z(j\omega)]$ at infinity.

$$N_{d.p.}(s = j\omega) = N_o \frac{s^m}{s^n} = N_o \quad \text{for} \quad n = m$$

(e) Because the origin is a point on the $j\omega$-axis, any poles or zeros located at the origin must be simple. Therefore, the lowest degree of the numerator polynomial can differ at most by one from the lowest degree of the denominator polynomial. That is, at very low frequency, one has

$$\begin{aligned}
N_{d.p.}(s) &= \frac{a_o}{b_o} \\
&= \frac{a_1 s}{b_o} \\
&= \frac{a_o}{b_1 s}
\end{aligned} \tag{10}$$

(f) When either the numerator or denominator polynomial has negative or missing coefficients, there must be poles or zeros in the right half-plane.

The zeros of $N(s)$ and $D(s)$ in the left half-plane show up in factors such as $(s + a)$ or $(s + a + jb)(s + a - jb) = s^2 + 2as + a^2 + b^2$. Note that the coefficients of both types of factors are positive. The product of factors of this form can only have positive coefficients with no missing terms.

An exception to this result occurs in the case when all the zeros of the polynomials lie on the jω-axis. In this case, the factor for a conjugate pair on the imaginary axis is

$$(s + jb)(s - jb) = s^2 + b^2$$

Hence, either all even powers or all odd powers of s are missing. The absence of all the odd powers occurs if there is no zero at the origin.

(g) It is important to keep in mind that even if a polynomial has only positive coefficients and no missing terms, there still may be zeros in the right half-plane. In other words, this is a necessary but not sufficient condition.

(h) It is easiest to check for the above properties. When these simple tests fail to show that the function lacks the positive-real property, the following properties establish a positive-real function.

(1) The function should be analytic in the right half-plane - check D(s) to determine whether it is Hurwitz.

(2) Poles on jω-axis must be simple and their residues positive and real.

(3) Re [N(s)] ≥ 0.

(i) The real part of N(s) is given by (3-4).

$$\text{Re}[N(j\omega)] = N_e(j\omega) = \left[\frac{q_1 d_1 - s^2 q_2 d_2}{d_1^2 - s^2 d_2^2}\right]_{s=j\omega}$$

$$= \frac{q_1 d_1 + \omega^2 q_2 d_2}{d_1^2 + \omega^2 d_2^2} \tag{11}$$

where q_1 and d_1 are functions of ω^2.

Since the denominator of (11) is always positive, it is therefore only necessary to investigate the numerator.

It is crucial to determine any existence of real zeros in the numerator of odd multiplicity.

Consider a possible zero, of multiplicity m, of U(s). In the neighborhood of this zero, the function can be approximated by

$$U_s \approx k(s - s_i)^m \tag{12}$$

The tangent of the function is given by its derivative with respect to s.

$$\frac{dU}{ds} = mk \, (\, s \, - \, s_i \,)^{m-1} \tag{13}$$

At $s = s^- \equiv s_i - \delta$, where δ is an arbitrarily small quantity, then

$$\left[\frac{dU}{ds} \right]_{s=s^-} = mk \, (\, s^- \, - \, s_i \,)^{m-1} = mk \, (- \, \delta)^{m-1} \tag{14}$$

At $s = s^+ \equiv s_i + \delta$, one has

$$\left[\frac{dU}{ds} \right]_{s=s^+} = mk\delta^{m-1} \tag{15}$$

For odd multiplicity, (14) and (15) yield

$$\left[\frac{dU}{ds} \right]_{s=s^-} > 0 \tag{16a}$$

$$\left[\frac{dU}{ds} \right]_{s=s^+} > 0 \tag{16b}$$

Hence, the function with a zero of odd multiplicity will change from negative to positive as the frequency passes the zero (Fig. 5-2).

For a zero with even multiplicity, (14) and (15) give

$$\left[\frac{dU}{ds} \right]_{s=s^-} < 0 \tag{17a}$$

$$\left[\frac{dU}{ds} \right]_{s=s^+} > 0 \tag{17b}$$

The function with a zero of even multiplicity has the shape shown in Fig. 5-3 in the neighborhood of the zero.

Since a positive-real function must remain positive on both sides of the zero, it follows that the zeros of the system function must be of even multiplicity.

(j) For the system function

$$N(j\omega) = N_r(j\omega) + jN_i(j\omega) \tag{18a}$$

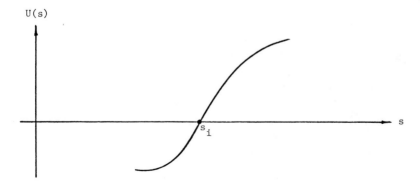

Figure 5-2: The frequency response near the zero of a function with zero of odd multiplicity.

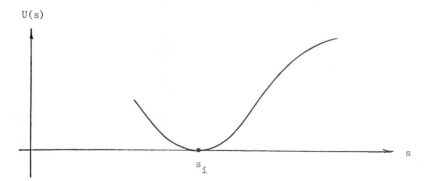

Figure 5-3: The frequency response near the zero of a function with zero of even multiplicity.

the argument is

$$\text{Arg } N(j\omega) = \tan^{-1} \frac{N_i(j\omega)}{N_r(j\omega)} \tag{18b}$$

where N_r and N_i are the real and the imaginary parts of $N(j\omega)$. Since $N_r(j\omega) \geq 0$, therefore

$$\frac{-\pi}{2} \leq \text{Arg } N(j\omega) \leq \frac{\pi}{2} \tag{19}$$

Equation (19) implies that the argument of $N(j\omega)$ must lie within this range so that

$$\text{Re } [N(j\omega)] \geq 0$$

6. Driving-Point Function of an LC Network

(a) An ideal LC network contains no resistive element and is lossless. Therefore, no real power can be absorbed, and hence,

$$P_{avg} = \frac{1}{2} I^2 \, Re \, [Z(j\omega)] = 0$$

or

$$Re \, [Z(j\omega)] = 0 \qquad (1)$$

(b) The impedance function can be expressed as [Eq. (3-4)]

$$Z(s) = \frac{q_1(s^2) + sq_2(s^2)}{d_1(s^2) + sd_2(s^2)} = Re \, [Z(s)] + j \, Im \, [Z(s)] \qquad (2a)$$

$$Re \, [Z(s)] = \frac{q_1 d_1 - s^2 q_2 d_2}{d_1^2 - s^2 d_2^2} \qquad (2b)$$

$$Im \, [Z(s)] = \frac{s(\, q_2 d_1 - q_1 d_2 \,)}{d_1^2 - s^2 d_2^2} \qquad (2c)$$

Equations (1) and (2b) yield

$$q_1 d_1 - s^2 q_2 d_2 = 0 \qquad (3a)$$

or

$$\frac{q_1}{sq_2} = \frac{sd_2}{d_1} \qquad (3b)$$

since the denominator of (2b) is always non-negative.

(c) The substitution of (3b) into (2a) gives

$$Z(s) = \frac{sq_2 \left[\dfrac{q_1}{sq_2} + 1 \right]}{d_1 \left[1 + \dfrac{sd_2}{d_1} \right]} = \frac{sq_2}{d_1} \qquad (4a)$$

or alternatively,

$$Z(s) = \frac{q_1 \left[1 + \dfrac{sq_2}{q_1} \right]}{sd_2 \left[\dfrac{d_1}{sd_2} + 1 \right]} = \frac{q_1}{sd_2} \tag{4b}$$

The driving-point impedance of an ideal LC network is an odd rational function.

(d) Any even function can be expressed in the form of $g(s)g(-s)$. It follows that the zeros of an even polynomial are symmetrical about both the real axis and the imaginary axis. Since the poles and zeros of $Z(s)$ cannot exist in the right half-plane, therefore, for the poles and zeros of (4) to be symmetrical with respect to the $j\omega$-axis they must remain on the $j\omega$-axis.

These poles and zeros on the $j\omega$-axis must be simple according to the stability condition. Since $Z(s)$ is an odd function, there must be either a pole or a zero at the origin and at infinity.

(e) Since the residues of $Z(s)$ at $j\omega$-axis poles should be positive and real, thus poles and zeros must alternate along the $j\omega$-axis.

Every odd positive-real function is purely imaginary when $s = j\omega$. Physically, such odd positive-real functions represent nothing more than a reactance or susceptance. As functions of ω, X and B have a series of poles. The sign of these functions changes as the frequency passes through each pole. In generic form,

$$\frac{k}{j\omega - j\omega_i} = \frac{jk}{\omega_i - \omega}$$

Hence, one deduces that the only behavior possible is that the zeros and the poles of an odd positive-real function alternate on the $j\omega$-axis (Fig. 6-1). Moreover, the derivatives of X and B with respect to ω are always positive.

The impedance function can be expanded in terms of partial fractions.

$$Z(s) = \frac{k_o}{s} + \frac{k_1}{s - j\omega_1} + \frac{k_1^*}{s + j\omega_1} + \cdots + k_\infty s \tag{5}$$

For the sake of generality, poles have been assumed to exist at both the origin and infinity. Since the residues are positive and real, $k_1 = k_1^*$, the contribution from the conjugate pair of poles is

$$\frac{k_1}{s - j\omega_1} + \frac{k_1^*}{s + j\omega_1} = \frac{2k_1 s}{s^2 + \omega_1^2} \tag{6}$$

Then (5) becomes

$$Z(s) = \frac{k_o}{s} + \frac{2k_1 s}{s^2 + \omega_1^2} + \frac{2k_2 s}{s^2 + \omega_2^2} + \cdots + k_\infty s \qquad \text{for } k_i = \text{real} \tag{7}$$

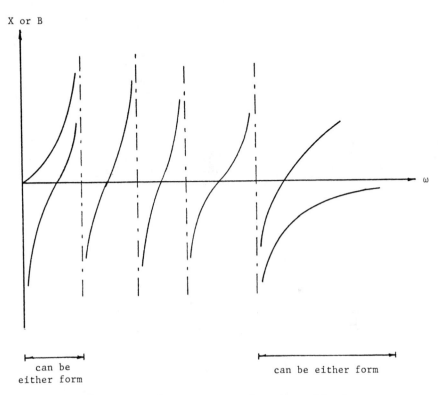

Figure 6-1: The frequency response of a positive real function.

When $s = j\omega$, (7) reduces to

$$Z(j\omega) = \frac{-jk_o}{\omega} + \frac{j2k_1\omega}{\omega_1^2 - \omega^2} + \cdots + jk_\infty\omega = jX(\omega) \tag{8}$$

The driving-point impedance function is purely imaginary and is known as the reactance function.

The derivative of $X(\omega)$ with respect to ω is

$$\frac{dX(\omega)}{d\omega} = \frac{k_o}{\omega^2} + \frac{2k_1(\omega_1^2 + \omega^2)}{(\omega_1^2 - \omega^2)^2} + \cdots + k_\infty > 0 \quad \text{for all } \omega \tag{9}$$

Since $dX/d\omega$ is always positive, $dZ/d\omega$ has the general shape shown in Fig. 6-1. Hence, poles and zeros are alternate on the $j\omega$-axis.

(f) The driving-point admittance of an LC network is the reciprocal of the corresponding driving-point impedance function, and both have the same mathematical form. The above analysis is valid for the admittance function. Hence, the driving-point admittance function is purely reactive and is known as the susceptance function $B(\omega)$.

$$Y(j\omega) = jB(\omega) \tag{10}$$

The susceptance function has properties similar to those of the reactance function.

7. Realization by Partial Fraction Expansion

(a) The properties of the driving-point impedance function of an LC network can be specified by either of the following:

(1) The function must be a positive-real, odd rational function.

(2) The function must have simple poles only along the $j\omega$-axis. The residues of the function at the poles must be positive and real.

(3) The function must have simple poles and simple zeros on the $j\omega$-axis. The poles and zeros must alternate, and there must be either a pole or a zero at the origin and at infinity.

A physical realization can be achieved if the driving-point impedance function has these properties.

(b) The partial fraction expression of the driving-point impedance function is [Eq. (6-7)],

$$Z(s) = \frac{k_o}{s} + \frac{2k_1 s}{s^2 + \omega_1^2} + \cdots + k_\infty s \tag{1}$$

A physical realization is accomplished if each term in (1) can be identified with an actual circuit. The impedance expressions of some simple circuit elements are shown in Fig. 7-1.

(c) The first term of (1) can be expressed as

$$\frac{k_o}{s} = \frac{1}{\dfrac{s}{k_o}} = \frac{1}{sC_o} \qquad C_o \equiv \frac{1}{k_o} \tag{2}$$

Therefore, the first term can be interpreted as the impedance of a capacitor which has the value defined by (2). This is shown in Fig. 7-2.

(d) The second term of (1) has the form of a parallel LC combination, Fig. 7-1c.

$$Z_{p1} = \frac{2k_1 s}{s^2 + \omega_1^2} = \frac{s(2k_1)}{s^2 + \omega_1^2} = \frac{s\dfrac{1}{C_1}}{s^2 + \dfrac{1}{L_1 C_1}} \tag{3a}$$

Therefore,

$$C_1 = \frac{1}{2k_1} \tag{3b}$$

$$\omega_1^2 = \frac{1}{L_1 C_1} \qquad \text{or} \qquad L_1 = \frac{1}{\omega_1^2 C_1} = \frac{2k_1}{\omega_1^2} \tag{3c}$$

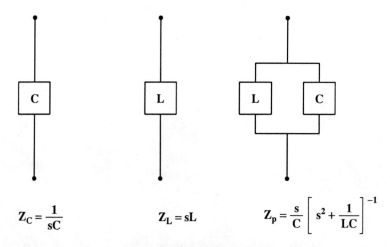

$$Z_C = \frac{1}{sC} \qquad\qquad Z_L = sL \qquad\qquad Z_p = \frac{s}{C}\left[s^2 + \frac{1}{LC} \right]^{-1}$$

Figure 7-1: The impedances of some simple circuit elements.

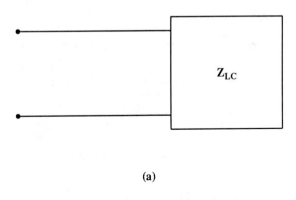

(a)

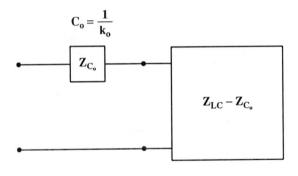

(b)

Figure 7-2: The extraction of a capacitive element from a network.

This is shown in Fig. 7-3.

(e) Similarly, the last term in (1) is identified as a series inductance (Fig. 7-4).

$$Z_{L\infty} = k_\infty s = sL_\infty \qquad L_\infty = k_\infty \tag{4}$$

(f) With real and positive residues, the network realization of each term should consist of positive elements.

(g) The driving-point admittance function can be realized similarly.

$$Y(s) = \frac{k_0'}{s} + \frac{2k_1's}{s^2 + \omega_1^2} + \cdots + k_\infty' s \tag{5}$$

The admittances of some simple circuit elements are shown in Fig. 7-5.

The realization of (5) is a parallel combination of admittance of LC and series LC elements as shown in Fig. 7-6.

It is to be noted that the number of coefficients, k_0, k_1, . . ., k_∞, is equal to the number of poles. The impedance functions are completely specified by this set of coefficients. The number of coefficients is the number of degrees of freedom. This number is also equal to the minimum number of circuit elements in the realization. A realization with this minimum number of elements is known as the canonical form.

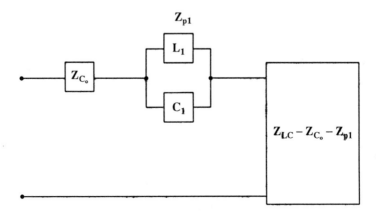

Figure 7-3: The extraction of a parallel LC-element from a network.

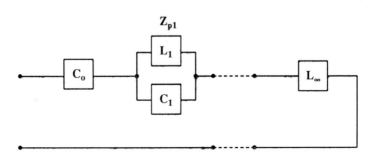

Figure 7-4: The general impedance representation of a network.

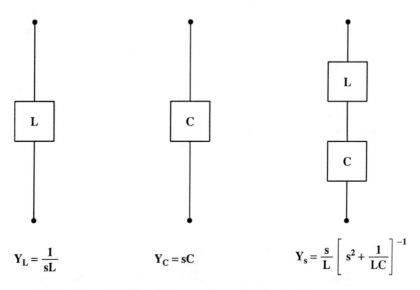

Figure 7-5: The admittances of some simple circuit elements.

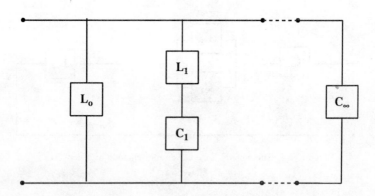

Figure 7-6: The general admittance representation of a network.

8. Realization by Continued Fraction

Instead of extracting impedance elements (or admittance elements) continuously, one could also extract impedance and admittance elements alternately. This method will be obvious if the reverse problem is considered first.

(a) Figure 8-1 shows a ladder network. The driving-point impedance $Z(s)$ can be obtained by the series and the parallel combinations of all the elements.

$$Z_a = Z_5 + \frac{1}{Y_6} \tag{1}$$

$$Y_a = \frac{1}{Z_a} = \frac{1}{Z_5 + \dfrac{1}{Y_6}} \tag{2}$$

$$Y_b = Y_4 + Y_a = Y_4 + \frac{1}{Z_5 + \dfrac{1}{Y_6}} \tag{3}$$

$$Z_c = Z_3 + \frac{1}{Y_b} = Z_3 + \frac{1}{Y_4 + \dfrac{1}{Z_5 + \dfrac{1}{Y_6}}} \tag{4}$$

$$Y_d = Y_2 + \frac{1}{Z_c} = Y_2 + \frac{1}{Z_3 + \dfrac{1}{Y_4 + \dfrac{1}{Z_5 + \dfrac{1}{Y_6}}}} \tag{5}$$

$$Z(s) = Z_1 + \frac{1}{Y_d} = Z_1 + \frac{1}{Y_2 + \dfrac{1}{Z_3 + \dfrac{1}{Y_4 + \dfrac{1}{Z_5 + \dfrac{1}{Y_6}}}}} \tag{6}$$

The form of $Z(s)$ as given by (6) is known as the continued fraction of the impedance.

(b) Consider the impedance function given by

$$Z(s) = \frac{a_n s^n + a_{n-2} s^{n-2} + \cdots + a_o}{b_{n-1} s^{n-1} + b_{n-3} s^{n-3} + \cdots + b_1 s} = \frac{Q_n(s)}{D_{n-1}(s)} \tag{7}$$

where $Q_n(s)$ is a polynomial of order n, and $D_{n-1}(s)$ is a polynomial of order $(n - 1)$. Equation (7) will be developed into a continued fraction by long division.

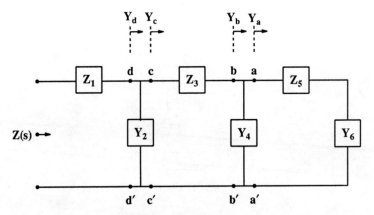

Figure 8-1: A ladder network.

Since the numerator is of higher order, thus

$$Z(s) = C_1 s + \frac{Q_{n-2}(s)}{D_{n-1}(s)} = C_1 s + \cfrac{1}{\cfrac{D_{n-1}(s)}{Q_{n-2}(s)}} \tag{8}$$

The second term is arranged so that it can be divided through again.

$$Z(s) = C_1 s + \cfrac{1}{C_2 s + \cfrac{D_{n-3}(s)}{Q_{n-2}(s)}}$$

$$= C_1 s + \cfrac{1}{C_2 s + \cfrac{1}{\cfrac{Q_{n-2}(s)}{D_{n-3}(s)}}} \tag{9}$$

$$Z(s) = C_1 s + \cfrac{1}{C_2 s + \cfrac{1}{C_3 s + \cfrac{1}{\cfrac{D_{n-3}}{Q_{n-4}}}}}$$

$$= C_1 s + \cfrac{1}{C_2 s + \cfrac{1}{C_3 s + \cfrac{1}{C_4 s + \cfrac{1}{\cfrac{Q_{n-4}}{D_{n-5}}}}}} \tag{10}$$

Comparing with (6), one can identify the physical meaning of each term.

$C_1 s$ is a series impedance term and corresponds to Z_1 in (6)

$C_2 s \equiv Y_2$

$C_3 s \equiv Z_3$

$C_4 s \equiv Y_4$

etc.

9. Magnitude and Frequency Normalization

Filter circuits are usually synthesized with a cutoff frequency of one radian per second and a load impedance of one ohm. Filter circuits designed with these restrictions are known to be normalized in both frequency and impedance level. The normalized filters can be converted into filters which meet the actual cutoff frequency and impedance level specifications by the process of denormalization.

Let the subscript n be used to denote the normalized frequency variable, ω_n, and the normalized circuit elements, L_n, C_n, and R_n. The corresponding denormalized parameters are denoted by the same notation without the subscript.

(a) Frequency Denormalization

The normalized frequency ω_n is related to the actual frequency ω by the relation

$$\omega = K_\omega \omega_n \tag{1}$$

where K_ω is the normalizing constant. The constant of normalization is dimensionless and is usually taken to be the value of the actual cutoff frequency.

The impedance of an element should remain invariant under frequency normalization. The actual value of each circuit element can be obtained from the normalized values by setting the impedances in the normalized and the denormalized cases equal to each other. For an inductor, one has

$$\omega_n L_n = \omega L = K_\omega \omega_n L \quad \text{or} \quad L = \frac{L_n}{K_\omega} \tag{2}$$

For a capacitor,

$$\frac{1}{\omega_n C_n} = \frac{1}{\omega C} \quad \text{or} \quad C = \frac{C_n}{K_\omega} \tag{3}$$

Since resistances are ideally independent of frequency, they are unaffected by frequency normalization.

(b) Magnitude Denormalization

Let the actual impedance level be K_z times the normalized value of one ohm. Then the denormalized impedance Z is related to the normalized impedance Z_n by the relation

$$Z = K_z Z_n \tag{4}$$

where K_z is a dimensionless constant of proportionality. Thus, for a normalized resistor R_n, the actual resistance R is

$$R = K_z R_n \tag{5}$$

For an inductance, the corresponding relationship is

$$\omega L = K_z (\omega L_n) \quad \text{or} \quad L = K_z L_n \tag{6}$$

Similarly, for a capacitor,

$$\frac{1}{\omega C} = K_z \frac{1}{\omega C} \quad \text{or} \quad C = \frac{C_n}{K_z} \tag{7}$$

(c) Frequency and Magnitude Denormalization

For combined frequency and magnitude denormalization, one simply combines the two sets of equations to give

$$R = K_z R_n \tag{8a}$$

$$C = \frac{C_n}{K_z K_\omega} \tag{8b}$$

$$L = L_n \frac{K_z}{K_\omega} \tag{8c}$$

9.1. Normalization

Normalization of circuit elements and frequency scale will simplify numerical calculations.

(a) When the elements R, L, and 1/C are divided by an impedance scale factor K_z, all the current values will become K_z times larger. However, the behavior of the network is essentially unchanged.

(b) When the elements L and C are multiplied by a frequency scale factor K_ω and s is simultaneously divided by the same factor K_ω, the resulting effects cancel, but this introduces the variable (s/K_ω), which is more convenient for the calculations.

For a given impedance function,

$$Z(s) = sL + R + \frac{1}{sC} \tag{1}$$

Then

$$\frac{Z(s)}{K_z} = s\frac{L}{K_z} + \frac{R}{K_z} + \frac{1}{sCK_z}$$

$$= \frac{s}{K_\omega} K_\omega \frac{L}{K_z} + \frac{R}{K_z} + \frac{1}{\frac{s}{K_\omega} K_\omega (CK_z)}$$

$$= s_n \frac{K_\omega L}{K_z} + \frac{R}{K_z} + \frac{1}{s_n (K_\omega K_z C)}$$

$$= s_n L_n + R_n + \frac{1}{s_n C_n} \tag{2a}$$

$$s_n \equiv \frac{s}{K_\omega} = \text{normalized complex frequency} \tag{2b}$$

$$L_n \equiv \frac{K_\omega}{K_z} L = \text{normalized inductance} \tag{2c}$$

$$R_n \equiv \frac{R}{K_z} = \text{normalized resistance} \tag{2d}$$

$$C_n \equiv K_\omega K_z C = \text{normalized capacitance} \tag{2e}$$

The process of normalization merely changes the scales in the response; the ratios of the responses do not change under normalization.

In actual construction of the network, the actual circuit elements are given by

$$L = L_n \frac{K_z}{K_\omega} \tag{3a}$$

$$R = R_n K_z \tag{3b}$$

$$C = \frac{C_n}{K_z K_\omega} \tag{3c}$$

10. Frequency Transformation

One can extend the utility of a network by means of a frequency transformation which translates the band of interest to a different location in the frequency spectrum. This is accomplished by a change of the frequency variable in the network function.

The resistive elements of a network are independent of frequency, but the inductive and capacitive elements are frequency dependent. If the parameter s' of the network function $H(s')$ is replaced by some other function, say, $s' = F(s)$ or $s = F^{-1}(s') \equiv G(s')$, where s is the new variable. This change of variable will modify the behavior of the network function. Physically, such a transformation will replace the original reactive elements by some different arrangement of reactive elements. The resultant network will be realizable if the transformation relation is properly chosen.

One type of transformation is to choose the function $F(s)$ or $G(s')$ to be a reactance function. This is equivalent to replacing the original independent variable s' by the reactance function of a combination of inductive and capacitive elements. This process actually replaces each element in the original network by another one-terminal LC-circuit. Consequently, the newly transformed network will be realizable. This transformation actually replaces the straight abscissa of ω' by a reactance curve. Since this alternates several times within the range $-\infty < \omega' < \infty$, the original characteristic repeats several times, at different frequency locations, with various scale distortions. Some of these transformations will be discussed in the following sections.

It is to be noted that the frequency transformations do not affect the particular network function - the dependent variable. In other words, a given ripple in the original network function will remain in the newly transformed network function.

A most general form of frequency transformation can be expressed as

$$\omega = G(\omega') \tag{1}$$

where ω' is the original frequency variable, ω is the newly transformed frequency variable, $G(\omega')$ is some reactance function which will map the original response in the ω'-domain into the desired response in the ω-domain. If the response of the original network is denoted by $H(j\omega')$, then the newly transformed response is

$$H(j\omega) = H(jG[\omega']) \tag{2}$$

The curves representing magnitude and the phase of the response $H(j\omega')$ have a one-to-one correspondence to the corresponding curves of the response $H(j\omega)$.

If the poles or zeros of $H(j\omega') \equiv H(p')$ are given as $(p' - p_j)$, then the corresponding poles or zeros of $H(j\omega)$ are given by

$$[G(p') - p_j]$$

which can be obtained by factoring the function $H(j\omega) \equiv H(p)$.

11. Frequency Scaling

Frequency scaling is one of the simplest frequency transformations. Frequency normalization and denormalization belong to this class of transformation. The function of transformation is

$$\omega = G(\omega') = K\omega' \tag{1}$$

where K is a constant factor of scaling. The function of transformation is a straight line with a slope of

$$\frac{d\omega'}{d\omega} = \frac{1}{K} = \frac{\omega'}{\omega} \tag{2}$$

The new frequency scale, ω, is expanded for $K < 1$. A graphical representation of frequency scaling is shown in Fig. 11-1, where a ficticious response composed of straight-line segments is used for the purpose of illustration.

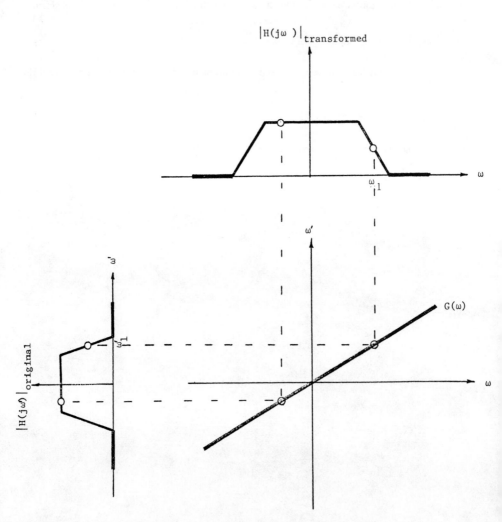

Figure 11-1: A graphical representation of frequency scaling.

12. Low-Pass to High-Pass Transformation

The ideal low-pass (LP) response is shown in Fig. 12-1, where ω_c' is the cutoff frequency. The ideal high-pass (HP) response is shown in Fig. 12-2. To convert a LP transfer function to a HP transfer function implies that the pass-band must be moved from the region between

$$\Omega' \equiv \frac{\omega'}{\omega_c'} = 0 \qquad \text{and} \qquad \Omega' = \pm 1 \tag{1}$$

to that between

$$\Omega \equiv \frac{\omega}{\omega_c} = \pm 1 \qquad \text{and} \qquad \Omega = \pm \infty \tag{2}$$

That is, the pass-band and the stop-band must be interchanged in the conversion from LP to HP operation. The frequency variables are normalized with respect to the cutoff frequencies.

In the concept of driving-point impedance, the pass-band infers zero input impedance so that all signals are transmitted without opposition. In the stop-band, the network provides infinite input impedance and no signal can pass through.

Hence, to accomplish the LP to HP transformation, the poles and zeros of the driving-point impedance function of the two networks must be interchanged. In a lossless network, this can be achieved by replacing capacitors by inductors and vice versa. Mathematically, this is equivalent to a frequency inversion. That is, the transformation is given by

$$j\Omega' = G(\Omega) = \frac{1}{j\Omega} \tag{3}$$

When non-normalized variables are used, the relation is

$$j\omega' = \frac{k_\omega^2}{j\omega} \tag{4}$$

where k_ω is a constant factor with the dimension of angular frequency. This factor is necessary so that the dimensions on both sides of (4) are consistent.

The normalized reactances are

$$X_L = j\Omega'L' = \frac{L'}{j\Omega} = \frac{1}{j\Omega C} \tag{5a}$$

$$C \equiv \frac{1}{L'} = \text{capacitance of HP circuit} \tag{5b}$$

$$X_C = \frac{1}{j\Omega'C'} = j\Omega L \tag{5c}$$

$$L \equiv \frac{1}{C'} = \text{inductance of HP circuit} \tag{5d}$$

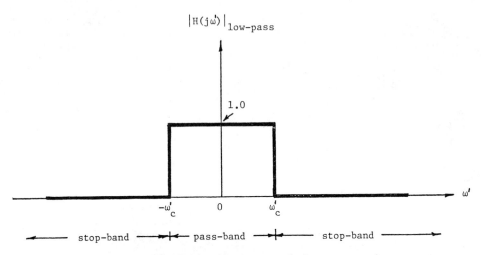

Figure 12-1: The ideal frequency response of a low-pass network.

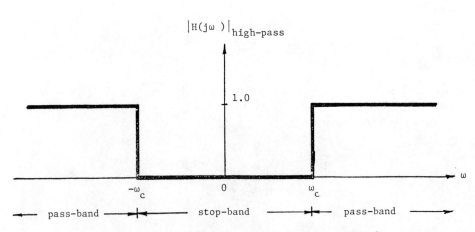

Figure 12-2: The ideal frequency response of a high-pass network.

Thus, the frequency inversion changes an inductor into a capacitor and vice versa, as illustrated in Fig. 12-3. The mapping of the transfer function from the Ω'-domain into the Ω-domain is shown in Fig. 12-4.

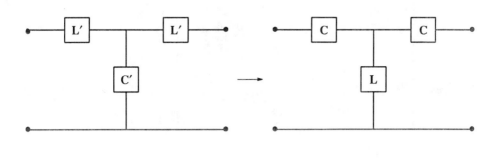

Figure 12-3: The transformation of circuit elements due to frequency inversion.

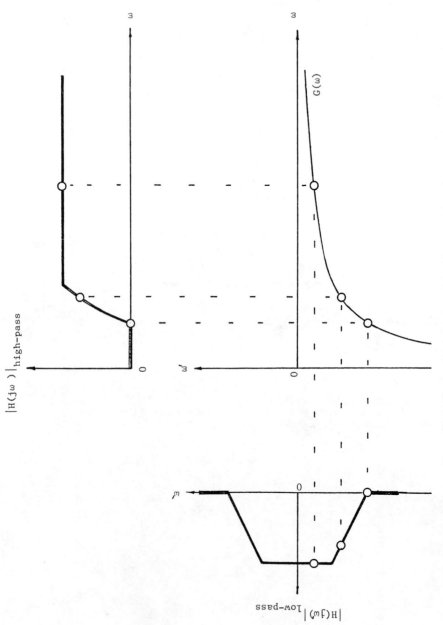

Figure 12-4: The graphical representation of frequency inversion.

13. Low-Pass to Band-Pass Transformation

The ideal band-pass (BP) response is shown in Fig. 13-1. A direct comparison with the ideal LP response, Fig. 12-1, reveals that the transformation function should have poles between

$$\omega = 0 \quad \text{and} \quad \omega = \omega_L \qquad \text{and between} \qquad \omega = \omega_U \quad \text{and} \quad \omega = \infty \qquad (1)$$

It should have zeros between

$$\omega = \omega_L \quad \text{and} \quad \omega = \omega_U \qquad (2)$$

In other words, the transformation function should have poles near $\omega = 0$ and near $\omega = \infty$ and a zero located in between these poles. It is convenient to use non-normalized frequency variables in the present analysis.

A series LC-reactance function provides poles at the frequency origin and at infinity and a zero at the resonant frequency.

$$j\omega' = j\omega L_o + \frac{1}{j\omega C_o} \qquad (3)$$

where L_o and C_o are scale factors. The transformation function $G(\omega')$ can be obtained by solving for ω from (3).

$$\omega^2 - \frac{\omega'}{L_o}\omega - \omega_o^2 = 0$$

$$\omega = \frac{\omega'}{2L_o} \pm \sqrt{\frac{\omega'^2}{4L_o^2} + \omega_o^2} = G(\omega') \qquad (4a)$$

$$\omega_o^2 = \frac{1}{L_o C_o} \qquad (4b)$$

The negative sign in front of the square root will be discarded since a negative frequency has little physical meaning.

At $\omega' = \omega_c'$, the corresponding cutoff in the new frequency ω-domain is

$$\omega = \frac{\omega_c'}{2L_o} + \sqrt{\frac{\omega_c'^2}{4L_o^2} + \omega_o^2} \equiv \omega_U \qquad (5a)$$

Similarly, at $\omega' = -\omega_c'$,

$$\omega_c = \frac{-\omega_c'}{2L_o} + \sqrt{\frac{\omega_c'^2}{4L_o^2} + \omega_o^2} \equiv \omega_L \qquad (5b)$$

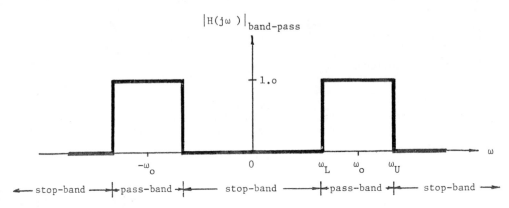

Figure 13-1: The ideal frequency response of a band-pass network.

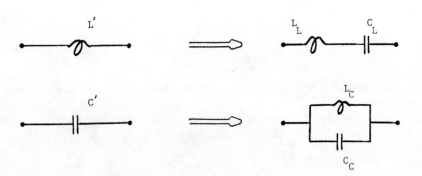

Figure 13-2: The transformation of circuit elements from low-pass to band-pass network.

These are the upper and lower cutoff frequencies of the band-pass network. The product of these two cutoff frequencies indicates that the mid-band frequency ω_o is the geometrical mean of these cutoff frequencies.

$$\omega_o = \sqrt{\omega_U \omega_L} \tag{6}$$

The bandwidth of the band-pass network is

$$BW = \omega_U - \omega_L = \frac{\omega_c'}{L_o} \tag{7}$$

With the transformation given by (3), the reactance of the inductor L' in the LP circuit is

$$X_L = j\omega'L' = \left[j\omega L_o + \frac{1}{j\omega C_o} \right] L' = j\omega L_L + \frac{1}{j\omega C_L} \tag{8a}$$

$$L_L \equiv L_o L' \qquad \text{and} \qquad C_L \equiv \frac{C_o}{L'} \tag{8b}$$

where L_L and C_L are the inductance and the capacitance, respectively, in the BP circuit replacing L' in the LP circuit. The transformation given by (3) converts the inductor in the LP network into a series LC-combination in the BP network (Fig. 13-2).

The capacitor in the LP circuit is similarly transformed.

$$X_C = \frac{1}{j\omega'C'} = \left[\left(j\omega L_o + \frac{1}{j\omega C_o} \right) C' \right]^{-1} = \left[\frac{1}{\dfrac{1}{j\omega C_C}} + \frac{1}{j\omega L_C} \right]^{-1} \tag{9a}$$

$$C_C \equiv L_o C' \qquad \text{and} \qquad L_C \equiv \frac{C_o}{C'} \tag{9b}$$

The reactance transformation converts the capacitor C' in the LP network into a parallel LC-combination in the BP network (Fig. 13-2).

The LP network is transformed by (3) into a BP network as shown in Fig. 13-3.

A graphical representation of the transformation can be obtained by rearranging (3) as

$$\omega' = \omega L_o - \frac{1}{\omega C_o} \equiv G(\omega) \tag{10}$$

This expression can be constructed graphically by adding the curve of ωL_o to that of $1/\omega C_o$ (Fig. 13-4). The graphical transformation of a LP network to a BP network with the transformation function $G(\omega)$ is shown in Fig. 13-5.

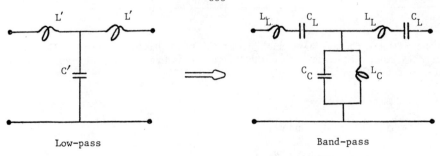

Figure 13-3: The transformation of a T-network from low-pass to band-pass
network.

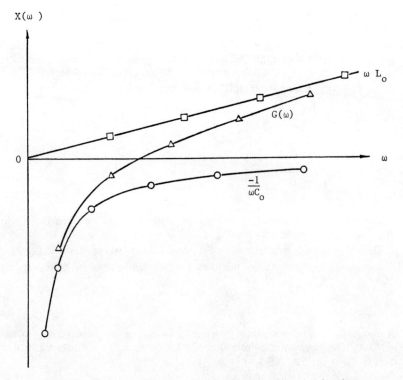

Figure 13-4: The graphical construction of the transformation function.

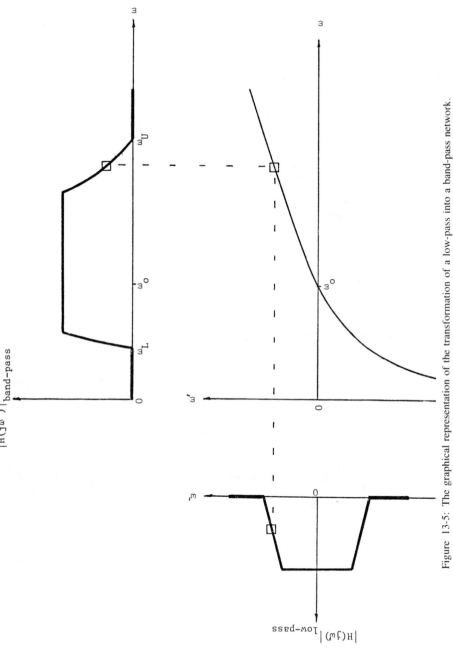

Figure 13-5: The graphical representation of the transformation of a low-pass into a band-pass network.

14. Low-Pass to Band-Stop Transformation

The ideal band-stop (BS) transfer function is shown in Fig. 14-1. This is an inversion of the BP transfer function, Fig. 13-1. The reactance function required for such a transformation is therefore the inverse of Eq. (13-3).

$$G(\omega) = j\omega' = \left[j\omega L_o + \frac{1}{j\omega C_o} \right]^{-1} \tag{1}$$

The new frequency variable is evaluated from (1).

$$\omega^2 + \frac{\omega}{\omega' L_o} - \omega_o^2 = 0$$

$$\omega = \sqrt{\frac{1}{(2\omega' L_o)^2} + \omega_o^2} - \frac{1}{2\omega' L_o} \tag{2a}$$

$$\omega_o^2 = \frac{1}{L_o C_o} \tag{2b}$$

At $\omega' = \omega_c'$,

$$\omega = \sqrt{\frac{1}{(2\omega_c' L_o)^2} + \frac{1}{2\omega_c' L_o}} - \frac{1}{2\omega_c' L_o} \equiv \omega_L \tag{3a}$$

At $\omega' = -\omega_c'$,

$$\omega = \sqrt{\frac{1}{(-2\omega_c' L_o)^2} + \omega_o^2} + \frac{1}{2\omega_c' L_o} = \omega_U \tag{3b}$$

The reactance function $G(\omega)$ can be obtain by inverting the reactance function for the LP-BP transformation. This is accomplished in Fig. 14-2. The graphical transformation of a LP response to a BS response is shown in Fig. 14-3.

The reactance of an inductive element in the LP network is

$$X_L = j\omega' L' = \frac{L'}{j\omega L_o + \dfrac{1}{j\omega C_o}} = \frac{1}{j\omega C_L + \dfrac{1}{j\omega L_L}} \tag{4a}$$

$$C_L \equiv \frac{L_o}{L'} \qquad L_L = C_o L' \tag{4b}$$

The inductive element of the LP network is transformed into a parallel LC-circuit (Fig. 14-4).

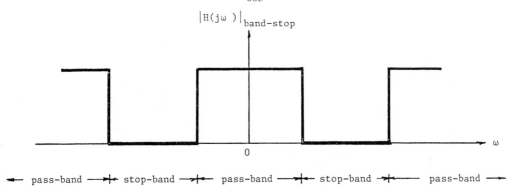

Figure 14-1: The ideal frequency response of a band-stop network.

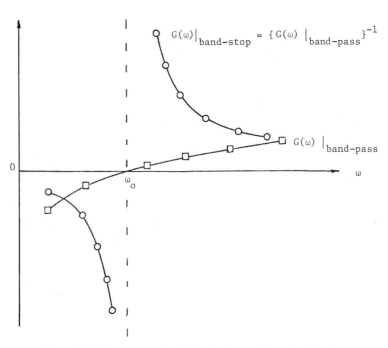

Figure 14-2: The construction of a band-stop transformation function.

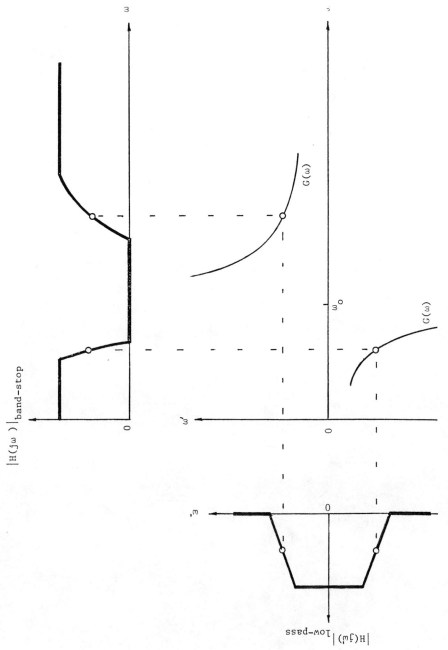

Figure 14-3: The graphical representation of a low-pass to band-stop transformation.

The reactance of a capacitive element is

$$X_C = \frac{1}{j\omega'C'} = j\omega L_C + \frac{1}{j\omega C_C} \tag{5a}$$

$$L_C = \frac{L_o}{C'} \qquad C_C = C_o C' \tag{5b}$$

The capacitive element in the LP network is transformed into a series connected LC-circuit (Fig. 14-4).

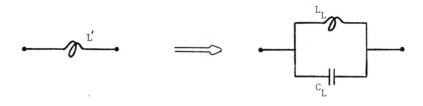

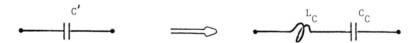

Figure 14-4: The transformation of circuit elements from low-pass to band-stop transformation.

15. Problems

1. Determine which of the following functions are possible system functions for a passive system.

(a) $F(p) = \dfrac{p - 1}{p\,(p - 3)}$

(b) $F(p) = \dfrac{p\,(p^2 + 9)}{p^2 + 5p + 4}$

(c) $F(p) = 5\,\dfrac{(p^2 + 4p + 8)\,(p^2 + 4p + 3)}{p^2 + 3p + 2}$

2. Determine the network whose driving-point impedance is

$$Z(p) = \frac{(2p + 3)\,p}{4p^2 + 6p}$$

3. Obtain the Forster equivalent network for the following driving-point impedance.

$$Z(p) = \frac{p\,(p^2 + 2)}{(p^2 + 1)\,(p^2 + 3)}$$

4. Determine the Cauer equivalent circuit for the driving-point impedance given in Problem 3.

5. A gyrator is terminated by a load Z_L. If the impedance matrix of the gyrator is given by

$$[z] = \begin{bmatrix} 0 & -k \\ k & 0 \end{bmatrix}$$

show that the gyrator input impedance is

$$Z_{in} = \frac{k^2}{Z_L}$$

A gyrator is therefore an impedance inverter.

CHAPTER XI   Normal Modes of a Waveguide

1. Orthogonality of Guided Waves

Solutions of wave equations generally have some properties of orthogonality. These are very important in problems when a general solution is expanded as a sum of normal modes.

The following orthogonality properties of the guided field solutions of wave equations will be verified without specifying the exact function of these fields.

(α) The product

$$e_{zp}e_{zq} \quad \text{or} \quad h_{zp}h_{zq}$$

integrated over the cross section of the guide is zero if $p \neq q$.

(β) The scalar product

$$\vec{e}_{tp} \cdot \vec{e}_{tq} \quad \text{or} \quad \vec{h}_{tp} \cdot \vec{h}_{tq}$$

integrated over the cross section of the guide is zero for $p \neq q$.

(γ) The quantity

$$\hat{z} \cdot (\vec{e}_{tp} \times \vec{h}_{tq})$$

integrated over the cross section of the guide is zero if $p \neq q$.

The above statements will now be proved separately.

(a) The surface integral

$$\int_S h_{zp}h_{zq}\, da = 0 \quad \text{for} \quad p \neq q \tag{1}$$

where S is the cross-sectional surface of the guide. Equation (1) can be proved by using the two-dimensional Green's theorem, (1.1-10),

$$\int_S (P \nabla^2 Q - Q \nabla^2 P)\, da = \oint_C \left[P \frac{\delta Q}{\delta n} - Q \frac{\delta P}{\delta n} \right] dL \tag{2a}$$

$$P \equiv P(x, y) \quad \text{and} \quad Q \equiv Q(x, y) \tag{2b}$$

Fields for TE waves are given by (Section II.4)

$$\nabla_t^2 h_{zk} + \gamma_k^2 h_{zk} = 0 \quad \text{or} \quad \nabla_t^2 h_{zk} = - \gamma_k^2 h_{zk} \tag{3a}$$

$$\vec{h}_{tk} = \frac{\mp j \gamma_{zk}}{\gamma_k^2} \nabla_t h_{zk} \quad \text{or} \quad \nabla_t h_{zk} = \frac{\gamma_k^2}{\mp j \gamma_{zk}} \vec{h}_{tk} \tag{3b}$$

$$\vec{e}_{tk} = Z_{TE_k} \hat{z} \times \vec{h}_{tk} \tag{3c}$$

$$\gamma_k^2 = \gamma_0^2 - \gamma_{zk}^2 \quad \text{and} \quad \gamma_0^2 = \omega^2 \mu \underline{\varepsilon} \tag{3d}$$

$$Z_{TE_k} = \frac{\omega \mu}{\mp \gamma_{zk}} \tag{3e}$$

The index k represents the kth mode. It consists of two digits in the case of a rectangular guide.

Let $P = h_{zp}$ and $Q = h_{zq}$; then (2) becomes

$$\int_S (h_{zp} \nabla^2 h_{zq} - h_{zq} \nabla^2 h_{zp}) \, da = \oint_C \left(h_{zp} \frac{\delta h_{zq}}{\delta n} - h_{zq} \frac{\delta h_{zp}}{\delta n} \right) dL \tag{4}$$

The boundary condition for TE waves is

$$\frac{\delta h_{zk}}{\delta n} = 0 \tag{5}$$

on the conducting surface, where $\hat{n}$ is the unit outward normal vector of the conducting wall. Equation (5) is equivalent to requiring the tangential components of the E-field to vanish at the conducting surface. The component of the E-field tangential to the conducting surface is given by

$$0 = (\hat{n} \times \hat{z}) \cdot \vec{e}_t = (\hat{n} \times \hat{z}) \cdot Z_{TE_k} (\hat{z} \times \vec{h}_{tk})$$

$$= (\hat{n} \times \hat{z}) \cdot Z_{TE_k} \frac{\mp j \gamma_{zk}}{\gamma_k^2} (\hat{z} \times \nabla_t h_{zk})$$

$$= \frac{j \omega \mu}{\gamma_k^2} \hat{n} \cdot (\hat{z} \times \hat{z} \times \nabla_t h_{zk})$$

$$= \frac{-j \omega \mu}{\gamma_k^2} \hat{n} \cdot \nabla_t h_{zk} = \frac{-j \omega \mu}{\gamma_k^2} \frac{\delta h_{zk}}{\delta n}$$

The use of (3a) and (5) in (4) yields

$$(\gamma_p^2 - \gamma_q^2) \int_S h_{zp} h_{zq} \, da = 0 \tag{6}$$

For $p \neq q$, the factor $(\gamma_p^2 - \gamma_q^2)$ is not zero; then one has

$$\int_S h_{zp} h_{zq} \, da = 0 \quad \text{for} \quad p \neq q \tag{7}$$

This proof fails when eigenvalues are degenerate. In the case of degeneracy, orthogonality of modes does not necessarily follow. Nevertheless, linear combinations of normal modes which are not orthogonal can be arranged to secure orthogonality.

The integral

$$\int_S e_{zp} e_{zq} \, da = 0 \tag{8}$$

can be verified similarly by using the relations for TM waves and imposing the boundary condition that $e_{zk} = 0$ on the conducting wall of the guide. Fields for TM waves are given by Section (II.5)

$$\nabla_t^2 e_{zk} + \gamma_k^2 e_{zk} = 0 \tag{9a}$$

$$\vec{e}_{tk} = \frac{\mp j \gamma_{zk}}{\gamma_k^2} \nabla_t e_{zk} \tag{9b}$$

$$\vec{h}_{tk} = \frac{-1}{Z_{TM_k}} \hat{z} \times \vec{e}_{tk} \tag{9c}$$

$$Z_{TM_k} = \frac{\mp \gamma_k}{\omega \underline{\varepsilon}} \tag{9d}$$

(b) The integral

$$\int_S \vec{e}_{tp} \cdot \vec{e}_{tq} \, da = 0 \qquad \text{for} \qquad p \ne q \tag{10}$$

can be verified from Green's first identity in two dimensions, (1.1-6),

$$\int_S (\nabla P \cdot \nabla Q + P \nabla^2 Q) \, da = \oint_C P \frac{\delta Q}{\delta n} \, dL \tag{11}$$

Let $P = e_{zp}$ and $Q = e_{zq}$, then

$$\int_S (\nabla e_{zp} \cdot \nabla e_{zq} + e_{zp} \nabla^2 e_{zq}) \, da = \oint_C e_{zp} \frac{\delta e_{zq}}{\delta n} \, dL = 0 \tag{12}$$

by virtue of the boundary condition imposed on the tangential component of the E-field. The substitution of (9a) and (9b) into (12) yields

$$\int_S \left[\frac{\gamma_p^2 \gamma_q^2}{(\mp j\gamma_{zp})(\mp j\gamma_{zq})} \, \vec{e}_{tp} \cdot \vec{e}_{tq} - \gamma_q^2 e_{zp} e_{zq} \right] da = 0$$

or

$$\int_S \vec{e}_{tp} \cdot \vec{e}_{tq} \, da = 0 \qquad \text{for} \qquad p \neq q \tag{13}$$

where (8) is imposed.

Similarly, by assuming $P = h_{zp}$ and $Q = h_{zq}$, one can show that

$$\int_S \vec{h}_{tp} \cdot \vec{h}_{tq} = 0 \qquad \text{for} \qquad p \neq q \tag{14}$$

by using the TE fields.

(c) The integral

$$\int_S \hat{z} \cdot (\vec{e}_{tp} \times \vec{h}_{tq}) \, da = 0 \qquad \text{for} \qquad p \neq q \tag{15}$$

can be verified by expressing the integrand in terms of the transverse components of magnetic fields. For TE waves

$$\vec{e}_{tp} = Z_{TE_p} \hat{z} \times \vec{h}_{tp}$$

then

$$\hat{z} \cdot (\vec{e}_{tp} \times \vec{h}_{tq}) = Z_{TE_p} \hat{z} \cdot [\,(\hat{z} \times \vec{h}_{tp}) \times \vec{h}_{tq}\,]$$

$$= -Z_{TE_p} \hat{z} \cdot [\,\hat{z}(\vec{h}_{tq} \cdot \vec{h}_{tp}) - \vec{h}_{tp}(\hat{z} \cdot \vec{h}_{tq})\,] = -Z_{TE_p} \vec{h}_{tp} \cdot \vec{h}_{tq} \tag{16}$$

Therefore,

$$\int_S \hat{z} \cdot (\vec{e}_{tp} \times \vec{h}_{tq}) \, da = -Z_{TE_p} \int_S \vec{h}_{tp} \cdot \vec{h}_{tq} \, da = 0 \tag{17}$$

by virtue of (14).

(d) The above proofs were carried out with the assumption that fields P and Q belong to the same type, i.e., both are TE waves or both are TM waves.

For the case when P and Q are of different types, the proof will be quite different.

The following integrals

$$\int_S e_{zp1} e_{zq2} \, da = 0 \qquad p \neq q \tag{18a}$$

$$\int_S h_{zp1} h_{zq2} \, da = 0 \qquad p \neq q \tag{18b}$$

are obviously satisfied when fields belong to different types, since e_z is always zero for TE waves while $h_z = 0$ for TM waves.

The integral

$$\int_S \vec{e}_{tp1} \cdot \vec{e}_{tq2} \, da = 0 \qquad p \neq q \tag{19}$$

can be verified when $\vec{e}_{tp1}$ belongs to TE waves and $\vec{e}_{tq2}$ belongs to TM waves or vice versa.

The $\vec{e}_{tp1}$ for TE waves is given by (3),

$$\vec{e}_{tp1} = Z_{TE_p} \hat{z} \times \vec{h}_{tp1} = \frac{\mp j\gamma_{zp}}{\gamma_p^2} Z_{TE_p} \hat{z} \times \nabla_t h_{zp1}$$

$$= -\frac{\mp j\gamma_{zp}}{\gamma_p^2} Z_{TE_p} \nabla_t \times \vec{h}_{zp1} \tag{20a}$$

The $\vec{e}_{tq2}$ for TM waves is given by (9b),

$$\vec{e}_{tq2} = \frac{\mp j\gamma_{zq}}{\gamma_q^2} \nabla_t e_{zq2} \tag{20b}$$

Then

$$\vec{e}_{tp1} \cdot \vec{e}_{tq2} = C \, [\, \nabla_t \times \vec{h}_{zp1} \,] \cdot \nabla_t e_{zq2} \tag{20c}$$

$$C = \frac{\gamma_{zp} \gamma_{zq} Z_{TE_p}}{\gamma_p^2 \gamma_q^2} \tag{20d}$$

The right-hand side of (20c) can be identified as

$$\nabla_t \cdot [\, e_{zq2} \nabla_t \times \vec{h}_{zp1} \,] = \nabla_t e_{zq2} \cdot \nabla_t \times \vec{h}_{zp1} + e_{zq2} \nabla_t \cdot [\, \nabla_t \times \vec{h}_{zp1} \,]$$

$$= \nabla_t \times \vec{h}_{zp1} \cdot \nabla_t e_{zq2} \tag{21}$$

Thus

$$\vec{e}_{tp1} \cdot \vec{e}_{tq2} = C \nabla \cdot [\, e_{zq2} \nabla_t \times \vec{h}_{zp1} \,] \tag{22}$$

The integration of (22) over the volume V of the guide of length L_o yields

$$\int_V \vec{e}_{tp1} \cdot \vec{e}_{tq2} \, dv = -C \int_V \nabla \cdot [\, e_{zq2} \nabla \times \vec{h}_{zp1} \,] \, dv$$

$$\int_S \int_0^{L_o} \vec{e}_{tp1} \cdot \vec{e}_{tq2} \, da \, dL = -C \oint_S [\, e_{zq2} \nabla \times \vec{h}_{zp1} \,] \cdot \hat{n} \, da$$

$$L_o \int_S \vec{e}_{tp1} \cdot \vec{e}_{tq2} \, da = 0 \tag{23}$$

The right-hand side of the above expression vanishes by virtue of $e_{zq2} = 0$ on the conducting guide wall; the contributions from the cross-sectional end surfaces cancel due to the oppositely directed normal and the two-dimensional nature of functions e_{zq2} and h_{zp1}.

The integral

$$\int_S \vec{h}_{tp1} \cdot \vec{h}_{tq2} \, da = 0 \qquad p \neq q \tag{24}$$

can be verified similarly (see Problem 6).

(e) The integral

$$\int_S \hat{z} \cdot (\, \vec{e}_{tp1} \times \vec{h}_{tq2} \,) \, da = 0 \qquad p \neq q \tag{25}$$

can be easily verified by substituting the appropriate fields into (25). Let $\vec{e}_{tp1}$ be the field of a TE wave and $\vec{h}_{tq2}$ be the field of a TM wave. Then from (3c), one has

$$\hat{z} \cdot (\, \vec{e}_{tp1} \times \vec{h}_{tq2} \,) = \hat{z} \cdot [\, Z_{TE_p} (\, \hat{z} \times \vec{h}_{tp1} \,) \times \vec{h}_{tq2} \,]$$

$$= -Z_{TE_p} \hat{z} \cdot [\, \vec{h}_{tq2} \times (\, \hat{z} \times \vec{h}_{tp1} \,) \,]$$

$$= -Z_{TE_p} \hat{z} \cdot [\, \hat{z} (\, \vec{h}_{tp1} \cdot \vec{h}_{tq2} \,) - (\, \hat{z} \cdot \vec{h}_{tq2} \,) \vec{h}_{tp1} \,]$$

$$= -Z_{TE_p} \vec{h}_{tp1} \cdot \vec{h}_{tq2} \tag{26}$$

Then (25) becomes

$$\int_S \hat{z} \cdot (\, \vec{e}_{tp1} \times \vec{h}_{tq2} \,) \, da = -Z_{TE_p} \int_S \vec{h}_{tp1} \cdot \vec{h}_{tq2} \, da = 0$$

by virtue of (24).

On the other hand, if $\vec{e}_{tp1}$ belongs to a TM wave and $\vec{h}_{tq2}$ belongs to a TE wave, then from (3c)

$$\hat{z} \times \vec{e}_{tq2} = Z_{TE_q} \hat{z} \times \hat{z} \times \vec{h}_{tq2} = -Z_{TE_q} \vec{h}_{tq2}$$

or

$$\vec{h}_{tq2} = \frac{-1}{Z_{TE_q}} \hat{z} \times \vec{e}_{tq2}$$

Then

$$\int_S \hat{z} \cdot (\vec{e}_{tp1} \times \vec{h}_{tq2})\, da = \frac{-1}{Z_{TE_q}} \int_S \hat{z} \cdot [\vec{e}_{tp1} \times (\hat{z} \times \vec{e}_{tq2})]\, da$$

$$= \frac{-1}{Z_{TE_q}} \int_S \vec{e}_{tp1} \cdot \vec{e}_{tq2}\, da = 0$$

by virtue of (23).

(f) In some applications, one needs the integrals of the squares of components of $\vec{E}$ or $\vec{H}$ and of the products of components with their conjugates. These will now be derived.

For TE waves

$$\int_S h_{zk} h_{zk}^*\, da = \frac{\gamma_k^2}{\gamma_{zk}\gamma_{zk}^*} \int_S \vec{h}_{tk} \cdot \vec{h}_{tk}^*\, da \tag{27}$$

This can be verified by using Green's first identity in two-dimensional form.

$$\int_S [\nabla_t P \cdot \nabla_t Q + P \nabla_t^2 Q]\, da = \oint_C P \frac{\delta Q}{\delta n}\, dL \tag{28}$$

Let $P = h_{zk}$ and $Q = h_{zk}^*$; then

$$\int_S [\nabla_t h_{zk} \cdot \nabla_t h_{zk}^* + h_{zk} \nabla_t^2 h_{zk}^*]\, da = \oint_C h_{zk} \frac{\delta h_{zk}^*}{\delta n}\, dL = 0 \tag{29}$$

by virtue of (5).

For a TE wave, (3b) gives

$$\nabla_t h_{zk} = \frac{\gamma_k^2}{\mp j\gamma_{zk}} \vec{h}_{tk} \tag{30a}$$

$$\nabla_t h_{zk}^* = \frac{\gamma_k^{*2}}{\pm j\gamma_{zk}^*} \vec{h}_{tk}^* \tag{30b}$$

$$\nabla_t^2 h_{zk}^* = -\gamma_k^{*2} h_{zk}^* \tag{30c}$$

Therefore,

$$\int_S \left[\frac{\gamma_k^2}{\mp j\gamma_{zk}} \frac{\gamma_k^{*2}}{\pm j\gamma_{zk}^*} \vec{h}_{tk} \cdot \vec{h}_{tk}^* + h_{zk} (-\gamma_k^{*2} h_{zk}^*) \right] da = 0$$

or

$$\int_S h_{zk} h_{zk}^* \, da = \frac{\gamma_k^2}{\gamma_{zk}\gamma_{zk}^*} \int_S \vec{h}_{tk} \cdot h_{tk}^* \, da \tag{31}$$

Similarly, for a TM wave, one has

$$\int_S e_{zk} e_{zk}^* \, da = \frac{\gamma_k^2}{|\gamma_{zk}|^2} \int_S \vec{e}_{tk} \cdot \vec{e}_{tk}^* \, da \tag{32}$$

(g) For both TE and TM waves,

$$\int_S \vec{e}_{tk} \cdot \vec{e}_{tk}^* \, da = Z_{T_k} Z_{T_k}^* \int_S \vec{h}_{tk} \cdot \vec{h}_{tk}^* \, da \tag{33a}$$

$$= -Z_{T_k}^* \int_S \hat{z} \cdot (\vec{e}_{tk} \times \vec{h}_{tk}^*) \, da \tag{33b}$$

$$Z_{T_k} = Z_{TE_k} \qquad \text{for TE wave} \tag{33c}$$

$$= Z_{TM_k} \qquad \text{for TM wave} \tag{33d}$$

The above expression can be verified by using the relation

$$\vec{e}_{tk} = Z_{T_k} \hat{z} \times \vec{h}_{tk} \tag{34}$$

then

$$\vec{e}_{tk} \cdot \vec{e}_{tk}^* = Z_{T_k} Z_{T_k}^* (\hat{z} \times \vec{h}_{tk}) \cdot (\hat{z} \times \vec{h}_{tk}^*)$$

$$= -|Z_{T_k}|^2 (\hat{z} \times \vec{h}_{tk}) \cdot (\vec{h}_{tk}^* \times \hat{z}) = -|Z_{T_k}|^2 \hat{z} \cdot [(\hat{z} \times \vec{h}_{tk}) \times \vec{h}_{tk}^*]$$

$$= -|Z_{T_k}|^2 [\hat{z} \times (\hat{z} \times \vec{h}_{tk})] \cdot \vec{h}_{tk}^* = |Z_{T_k}|^2 \vec{h}_{tk} \cdot \vec{h}_{tk}^*$$

thus proving Eq. (33a).

The relation (33b) can be verified by using the following identity.

$$\vec{e}_{tk} \cdot \vec{e}_{tk}^* = \vec{e}_{tk} \cdot Z_{T_k}^* \hat{z} \times \vec{h}_{tk}^* = Z_{T_k}^* \hat{z} \times \vec{h}_{tk}^* \cdot \vec{e}_{tk}$$

$$= Z_{T_k}^* \hat{z} \cdot \vec{h}_{tk}^* \times \vec{e}_{tk}$$

1.1. Green's Theorem

(a) Green's theorem can be obtained by applying the divergence theorem to an appropriate vector function $\vec{Q}(\vec{r})$.

$$\int_V \nabla \cdot \vec{Q} \, dv = \oint_S \vec{Q} \cdot \hat{n} \, da \tag{1}$$

Let

$$\vec{Q} \equiv f \nabla g \tag{2a}$$

$$f \equiv f(\vec{r}) \qquad \text{and} \qquad g \equiv g(\vec{r}) \tag{2b}$$

and by direct expansion, one obtains

$$\nabla \cdot \vec{Q} = f \nabla^2 g + \nabla f \cdot \nabla g \tag{3}$$

The substitution of (2) and (3) into (1) yields

$$\int_V [\, f \nabla^2 g + \nabla f \cdot \nabla g \,] \, dv = \oint_S (\, f \nabla g \,) \cdot \hat{n} \, da = \oint_S f \frac{\delta g}{\delta n} \, da \tag{4}$$

This is known as Green's first identity.

Let

$$\vec{Q}' \equiv g \nabla f \tag{5}$$

The application of the divergence theorem to $\vec{Q}'$ gives

$$\int_V [\, g \nabla^2 f + \nabla f \cdot \nabla g \,] \, dv = \oint_S g \frac{\delta f}{\delta n} \, da \tag{6}$$

The difference between (4) and (6) is

$$\int_V [\, f \nabla^2 g - g \nabla^2 f \,] \, dv = \oint_S \left[\, f \frac{\delta g}{\delta n} - g \frac{\delta f}{\delta n} \,\right] da \tag{7}$$

This is known as Green's second identity or Green's theorem.

(b) In two-dimensional problems, functions f and g are independent of the longitudinal coordinate, z, i.e.,

$$f = f(x, y) \qquad \text{and} \qquad g = g(x, y) \tag{8}$$

The volume integral can then be written as

$$\int_V dv = \int_{z_r}^{z_r+z_o} dz \int_S dx\, dy = z_o \int_{S_c} dx\, dy \tag{9a}$$

where z_r is the reference location and z_o is any arbitrary distance from the reference point, and S_c is the cross-sectional area of the volume V. The closed surface integral can be expressed as

$$\oint_S \hat{n}\, da = \int_{z_r}^{z_r+z_o} dz \oint_C \hat{n}\, dL = z_o \oint_C \hat{n}\, dL \tag{9b}$$

$$S = S_{cylindrical} + S_{end} \tag{9c}$$

where C is the contour of the cross-sectional area of the volume V, and the contributions from two end surfaces cancel.

With (9), expressions (4) and (7) become

$$\int_S [\, f \nabla^2 g + \nabla f \cdot \nabla g \,]\, da = \oint_C f \frac{\delta g}{\delta n}\, dL \tag{10a}$$

$$\int_S [\, f \nabla^2 g - g \nabla^2 f \,]\, da = \oint_C \left[\, f \frac{\delta g}{\delta n} - g \frac{\delta f}{\delta n} \,\right] dL \tag{10b}$$

where S in (10) is understood to be the cross-sectional area of the volume V and is bounded by the contour C. Equation (10) is known as Green's identity in two-dimensional form.

2. Guided Wave Theory

Fields in a medium characterized by μ, ε, and σ and in a region which excludes all sources are governed by Maxwell's equations.

$$\nabla \times \vec{E} = -j\omega\mu\vec{H} \tag{1a}$$

$$\nabla \times \vec{H} = j\omega\underline{\varepsilon}\vec{E} \tag{1b}$$

$$\underline{\varepsilon} = \varepsilon \left(1 + \frac{\sigma}{j\omega\varepsilon} \right) \tag{1c}$$

The divergence equations follow automatically from (1). These equations assume that all quantities vary with time as a function $e^{j\omega t}$.

$$\underline{\vec{G}} \equiv \underline{\vec{G}}(\vec{r}, t) = \vec{G}(\vec{r})e^{j\omega t} \tag{2}$$

where $\underline{\vec{G}}$ stands for either $\vec{E}$ or $\underline{\vec{H}}$.

The coupled set of equations, (1), can be decoupled to have the form, (Section II.2)

$$\nabla^2\vec{G} + \gamma_0^2\vec{G} = 0 \tag{3a}$$

$$\gamma_0^2 \equiv \omega^2\mu\underline{\varepsilon} \tag{3b}$$

Both $\vec{E}$ and $\vec{H}$ satisfy this vector wave equation.

A waveguide is defined to be a cylindrical region with uniform cross section, bounded by a conducting surface and filled with a dielectric. The problem is to find $\vec{E}$ and $\vec{H}$ which satisfy Maxwell's equations within the guide and the associated boundary conditions at the boundary of the guide.

Assume a trial solution of the form

$$\vec{G}(x, y, z) = \vec{g}(x, y) \, e^{\mp j\gamma_z z} \tag{4}$$

where γ_z is the propagation constant of the guided wave, which is yet to be determined. The $\mp$ signs in front of γ_z indicate two possible solutions. Then (3) becomes

$$\nabla_t^2\vec{g} + \gamma^2\vec{g} = 0 \tag{5a}$$

$$\gamma^2 = \gamma_0^2 - \gamma_z^2 \tag{5b}$$

$$\nabla_t^2 \equiv \frac{\delta^2}{\delta x^2} + \frac{\delta^2}{\delta y^2} = \text{Laplacian operator in the transverse coordinates} \tag{5c}$$

By decomposing the field vectors into their axial and transverse components, one has

$$\vec{g} = \vec{g}_z + \vec{g}_t \tag{6a}$$

In Cartesian coordinates,

$$\vec{g}_z = \hat{z}g_z \quad \text{and} \quad \vec{g}_t = \hat{x}g_x + \hat{y}g_y \tag{6b}$$

Then (5) becomes

$$\nabla_t^2 g_z + \gamma^2 g_z = 0 \tag{7a}$$

$$\nabla_t^2 \vec{g}_t + \gamma^2 \vec{g}_t = 0 \tag{7b}$$

It can further be shown (Section II.6) that the transverse field components can be expressed in terms of axial field components.

$$\vec{e}_t = \frac{\mp \gamma_z}{\gamma^2} \nabla_t e_z + \frac{j\gamma_o}{\gamma^2} \hat{z} \times \nabla_t (\eta h_z) \tag{8a}$$

$$\eta \vec{h}_t = \frac{\mp \gamma_z}{\gamma^2} \nabla_t (\eta h_z) - \frac{j\gamma_o}{\gamma^2} \hat{z} \times \nabla_t e_z \tag{8b}$$

$$\eta = \sqrt{\frac{\mu}{\varepsilon}}, \quad \gamma_o = \omega\sqrt{\mu\varepsilon}, \quad \gamma^2 = \gamma_o^2 - \gamma_z^2 \tag{8c}$$

The problem is now reduced to finding the axial field components, e_z and h_z, which satisfy (7) and the associated boundary conditions.

One of the most powerful techniques in solving partial differential equations is the method of separation of variables (Section II.11). The solution of (7a) in rectangular coordinates for the traveling wave in the positive z-direction is [Eq. (II.11-10)].

$$G_z(x, y, z) = (X^+ e^{-j\gamma_x x} + X^- e^{j\gamma_x x})(Y^+ e^{-j\gamma_y y} + Y^- e^{j\gamma_y y}) e^{-j\gamma_z z} \tag{9a}$$

$$\gamma^2 = \gamma_x^2 + \gamma_y^2 \tag{9b}$$

The constants of integration, $X^{\pm}$ and $Y^{\pm}$, are to be determined by imposing boundary conditions.

3. Normal Modes

Since both Maxwell's equations and the boundary conditions are linear, therefore a most general solution can be constructed by the superposition of all possible modes with appropriate coefficients. In waveguide problems, this general solution will consist of traveling waves in both the positive z-direction and the negative z-direction. It is to be noted from (1-3) and (1-9) that to reverse the direction of propagation, one must change the sign of γ_{zk}. This will also change the sign of the wave impedances, Z_{TE_k} or Z_{TM_k} [(II.4-6c) and (II.5-4b)]. It thus brings a change in sign in the relation between transverse components of $\vec{E}$ and $\vec{H}$ and in the relation between $\vec{e}_{tk}$ and e_{zk} (or between $\vec{h}_{tk}$ and h_{zk}).

Let

$\vec{e}_{tk}(x, y)\, E_k^+$ = magnitude of kth mode of traveling wave in positive z–direction

$\vec{e}_{tk}(x, y)\, E_k^-$ = magnitude of kth mode of traveling wave in negative z–direction

Then the complete field can be expressed as

$$\underline{\vec{E}} = \sum_k \{ \vec{e}_{tk} [\, E_k^+ e^{j(\omega t - \gamma_{zk} z)} + E_k^- e^{j(\omega t + \gamma_{zk} z)} \,]$$

$$+ \hat{z} e_{zk} [\, E_k^+ e^{j(\omega t - \gamma_{zk} z)} - E_k^- e^{j(\omega t + \gamma_{zk} z)} \,] \} \tag{1a}$$

$$\underline{\vec{H}} = \sum_k \{ \vec{h}_{tk} [\, E_k^+ e^{j(\omega t - \gamma_{zk} z)} - E_k^- e^{j(\omega t + \gamma_{zk} z)} \,] \qquad \vec{h}_{tk} \equiv \vec{h}_{tk}(x, y)$$

$$+ \hat{z} h_{zk} [\, E_k^+ e^{j(\omega t - \gamma_{zk} z)} + E_k^- e^{j(\omega t + \gamma_{zk} z)} \,] \} \qquad h_{zk} \equiv h_{zk}(x, y) \tag{1b}$$

Unique solution may be obtained by specifying the values of the tangential components of both $\vec{E}$ and $\vec{H}$ as a function of x and y across a single cross section of the guide.

If $z = 0$ is chosen at the cross section where the tangential $\underline{\vec{E}}$ and $\underline{\vec{H}}$ are given, then one has at $z = 0$:

$$\underline{\vec{E}}_t \equiv \vec{E}_t e^{j\omega t} = \sum_k \vec{e}_{tk} (E_k^+ + E_k^-)\, e^{j\omega t} \quad \text{or} \quad \vec{E}_t = \sum_k \vec{e}_{tk} (E_k^+ + E_k^-) \tag{2a}$$

and

$$\vec{H}_t = \sum_k \vec{h}_{tk} (E_k^+ - E_k^-) \tag{2b}$$

where $\vec{E}_t(x, y)$ and $\vec{H}_t(x, y)$ are given functions. The coefficients $E_k^\pm$ can be evaluated by multiplying (2a) by $\vec{e}_{tj}$ and integrating over the cross section of the guide.

$$\int_S \vec{E}_t \cdot \vec{e}_{tj}\, da = \int_S \sum_k \vec{e}_{tk} \cdot \vec{e}_{tj} (E_k^+ + E_k^-)\, da = (E_j^+ + E_j^-) \int_S \vec{e}_{tj} \cdot \vec{e}_{tj}\, da \tag{3}$$

where the property of orthogonality has been used.

$$\int_S \vec{e}_{tk} \cdot \vec{e}_{tj} \, da = 0 \qquad \text{for} \qquad k \neq j \tag{1-10}$$

It is convenient to normalize the normal-mode function such that

$$\int_S \vec{e}_{tk} \cdot \vec{e}_{tk}^* \, da = 1 \tag{4}$$

For the normalized normal-mode function, (3) becomes

$$E_j^+ + E_j^- = \int_S \vec{E}_t \cdot \vec{e}_{tj} \, da \tag{5a}$$

But

$$\int_S \vec{H}_t \cdot \vec{h}_{tj} \, da = \int_S \sum_k \vec{h}_{tk} \cdot \vec{h}_{tj} \, (E_k^+ - E_k^-) \, da = (E_j^+ - E_j^-) \int_S \vec{h}_{tj} \cdot \vec{h}_{tj} \, da$$

But by (1-33a), one has

$$\int_S \vec{H}_t \cdot \vec{h}_{tj} \, da = \frac{(E_j^+ - E_j^-)}{Z_{T_j}^2} \int_S \vec{e}_{tj} \cdot \vec{e}_{tj} \, da = \frac{E_j^+ - E_j^-}{Z_{T_j}^2}$$

or

$$E_j^+ - E_j^- = Z_{T_j}^2 \int_S \vec{H}_t \cdot \vec{h}_{tj} \, da \tag{5b}$$

The coefficients $E_j^{\pm}$ can be determined from (5) once the functions $\vec{E}_t$ and $\vec{H}_t$ are specified.

A general expression for the field within a waveguide has been given in terms of the normal modes with appropriate coefficients. However, the possibility of expressing a function in terms of orthogonal functions is much more general.

4. Orthogonal Functions for Electromagnetic Fields

Electromagnetic fields in a homogeneous medium which contains no current densities or charge densities of the sources satisfy Maxwell's equations.

$$\nabla \times \vec{E} = -j\omega\sqrt{\mu\varepsilon} \left[\sqrt{\frac{\mu}{\varepsilon}} \, \vec{H} \right] \tag{1a}$$

$$\nabla \cdot \vec{E} = \frac{\rho}{\varepsilon} \tag{1b}$$

$$\nabla \times \left[\sqrt{\frac{\mu}{\varepsilon}} \, \vec{H} \right] = j\omega\sqrt{\mu\varepsilon} \, \vec{E} \tag{2a}$$

$$\nabla \cdot \left[\sqrt{\frac{\mu}{\varepsilon}} \, \vec{H} \right] = 0 \tag{2b}$$

From the theory of vector analysis, any vector function whose flow sources and vortex sources are specified can be uniquely represented as the sum of the gradient of a scalar potential and the curl of a vector potential. In other words, any vector function can be decomposed into an irrotational field (zero curl) and a solenoidal field (zero divergence). In particular cases, one of these fields could be zero.

Since the $\vec{E}$-field has both vortex and flow sources, it is convenient to subdivide the $\vec{E}$-field as follows. Let

$$\vec{E} = \vec{U} + \vec{V} \tag{3}$$

where $\vec{U}$ and $\vec{V}$ are defined by the following relations.

$$\nabla \times \vec{U} = -j\omega\sqrt{\mu\varepsilon} \left[\sqrt{\frac{\mu}{\varepsilon}} \, \vec{H} \right] \tag{4a}$$

$$\nabla \cdot \vec{U} = 0 \tag{4b}$$

$$\nabla \times \vec{V} = 0 \tag{5a}$$

$$\nabla \cdot \vec{V} = \frac{\rho}{\varepsilon} \tag{5b}$$

That is, the $\vec{U}$-function is solenoidal and the $\vec{V}$-function is irrotational.

Similarly, the $(\sqrt{\mu/\varepsilon} \, \vec{H})$-field can be represented by a solenoidal function $\vec{W}$, since the $\vec{H}$-field does not have any flow source. The $\vec{W}$-function is defined by

$$\nabla \times \vec{W} = j\omega\sqrt{\mu\varepsilon} \, \vec{E} \tag{6a}$$

$$\nabla \cdot \vec{W} = 0 \tag{6b}$$

(a) Potential Functions $\vec{U}$ and $\vec{W}$:

$\vec{U}$ and $\vec{W}$ are potential functions and can be defined more explicitly as

$$\nabla \times \vec{U}_k = c_k^* \vec{W}_k \tag{7a}$$

$$\nabla \times \vec{W}_k = c_k \vec{U}_k \qquad c_k = j\omega \sqrt{\mu\varepsilon} \tag{7b}$$

where c_k is a constant, as suggested by (4a) and (6a). $\vec{U}_k$ and $\vec{W}_k$ satisfy the following boundary conditions.

$$\hat{n} \times \vec{U}_k = 0 \qquad \text{on } S_{sc} \qquad \text{short-circuited boundary condition} \tag{8a}$$

$$\hat{n} \times \vec{W}_k = 0 \qquad \text{on } S_{oc} \qquad \text{open-circuited boundary condition} \tag{8b}$$

where $\hat{n}$ is the outward normal vector to the surface. These boundary conditions imply that the tangential components of $\vec{U}_k$ vanish on the surface S_{sc} and the tangential components of $\vec{W}_k$ vanish on the surface S_{oc}.

The integral of $\vec{U}_k$ over a contour on surface S_{sc} yields

$$\oint_C \vec{U}_k \cdot d\vec{L} = 0 = \int_{S_{sc}} \nabla \times \vec{U}_k \cdot \hat{n}\, da = c_k^* \int_{S_{sc}} \vec{W}_k \cdot \hat{n}\, da \tag{9}$$

The contour integral on surface S_{sc} is zero since the components of $\vec{U}_k$ along the contour do not exist, (8a). The surface integral is obtained by Stokes' theorem, and (7a) is used to obtain the last expression. Thus (8a) also implies

$$\hat{n} \cdot \vec{W}_k = 0 \qquad \text{on } S_{sc} \tag{10a}$$

Similarly, one concludes that (8b) also implies

$$\hat{n} \cdot \vec{U}_k = 0 \qquad \text{on } S_{oc} \tag{10b}$$

Equation (7) can be decoupled by taking the curl of (7a) and using (7b) to eliminate $\vec{W}_k$.

$$\nabla \times \nabla \times \vec{U}_k = c_k^2 \vec{U}_k \tag{11a}$$

Similarly, one obtains

$$\nabla \times \nabla \times \vec{W}_k = c_k^2 \vec{W}_k \tag{11b}$$

Equation (11) can be reduced to the familiar wave equation.

$$\nabla^2 \vec{U}_k + c_k^2 \vec{U}_k = 0 \tag{12a}$$

$$\nabla^2 \vec{W}_k + c_k^2 \vec{W}_k = 0 \tag{12b}$$

These equations have an infinite set of solutions corresponding to different values of c_k and satisfy the boundary conditions, (8).

These potential functions, $\vec{U}_k$ and $\vec{W}_k$, have the following orthogonality properties.

$$\int_V \vec{U}_j \cdot \vec{U}_k \, dv = 0 \qquad \text{for } j \neq k \tag{13a}$$

$$\int_V \vec{W}_j \cdot \vec{W}_k \, dv = 0 \qquad \text{for } j \neq k \tag{13b}$$

The verification of (13) is left as an example (Section 3.1).

The potential functions $\vec{U}_k$ and $\vec{W}_k$ can be normalized.

$$\int_V \vec{U}_k \cdot \vec{U}_k^* \, dv = 1 \tag{14a}$$

$$\int_V \vec{W}_k \cdot \vec{W}_k^* \, dv = 1 \tag{14b}$$

These expressions are consistent with each other and can be verified by the follow expansion.

$$\int_V \nabla \cdot [\, \vec{U}_k \times \nabla \times \vec{U}_k^* \,] \, dv = \int_V [\, (\nabla \times \vec{U}_k) \cdot (\nabla \times \vec{U}_k^*) - \vec{U}_k \cdot \nabla \times \nabla \times \vec{U}_k^* \,] \, dv$$

$$\oint_S [\, \vec{U}_k \times \nabla \times \vec{U}_k^* \,] \cdot \hat{n} \, da = \int_V [\, |c_k|^2 \, \vec{W}_k \cdot \vec{W}_k^* - |c_k|^2 \, \vec{U}_k \cdot \vec{U}_k^* \,] \, dv$$

$$0 = |c_k|^2 \int_V [\, \vec{W}_k \cdot \vec{W}_k^* - \vec{U}_k \cdot \vec{U}_k^* \,] \, dv$$

or

$$\int_V \vec{W}_k \cdot \vec{W}_k^* \, dv = \int_V \vec{U}_k \cdot \vec{U}_k^* \, dv \tag{15}$$

(b) Potential Function $\vec{V}$

Because of the irrotational property of $\vec{V}$ [Eq. (5)], this function can be represented by the gradient of a scalar function.

$$\vec{V}_k d_k = \nabla \phi_k \tag{16a}$$

where d_k is a constant. It is further assumed that ϕ_k satisfies the wave equation.

$$\nabla^2 \phi_k + d_k^2 \phi_k = 0 \tag{16b}$$

Equation (16) implies that $\vec{V}_k$ also satisfies the wave equation since the gradient of (16b) is

$$\nabla^2 (\nabla \phi_k) + d_k^2 \nabla \phi_k = 0 \quad \text{or} \quad \nabla^2 \vec{V}_k + d_k^2 \vec{V}_k = 0 \tag{17}$$

The function ϕ_k is assumed to satisfy the homogeneous boundary condition.

$$\phi_k = 0 \quad \text{on } S_{sc} \text{ and } S_{oc} \tag{18a}$$

Since $\phi_k = 0 = $ constant on the bounding surface, hence its derivatives tangential to the surface vanish, i.e.,

$$\hat{n} \times \nabla \phi_k = 0 = d_k \hat{n} \times \vec{V}_k \quad \text{or} \quad \hat{n} \times \vec{V}_k = 0 \quad \text{on boundary} \tag{18b}$$

The tangential components of $\vec{V}_k$ vanish along the bounding surface.

The orthogonality properties

$$\int_V \phi_j \phi_k \, dv = 0 \qquad \text{for } j \neq k \tag{19a}$$

$$\int_V \vec{V}_j \cdot \vec{V}_k \, dv = 0 \qquad \text{for } j \neq k \tag{19b}$$

can be confirmed by invoking Green's theorem

$$\int_V (P \, \nabla^2 Q - Q \, \nabla^2 P) \, dv = \oint_S \left[P \frac{\delta Q}{\delta n} - Q \frac{\delta P}{\delta n} \right] da$$

With

$$P = \phi_j \quad \text{and} \quad Q = \phi_k$$

then

$$\int_V (\phi_j \, \nabla^2 \phi_k - \phi_k \, \nabla^2 \phi_j) \, dv = \oint_S \left[\phi_j \frac{\delta \phi_k}{\delta n} - \phi_k \frac{\delta \phi_j}{\delta n} \right] da = 0$$

by virtue of (18a). Hence

$$\int_V (- d_k^2 \phi_j \phi_k + d_j^2 \phi_j \phi_k) \, dv = 0 \quad \text{or} \quad (d_j^2 - d_k^2) \int_V \phi_j \phi_k \, dv = 0$$

which proves (19a) since $d_k \neq d_j$.

Equation (19b) can be verified by Green's first identity.

$$\int_V (P \, \nabla^2 Q + \nabla P \cdot \nabla Q) \, dv = \oint_S P \frac{\delta Q}{\delta n} da$$

$$\int_V (\phi_j \, \nabla^2 \phi_k + \nabla\phi_j \cdot \nabla\phi_k) \, dv = \oint_S \phi_j \frac{\delta\phi_k}{\delta n} \, da = 0$$

by virtue of the boundary condition and hence

$$\int_V (-d_k^2 \phi_j \phi_k + d_k d_j \vec{V}_j \cdot \vec{V}_k) \, dv = 0$$

or

$$d_j d_k \int_V \vec{V}_j \cdot \vec{V}_k \, dv = d_k^2 \int_V \phi_j \phi_k \, dv = 0 \qquad j \neq k \tag{20}$$

by virtue of (19a) thus proving the othogonality properties, (19).

(c) Normalization

The normalized functions obey the relations

$$\int_V \vec{V}_k \cdot \vec{V}_k \, dv = 1 \qquad \text{and} \qquad \int_V \phi_k \phi_k \, dv = 1 \tag{21}$$

The consistency of these relations can be substantiated from (20) by setting $j = k$.

(d) Orthogonality

Orthogonality exists between $\vec{U}_k$ and $\vec{V}_k$.

$$\int_V \vec{U}_j \cdot \vec{V}_k \, dv = 0 \tag{22}$$

This can be established by the identity

$$\nabla \cdot (\phi_j \vec{U}_j) = \phi_k \, \nabla \cdot \vec{U}_j + \nabla\phi_k \cdot \vec{U}_j = d_k \vec{V}_k \cdot \vec{U}_j \tag{23}$$

in accordance with (4b) and (16a). Integrating (23) over the volume V creates

$$\oint_S \phi_k \vec{U}_j \cdot \hat{n} \, da = 0 = d_k \int_V \vec{V}_k \cdot \vec{U}_j \, dv \tag{24}$$

which is equivalent to (22).

(e) Expansion of Fields in Terms of Potential Functions

A set of potential functions $\vec{U}_k$, $\vec{V}_k$, and $\vec{W}_k$ has been established. Any arbitrary physical fields within the volume V can be expanded in terms of these potentials. Here the term physical fields implies well-behaved fields with reasonable continuity.

Since $\vec{U}_k$ and $\vec{W}_k$ are related by (7), it is unnecessary to use both of these for a given expansion. If the field to be expanded has boundary conditions which more closely resemble the boundary conditions satisfied by $\vec{U}_k$ than by $\vec{W}_k$, then its expansion in $\vec{U}_k$'s will converge better than expansion in terms of $\vec{W}_k$'s, and vice versa. As in Fourier expansion, one can choose either a sine or a cosine series. If the function to be expanded satisfies the same type of boundary condition as the cosine function, the expansion in a cosine series will converge better than that in a sine series.

The procedure of expanding an arbitrary field vector function $\vec{G}$ in terms of potentials $\vec{U}_k$ and $\vec{V}_k$ is as follows:

(α) Separate the solenoidal and irrotational portions of the field. Let

$$\vec{G} = \vec{G}_a + \vec{G}_b \tag{25a}$$

$$\vec{G}_a = \text{solenoidal field} \tag{25b}$$

$$\vec{G}_b = \text{irrotational field} \tag{25c}$$

(β) Express each portion of $\vec{G}$ in terms of a potential function.

$$\vec{G}_a = \sum_k u_k \vec{U}_k \tag{26a}$$

$$\vec{G}_b = \sum_k v_k \vec{V}_k \tag{26b}$$

(γ) Combine (26) and (25) to obtain $\vec{G}$.

$$\vec{G} = \sum_k u_k \vec{U}_k + \sum_k v_k \vec{V}_k \tag{27}$$

The coefficients u_k and v_k may be evaluated by multiplying (27) by $\vec{U}_j$ and integrating over the volume V.

$$\int_V \vec{U}_j \cdot \vec{G} \, dv = \int_V [\sum_k u_k \vec{U}_k \cdot \vec{U}_j + \sum_k v_k U_j \cdot \vec{V}_k] \, dv$$

$$= u_j \int_V \vec{U}_j \cdot \vec{U}_j \, dv = u_j \tag{28a}$$

by virtue of (22) and (14). Similarly, v_j can be determined by

$$\int_V \vec{V}_j \cdot \vec{G} \, dv = v_j \tag{28b}$$

Hence, Eq. (27) becomes

$$\vec{G} = \sum_k \vec{U}_k \int_V \vec{U}_k \cdot \vec{G} \, dv + \sum_k \vec{V}_k \int_V \vec{V}_k \cdot \vec{G} \, dv \tag{29}$$

5. Expansion of Fields in Normal Modes

It was shown that within the waveguide bounded by perfectly conducting walls, Maxwell's equations have an infinite number of solutions, known as normal modes. Since Maxwell's equations as well as boundary conditions are linear, the superposition of various fields, each of which satisfies Maxwell's equations and the same set of boundary conditions, will also be a solution. Therefore, a general solution has the form

$$\vec{E} = \sum_k \{ \vec{e}_{tk} [E_k^+ e^{j(\omega t - \gamma_{zk} z)} + E_k^- e^{j(\omega t + \gamma_{zk} z)}]$$

$$+ \hat{z} e_{zk} [E_k^+ e^{j(\omega t - \gamma_{zk} z)} - E_k^- e^{j(\omega t + \gamma_{zk} z)}] \} \tag{1a}$$

$$\vec{H} = \sum_k \{ \vec{h}_{tk} [E_k^+ e^{j(\omega t - \gamma_{zk} z)} - E_k^- e^{j(\omega t + \gamma_{zk} z)}]$$

$$+ \hat{z} h_{zk} [E_k^+ e^{j(\omega t - \gamma_{zk} z)} + E_k^- e^{j(\omega t + \gamma_{zk} z)}] \} \tag{1b}$$

where $E_k^\pm$ are constant coefficients to be determined by imposing the associated boundary conditions. The sign of the traveling wave in the negative z-direction is determined from (1-3) and (1-9).

(a) The reversing of a wave is accompanied by a change in the sign of γ_{zk}.

(b) This also changes the signs of the wave impedances.

(c) Fields in the negative wave of a TE mode:

$\vec{h}_{tk}$ changes sign as a consequence of (a).

$\vec{e}_{tk}$ remains unchanged because the sign of γ_{zk} changes twice, once in $\vec{h}_{tk}$ and again in Z_{TE_k}.

(d) Fields in the negative wave of a TM mode:

$\vec{e}_{tk}$ changes sign because of (a).

$\vec{h}_{tk}$ remains unchanged.

The sign convention in this analysis is chosen in accordance with (II.4-8).

The coefficients $E_k^\pm$ can be determined if the fields are known at a cross-sectional surface. If the origin of coordinates is chosen at this surface, then at $z = 0$,

$$\vec{E}_t = \sum_k \vec{e}_{tk} (E_k^+ + E_k^-) \tag{2a}$$

$$\vec{H}_t = \sum_k \vec{h}_{tk} (E_k^+ - E_k^-) \tag{2b}$$

The coefficients can be evaluated by multiplying (2a) by $\vec{e}_{tj}$ and integrating over the cross section.

$$\int_S \vec{E}_t \cdot \vec{e}_{tj} \, da = \int_S \sum_k \vec{e}_{tk} \cdot \vec{e}_{tj} \, (E_k^+ + E_k^-) \, da$$

or

$$E_j^+ + E_j^- = \int_S \vec{E}_t \cdot \vec{e}_{tj} \, da \tag{3a}$$

since the mode functions are orthogonal. Similarly,

$$E_j^+ - E_j^- = \int_S \vec{H}_t \cdot \vec{h}_{tj} \, da = Z_{f_j}^2 \int_S \vec{E}_t \cdot \vec{e}_{tj} \, da \tag{3b}$$

Hence, the coefficients $E_j^\pm$ can be determined from (3) once the fields at the surface S are specified.

6. Power and Energy Relations

The energy density or Poynting vector is an important quantity from an experimental point of view. Generally, it is not convenient to measure the electric or magnetic field at microwave frequencies. However, one can measure directly the standing-wave ratio and the power flow. With these types of measurements, one can then deduce the actual electric and magnetic fields.

(a) It was shown in Eq. (II.10-13) that the time-average Poynting vector is given by

$$< \vec{\underline{S}} > = \frac{1}{2} \, \text{Re} \, [\, \vec{\underline{E}} \times \vec{\underline{H}}^* \,] \tag{1}$$

For fields expressed in terms of normal modes [Eq. (4-1)],

$$\vec{\underline{E}} = \sum_k [\, \vec{e}_{tk} \, (\, \underline{E}_k^+ + \underline{E}_k^- \,) + \hat{z} e_{zk} \, (\, \underline{E}_k^+ - \underline{E}_k^- \,) \,] \tag{2a}$$

$$= \sum_k [\, \vec{\underline{E}}_{tk} + \hat{z} \underline{E}_{zk} \,] \tag{2α}$$

$$\vec{\underline{H}} = \sum_k [\, \vec{h}_{tk} \, (\, \underline{E}_k^+ - \underline{E}_k^- \,) + \hat{z} h_{zk} \, (\, \underline{E}_k^+ + \underline{E}_k^- \,) \,] \tag{2b}$$

$$= \sum_k [\, \vec{\underline{H}}_{tk} + \hat{z} \underline{H}_{zk} \,] \tag{2β}$$

$$\vec{\underline{E}}_k^\pm = E_k^\pm e^{j\theta_k^\pm} \qquad E_k^\pm = \text{scalar factor} \tag{2c}$$

$$\theta_k^\pm = \omega t \mp \gamma_{zk} z \tag{2d}$$

$$\gamma_{zk} = \beta_{zk} - j\alpha_{zk} \tag{2e}$$

The substitution of (2α) and (2β) into (1) yields

$$< \vec{\underline{S}} > = \frac{1}{2} \, \text{Re} \, [\, \{ \sum_k (\vec{\underline{E}}_{tk} + \hat{z} \underline{E}_{zk}) \} \times \{ \sum_j (\vec{\underline{H}}_{tj}^* + \hat{z} \underline{H}_{zj}^*) \} \,]$$

$$= \frac{1}{2} \, \text{Re} \, [\, \sum_k \sum_j (\vec{\underline{E}}_{tk} \times \vec{\underline{H}}_{tj}^* + \vec{\underline{E}}_{tk} \times \hat{z} \, \underline{H}_{zj}^* + \hat{z} \underline{E}_{zk} \times \vec{\underline{H}}_{tj}^*) \,] \tag{3}$$

The component of $< \vec{\underline{S}} >$ in the longitudinal direction, $\hat{z}$, of the guide is

$$< \underline{S}_z > = \hat{z} \cdot < \vec{\underline{S}} > = \frac{1}{2} \, \text{Re} \, [\, \sum_k \sum_j \vec{\underline{E}}_{tk} \times \vec{\underline{H}}_{tj}^* \,] \tag{4}$$

The total power transfer along the guide is the integral of $< \underline{S}_z >$ over the cross-sectional area S_c of the guide.

$$P = \frac{1}{2} \operatorname{Re} \int_{S_C} \sum_k \sum_j \hat{z} \cdot (\vec{e}_{tk} \times \vec{h}_{tj}^*) \, F_{kj} \, da$$

$$= \frac{1}{2} \operatorname{Re} [\sum_k \sum_j F_{kj} \int_{S_C} \vec{e}_{tk} \times \vec{h}_{tj}^* \cdot \hat{z} \, da \tag{5a}$$

$$F_{kj} = (\underline{E}_k^+ + \underline{E}_k^-)(\underline{E}_j^{+*} - \underline{E}_j^{-*}) \tag{5b}$$

The surface integral can be transformed by (1-33b).

$$\int_{S_C} \vec{e}_{tk} \times \vec{h}_{tj}^* \cdot \hat{z} \, da = \frac{1}{Z_{T_k}^*} \int_{S_C} \vec{e}_{tk} \cdot \vec{e}_{tj}^* \, da = \frac{1}{Z_{T_k}^*} \qquad \text{for } j = k \tag{1-33b}$$

for normalized orthogonal mode functions. Therefore, (5a) becomes

$$P = \frac{1}{2} \operatorname{Re} \sum_k \frac{F_{kk}}{Z_{T_k}^*} = \frac{1}{2} \operatorname{Re} \sum_k P_k \tag{6a}$$

$$F_{kk} = (\underline{E}_k^+ + \underline{E}_k^-)(\underline{E}_k^{+*} - \underline{E}_k^{-*}) = |\underline{E}_k^+|^2 - |\underline{E}_k^-|^2 + 2j \operatorname{Im} [\underline{E}_k^- \underline{E}_k^{+*}]$$

$$= |E_k^+|^2 e^{j(\theta_k^+ - \theta_k^{+*})} - |E_k^-|^2 e^{j(\theta_k^- - \theta_k^{-*})} + 2j \operatorname{Im} [E_k^- E_k^{+*} e^{j(\theta_k^- - \theta_k^{+*})}]$$

$$= |E_k^+|^2 \left[e^{-2\alpha_{zk}z} - \frac{|E_k^-|^2}{|E_k^+|^2} e^{2\alpha_{zk}z} + 2j \operatorname{Im} \left[\frac{E_k^-}{E_k^+} e^{j2\beta_{zk}z} \right] \right]$$

$$= |E_k^+|^2 [e^{-2\alpha_{zk}z} - |\Gamma_k|^2 e^{2\alpha_{zk}z} + 2j |\Gamma_k| \sin(2\beta_{zk}z + \phi_k)] \tag{6b}$$

$$\Gamma_k = \frac{E_k^-}{E_k^+} = |\Gamma_k| e^{j\phi_k} \tag{6c}$$

$$P_k = \frac{F_{kk}}{Z_{T_k}^*} = \frac{1}{Z_{T_k}^*} [p_k^+ - p_k^- + j p_{kr}] = P_k^+ - P_k^- + j P_{kr} \tag{6d}$$

$$P_k^\pm = \frac{p_k^\pm}{Z_{T_k}^*} = \text{power of } \pm z \text{ traveling waves} \tag{6e}$$

$$p_k^{\pm} = |E_k^{\pm}|^2 e^{\mp 2\alpha_{zk}z} \tag{6eα}$$

$$P_{kr} = \frac{2|\Gamma_k| \, |E_k^+|^2}{Z_{T_k^*}} \sin(2\beta_{zk}z + \phi_k) = \text{reactive voltage-amperes} \tag{6f}$$

Equation (6a) indicates that the power flows of separate modes are independent of each other on account of orthogonal properties between modes. Hence, the power flow of various modes can be treated separately and then summed up to get the total result.

Let the wave admittance be represented by

$$Y_{T_k}^* = \frac{1}{Z_{T_k}^*} = G_k - jB_k \tag{7}$$

Then (6d) becomes

$$P_k = (G_k - jB_k)[\, p_k^+ - p_k^- + jp_{kr}\,]$$

$$= [\, G_k(\, p_k^+ - p_k^-\,) + B_k p_{kr}\,] - j[\, B_k(\, p_k^+ - p_k^-\,) - G_k p_{kr}\,] \tag{8}$$

and the substitution of (8) into (6a) yields

$$P = \frac{1}{2} \sum_k [\, G_k(\, p_k^+ - p_k^-\,) + B_k p_{kr}\,] \tag{9}$$

(b) Above cutoff in a lossless medium, the wave impedance Z_{T_k} is purely real and the attenuation constant α_{zk} is zero. Then (9) becomes

$$P = \frac{1}{2} \sum_k G_k(\, p_k^+ - p_k^-\,) = \frac{1}{2} \sum_k G_k(\, |E_k^+|^2 - |E_k^-|^2\,) \tag{10}$$

The net power flow is the difference between the power of the positive z traveling wave and that of the negative z traveling wave. It is independent of the longitudinal position z.

(c) Below cutoff in a lossless medium, the propagation and the wave impedance are purely imaginary.

$$\gamma_{zk} = \beta_{zk} - j\alpha_{zk} = -j\alpha_{zk} \tag{11a}$$

$$Z_{T_k} = \frac{1}{G_k - jB_k} = \frac{1}{-jB_k} \tag{11b}$$

The power flow, (9), becomes

$$P = \frac{1}{2} \sum_k B_k P_{kr} = \sum_k B_k |E_k^+|^2 |\Gamma_k| \sin \phi_k \tag{12}$$

The power flow is again independent of the position coordinate z.

Beyond cutoff, a purely attenuated wave carries no power in a lossless medium. This is because at $z = \infty$, the amplitude of the wave is attenuated to zero; consequently, no power can exist there. Since the medium is lossless, all power that passes through any surface transverse to the direction of propagation must flow across any other such surface at a different location. Hence, there exists no net power flow along the direction of propagation anywhere.

However, when both a positive attenuated disturbance $e^{\alpha_{zk} z}$ and a negative attenuated disturbance $e^{-\alpha_{zk} z}$ exist simultaneously, then one must only consider finite values of z. This is because one of these disturbances would always be large for large values of z. There is no location where the field can be zero and hence power flow exists everywhere.

In a waveguide beyond cutoff, one will find power flow exists within a short interval along the axial direction. The amount of flow decreases exponentially as the length of the guide increases. This is a similar phenomenon to total internal reflection. The total internally reflected wave in the denser medium is accompanied by an exponentially damped disturbance in the rarer medium. When the rarer medium extends to infinity, the exponential wave carries no energy and all energy is reflected.

If, on the other hand, the thin rare medium is followed by another dense medium, then a positive and a negative exponential disturbance exist in the rare medium and power will be transmitted through the rare medium to the third medium.

(d) In a practical medium, the wave impedance can be approximated to be real and the propagation coefficient has a small imaginary component.

$$Z_{T_k} \approx \frac{1}{G_k} \tag{13a}$$

$$\gamma_{zk} = \beta_{zk} - j\alpha_{zk} \tag{13b}$$

The power flow in such a practical medium is

$$P = \frac{1}{2} \sum_k G_k (p_k^+ - p_k^-) = \frac{1}{2} \sum_k G_k |E_k^+|^2 (e^{-2\alpha_{zk} z} - |\Gamma_k|^2 e^{2\alpha_{zk} z}) \tag{14}$$

The power flow is the difference between the power in the oppositely traveling waves at the location. This power flow is a function of z along the direction of propagation.

7. Attenuation in Lossy Dielectric

The attenuation coefficient can be evaluated from the decrease of power flow per unit distance along the direction of flow. The Poynting theorem states that the rate of energy inflow is equal to the sum of: (i) the rate of Joulean dissipation, (ii) the rate of change of electromagnetic energy, and (iii) the rate of energy density created, within the volume.

$$-\oint_S \underline{\vec{S}} \cdot \hat{n} \, da = \int_V \underline{\vec{E}} \cdot \underline{\vec{J}} \, dv + \frac{\partial}{\partial t} \int_V \underline{w} \, dv \qquad \text{(II.10–6)} \qquad (1a)$$

$$\underline{w} = \underline{w}_e + \underline{w}_m = \frac{1}{2} \left(\epsilon \underline{E}^2 + \mu \underline{H}^2 \right) \qquad (1b)$$

The time-average expression of (1a) is

$$-\oint_S <\underline{\vec{S}}> \cdot \hat{n} \, da = \int_V <\underline{\vec{E}} \cdot \underline{\vec{J}}> dv + \frac{\partial}{\partial t} \int_V <\underline{w}> dv \qquad (1c)$$

As shown by (6-6a), the integral of energy density over the cross section contains no cross product terms by virtue of orthogonality of the mode functions. Thus, one can consider each mode separately and sum up the results at the end.

For fields varying as $e^{j\omega t}$ expressed as the real part of the complex exponential, the time-average energy density is

$$<w> = \frac{1}{4} \, \text{Re} \, [\, \epsilon \underline{\vec{E}} \cdot \underline{\vec{E}}^* + \mu \underline{\vec{H}} \cdot \underline{\vec{H}}^* \,] \qquad (2)$$

The time-average value of the electromagnetic energy density, $\underline{w}$, is independent of time and disappears from (1c). Hence, the Poynting theorem for this case is

$$-\oint_S <\underline{\vec{S}}> \cdot \hat{n} \, da = \frac{1}{2} \, \text{Re} \int_V \sigma_d \underline{\vec{E}} \cdot \underline{\vec{E}}^* \, dv \qquad \text{in region excluding sources} \qquad (3a)$$

$$<\underline{\vec{E}} \cdot \underline{\vec{J}}> = \frac{1}{2} \, \text{Re} \, [\, \underline{\vec{E}} \cdot \underline{\vec{J}}^* \,] = \frac{1}{2} \, \text{Re} \, [\, \sigma_d \underline{\vec{E}} \cdot \underline{\vec{E}}^* \,] \qquad (3b)$$

where σ_d is the conductivity of the dielectric medium filling the guide.

(a) Apply (3) to a short length Δz of guide with uniform cross section S_c (Fig. 7-1). The guide is made of a perfect conductor. The closed surface integral of (3) is carried over a surface which is composed of $S_1 = S_c$ at $z = z_0$, $S_2 = S_c$ at $z = z_0 + \Delta z$, and the surface of the guide wall between z_0 and $z_0 + \Delta z$. Since the guide is made of a perfect conductor, there will be no contribution from the wall surface.

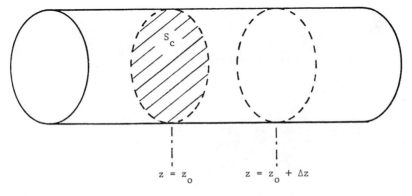

Figure 7-1: A short section of waveguide

$$-\oint_{S} <\vec{\underline{S}}> \cdot \hat{n} \, da = -\int_{S_1} <\vec{\underline{S}}> \cdot (-\hat{z}) \, da - \int_{S_2} <\vec{\underline{S}}> \cdot \hat{z} \, da \equiv P_1 - P_2$$

$$= -\frac{P_2 - P_1}{\Delta z} \Delta z = -\frac{\partial P}{\partial z} \Delta z \tag{4a}$$

The right-hand side of (3a) is

$$\frac{1}{2} \, \mathrm{Re} \int_{V} \sigma_d \vec{\underline{E}} \cdot \vec{\underline{E}}^* \, dv = \frac{1}{2} \, \mathrm{Re} \int_{z_o}^{z_o+\Delta z} dz \int_{S_c} \sigma_d \vec{\underline{E}} \cdot \vec{\underline{E}}^* \, da = \frac{\Delta z}{2} \, \mathrm{Re} \int_{S_c} \sigma_d \vec{\underline{E}} \cdot \vec{\underline{E}}^* \, da \tag{4b}$$

With (4), (3a) becomes

$$-\frac{\partial P}{\partial z} = \frac{1}{2} \, \mathrm{Re} \int_{S_c} \sigma_d \vec{\underline{E}} \cdot \vec{\underline{E}}^* \, da \tag{5}$$

For TE waves traveling in the positive z-direction, one has

$$\vec{\underline{E}} = \sum_{k} \vec{e}_{tk} E_k e^{j\theta_k} \tag{6a}$$

$$\vec{\underline{H}} = \sum_{k} (\vec{h}_{tk} + \hat{z} h_{zk}) E_k e^{j\theta_k} \tag{6b}$$

$$E_k \equiv E_k^+ \qquad \text{and} \qquad \theta_k \equiv \theta_k^+ = \omega t - \gamma_{zk} z \tag{6c}$$

$$\gamma_{zk} = \beta_{zk} - j\alpha_{zk} = \sqrt{\gamma_o^2 - \gamma_{kc}^2} \,, \qquad \gamma_o^2 = \omega^2 \mu \underline{\varepsilon} \,, \qquad \gamma_{ck}^2 = \omega_{ck}^2 \mu \underline{\varepsilon} \tag{6d}$$

$$\vec{e}_{tk} = Z_{TE_k} \hat{z} \times \vec{h}_{tk} \qquad \text{and} \qquad Z_{TE_k} = \frac{\omega \mu}{-\gamma_{zk}} \tag{6e}$$

where the superscripts + have been omitted. Only the positive traveling wave will be used in the present analysis since the addition of the reflected wave will complicate the results and will help little in the understanding of the situation.

The flow of Poynting flux through any cross-sectional area S_c is

$$P = -\int_{S_c} <\vec{\underline{S}}> \cdot \hat{n} \, da = \frac{1}{2} \, \mathrm{Re} \, [\sum_{k} \sum_{j} F_{kj} \int_{S_c} \vec{e}_{tk} \times \vec{h}_{tk}^* \cdot \hat{z} \, da \,]$$

$$= \frac{1}{2} \, \mathrm{Re} \sum_{k} \frac{F_{kk}}{Z_{T_k}^*} = \frac{1}{2} \, \mathrm{Re} \left[\sum_{k} \frac{|E_k|^2}{Z_{TE_k}^*} e^{-2\alpha_{zk} z} \right] \tag{7a}$$

$$F_{kk} = |E_k|^2 e^{j(\theta_k - \theta_k^*)} = |E_k|^2 e^{-2\alpha_{zk} z} \tag{7b}$$

The left-hand side of (5) is then obtained by the differentiation of (7a)

$$-\frac{\partial P}{\partial z} = \mathrm{Re} \sum_k \frac{\alpha_{zk} |E_k|^2}{Z_{TE_k}^*} e^{-2\alpha_{zk}z} \tag{8}$$

The right-hand side of (5) is

$$\frac{1}{2} \mathrm{Re} \int_{S_C} \sigma_d \vec{\underline{E}} \cdot \vec{\underline{E}}^* \, da = \frac{1}{2} \mathrm{Re} \sum_k \sigma_d |E_k|^2 e^{-2\alpha_{zk}z} \tag{9}$$

The substitution of (8) and (9) into (5) yields

$$\frac{\alpha_{zk}}{Z_{TE_k}^*} = \frac{\sigma_d}{2} \qquad \text{or} \qquad \alpha_{zk} = \frac{\sigma_d Z_{TE_k}^*}{2} \tag{10}$$

This is the attenuation constant of the TE waves due to the losses in the dielectric medium filling the guide.

8. Relation Between Energy Densities

The time-average energy densities are

$$< w_e > = \frac{1}{4} \text{Re} [\, \varepsilon \underline{\vec{E}} \cdot \underline{\vec{E}}^* \,] \tag{1a}$$

$$< w_m > = \frac{1}{4} \text{Re} [\, \mu \underline{\vec{H}} \cdot \underline{\vec{H}}^* \,] \tag{1b}$$

For a TE wave traveling in the positive z-direction [Eq. (7-6)], the total electric and magnetic energies within the volume of short length, Δz, of guide (Fig. 7-1) is

$$< \Delta w_e > = \int_V < w_e > dv = \frac{1}{4} \text{Re} \int_{z_o}^{z_o + \Delta z} dz \int_{S_C} \varepsilon \sum_k \vec{e}_{tk} \cdot \vec{e}_{tk}^* F_{kk} \, da$$

$$= \frac{1}{4} \Delta z \, \varepsilon \sum_k |E_k|^2 \, e^{-2\alpha_{zk} z} \tag{2a}$$

$$< \Delta w_m > = \int_V < w_m > dv$$

$$= \frac{1}{4} \text{Re} \int_{z_o}^{z_o + \Delta z} dz \int_{S_C} \mu \sum_k (\vec{h}_{tk} \cdot \vec{h}_{tk}^* + h_{zk} h_{zk}^*) F_{kk} \, da$$

$$= \frac{\mu \Delta z}{4} \text{Re} \sum_k \frac{1}{Z_{TE_k} Z_{TE_k}^*} \left[1 + \frac{\gamma_{ck}^2}{\gamma_{zk} \gamma_{zk}^*} \right] \int_{S_C} \vec{e}_{tk} \cdot \vec{e}_{tk}^* F_{kk} \, da \tag{2b}$$

where F_{kk} is given by (6-6b) and

$$\int_S h_{zk} h_{zk}^* \, da = \frac{\gamma_{ck}^2}{\gamma_{zk} \gamma_{zk}^*} \int_S \vec{h}_{tk} \cdot \vec{h}_{tk}^* \, da \tag{1-27}$$

$$\int_S \vec{h}_{tk} \cdot \vec{h}_{tk}^* \, da = \frac{1}{Z_{TE_k} Z_{TE_k}^*} \int_S \vec{e}_{tk} \cdot \vec{e}_{tk}^* \, da \tag{1-33a}$$

The wave impedance and various γ's are given in (7-6).

The ratio of energy densities for each mode is therefore

$$\frac{< \Delta w_{m_k} >}{< \Delta w_{e_k} >} = \frac{\mu}{\varepsilon} \frac{1}{Z_{TE_k} Z_{TE_k}^*} \left[1 + \frac{\gamma_{ck}^2}{\gamma_{zk} \gamma_{zk}^*} \right] = \frac{\gamma_{zk} \gamma_{zk}^*}{\omega^2 \mu \varepsilon} \left[1 + \frac{\gamma_{ck}^2}{\gamma_{zk} \gamma_{zk}^*} \right] \tag{3}$$

$$Z_{TE_k} = \frac{\mp \omega \mu}{\gamma_{zk}} \qquad \text{(II.4-6c)}$$

The propagation coefficient is

$$\gamma_{zk} = \sqrt{\omega^2 \mu \varepsilon \left[1 - j \frac{\sigma_d}{\omega \varepsilon} \right] - \gamma_{ck}^2} \qquad \text{(7-6d)}$$

$$= \sqrt{(\omega^2 \mu \varepsilon - \gamma_{ck}^2) - j \omega \mu \sigma_d} \qquad \text{(4a)}$$

$$\gamma_{zk} \gamma_{zk}^* = \sqrt{(\omega^2 \mu \varepsilon - \gamma_{ck}^2)^2 + (\omega \mu \sigma_d)^2}$$

$$= \omega^2 \mu \varepsilon \sqrt{\left[1 - \frac{\gamma_{ck}^2}{\omega^2 \mu \varepsilon} \right]^2 + \left[\frac{\sigma_d}{\omega \varepsilon} \right]^2} \qquad \text{(4b)}$$

The ratio of time-average energies then becomes

$$\frac{<\Delta w_{m_k}>}{<\Delta w_{e_k}>} = \sqrt{\left[1 - \frac{\gamma_{ck}^2}{\omega^2 \mu \varepsilon} \right]^2 + \left[\frac{\sigma_d}{\omega \varepsilon} \right]^2} + \frac{\gamma_{ck}^2}{\omega^2 \mu \varepsilon} \qquad \text{(5)}$$

(a) In a lossless medium, $\sigma_d = 0$, (5) becomes

$$\frac{<\Delta w_{m_k}>}{<\Delta w_{e_k}>} = 1 \qquad \text{(6)}$$

The time-average value of electric energy density is equal to that of the magnetic energy density in a lossless medium.

(b) In a good conducting medium, the term containing σ_d becomes much larger and (5) can be approximated as

$$\frac{<\Delta w_{m_k}>}{<\Delta w_{e_k}>} \approx \frac{\sigma_d}{\omega \varepsilon} \gg 1 \qquad \text{(7)}$$

The time-average value of the magnetic energy density is much larger than that of the electric energy density in a conducting medium.

9. Velocity of Energy Transport

The fields in a waveguide are represented by traveling waves. It is reasonable to interpret the energy flow as the energy density multiplied by a velocity factor.

It was shown in (8-6) that the electric energy density is equal to the magnetic energy density in a lossless medium. The total energy density for a positive traveling wave is

$$< w_k > \; = \; \frac{1}{2} \sum_k \varepsilon \, |E_k|^2 \qquad E_k \equiv E_k^+ \tag{1}$$

Let the power flow in a lossless medium be represented by

$$P \; = \; \sum_k < w_k > v_k \; = \; \frac{1}{2} \sum_k \varepsilon \, |E_k|^2 \, v_k \tag{2}$$

where v_k is a velocity factor to be determined. The general expression of the inflow of Poynting flux is given by (6-6). Specializing (6-6) for the positive z-traveling TE wave in a lossless medium, one has

$$P \; = \; \frac{1}{2} \sum_k \frac{1}{Z_{TE_k}} \, |E_k|^2 \tag{3a}$$

$$Z_{TE_k} \; = \; \frac{\omega\mu}{\gamma_{zk}} \tag{3b}$$

$$\gamma_{zk} \; = \; \sqrt{\omega^2\mu\varepsilon - \gamma_{ck}^2 - j\omega\mu\sigma_d} \; = \; \sqrt{\omega^2\mu\varepsilon - \gamma_{ck}^2} \tag{3c}$$

Equating (2) and (3a) yields

$$v_k \; = \; \frac{1}{\varepsilon Z_{TE_k}} \; = \; \frac{\gamma_{zk}}{\omega\mu\varepsilon} \tag{4a}$$

$$= \; \frac{1}{\sqrt{\mu\varepsilon}} \sqrt{1 - \frac{\gamma_{ck}^2}{\omega^2\mu\varepsilon}} \tag{4b}$$

The phase velocity of the kth mode is

$$v_{pk} \; = \; \left[\frac{\omega}{\gamma_{zk}} \right]_{\alpha_{zk}=0} \; = \; \frac{\dfrac{1}{\sqrt{\mu\varepsilon}}}{\sqrt{1 - \dfrac{\gamma_{ck}^2}{\omega^2\mu\varepsilon}}} \tag{5}$$

The geometrical mean of the phase velocity and the velocity factor v_k is the velocity of propagation of a plane wave.

The group velocity of the kth mode is given by

$$v_{gk} = \frac{\partial \omega}{\partial \beta_{zk}} = \lim_{\Delta\omega \to 0} \left[\frac{\Delta\beta_{zk}}{\Delta\omega} \right]^{-1} = \left[\frac{\partial \beta_{zk}}{\partial \omega} \right]^{-1} \qquad \text{(II.7-10)}$$

$$\frac{\partial \beta_{zk}}{\partial \omega} = \frac{\partial}{\partial \omega} \sqrt{\omega^2 \mu\varepsilon - \gamma_{ck}^2} = \frac{\sqrt{\mu\varepsilon}}{\sqrt{1 - \dfrac{\gamma_{ck}^2}{\omega^2 \mu\varepsilon}}}$$

or

$$v_{gk} = \frac{1}{\sqrt{\mu\varepsilon}} \sqrt{1 - \frac{\gamma_{ck}^2}{\omega^2 \mu\varepsilon}} \qquad (6)$$

This is identical to (4b), or the group velocity is the velocity of energy transport.

10. Complex Frequency

In transient problems, fields vary sinusoidally as well as exponentially, increasing or decaying, in the time domain. Such cases can be easily analyzed by means of complex frequencies.

(a) Series RLC-Circuit

The differential equation for a series RLC-circuit, Fig. 10-1, is

$$L \frac{di}{dt} + Ri + \frac{1}{C} \int i\, dt = v(t) \tag{1}$$

The natural response of this circuit is the solution to the corresponding homogeneous equation.

$$L \frac{di}{dt} + Ri + \frac{1}{C} \int i\, dt = 0 \tag{2}$$

Let the trial solution be

$$i(t) = I e^{j\omega t} \tag{3}$$

where I and ω are some constants to be determined. The substitution of (3) into (2) gives

$$\left[j\omega L + R + \frac{1}{j\omega C} \right] I e^{j\omega t} = 0$$

or

$$\omega^2 - \frac{jR}{L}\omega - \frac{1}{LC} = 0 \tag{4}$$

This is the characteristic equation of this circuit. In circuit theory, it is conventional to define

$$\omega_o = \frac{1}{\sqrt{LC}} \qquad\qquad Q_s = \frac{\omega_o L}{R} = Q \text{ for series RLC--circuit} \tag{5}$$

Then (4) becomes

$$\omega^2 - \frac{jR}{L}\omega - \omega_o^2 = 0 \qquad \text{or} \qquad \omega^2 - \frac{jR}{\omega_o L}\omega_o\omega - \omega_o^2 = 0$$

$$\omega^2 - \frac{j\omega_o}{Q_s}\omega - \omega_o^2 = 0 \tag{4b}$$

and the characteristic roots are

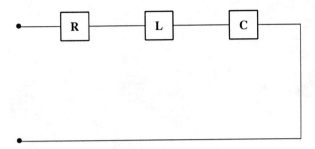

Figure 10-1: A series RLC circuit

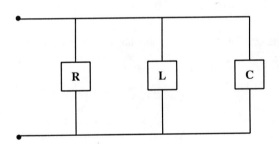

Figure 10-2: A parallel RLC circuit

$$\omega = \pm\omega_o \sqrt{1 - \frac{1}{(2Q_s)^2}} + \frac{j\omega_o}{2Q_s} \equiv \omega_1 + j\omega_2 \tag{6a}$$

$$\omega_1 = \pm\omega_o \sqrt{1 - \frac{1}{(2Q_s)^2}} \tag{6b}$$

$$\omega_2 = \frac{\omega_o}{2Q_s} \tag{6c}$$

$$\omega_o^2 = \omega_1^2 + \omega_2^2 \tag{6d}$$

$$Q_s = \frac{\omega_o}{2\omega_2} = \frac{[\omega_1^2 + \omega_2^2]^{\frac{1}{2}}}{2\omega_2} = \frac{1}{2}\left[\left(\frac{\omega_1}{\omega_2}\right)^2 + 1\right]^{\frac{1}{2}} \tag{6e}$$

If ω_2 is small compared to ω_1, or if Q_s is large compared to unity, then $\omega_o \approx \omega_1$ and $Q_s \approx \omega_1/(2\omega_2)$.

(b) Parallel RLC-Circuit

The differential equation for a parallel RLC-circuit, Fig. 10-2, is

$$C\frac{dv}{dt} + \frac{v}{R} + \frac{1}{L}\int v\,dt = i(t) \tag{7}$$

The natural response satisfies the corresponding homogeneous equation.

$$C\frac{dv}{dt} + \frac{v}{R} + \frac{1}{L}\int v\,dt = 0 \tag{8}$$

Let the trial solution be

$$v(t) = Ve^{j\omega t} \tag{9}$$

The characteristic equation is

$$\omega^2 - \frac{j\omega_o}{Q_p}\omega - \omega_o^2 = 0 \qquad Q_p = \omega_o RC, \qquad \omega_o = \frac{1}{\sqrt{LC}} \tag{10}$$

where Q_p is the Q factor for the parallel circuit. The characteristic roots are

$$\omega = \omega_1 + j\omega_2 \tag{11a}$$

$$\omega_1 = \omega_o \sqrt{1 - \frac{1}{(2Q_p)^2}} \tag{11b}$$

$$\omega_2 = \frac{\omega_o}{2Q_p} \tag{11c}$$

$$\omega_o^2 = \omega_1^2 + \omega_2^2 \tag{11d}$$

$$Q_p = \frac{\omega_o}{2\omega_2} = \frac{1}{2}\left[\left[\frac{\omega_1}{\omega_2}\right]^2 + 1\right]^{\frac{1}{2}} \tag{11e}$$

While the definitions for Q_s and Q_p in terms of circuit parameters are different in (5) and (10), their definitions in terms of the real and imaginary parts of the frequency are the same for both. It is more useful to define the Q factor in the latter way in problems at microwave frequencies.

(c) Physical Significance of Q

The physical significance of Q can be ascertained from the energy relation. The voltage, (9), is given by

$$v(t) = Ve^{j(\omega_1 + j\omega_2)t} = Ve^{-\omega_2 t + j\omega_1 t} \tag{12a}$$

The current in the resistor is

$$i_R = \frac{v(t)}{R} = \frac{V}{R}e^{-\omega_2 t + j\omega_1 t} \tag{12b}$$

The power dissipated in the resistor is

$$P = \frac{1}{2}\text{Re}\,[\,v(t)i_R^*(t)\,] = \frac{1}{2}\frac{V^2}{R}e^{-2\omega_2 t} = P_o\,e^{-\frac{\omega_o}{Q_p}t} \tag{12c}$$

$$P_o = \frac{1}{2}\frac{V^2}{R} \quad\text{and}\quad \omega_2 = \frac{\omega_o}{2Q_p} \qquad \text{for parallel RLC-circuit} \tag{12d}$$

The decrease of energy per unit time is

$$-\frac{dP}{dt} = \frac{\omega_o}{Q_p}P$$

or

$$Q_p = \frac{\omega_o P}{-\dfrac{dP}{dt}} = \frac{\omega_o \times \text{total energy}}{\text{decrease of energy per second}} \tag{13}$$

Since a series circuit can be transformed into an equivalent parallel circuit, and vice versa, the above result is valid for either a parallel or a series circuit; hence, the subscript on Q may be dropped.

11. Propagation with Complex Frequency

The propagation coefficient is given by

$$\gamma_{zk} \equiv \sqrt{\omega^2\mu \left[\varepsilon - \frac{j\sigma_d}{\omega} \right] - \gamma_{ck}^2} \tag{8-4a}$$

Then

$$\gamma_{zk}^2 = \omega^2\mu\varepsilon - \gamma_{ck}^2 - j\omega\mu\sigma_d \tag{1}$$

Let the complex frequency be

$$\omega = \omega_1 + j\omega_2 \tag{2}$$

One then has

$$\gamma_{zk}^2 = (\omega_1 + j\omega_2)^2 \mu\varepsilon - \gamma_{ck}^2 - j(\omega_1 + j\omega_2)\mu\sigma_d$$

$$= [(\omega_1^2 - \omega_2^2)\mu\varepsilon + \omega_2\mu\sigma_d - \gamma_{ck}^2] + j\omega_1\mu(2\omega_2\varepsilon - \sigma_d)$$

or the propagation coefficient in terms of the complex frequency is

$$\gamma_{zk} = \sqrt{(\omega_1^2 - \omega_2^2)\mu\varepsilon + \omega_2\mu\sigma_d - \gamma_{ck}^2 + j\omega_1\mu(2\omega_2\varepsilon - \sigma_d)} \tag{3}$$

(a) Attenuation in the space domain: When $\omega_2 = 0$ or $\omega = \omega_1$, that is, in the case of a real frequency, (3) takes the form

$$\gamma_{zk} = \sqrt{\omega_1^2\mu\varepsilon - \gamma_{ck}^2 - j\omega_1\mu\sigma_d} = \beta_{zk} - j\alpha_{zk} \tag{4}$$

The propagation coefficient is, in general, complex; both propagation and attenuation exist in the direction of propagation. However, there is no damping as a function of time.

(b) Attenuation in the time-domain: When $\omega_2 = \sigma_d/2\varepsilon$, this makes the imaginary term in (3) disappear; then

$$\gamma_{zk} = \sqrt{(\omega_1^2 - \omega_2^2)\mu\varepsilon + \omega_2\mu\sigma_d - \gamma_{ck}^2} \tag{5}$$

If the frequency ω_1 is above cutoff, then there will be propagation without attenuation in the direction of propagation. However, the wave amplitude will be attenuated exponentially with time as $e^{-\omega_2 t}$ since energy is absorbed by the medium.

For the low-loss case, the Q factor is

$$Q \approx \frac{\omega_1}{2\omega_2} \tag{10-6e}$$

$$= \frac{\omega_1}{2[\,\sigma_d/(\,2\varepsilon\,)\,]} \qquad \text{since} \qquad \omega_2 = \frac{\sigma_d}{2\varepsilon} \tag{6}$$

These are two extreme cases; the intermediate cases, where waves are attenuated both in time and space, can best be demonstrated by way of the Poynting theorem.

When the time-average electromagnetic energy density at any location is independent of time, the Poynting theorem is expressed as (7-5), which implies that the decrease of time-average power flow in the direction of propagation is equal to the time-average Joulean dissipation.

In the general case of complex frequency, where the energy density can vary with time, the total decrease of power flow should include both the change of energy density as a function of position and as a function of time. The Poynting theorem in the complex-frequency domain is therefore a modified version of (7-1c).

$$\frac{\partial}{\partial z} \int_{S_c} <\vec{\underline{S}}> \cdot \hat{z} \; da \; + \; \frac{\partial}{\partial t} \int_{S_c} <\underline{w}> da \; = \; -\frac{1}{2} \, \text{Re} \int_{S_c} \sigma_d \, \vec{\underline{E}} \cdot \vec{\underline{E}}^* \; da \tag{7a}$$

where (Fig. 11-1)

$$-\oint_S <\vec{\underline{S}}> \cdot \hat{n} \; da \; = \; -\left[\int_{S_1} <\vec{\underline{S}}> \cdot (-\hat{z}) \; da \; + \; \int_{S_2} <\vec{\underline{S}}> \cdot \hat{z} \; da \right]$$

$$= \; P_1 - P_2 \; = \; -\frac{P_2 - P_1}{\Delta z} \, \Delta z$$

$$= \; -\frac{\partial P}{\partial z} \, \Delta z \; = \; -\Delta z \, \frac{\partial}{\partial z} \int_{S_c} <\vec{\underline{S}}> \cdot \hat{z} \; da \tag{7b}$$

To provide a better interpretation, Eq. (7a) is arranged in terms of a common integral,

$$\int_s <\underline{w}> da$$

Since for any mode, it has been shown [Eq. (9-2)] that

$$P \; = \; <\underline{S}_z> \; = \; <\underline{w}> v_g \tag{6-2}$$

and in a lossless medium, $<\underline{w}_e> \; = \; <\underline{w}_m>$ [Eq. (8-6)], therefore

$$<\underline{w}> \; = \; <\underline{w}_e> \; + \; <\underline{w}_m> \; = \; 2 <\underline{w}_e> \; = \; \frac{1}{2} \, \varepsilon \, \text{Re} \, [\, \vec{\underline{E}} \cdot \vec{\underline{E}}^* \,]$$

or

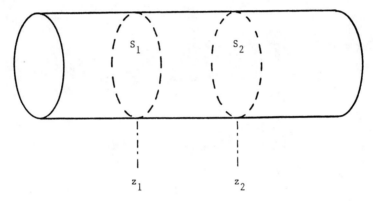

Figure 11-1: A section of waveguide

$$\text{Re} \, [\, \vec{\underline{E}} \cdot \vec{\underline{E}}^* \,] \; = \; \frac{2}{\varepsilon} < \underline{w} >$$

Equation (7a) can be expressed in terms of $< \underline{w} >$.

$$v_g \frac{\partial}{\partial z} \int_{S_c} < \underline{w} > da \; + \; \frac{\partial}{\partial t} \int_{S_c} < \underline{w} > da \; = \; - \frac{\sigma_d}{\varepsilon} \int_{S_c} < \underline{w} > da \tag{8}$$

This is a general expression of the Poynting theorem. It states that the time rate of increase of energy is the inflow of energy diminished by the Joulean dissipation. The two extreme cases (a) and (b) will now be reexamined by means of Eq. (8).

(α) When $< \underline{w} >$ is independent of z and

$$< \underline{w} > \; = \; w_0(x, y) \, e^{-2\omega_2 t} \tag{9}$$

Eq. (8) then gives

$$-2\omega_2 \int_{S_c} < \underline{w} > da \; = \; - \frac{\sigma_d}{\varepsilon} \int_{S_c} < \underline{w} > da \qquad \text{or} \qquad \omega_2 \; = \; \frac{\sigma_d}{2\varepsilon} \tag{10}$$

which is identical to (6).

(β) When $< \underline{w} >$ is independent of t and

$$< \underline{w} > \; = \; w_0 e^{-2\alpha_z z} \tag{11}$$

then Eq. (8) yields

$$-2\alpha_z v_g \; = \; - \frac{\sigma_d}{\varepsilon} \qquad \text{or} \qquad \alpha_z \; = \; \frac{\sigma_d}{2 v_g \varepsilon} \tag{12a}$$

$$= \; \frac{\omega_1}{2 v_g Q} \tag{12b}$$

by virtue of (6). This expression for α_z can also be obtained from (3) by setting $\omega_2 = 0$; see Section 11.1.

(γ) For the general case,

$$< \underline{w} > \; = \; w_0 e^{-2\alpha z \, - \, 2\omega_2 t} \tag{13}$$

Then (8) gives

$$-2\alpha v_g \; - \; 2\omega_2 \; = \; - \frac{\sigma_d}{\varepsilon} \qquad \text{or} \qquad \alpha \; = \; \frac{\sigma_d}{2\varepsilon v_g} \; - \; \frac{\omega_2}{v_g} \tag{14}$$

This is a relation between the attenuation coefficient in the z-domain and the damping constant in the time domain for the low-loss case.

11.1 Example: Determination of Attenuation Coefficient - Real Frequency Case.

Determine the attenuation coefficient from Eq. (11-4)

Solution :

For the case when the frequency is real, $\omega_2 = 0$, (11-3) is simplified to (11-4).

$$\gamma = \sqrt{\omega_1^2\mu\varepsilon - \gamma_{ck}^2 - j\omega_1\mu\sigma_d} = \beta + j\alpha \tag{1}$$

where the subscript z is omitted. Squaring both sides of (1) yields

$$\beta^2 - \alpha^2 + j2\alpha\beta = m - j2n \tag{2a}$$

$$m = \omega_1^2\mu\varepsilon - \gamma_{ck}^2 \quad \text{and} \quad n = \frac{1}{2}\omega_1\mu\sigma_d \tag{2b}$$

or

$$\beta^2 - \alpha^2 = m \tag{3a}$$

$$2\alpha\beta = -2n \quad \text{or} \quad \beta = \frac{-n}{\alpha} \tag{3b}$$

β can be eliminated from (3a) by means of (3b), and one obtains

$$\alpha^4 + m\alpha^2 - n^2 = 0 \tag{4}$$

The solution of (4) is

$$\alpha = \sqrt{\frac{-m}{2}}\sqrt{1 \pm \sqrt{1 + \frac{4n^2}{m^2}}} \tag{5}$$

The last term in the radical is

$$\frac{2n}{m} = \frac{\dfrac{\sigma_d}{\omega_1\varepsilon}}{1 - \dfrac{\gamma_{ck}^2}{\omega_1^2\mu\varepsilon}} \tag{6}$$

At frequencies away from cutoff, this is a small quantity for the low-loss case and (5) may be approximated by binomial expansion.

$$\alpha \approx \sqrt{\frac{-m}{2}}\sqrt{1 \pm \left[1 + \frac{1}{2}\frac{4n^2}{m^2}\right]} \tag{7}$$

The plus sign under the root yields

$$\alpha_+ = j \sqrt{\frac{m}{2}} \sqrt{2 + \frac{2n^2}{m^2}} \approx j\sqrt{m} \tag{8}$$

This result is not acceptable since α is defined to be real by (1).

When the negative sign in (7) is used, one has

$$\alpha = \sqrt{\frac{-m}{2}} \sqrt{\frac{-1}{2} \frac{4n^2}{m^2}} = \frac{n}{\sqrt{m}} = \frac{\frac{1}{2} \omega_1 \mu \sigma_d}{\sqrt{\omega_1^2 \mu \varepsilon - \gamma_{ck}^2}} = \frac{1}{v_g} \frac{\sigma_d}{2\varepsilon} \tag{9a}$$

$$v_g = \frac{1}{\sqrt{\mu \varepsilon}} \sqrt{1 - \frac{\gamma_{ck}^2}{\omega_1^2 \mu \varepsilon}} \tag{9b}$$

12. Losses in the Guide Wall

In all previous sections, a waveguide made of perfect conductor was considered. Losses in the dielectric filling the guide were investigated.

At microwave frequencies, waveguides are usually constructed of good conductors, and the attenuation is ordinarily very small. As a result, an approximate method is justified for the determination of the attenuation.

The Poynting theorem of (11-8) can be modified to the following form.

$$v_g \frac{\partial}{\partial z} \int_{S_c} <\underline{w}> da \; + \; \frac{\partial}{\partial t} \int_{S_c} <\underline{w}> da \; = \; -\frac{\omega}{Q} \int_{S_c} <\underline{w}> da \tag{1}$$

where the coefficient of the right-hand term is replaced by ω/Q since losses are proportional to the inverse of Q. This replacement of the coefficient is justified as follows:

(a) If $<\underline{w}>$ is independent of z, and

$$<\underline{w}> \; = \; w_o e^{-2\omega_2 t}$$

then (1) yields

$$-2\omega_2 \; = \; -\frac{\omega}{Q} \qquad \text{or} \qquad Q \approx \frac{\omega_1}{2\omega_2} \qquad \text{for } \omega_2 \ll \omega_1 \tag{2}$$

which is identical to (10-6e).

(b) If $<\underline{w}>$ is independent of t and

$$<\underline{w}> \; = \; w_o e^{-2\alpha_z z}$$

then

$$2\alpha_z v_g \; = \; \frac{\omega}{Q} \qquad \text{or} \qquad 2\alpha_z \; = \; \frac{\omega}{v_g Q} \qquad \omega \approx \omega_1 \tag{3}$$

This is (11-12b) but has more general implications since the Q in (3) accounts for losses in either the dielectric or the wall or both.

The problem of determining the attenuation coefficient is now reduced to that of evaluating Q. The definition of Q is

$$Q \; = \; \frac{\omega_o \times \text{total energy}}{\text{decrease of energy per second}} \tag{10-13}$$

The ohmic loss per unit length in the wall can be evaluated from the inflow of Poynting's flux into the wall. The tangential components of electromagnetic fields are related by the surface impedance of the conductor, and the latter is defined by

$$\vec{\underline{E}}_{tan} \; = \; Z_s \vec{\underline{I}}_s \tag{4a}$$

$$Z_s = (1 + j) \sqrt{\frac{\omega\mu}{2\sigma_c}} = (1 + j) \frac{1}{\sigma_c \delta_s} \tag{4b}$$

$$\delta_s = \sqrt{\frac{2}{\omega\mu\sigma_c}} = \text{skin depth} \tag{4c}$$

$\vec{E}_{tan}$ = tangential components of E–field on conductor surface

$\vec{J}_s = \hat{n} \times \vec{H}$ = surface current density

The time-average value of the ohmic losses, $< \underline{w}_L >$, in a guide of length Δz is given by

$$< \underline{w}_L > = \frac{1}{2} \text{Re} \int_{\text{wall}} -\hat{n} \cdot \underline{\vec{E}} \times \underline{\vec{H}}^* \, da = \frac{1}{2} \text{Re} \int_{z_o}^{z_o + \Delta z} dz \oint_C Z_s H_{tan}^2 \, dL$$

$$= \frac{\Delta z}{2} \text{Re} \oint_C Z_s H_{tan}^2 \, dL \tag{5}$$

where $\hat{n}$ is the outward normal of the wall and C is the contour bounding the cross section of the guide.

For the case of low loss, the total stored energy will be twice the magnetic energy [Eq. (8-6)]

$$< \underline{w} > = 2 < \underline{w}_m > = \frac{1}{2} \text{Re} \int_V [\mu \underline{\vec{H}} \cdot \underline{\vec{H}}^*] \, dv$$

$$= \frac{\mu \Delta z}{2} \int_{S_C} \text{Re} [\underline{\vec{H}} \cdot \underline{\vec{H}}^*] \, da \tag{6}$$

Thus, the Q_c factor for the conducting wall is

$$\frac{1}{Q_c} = \frac{< \underline{w}_L >}{\omega < \underline{w} >} = \frac{\text{Re} \oint_C Z_s \underline{\vec{H}}_{tan} \cdot \underline{\vec{H}}_{tan}^* \, dL}{\omega\mu \int_{S_C} \text{Re} [\underline{\vec{H}} \cdot \underline{\vec{H}}^*] \, da}$$

$$= \frac{\delta_s}{2} \frac{\oint_C \underline{\vec{H}}_{tan} \cdot \underline{\vec{H}}_{tan}^* \, dL}{\int_{S_C} \underline{\vec{H}} \cdot \underline{\vec{H}}^* \, da} \tag{7}$$

The total Q for a guide with losses in both the dielectric and the conducting wall is given by

$$\frac{1}{Q} = \frac{1}{Q_c} + \frac{1}{Q_d} \qquad\qquad (8a)$$

$$Q_d = \frac{\varepsilon\omega}{\sigma_d} \qquad (11\text{--}6) \qquad\qquad (8b)$$

    It is to be noted that this approximate method for the determination of Q is only good when losses are small and fields are not greatly disturbed by the presence of losses. This method is valid when only the direct wave (excluding the reflected wave) is considered. The power is therefore directly proportional to the square of the amplitude of the direct wave.

    When the reflected wave is present, the resultant wave is the sum of the direct and the reflected waves. Hence, the total losses from both waves are not the sum of the losses from the individual waves. The presence of the reflected wave complicates the problem, and there does not seem to be any simple technique that can be applied to handle such cases.

13. Problems

1. Two parallel-plate guides with different heights b and d are joined together as shown in Fig. 13-1. If an incident wave with a magnetic field H_{io} approaches guide 1 from the left, determine the total field in both guides.

Hint: Assume the field in each guide is as follows:

$$H_{y1} = H_{io}e^{-j\beta_1 z} + \sum_{m=1}^{\infty} H_{mr}e^{\gamma_{m1} z}$$

$$H_{y2} = H_{to}e^{-j\beta_2 z} + \sum_{m=1}^{\infty} H_{mt}e^{\gamma_{m2} z}$$

2. A cylindrical perfectly conducting post of radius r_o is located at a distance d from the side-wall of a rectangular guide, Fig. 13-2. The E-field of the incident wave is given by

$$\vec{E}(\vec{r}, t) = \hat{x}E_o \sin \frac{\pi y}{a} e^{j(\omega t - \beta z)} \qquad \beta = \sqrt{\omega^2 \mu \varepsilon - (\pi/a)^2}$$

To simplify the problem, consider a thin post (very small r_o) which may be represented by a line. Determine the approximate E-field in the guide.

3. The fields in an air-filled rectangular waveguide are

$$\vec{H} = H_o \left[\hat{z} \cos \frac{\pi x}{a} + \hat{x} \frac{j\beta_{10}}{k_{c10}^2} \frac{\pi}{a} \sin \frac{\pi x}{a} \right]$$

$$\vec{E} = \hat{y}Z_{h10}H_o \frac{j\beta_{10}}{k_{c10}^2} \sin \frac{\pi x}{a}$$

$$k_{cmn} = \sqrt{(m\pi/a)^2 + (n\pi/b)^2} \quad \text{and} \quad \gamma_{mn} = j\beta_{mn} = j\sqrt{k_o^2 - k_{cmn}^2}$$

$$Z_{hmn} = Z_o \frac{k_o}{\beta_{mn}} \qquad Z_o = \sqrt{\frac{\mu_o}{\varepsilon_o}} \qquad k_o = \sqrt{\mu_o\varepsilon_o}$$

The dimensions of the guide are a = 0.9 in. and b = 0.4 in.

Determine the rate of energy flow in the guide.

4. Find the power loss in the guide described in Problem 3, assuming the conductivity of the conducting wall is 6.1×10^7 mhos/m.

5. Determine the Q of the guide in Problem 4.

6. Show that

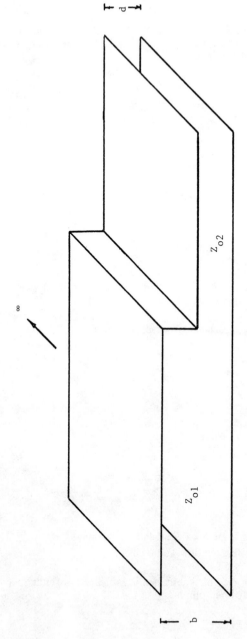

Figure 13-1: Parallel-plate lines in cascade

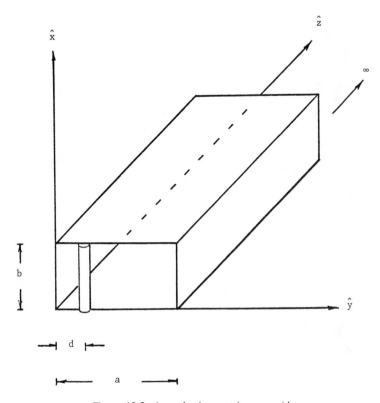

Figure 13-2: A conducting post in waveguide

$$\int_S \vec{h}_{tp} \cdot \vec{h}_{tq} \, da \ = \ 0 \qquad \text{for } p \ne q$$

where $\vec{h}_{tp}$ and $\vec{h}_{tq}$ are the transverse field components which belong to different waves, i.e., $\vec{h}_{tp}$ belongs to TE wave and $\vec{h}_{tq}$ belongs to TM wave or vice versa. Hints: By using Eqs. (1-3) and (1-9), the product

$$\vec{h}_{tp} \cdot \vec{h}_{tq} \ = \ c_o \nabla_t \cdot (\, e_{zq} \, \nabla_t \times \vec{h}_{zp} \,)$$

$$c_o \ = \ \frac{\gamma_{zp} \gamma_{zq}}{\gamma_p^2 \gamma_q^2 Z_{TM_q}}$$

$$\int_V \vec{h}_{tp} \cdot \vec{h}_{tq} \, dv \ = \ c_o \int_S e_{zq} \, \nabla_t \times \vec{h}_{zp} \cdot \hat{n} \, da \ = \ 0$$

Since $e_{zq} = 0$ on the cylindrical conducting wall, the contributions from both end surfaces cancel.

But

$$\int_V \vec{h}_{tp} \cdot \vec{h}_{tq} \, dv \ = \ \int_{L_1}^{L_1+L_o} dL \int_S \vec{h}_{tp} \cdot \vec{h}_{tq} \, da \ = \ 0$$

Therefore,

$$\int_S \vec{h}_{tp} \cdot \vec{h}_{tq} \, da \ = \ 0$$

which is the desired result to be proved.

CHAPTER XII   Resonant Cavity

1. Introduction

The properties of a general resonant cavity will be investigated in this chapter. It will be demonstrated that a waveguide terminated at both ends is a special case of a resonant cavity. The subjects examined in this and the next chapter are available in advanced textbooks such as Altman [C-1], Collin [C-2] and Slater [C-9].

A cavity is defined to be a region completely enclosed by a perfect conductor or a perfect insulator. A cavity enclosed by a practical conductor will also be investigated.

The object of this chapter is to investigate the properties of a resonant cavity. This is accomplished by expressing the electromagnetic fields inside the cavity in terms of orthogonal waveguide mode functions. This involves the following processes:

(a) Expansion of Maxwell's equations in terms of orthogonal mode functions with appropriate coefficients of expansion.

(b) Determination of the coefficients of expansion.

The analysis will cover various practical cases ranging from the ideal lossless case to cases with practical losses.

2. Normal-Mode Functions for the Cavity

The problem is to obtain solutions for Maxwell's equations within the cavity subject to appropriate boundary conditions. Potential function techniques will be employed. The electromagnetic fields in the cavity will be represented by scalar and vector potential functions. These potential functions should satisfiy certain boundary conditions consistent with those imposed upon the fields.

Orthogonal functions can be set up to satisfy two types of boundary conditions:

(i) Short-circuit boundary conditions - The tangential components of $\vec{E}$ and the normal component of $\vec{H}$ are required to be zero on the bounding surface of the short circuit, S_{sc}.

(ii) Open-circuit boundary conditions - The normal component of $\vec{E}$ and the tangential components of $\vec{H}$ are required to vanish on the surface of the open circuit, S_{oc}.

(a) In accordance with (XI.4-29), the $\underline{\vec{E}}$- and $\underline{\vec{H}}$-fields can be expressed as

$$\underline{\vec{E}} = \sum_k \left[\vec{U}_k \int_V \vec{U}_k \cdot \underline{\vec{E}} \, dv + \vec{V}_k \int_V \vec{V}_k \cdot \underline{\vec{E}} \, dv \right] \tag{1a}$$

$$\underline{\vec{H}} = \sum_k \vec{W}_k \int_V \vec{W}_k \cdot \underline{\vec{H}} \, dv \tag{1b}$$

$$\underline{\vec{J}} = \sum_k \left[\vec{U}_k \int_V \vec{U}_k \cdot \underline{\vec{J}} \, dv + \vec{V}_k \int_V \vec{V}_k \cdot \underline{\vec{J}} \, dv \right] \tag{1c}$$

$$\underline{\rho} = \sum_k \phi_k \int_V \phi_k \, \underline{\rho} \, dv \tag{1d}$$

$\underline{\vec{J}}$ can be expanded similarly to $\underline{\vec{E}}$ since $\underline{\vec{J}} = \sigma \underline{\vec{E}}$; $\underline{\rho}$ is expanded in terms of ϕ_k because $\underline{\rho} = \operatorname{div} \underline{\vec{D}}$.

(b) With

$$\nabla \times \underline{\vec{E}} = -\mu \frac{\partial \underline{\vec{H}}}{\partial t}$$

curl $\underline{\vec{E}}$ can be expanded in terms of $\vec{W}_k$.

$$\nabla \times \underline{\vec{E}} = \sum_k \vec{W}_k \int_V \vec{W}_k \cdot \nabla \times \underline{\vec{E}} \, dv \tag{2}$$

The integral may be evaluated by techniques of integration by parts. The integrand of (2) can be expressed as follows.

$$\nabla \cdot (\vec{\underline{E}} \times \nabla \times \vec{U}_k) = (\nabla \times \vec{\underline{E}}) \cdot (\nabla \times \vec{U}_k) - \vec{\underline{E}} \cdot \nabla \times \nabla \times \vec{U}_k$$

or

$$c_k^* \nabla \cdot (\vec{\underline{E}} \times \vec{W}_k) = c_k^* \vec{W}_k \cdot \nabla \times \vec{\underline{E}} - c_k^2 \vec{\underline{E}} \cdot \vec{U}_k \qquad (2\alpha)$$

since

$$\nabla \times \vec{U}_k = c_k^* \vec{W}_k \qquad \text{and} \qquad \nabla \times \vec{W}_k = c_k \vec{U}_k \qquad (XI.4\text{-}7)$$

The integral on the right-hand side of (2) can now be evaluated. With $c_k^* = -c_k$, one has

$$\int_V \vec{W}_k \cdot \nabla \times \vec{\underline{E}}\, dv = c_k \int_V \vec{\underline{E}} \cdot \vec{U}_k\, dv + \oint_S \vec{\underline{E}} \times \vec{W}_k \cdot \hat{n}\, da$$

$$= c_k \int_V \vec{\underline{E}} \cdot \vec{U}_k\, dv + \oint_S \hat{n} \times \vec{\underline{E}} \cdot \vec{W}_k\, da \qquad (3)$$

The closed surface integral is evaluated over the surface S, which is, in general, composed of S_{sc} and S_{oc}; and since

$$\vec{\underline{E}} \times \vec{W}_k \cdot \hat{n} = \vec{\underline{E}} \cdot \vec{W}_k \times \hat{n} = 0 \qquad \text{on } S_{oc}$$

it is therefore only necessary to evaluate the integral over S_{sc}. There are situations where $\hat{n} \times \vec{\underline{E}} \neq 0$ on S_{sc}, such as when the short circuit is not perfect; therefore, this integral is retained for generality. Thus,

$$\nabla \times \vec{\underline{E}} = \sum_k \vec{W}_k \left[c_k \int_V \vec{\underline{E}} \cdot \vec{U}_k\, dv + \int_{S_{sc}} \hat{n} \times \vec{\underline{E}} \cdot \vec{W}_k\, da \right] \qquad (4)$$

By analogous reasoning, curl $\vec{\underline{H}}$ is expanded in terms of $\vec{U}_k$ as

$$\nabla \times \vec{\underline{H}} = \sum_k \vec{U}_k \int_V \vec{U}_k \cdot \nabla \times \vec{\underline{H}}\, dv$$

$$= \sum_k \vec{U}_k \left[c_k^* \int_V \vec{\underline{H}} \cdot \vec{W}_k\, dv + \int_{S_{oc}} \hat{n} \times \vec{\underline{H}} \cdot \vec{U}_k\, da \right] \qquad (5)$$

The closed surface integral is evaluated over S_{oc} only since $\hat{n} \times \vec{U}_k = 0$ on S_{sc}.

(c) The term div $\underline{\vec{D}}$ can be expanded in terms of the scalar potential functions.

$$\nabla \cdot \underline{\vec{D}} = \sum_k \phi_k \int_V \phi_k \nabla \cdot \underline{\vec{D}} \, dv \tag{6}$$

By techniques of integration by parts, one obtains

$$\int_V \phi_k \nabla \cdot \underline{\vec{D}} \, dv = \int_V [\nabla \cdot (\phi_k \underline{\vec{D}}) - \underline{\vec{D}} \cdot \nabla \phi_k] \, dv$$

$$= \oint_S \phi_k \underline{\vec{D}} \cdot \hat{n} \, da - \int_V \underline{\vec{D}} \cdot \vec{V}_k d_k \, dv \tag{7}$$

by virtue of (XI.4-16a). Hence,

$$\nabla \cdot \underline{\vec{D}} = \sum_k \phi_k \left[-d_k \int_V \underline{\vec{D}} \cdot \vec{V}_k \, dv + \oint_S \phi_k \hat{n} \cdot \underline{\vec{D}} \, da \right] \qquad S = S_{oc} + S_{sc}$$

$$= \sum_k -\phi_k d_k \int_V \underline{\vec{D}} \cdot \vec{V}_k \, dv \tag{8}$$

since $\phi_k = 0$ on S_{oc} and S_{sc}, and thus the surface integral in (8) is equal to zero.

(d) Maxwell's equations for potential functions can now be obtained by using the above defined quantities.

$$\nabla \times \underline{\vec{E}} + \mu \frac{\partial \underline{\vec{H}}}{\partial t} = 0 \tag{9}$$

The substitution of (4) and (1b) into (9) yields

$$\sum_k \vec{W}_k \left[c_k \int_V \underline{\vec{E}} \cdot \vec{U}_k \, dv + \mu \frac{\partial}{\partial t} \int_V \underline{\vec{H}} \cdot \vec{W}_k \, dv + \int_{S_{sc}} \hat{n} \times \underline{\vec{E}} \cdot \vec{W}_k \, da \right] = 0$$

or

$$c_k \int_V \underline{\vec{E}} \cdot \vec{U}_k \, dv + \mu \frac{\partial}{\partial t} \int_V \underline{\vec{H}} \cdot \vec{W}_k \, dv = - \int_{S_{sc}} \hat{n} \times \underline{\vec{E}} \cdot \vec{W}_k \, da \tag{10}$$

The substitution of (1a), (1c), and (5) into

$$\nabla \times \underline{\vec{H}} = \varepsilon \frac{\partial \underline{\vec{E}}}{\partial t} + \underline{\vec{J}}$$

produces

$$\sum_k \vec{U}_k \left[c_k^* \int\limits_V \underline{\vec{H}} \cdot \vec{W}_k \, dv + \int\limits_{S_{\infty}} \hat{n} \times \underline{\vec{H}} \cdot \vec{U}_k \, da \right]$$

$$= \left\{ \sum_k \left[\vec{U}_k \, \varepsilon \frac{\partial}{\partial t} \int\limits_V \vec{U}_k \cdot \underline{\vec{E}} \, dv + \vec{V}_k \, \varepsilon \frac{\partial}{\partial t} \int\limits_V \vec{V}_k \cdot \underline{\vec{E}} \, dv \right] \right\}$$

$$+ \sum_k \left[\vec{U}_k \int\limits_V \vec{U}_k \cdot \underline{\vec{J}} \, dv + \vec{V}_k \int\limits_V \vec{V}_k \cdot \underline{\vec{J}} \, dv \right]$$

Since U_k and V_k are independent potential functions, the above relation may be separated as follows.

$$c_k^* \int\limits_V \underline{\vec{H}} \cdot \vec{W}_k \, dv + \int\limits_{S_{\infty}} \hat{n} \times \underline{\vec{H}} \cdot \vec{U}_k \, da = \varepsilon \frac{\partial}{\partial t} \int\limits_V \vec{U}_k \cdot \underline{\vec{E}} \, dv + \int\limits_V \vec{U}_k \cdot \underline{\vec{J}} \, dv$$

or

$$c_k^* \int\limits_V \underline{\vec{H}} \cdot \vec{W}_k \, dv - \varepsilon \frac{\partial}{\partial t} \int\limits_V \underline{\vec{E}} \cdot \vec{U}_k \, dv = \int\limits_V \vec{U}_k \cdot \underline{\vec{J}} \, dv - \int\limits_{S_{\infty}} \hat{n} \times \underline{\vec{H}} \cdot \vec{U}_k \, da \qquad (11a)$$

and

$$\varepsilon \frac{\partial}{\partial t} \int\limits_V \vec{V}_k \cdot \underline{\vec{E}} \, dv + \int\limits_V \vec{V}_k \cdot \underline{\vec{J}} \, dv = 0 \qquad (11b)$$

The substitution of (1d) and (8) into

$$\nabla \cdot \underline{\vec{D}} = \rho$$

yields

$$\sum_k \phi_k \left[-d_k \int\limits_V \underline{\vec{D}} \cdot \vec{V}_k \, dv \right] = \sum_k \phi_k \int\limits_V \phi_k \rho \, dv$$

or

$$-d_k \varepsilon \int\limits_V \underline{\vec{E}} \cdot \vec{V}_k \, dv = \int\limits_V \phi_k \rho \, dv \qquad (12)$$

It is to be noted that (11b) and (12) are equivalent (see Section 2.1)

Equations (10) and (11a) determine the solenoidal (zero divergence) part of the electromagnetic fields, $\vec{\underline{E}}$ and $\vec{\underline{H}}$. It is this part of the field that shows properties of wave propagation. The coefficients

$$e \equiv \int\limits_V \vec{U}_k \cdot \vec{\underline{E}} \, dv \qquad \text{and} \qquad h \equiv \int\limits_V \vec{W}_k \cdot \vec{\underline{H}} \, dv \tag{13}$$

are determined from these equations as functions of time in terms of the source integrals.

The coefficient e can be eliminated from (10) by differentiating (10) with respect to time and then replacing $\dfrac{\partial e}{\partial t}$ by (11a); hence, one has

$$\mu\varepsilon \frac{\partial^2 h}{\partial t^2} + c_k^2 h = c_k \int\limits_V \vec{U}_k \cdot \vec{\underline{J}} \, dv$$

$$- c_k \int\limits_{S_{oc}} \hat{n} \times \vec{\underline{H}} \cdot \vec{U}_k \, da - \varepsilon \frac{\partial}{\partial t} \int\limits_{S_{sc}} \hat{n} \times \vec{\underline{E}} \cdot \vec{W}_k \, da \tag{14a}$$

A similar relation may be obtained for the coefficient e by differentiating (11a) with respect to time and using (10) to eliminate the term $\dfrac{\partial h}{\partial t}$. This is found to be

$$\mu\varepsilon \frac{\partial^2 e}{\partial t^2} + c_k^2 e = -\mu \frac{\partial}{\partial t} \int\limits_V \vec{U}_k \cdot \vec{\underline{J}} \, dv$$

$$- c_k^* \int\limits_{S_{sc}} \hat{n} \times \vec{\underline{E}} \cdot \vec{W}_k \, da + \mu \frac{\partial}{\partial t} \int\limits_{S_{oc}} \hat{n} \times \vec{\underline{H}} \cdot \vec{U}_k \, da \tag{14b}$$

Equations (14) are the differential equations for the expansion coefficients e and h. The terms on the right-hand side of (14) depend upon the specific features of the problem such as the existence of sources and losses. Special cases will be taken up in the following sections.

The remaining relation, either (11b) or (12), prescribes the irrotational (zero curl) part of $\vec{\underline{E}}$-field. Since this part of $\vec{\underline{E}}$ has zero curl, it is hence derivable from a scalar function.

The above result can be summarized as follows. The $\vec{\underline{E}}$-field and the current density $\vec{\underline{J}}$ can be represented as sum of their solenoidal parts, $\vec{\underline{E}}_1$ and $\vec{\underline{J}}_1$, and their irrotational parts, $\vec{\underline{E}}_2$ and $\vec{\underline{J}}_2$.

$$\vec{\underline{E}} = \vec{\underline{E}}_1 + \vec{\underline{E}}_2 \tag{15a}$$

$$\vec{\underline{J}} = \vec{\underline{J}}_1 + \vec{\underline{J}}_2 \tag{15b}$$

Then Maxwell's equations become

$$\nabla \times \vec{\underline{E}} = \nabla \times \vec{\underline{E}}_1 + \nabla \times \vec{\underline{E}}_2 = -\dot{\vec{\underline{B}}}$$

or

$$\nabla \times \vec{\underline{E}}_1 = -\dot{\vec{\underline{B}}} \qquad (16a)$$

$$\nabla \times \vec{\underline{E}}_2 = 0 \qquad (16b)$$

$$\nabla \times \vec{\underline{H}} = \dot{\vec{\underline{D}}} + \vec{\underline{J}} = \dot{\vec{\underline{D}}}_1 + \dot{\vec{\underline{D}}}_2 + \vec{\underline{J}}_1 + \vec{\underline{J}}_2$$

or

$$\nabla \times \vec{\underline{H}} = \dot{\vec{\underline{D}}}_1 + \vec{\underline{J}}_1 \qquad (17a)$$

$$0 = \dot{\vec{\underline{D}}}_2 + \vec{\underline{J}}_2 \qquad (17b)$$

$$\nabla \cdot \vec{\underline{D}} = \nabla \cdot \vec{\underline{D}}_1 + \nabla \cdot \vec{\underline{D}}_2 = \rho$$

or

$$\nabla \cdot \vec{\underline{D}}_1 = 0 \qquad (18a)$$

$$\nabla \cdot \vec{\underline{D}}_2 = \rho \qquad (18b)$$

$$\nabla \cdot \vec{\underline{B}} = 0 \qquad (19)$$

Equations (17b) and (18b) are equivalent by virtue of the equation of continuity.

Because of (16b), $\vec{\underline{E}}_2$ can be expressed as

$$\vec{\underline{E}}_2 = -\nabla \phi \qquad (20)$$

and

$$\nabla \cdot \vec{\underline{E}}_2 = -\nabla^2 \phi = \frac{\rho}{\varepsilon} \qquad (21)$$

Therefore, $\vec{\underline{E}}_2$ is derivable from a scalar potential which satisfies Poisson's equation. The determination of $\vec{\underline{E}}_2$ is similar to an electrostatic problem with the exception that the charge distribution, and hence $\vec{\underline{E}}_2$, varies with time.

2.1 Example: Equivalence of Eqs. (2-11b) and (2-12)

Show that Eq. (2-11b) is equivalent to Eq. (2-12).

Solution :

These relations may be shown to be equivalent by the condition of continuity. Multiplying Eq. (2-11b) by d_k yields

$$-d_k\varepsilon \frac{\partial}{\partial t} \int_V \vec{V}_k \cdot \vec{E} \, dv = d_k \int_V \vec{V}_k \cdot \vec{J} \, dv \tag{1}$$

Differentiating Eq. (2-12) with respect to time gives

$$-d_k\varepsilon \frac{\partial}{\partial t} \int_V \vec{V}_k \cdot \vec{E} \, dv = \frac{\partial}{\partial t} \int_V \phi_k \rho \, dv \tag{2}$$

Equating (1) and (2) produces

$$d_k \int_V \vec{J} \cdot \vec{V}_k \, dv = \frac{\partial}{\partial t} \int_V \phi_k \rho \, dv \tag{3}$$

The integrand on the left-hand side of (3) can be obtained from the following identity.

$$\nabla \cdot (\phi_k \underline{\vec{J}}) = \underline{\vec{J}} \cdot \nabla \phi_k + \phi_k \nabla \cdot \underline{\vec{J}} = d_k \underline{\vec{J}} \cdot \vec{V}_k + \phi_k \nabla \cdot \underline{\vec{J}} \tag{4}$$

Since

$$\vec{V}_k d_k = \nabla \phi_k \tag{XI.4-16a}$$

then (3) becomes

$$d_k \int_V \underline{\vec{J}} \cdot \vec{V}_k \, dv = \oint_S \phi_k \underline{\vec{J}} \cdot \hat{n} \, da - \int_V \phi_k \nabla \cdot \underline{\vec{J}} \, dv = -\int_V \phi_k \nabla \cdot \underline{\vec{J}} \, dv \tag{5}$$

since $\phi_k = 0$ on surface S. Therefore,

$$\int_V \phi_k \left[\nabla \cdot \underline{\vec{J}} + \frac{\partial \rho}{\partial t} \right] dv = 0 \tag{6}$$

Since the integrand of (6) vanishes by virtue of the condition of continuity, (2-11b) and (2-12) are identical.

3. Free Oscillations of a Cavity

(a) It was shown in Eq. (2-1) that the field vectors can be expressed in terms of the normal modes.

$$\underline{\vec{E}} = \sum_k \left[\vec{U}_k \int_V \vec{U}_k \cdot \underline{\vec{E}} \, dv + \vec{V}_k \int_V \vec{V}_k \cdot \underline{\vec{E}} \, dv \right] \tag{1a}$$

$$\underline{\vec{H}} = \sum_k \vec{W}_k \int_V \vec{W}_k \cdot \underline{\vec{H}} \, dv \tag{1b}$$

With these expansions, Maxwell's equations become [Eqs. (2-10) - (2-12)]

$$c_k \int_V \vec{U}_k \cdot \underline{\vec{E}} \, dv + \mu \frac{\partial}{\partial t} \int_V \vec{W}_k \cdot \underline{\vec{H}} \, dv = - \int_{S_{sc}} \hat{n} \times \underline{\vec{E}} \cdot \vec{W}_k \, da \tag{2a}$$

$$c_k \int_V \vec{W}_k \cdot \underline{\vec{H}} \, dv - \varepsilon \frac{\partial}{\partial t} \int_V \vec{U}_k \cdot \underline{\vec{E}} \, dv = \int_V \vec{U}_k \cdot \underline{\vec{J}} \, dv - \int_{S_{oc}} \hat{n} \times \underline{\vec{H}} \cdot \vec{U}_k \, da \tag{2b}$$

$$-\varepsilon \frac{\partial}{\partial t} \int_V \vec{V}_k \cdot \underline{\vec{E}} \, dv = \int_V \vec{V}_k \cdot \underline{\vec{J}} \, dv \tag{2c}$$

$$-d_k \varepsilon \int_V \vec{V}_k \cdot \underline{\vec{E}} \, dv = \int_V \phi_k \rho \, dv \tag{2d}$$

The coupled set, (2a) and (2b), may be decoupled [Eq. (2-14)].

$$\mu\varepsilon \frac{\partial^2 h}{\partial t^2} + c_k^2 h = c_k \int_V \vec{U}_k \cdot \underline{\vec{J}} \, dv$$

$$- c_k \int_{S_{oc}} \hat{n} \times \underline{\vec{H}} \cdot \vec{U}_k \, da - \varepsilon \frac{\partial}{\partial t} \int_{S_{sc}} \hat{n} \times \underline{\vec{E}} \cdot \vec{W}_k \, da \tag{3a}$$

$$\mu\varepsilon \frac{\partial^2 e}{\partial t^2} + c_k^2 e = -\mu \frac{\partial}{\partial t} \int_V \vec{U}_k \cdot \underline{\vec{J}} \, dv$$

$$- c_k^* \int_{S_{sc}} \hat{n} \times \underline{\vec{E}} \cdot \vec{W}_k \, da + \mu \frac{\partial}{\partial t} \int_{S_{oc}} \hat{n} \times \underline{\vec{H}} \cdot \vec{U}_k \, da \tag{3b}$$

$$e \equiv \int_V \vec{U}_k \cdot \underline{\vec{E}} \, dv \qquad h \equiv \int_V \vec{W}_k \cdot \underline{\vec{H}} \, dv \tag{3c}$$

(b) Consider a cavity filled with a lossless dielectric and enclosed by perfectly conducting walls. There is no current density contained within the cavity, $\vec{J} = 0$. The tangential components of $\underline{\vec{E}}$ and $\vec{U}_k$ are zero on surface S_{sc} (conducting walls) and the tangential components of $\underline{\vec{H}}$ and $\vec{W}_k$ are zero on surface S_{oc} (open-circuited surface). Equation (3) becomes

$$\mu\varepsilon \frac{\partial^2 h}{\partial t^2} + c_k^2 h = 0 \tag{4a}$$

$$\mu\varepsilon \frac{\partial^2 e}{\partial t^2} + c_k^2 e = 0 \tag{4b}$$

The solutions are

$$h = \int_V \vec{W}_k \cdot \underline{\vec{H}} \, dv = K_{hk} e^{j\omega_{ok}t} \tag{5a}$$

$$e = \int_V \vec{U}_k \cdot \underline{\vec{E}} \, dv = K_{ek} e^{j\omega_{ok}t} \tag{5b}$$

$$\omega_{ok}^2 \mu\varepsilon = c_k^2 \tag{5c}$$

where K_{hk} and K_{ek} are arbitrary constants, and ω_{ok} is the angular frequency of the kth mode for the lossless case. The general solution of a free oscillator is given by the superposition of all normal modes with appropriate coefficients. The relation between e and h can be determined from (2a) or (2b). The substitution of (5) into (2a) yields

$$c_k e + j\mu\omega_{ok} K_{hk} e^{j\omega_{ok}t} = c_k e + j\omega_{ok}\mu h = 0$$

or

$$\frac{e}{h} = \frac{\int_V \vec{U}_k \cdot \underline{\vec{E}} \, dv}{\int_V \vec{W}_k \cdot \underline{\vec{H}} \, dv} = -j\sqrt{\frac{\mu}{\varepsilon}} \tag{6}$$

The magnitude of the ratio of coefficients on the left-hand side of (6) is equal to the ratio of $|\underline{\vec{E}}|$ to $|\underline{\vec{H}}|$ in a uniform plane wave, but there is a 90° phase difference.

The free oscillation of an ideal lossless cavity is analogous to the oscillation in a lumped LC-circuit.

(c) Consider the same cavity as described in (b) except that it is now filled with a dielectric with losses, i.e., it has conductivity σ_d. Then the current density $\vec{J} = \sigma_d \underline{\vec{E}}$, and (3b) becomes

$$\mu\varepsilon \frac{\partial^2 e}{\partial t^2} + \sigma_d \mu \frac{\partial e}{\partial t} + c_k^2 e = 0 \tag{7}$$

The solution is

$$e = \int_V \vec{U}_k \cdot \underline{\vec{E}} \, dv = K_{ek} e^{j\omega_k t} \tag{8a}$$

$$\omega_k = j\omega_{ok} \left[\pm \sqrt{1 - \frac{1}{(2Q_d)^2}} + \frac{j}{2Q_d} \right] \equiv \omega_{1k} + j\omega_{2k} \tag{8b}$$

$$Q_d = \frac{\omega_{ok}\varepsilon}{\sigma_d} = Q \text{ of dielectric} \qquad \omega_{ok} = \frac{c_k^2}{\mu\varepsilon} \tag{8c}$$

With a complex frequency ω_k, the oscillation is damped and is similar to that in an RLC-circuit [Eq. XI.10-6].

Following the same procedure, one can show that the coefficient h behaves the same as e by using (3a).

$$\mu\varepsilon\frac{\partial^2 h}{\partial t^2} + c_k^2 h = c_k\sigma_d e = \sigma_d \left[-\mu\frac{\partial h}{\partial t} \right]$$

or

$$\frac{\partial^2 h}{\partial t^2} + \sigma_d\mu\frac{\partial h}{\partial t} + c_k^2 h = 0 \tag{7a}$$

by virtue of (2a) and the boundary conditions, $\hat{n} \times \vec{E} = 0$ on S_{sc} and $\hat{n} \times \vec{H} = 0$ on S_{oc}. This is identical to (7) with e replaced by h.

(d) The problem of additional losses in the conducting walls can be investigated by including the tangential component of $\vec{E}$ over the surface S_{sc}. The electromagnetic fields in a conducting medium with conductivity σ_c are related by

$$\hat{n} \times \vec{E} = \vec{H}(1 + j)\sqrt{\frac{\omega\mu}{2\sigma_c}} \qquad \omega \equiv \omega_{ok} \tag{9a}$$

$$\eta = \sqrt{\frac{\mu}{\varepsilon}} = \left[\frac{\mu}{\varepsilon\left[1 - \dfrac{j\sigma_c}{\omega\varepsilon}\right]} \right]^{\frac{1}{2}} \approx (1 + j)\sqrt{\frac{\omega\mu}{2\sigma_c}} \tag{9b}$$

where η is the wave impedance of a good conductor, $\sigma_c/\omega\varepsilon \gg 1$. The substitution of (9) into (3a) produces

$$\mu\varepsilon\frac{\partial^2 h}{\partial t^2} + (\eta W + \mu\sigma_d)\frac{\partial h}{\partial t} + c_k^2 h = 0 \tag{10a}$$

where, from (2a),

$$c_k e = -\mu \frac{\partial h}{\partial t} - \eta \int_{S_{sc}} \underline{\vec{H}} \cdot \vec{W}_k \, da$$

$$= -\mu \frac{\partial h}{\partial t} - \eta \int_{S_{sc}} \vec{W}_k \cdot \left[\vec{W}_k \int_V \vec{W}_k \cdot \underline{\vec{H}} \, dv \right] da$$

$$= -\mu \frac{\partial h}{\partial t} - \eta \left[\int_V \vec{W}_k \cdot \underline{\vec{H}} \, dv \right] \int_{S_{sc}} \vec{W}_k \cdot \vec{W}_k \, da = -\mu \frac{\partial h}{\partial t} - \eta \dot{W} h$$

$$\approx -\mu \frac{\partial h}{\partial t} \tag{10b}$$

$$\varepsilon \frac{\partial}{\partial t} \int_{S_{sc}} \hat{n} \times \underline{\vec{E}} \cdot \vec{W}_k \, da = \varepsilon \frac{\partial}{\partial t} \int_{S_{sc}} \eta \underline{\vec{H}} \cdot \vec{W}_k \, da$$

For good conductor, $\eta \ll 1$, the last term is neglected.

$$= \varepsilon \eta \frac{\partial}{\partial t} \int_{S_{sc}} \vec{W}_k \cdot \left[\vec{W}_k \int_V \vec{W}_k \cdot \underline{\vec{H}} \, dv \right] da$$

$$= \varepsilon \eta W \frac{\partial h}{\partial t} \tag{10c}$$

$$W \equiv \int_{S_{sc}} \vec{W}_k \cdot \vec{W}_k \, da \tag{10d}$$

Let the solution be proportional to $\exp(j\omega_k t)$. The angular frequency is determined from

$$\omega_k^2 - j\omega_k \left(\frac{\eta W}{\mu} + \frac{\sigma_d}{\varepsilon} \right) - \omega_{ok}^2 = 0 \tag{11}$$

where $c_k^2 \approx \omega_{ok}^2 \mu \varepsilon$ as in the lossless case.

$$\omega_k = \omega_{ok} \left[\pm \sqrt{1 - \left[\frac{1+j}{2Q_c} + \frac{1}{2Q_d} \right]^2} + j \left[\frac{1+j}{2Q_c} + \frac{1}{2Q_d} \right] \right] \equiv \omega_{1k} + j\omega_{2k} \tag{12a}$$

$$\frac{\eta W}{2\omega_{ok}\mu} = W \frac{1+j}{2\omega_{ok}\mu} \sqrt{\frac{\omega_{ok}\mu}{2\sigma_c}} = W \frac{(1+j)\delta_s}{2} = \frac{1+j}{2Q_c} \tag{12b}$$

$$\delta_s \equiv \frac{1}{\sqrt{2\omega_{ok}\mu\sigma_c}} = \text{skin depth} \tag{12c}$$

$$Q_c = \frac{1}{W\delta_s} \qquad Q \text{ factor of conductor} \qquad Q_d = \frac{\omega_{ok}\varepsilon}{\sigma_d} \tag{12d}$$

The real and imaginary parts of ω can be approximately obtained by binomial expansion.

$$\frac{\omega_k}{\omega_{ok}} \approx 1 - \frac{1}{2}\left[\frac{1+j}{2Q_c} + \frac{1}{2Q_d}\right]^2 + j\left[\frac{1+j}{2Q_c} + \frac{1}{2Q_d}\right]$$

$$= 1 - \frac{1}{2Q_c} - \frac{1}{8}\left[\frac{2}{Q_cQ_d} + \frac{1}{Q_d^2}\right]$$

$$+ j\left[\left[\frac{1}{2Q_c} + \frac{1}{2Q_d}\right] - \frac{1}{4}\left[\frac{1}{Q_c^2} + \frac{1}{Q_cQ_d}\right]\right]$$

$$\approx \left[1 - \frac{1}{2Q_c}\right] + \frac{j}{2}\left[\frac{1}{Q_c} + \frac{1}{Q_d}\right]$$

or

$$\omega_k = \omega_{1k} + j\omega_{2k} \tag{13a}$$

$$\omega_{1k} = \omega_{ok}\left[1 - \frac{1}{2Q_c}\right] \tag{13b}$$

$$\omega_{2k} = \frac{\omega_{ok}}{2}\left[\frac{1}{Q_c} + \frac{1}{Q_d}\right] \tag{13c}$$

It is to be noted that ω_{1k} has a first-order correction on account of Q_c. The effect of the wall is usually small for a cavity made of good conductors. Nevertheless, it plays an important role in accurate frequency determination.

The total Q for the case with losses in both the dielectric and the wall is given by

$$\frac{1}{Q} = \frac{1}{Q_d} + \frac{1}{Q_c} \tag{14}$$

The determination of Q_c is based upon the power losses produced by a single mode of oscillation. In cases where several modes are excited simultaneously, these losses must be calculated from the complete field. The sum of the losses resulting from different modes individually will not yield the correct value. The calculation of Q for such cases is complicated and is rarely carried out. This is because one has little interest in Q except near mode resonance. In practice, the resonance mode is excited with an amplitude much greater than any other modes and the above evaluation is approximately correct.

3.1. Impedance of a Resonant Line Section

The input impedance of a transmission line terminated by a short circuit is given by

$$Z = Z_0 \tanh (\alpha + j\beta)L \tag{1}$$

where L is the length of the line.

The input impedance of a parallel RLC-circuit is

$$Z_{RLC} = \cfrac{1}{\cfrac{1}{R} + \cfrac{1}{j\omega L} + j\omega C} = \cfrac{1}{\cfrac{1}{R} + j\omega_0 C \left[\cfrac{\omega}{\omega_0} - \cfrac{\omega_0}{\omega} \right]}$$

$$= \cfrac{\cfrac{1}{\omega_0 C}}{\cfrac{1}{Q} + j \left[\cfrac{\omega}{\omega_0} - \cfrac{\omega_0}{\omega} \right]} \tag{2a}$$

$$\omega_0 = \frac{1}{\sqrt{LC}} \qquad\qquad Q = \omega_0 CR \tag{2b}$$

It will now be shown that (1) can be expressed in a form similar to that of (2a) when the length L is adjusted to approximately the resonant length. Near resonance, the phase constant can be expressed as

$$\beta = \frac{\omega}{v} = \frac{\omega_0}{v} \frac{\omega}{\omega_0} = \beta_0 \left[1 + \frac{\omega - \omega_0}{\omega_0} \right] \qquad \beta_0 \equiv \frac{\omega_0}{v}$$

$$\approx \beta_0 \left[1 + \frac{\omega - \omega_0}{\omega_0} \frac{\omega + \omega_0}{2\omega} \right] \qquad \text{for } \omega \approx \omega_0$$

$$= \beta_0 \left[1 + \frac{1}{2} \frac{\omega^2 - \omega_0^2}{\omega\omega_0} \right]$$

$$= \beta_0 \left[1 + \frac{1}{2} \left[\frac{\omega}{\omega_0} - \frac{\omega_0}{\omega} \right] \right] = \beta_0 + \Delta \tag{3a}$$

$$\Delta = \frac{\beta_0}{2} \left[\frac{\omega}{\omega_0} - \frac{\omega_0}{\omega} \right] \tag{3b}$$

The hyperbolic function can be approximated as follows.

$$\coth (\alpha + j\beta)L = \frac{\sinh 2\alpha L - j \sin 2\beta L}{\cosh 2\alpha L - \cos 2\beta L}$$

For a high-Q circuit, $2\alpha L \ll 1$,

$$\sinh 2\alpha L \approx 2\alpha L \tag{4a}$$

$$\cosh 2\alpha L \approx 1 \tag{4b}$$

$$\sin 2\beta L = \sin (2\beta_0 L + 2\Delta L) = \sin 2\beta_0 L \cos 2\Delta L + \cos 2\beta_0 L \sin 2\Delta L$$

$$= -\sin 2\Delta L \approx -2\Delta L \quad \text{for} \quad \beta_0 L = \frac{\pi}{2} \tag{4c}$$

$$\cos 2\beta L \approx -1 \quad \text{for} \quad \beta_0 L = \frac{\pi}{2} \tag{4d}$$

Therefore,

$$\coth (\alpha + j\beta)L \approx \frac{1}{2} \left[2\alpha L + j\frac{\pi}{2} \left[\frac{\omega}{\omega_0} - \frac{\omega_0}{\omega} \right] \right] = \alpha L + j\frac{\pi}{4} \left[\frac{\omega}{\omega_0} - \frac{\omega_0}{\omega} \right]$$

$$= \alpha L + j\frac{\beta_0 L}{2} \left[\frac{\omega}{\omega_0} - \frac{\omega_0}{\omega} \right] \tag{5}$$

The substitution of (5) into (1) yields

$$Z = \frac{Z_0}{\alpha L + j\dfrac{\beta_0 L}{2} \left[\dfrac{\omega}{\omega_0} - \dfrac{\omega_0}{\omega} \right]} = \frac{\dfrac{2Z_0}{\beta_0 L}}{\dfrac{2\alpha}{\beta_0} + j\left[\dfrac{\omega}{\omega_0} - \dfrac{\omega_0}{\omega} \right]} \tag{6}$$

Direct comparison with (2) gives

$$\frac{1}{R} = \frac{\alpha L}{Z_0} \quad \text{or} \quad R = \frac{Z_0}{\alpha L} \tag{7a}$$

$$\omega_0 C = \frac{\beta_0 L}{2Z_0} \quad \text{or} \quad Q = \frac{\beta_0}{2\alpha} \tag{7b}$$

The impedance of a resonance section of transmission line can be expressed as the impedance of a parallel resonance network. The resonance frequency will be that for the loss-less section.

4. Input Impedance of a Cavity

When a waveguide is connecting a cavity to another device, coupling exists between the cavity and the device. The effect of coupling between the cavity and the other circuit will now be investigated. Let the cross-sectional surface S_r of the guide be the input plane to the cavity or the output plane from the cavity. The volume of interest, V, includes both the cavity and the portion of the guide between the reference surface S_r and the cavity. The problem is to determine the field at S_r in terms of the normal resonance modes. From (2-14),

$$\mu\varepsilon \frac{\partial^2 h}{\partial t^2} + c_k^2 h = c_k \int_V \vec{U}_k \cdot \underline{\vec{J}} \, dv$$

$$- c_k \int_{S_{oc}} \hat{n} \times \underline{\vec{H}} \cdot \vec{U}_k \, da - \varepsilon \frac{\partial}{\partial t} \int_{S_{sc}} \hat{n} \times \underline{\vec{E}} \cdot \vec{W}_k \, da \tag{1a}$$

$$\mu\varepsilon \frac{\partial^2 e}{\partial t^2} + c_k^2 e = -\mu \frac{\partial}{\partial t} \int_V \vec{U}_k \cdot \underline{\vec{J}} \, dv$$

$$- c_k^* \int_{S_{sc}} \hat{n} \times \underline{\vec{E}} \cdot \vec{W}_k \, da + \mu \frac{\partial}{\partial t} \int_{S_{oc}} \vec{U}_k \cdot \hat{n} \times \underline{\vec{H}} \, da \tag{1b}$$

$$h \equiv \int_V \vec{W}_k \cdot \underline{\vec{H}} \, dv \qquad e \equiv \int_V \vec{U}_k \cdot \underline{\vec{E}} \, dv \tag{1c}$$

The normal resonance mode is defined to be the oscillating mode when an open circuit is maintained at the surface S_r. For a lossless passive system, the coefficients for the normal resonance mode are solutions of (1) with terms on the right-hand side set equal to zero. The surface integral over S_r has no contribution since its integrand $\hat{n} \times \underline{\vec{H}} = 0$ over the plane of the open circuit. This is the case of free oscillation for a lossless system.

When the boundary condition over S_r is not an open circuit, the integral over S_r will be different from zero. Its contribution to (1) will produce a shift in the frequency of resonance and a change in the Q of the system [Section 4(c)].

The problem is to find the fields, both $\underline{\vec{E}}$ and $\underline{\vec{H}}$, within the waveguide-cavity system, particularly, within the guide and over the surface S_r. The ratio of the $\underline{\vec{E}}$-field to the $\underline{\vec{H}}$-field provides the input impedance at S_r. This is important information for many applications.

(a) Lossless Case

An arbitrary field function can be expanded in terms of orthogonal functions [Eq. (2-1a)].

$$\underline{\vec{E}} = \sum_k \vec{U}_k \int_V \vec{U}_k \cdot \underline{\vec{E}} \, dv \tag{2}$$

With the goal of determining the fields over S_r, the function $\vec{U}_k$ is expanded in terms of the eigenfunctions of the waveguide.

$$\vec{U}_k = \sum_m v_{km} \vec{e}_{tm} \tag{3}$$

where the v_{km}'s are coefficients independent of time, and the $\vec{e}_{tm}$'s are the normalized eigenfunctions of the transverse components of $\vec{E}$.

Let the tangential $\vec{H}$-field imposed over the input surface S_r be

$$\underline{\vec{H}} = \sum_n \underline{i}_n Z_{T_n} \vec{h}_{tn} \tag{4}$$

where the $\vec{h}_{tn}$'s are the normalized eigenfunctions of the transverse component of $\underline{\vec{H}}$, the $\underline{i}_n$'s are coefficients which have the time variation $\exp(j\omega t)$, and the Z_{T_n}'s are the wave impedances, E_{TE_k} or Z_{TM_k}.

With the above representation of the fields, the surface integral over $S_{oc} = S_r$ in (1a) becomes

$$\int_{S_r} \hat{n} \times \underline{\vec{H}} \cdot \vec{U}_k \, da = \int_{S_r} [-\hat{z} \times \sum_n \underline{i}_n Z_{T_n} \vec{h}_{tn}] \cdot [\sum_m v_{km} \vec{e}_{tm}] \, da$$

$$= \sum_m \sum_n \underline{i}_n v_{km} \int_{S_r} \vec{e}_{tn} \cdot \vec{e}_{tm} \, da$$

$$= \sum_n \underline{i}_n v_{kn} \tag{5}$$

$$\oint_S \vec{e}_{tk} \cdot \vec{e}_{tk} \, da = Z_{T_k} \oint_S \hat{z} \cdot \vec{e}_{tk} \times \vec{h}_{tk} \, da \tag{XI.1-33}$$

The surface S enclosing the volume of interest V is composed of the surface S_r and the conducting walls. Since there will be no contribution from the conducting walls, the closed surface integral is carried over S_r only. The unit normal $\hat{n}$ of S_r is equal to $-\hat{z}$ (Fig. 4-1). The relation [Eq. (XI.1-3) or (XI.1-9)]

$$Z_{T_n} \vec{h}_{tn} = -\hat{z} \times \vec{e}_{tn} \tag{6}$$

and the property of orthogonality of the eigenfunctions are employed to obtain (5).

The substitution of (5) into (1b) yields

$$\mu\varepsilon \frac{\partial^2 e}{\partial t^2} + c_k^2 e = \mu \frac{\partial}{\partial t} \sum_n \underline{i}_n v_{kn} = j\omega\mu \sum_n \underline{i}_n v_{kn} \tag{7a}$$

$$\underline{i}_n = i_n e^{j\omega t} \tag{7b}$$

Let the solution be

$$e \equiv \int_V \vec{U}_k \cdot \underline{\vec{E}} \, dv = K_{ek} e^{j\omega t} \tag{8}$$

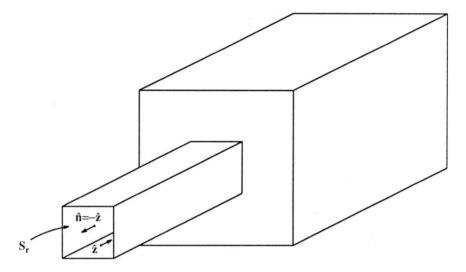

Figure 4-1: A general resonator

Then

$$(-\omega^2\mu\varepsilon + c_k^2) K_{ek}e^{j\omega t} = j\omega\mu \sum_n \underline{i}_n v_{kn}$$

$$K_{ek}e^{j\omega t} = \frac{j\omega\mu \sum\limits_n \underline{i}_n v_{kn}}{\omega_k^2\mu\varepsilon - \omega^2\mu\varepsilon} = \sum_n \frac{1}{j\Omega_\Delta} \frac{\underline{i}_n v_{kn}}{\omega_k\varepsilon} = \int_V \vec{U}_k \cdot \underline{\vec{E}} \, dv = e \tag{9a}$$

$$\Omega_\Delta \equiv \frac{\omega}{\omega_k} - \frac{\omega_k}{\omega} \tag{9b}$$

The $\underline{\vec{E}}$-field is obtained by substituting (3) and (9) into (2).

$$\underline{\vec{E}} = \sum_k [\sum_m v_{km}\vec{e}_{tm}] \left[\sum_n \frac{1}{j\Omega_\Delta} \frac{\underline{i}_n v_{kn}}{\omega_k\varepsilon} \right]$$

$$= \sum_m \vec{e}_{tm} \sum_n \underline{i}_n \sum_k \frac{1}{j\Omega_\Delta} \frac{v_{km} \, v_{kn}}{\omega_k\varepsilon}$$

$$= \sum_m \vec{e}_{tm} \sum_n \underline{i}_n Z_{nm} = \sum_m \vec{e}_{tm}\underline{V}_m \tag{10a}$$

$$Z_{nm} \equiv \sum_k \frac{1}{\Omega_\Delta} \frac{v_{km}v_{kn}}{\omega_k\varepsilon} \tag{10b}$$

$$\underline{V}_m \equiv \sum_n \underline{i}_n Z_{nm} \tag{10c}$$

The coefficients $\underline{V}_m$ and $\underline{i}_n$ can be interpreted as voltage and current, respectively, since the ratio

$$\frac{\underline{V}_m}{\underline{i}_n} = Z_{nm}$$

and

$$\frac{1}{2} \operatorname{Re} \oint_S \underline{\vec{E}} \times \underline{\vec{H}}^* \cdot \hat{n} \, da = \frac{1}{2} \operatorname{Re} \sum_m \sum_n \oint_S \vec{e}_{tm} \times \vec{h}_{tn}^* \cdot \hat{n} \, da \, [\, \underline{V}_m \, \underline{i}_n^* \, Z_{T_n}]$$

$$= \frac{1}{2} \operatorname{Re} \sum_m \sum_n \left[\frac{1}{Z_{T_n}} \oint_S \vec{e}_{tm} \cdot \vec{e}_{tn} \, da \right] [\, \underline{V}_m \underline{i}_n^* \, Z_{T_n}] = \frac{1}{2} \operatorname{Re} [\, \underline{V}_m \underline{i}_n^*]$$

where (XI.1-33) is used to obtain the final expression.

(b) Single-Mode Propagation (Lossless Case)

In most practical applications, propagation is restricted to a dominant mode. Let this be the first mode; then (10) becomes

$$\underline{V}_1 = \underline{i}_1 Z_{11} \tag{11a}$$

$$Z_{11} = \frac{\underline{V}_1}{\underline{i}_1} = \sum_k \frac{1}{j\Omega_\Delta} \frac{v_{k1}^2}{\omega_k \varepsilon} \tag{11b}$$

This is the input impedance at surface S_r for a guide which propagates only the dominant mode. It is made up of a sum of resonance terms. The impedance is infinite at the resonance frequency. At frequencies near ω_i, the ith term is much greater and varies rapidly with frequency, whereas all other terms are rather small and slow varying; those terms are lumped together as Z_Δ for convenience.

$$Z_{11} = \frac{1}{j\Omega_\Delta} \frac{v_{k1}^2}{\omega_k \varepsilon} + Z_\Delta \qquad \text{for } \omega \approx \omega_k \tag{11c}$$

$$Z_\Delta = \sum_{j \neq k} \frac{1}{j\Omega_j} \frac{v_{j1}^2}{\omega_j \varepsilon} \qquad \Omega_j = \frac{\omega}{\omega_j} - \frac{\omega_j}{\omega} \tag{11d}$$

(c) Terminated Guide

In the above analysis, the $\vec{H}$-field is imposed over the reference surface S_r without specifying how this is done. Let this be accomplished by terminating S_r by an impedance Z_1. Z_{11} is the input impedance looking into the guide toward the cavity, so it is the negative of Z_1, the impedance looking away from the cavity. Therefore,

$$Z_1 = -Z_{11} = -\left[\frac{1}{j\Omega_\Delta} \frac{v_{k1}^2}{\omega_k \varepsilon} + Z_\Delta \right]$$

or

$$j\,\Omega_\Delta + \frac{\dfrac{v_{k1}^2}{\omega_k\,\varepsilon}}{Z_1 + Z_\Delta} = 0 \tag{12}$$

Let the normalized admittance be

$$g + jb = \frac{1}{\dfrac{Z_1 + Z_\Delta}{Z_{o1}}} = \frac{Z_{o1}}{Z_1 + Z_\Delta} \tag{13}$$

where Z_{o1} is the characteristic impedance of the first mode. Then

$$j\Omega_\Delta + \frac{(g + jb)\, v_{k1}^2}{\omega_k \varepsilon Z_{o1}} = 0 \quad \text{or} \quad \Omega_\Delta + \frac{g + jb}{Q_{ek1}} = 0 \tag{14a}$$

$$j\left[\Omega_\Delta + \frac{b}{Q_{ek1}}\right] + \frac{g}{Q_{ek1}} = 0 \tag{14b}$$

$$Q_{ek1} \equiv \frac{\omega_k \varepsilon Z_{o1}}{v_{k1}^2} \tag{14c}$$

where Q_{ek1} is the external Q for the kth mode. The complex frequency ω can be solved from (14a).

$$\omega^2 + \frac{\omega_k\,(b - jg)}{Q_{ek1}}\, \omega - \omega_k^2 = 0 \tag{15a}$$

$$\omega = \omega_k\left[\pm\sqrt{1 + \left[\frac{b - jg}{2Q_{ek1}}\right]^2} - \frac{b - jg}{2Q_{ek1}}\right]$$

$$\approx \omega_k\left\{\pm\left[1 + \frac{1}{2}\left[\frac{b - jg}{2Q_{ek1}}\right]^2\right] - \frac{b}{2Q_{ek1}} + \frac{jg}{2Q_{ek1}}\right\} = \omega_1 + j\omega_2 \tag{15b}$$

$$\omega_1 \approx \pm\omega_k\left[1 - \frac{b}{2Q_{ek1}}\right] \equiv \pm\omega_k \mp \Delta\omega_{1k} \qquad \Delta\omega_{1k} = \frac{b\omega_k}{2Q_{ek1}} \tag{15c}$$

$$\omega_2 \approx \frac{g\omega_k}{2Q_{ek1}} \tag{15d}$$

$$Q = \frac{\omega_1}{2\omega_2} \approx \frac{Q_{ek1}}{g} \tag{15e}$$

The second-order terms were neglected in the above approximation. The terminating impedance Z_1 produces a shift in resonance frequency, $\Delta\omega_{1k}$, which is directly related to the reactive component b of the normalized admittance. The resulting Q is different from the external Q_{ek1} by a factor of $1/g$, where g is the normalized conductance of the external load. Q_{ek1} is the external Q of the kth mode for the first guided mode. If the terminating impedance is adjusted so that

$$Z_1 + Z_\Delta = Z_{o1} = \text{characteristic impedance}$$

then

$$g = 1 \quad \text{and} \quad Q = Q_{ek1}$$

This is the Q for the resonant mode. If Z_Δ is neglected, this implies that the guide is matched. Then Q_{ek1} is equal to the Q with a matched load. By an appropriate choice of the location of the surface S_r, Z_Δ can be made equal to zero, and $Q_{ek1} = Q_{match}$. Thus, Q_{ek1} is a measure of the coupling of the kth mode through the output. Equation (11) can be expressed in terms of Q_{ek1} by dividing through by Z_{o1}.

$$\frac{Z_{11}}{Z_{o1}} = \frac{1}{j\Omega_\Delta} \frac{1}{Q_{ek1}} + \frac{Z_\Delta}{Z_{o1}} = \sum_k \frac{1}{j\Omega_\Delta} \frac{1}{Q_{ek1}} \tag{16}$$

Equation (11d) is used to obtain the final result.

(d) Cavity with Losses

When the cavity is constructed of practical materials, losses are incurred in both the dielectric and the conducting walls. The integrals on the right-hand side of (1) must be included in the analysis. It is more convenient to use (1b).

$$\mu\varepsilon \frac{\partial^2 e}{\partial t^2} + c_k^2 e = \mu \frac{\partial}{\partial t} \left[\int_{S_{oc}} \vec{U}_k \cdot \hat{n} \times \underline{\vec{H}} \, da - \int_V \vec{U}_k \cdot \underline{\vec{J}} \, dv \right]$$

$$- c_k^* \int_{S_{sc}} \hat{n} \times \underline{\vec{E}} \cdot \vec{W}_k \, da \tag{1b}$$

$$\equiv I_1 + I_2 + I_3 \tag{17}$$

Each term on the right-hand side of (17) will be evaluated separately.

$$I_1 = \mu \frac{\partial}{\partial t} \int_{S_{oc}} \vec{U}_k \cdot \hat{n} \times \underline{\vec{H}} \, da = \mu \frac{\partial}{\partial t} \sum_n i_n v_{kn} \tag{18a}$$

by virtue of (5).

$$I_2 = -\mu \frac{\partial}{\partial t} \int_V \sigma_d \underline{\vec{E}} \cdot \vec{U}_k \, dv = -\mu\sigma_d \frac{\partial e}{\partial t} \tag{18b}$$

$$I_3 = -c_k^* \int\limits_{S_{sc}} Z_s \underline{\vec{H}} \cdot \vec{W}_k \, da = -c_k^* Z_s \int\limits_{S_{sc}} \vec{W}_k \cdot \vec{W}_k \, da \int\limits_V \underline{\vec{H}} \cdot \vec{W}_k \, dv$$

$$= \frac{-c_k^* Z_s We}{-j\sqrt{\dfrac{\mu}{\varepsilon}}} = -j\omega_k \varepsilon Z_s We \qquad (18c)$$

$$W \equiv \int\limits_{S_{sc}} \vec{W}_k \cdot \vec{W}_k \, da \qquad (18d)$$

$$\hat{n} \times \underline{\vec{E}} = Z_s \underline{\vec{H}} \qquad Z_s = (1+j)\sqrt{\frac{\omega\mu}{2\sigma_c}} = \text{surface impedance} \qquad (XI.12\text{-}4)$$

$$\int\limits_V \vec{W}_k \cdot \underline{\vec{H}} \, dv = j\sqrt{\frac{\varepsilon}{\mu}} \int\limits_V \vec{U}_k \cdot \underline{\vec{E}} \, dv \qquad (3\text{-}6)$$

The substitution of (18) into (17) yields

$$\mu\varepsilon \frac{\partial^2 e}{\partial t} + \mu\sigma_d \frac{\partial e}{\partial t} + j\omega_k \varepsilon Z_s We + c_k^2 e = \mu \frac{\partial}{\partial t} \sum_n \underline{i}_n v_{kn} \qquad (19)$$

Let the solution of (19) be

$$e = e_o e^{j\omega t} \qquad \text{for} \qquad \underline{i}_n = i_{no} e^{j\omega t} \qquad (20a)$$

Then

$$-\omega^2 \mu\varepsilon e + j(\omega\mu\sigma_d + \omega_k \varepsilon Z_s W)e + \omega_k^2 \mu\varepsilon e = j\omega\mu \sum_n \underline{i}_n v_{kn} \qquad (20b)$$

or

$$e = \int\limits_V \vec{U}_k \cdot \underline{\vec{E}} \, dv = \frac{1}{D} j\omega\mu \sum_n \underline{i}_n v_{kn}$$

$$= \frac{\dfrac{1}{\omega_k \varepsilon} \sum\limits_n \underline{i}_n v_{kn}}{j\Omega_\Delta + \left[\dfrac{1}{Q_d} + \dfrac{1+j}{Q_c} \right]} \qquad (21a)$$

$$D \equiv -\mu\varepsilon \left[\omega_k \omega \Omega_\Delta - j \left[\frac{\omega \sigma_d}{\varepsilon} + \frac{\omega_k Z_s W}{\mu} \right] \right]$$

$$= -j\omega_k \omega\mu\varepsilon \left[j\Omega_\Delta + \left[\frac{\sigma_d}{\omega_k \varepsilon} + \frac{Z_s W}{\omega\mu} \right] \right]$$

$$= -j\omega_k \omega\mu\varepsilon \left[j\Omega_\Delta + \left[\frac{1}{Q_d} + \frac{1}{Q_w} \right] \right] \tag{21b}$$

$$Q_d = \frac{\omega_k \varepsilon}{\sigma_d} \tag{21c}$$

$$\frac{1}{Q_w} = \frac{Z_s W}{\omega\mu} = \frac{1+j}{Q_c} \qquad Q_c = \frac{2}{W\delta_s} \tag{21d}$$

where Q_w is the Q for the wall of the guide, and the skin depth, δ_s, is given by Eq. (II.12-4c).

(e) Resonance Frequency for a Cavity with Losses

The resonance frequency for a cavity with losses can be determined from (20b) by setting terms on the right-hand side to zero.

$$\omega^2 - j\omega\omega_k \left[\frac{\sigma_d}{\omega_k \varepsilon} + \frac{W Z_s}{\omega\mu} \right] - \omega_k^2 = 0$$

or

$$\omega^2 - j\frac{\omega_k}{Q_t}\omega - \omega_k^2 = 0 \tag{22a}$$

$$\frac{1}{Q_t} = \frac{1}{Q_d} + \frac{1}{Q_w} \tag{22b}$$

The solution of (22a) is

$$\omega = \omega_k \left[\frac{j}{2Q_t} \pm \sqrt{1 - \frac{1}{4Q_t^2}} \right] \approx \omega_k \left[1 - \frac{1}{8Q_t^2} + \frac{j}{2Q_t} \right] \tag{23}$$

The negative sign in front of the radical is discarded since negative frequency is of little interest. Equation (24) can be further approximated for the low-loss case by keeping only the first power of Q's.

$$\omega \approx \omega_k \left[1 + \frac{j}{2Q_t} \right]$$

$$= \omega_k \left[\left[1 - \frac{1}{2Q_c} \right] + \frac{j}{2} \left[\frac{1}{Q_d} + \frac{1}{Q_c} \right] \right] = \omega_1 + j\omega_2 \tag{24a}$$

$$\omega_1 = \omega_k \left[1 - \frac{1}{2Q_c} \right] \tag{24b}$$

$$\omega_2 = \frac{\omega_k}{2} \left[\frac{1}{Q_d} + \frac{1}{Q_c} \right] \tag{24c}$$

5. Multiport Cavity

When a cavity has more than one connecting port, the last term on the right-hand side of (4-1b), or equivalently (4-5), should be evaluated over all ports. Equation (4-5) then has the form

$$\oint_S \hat{n} \times \vec{H} \cdot \hat{n} \, da = \sum_m \sum_n \sum_p i_n v_{km} \int_{S_{cp}} \vec{e}_{tn} \cdot \vec{e}_{tm} \, da$$

$$= \sum_{n_p} i_{n_p} v_{kn_p} \tag{1a}$$

$$\sum_{n_p} \equiv \sum_n \sum_p \tag{1b}$$

where S_{cp} is the cross section of the output plane of the pth port. Equation (1b) implies that the summation is taken over all ports; this consists of the summation of all propagating modes in each of the ports. Mathematically, the results obtained for a single-port cavity are valid for a multiport cavity by simply replacing the summation over n by a summation over n_p as defined by (1b).

Beginning with the general wave equation, (4-1b) or (4-17), or equivalently the relation (4-19) after all integrations have been carried out, the corresponding expression for a multiport cavity is

$$\mu\varepsilon \frac{\partial^2 e}{\partial t^2} + \mu\sigma_d \frac{\partial e}{\partial t} + j\omega_k \varepsilon Z_s W e + c_k^2 e = \mu \frac{\partial}{\partial t} \sum_{n_p} i_{n_p} v_{kn_p} \tag{2}$$

Let (2) have the same type of solution as given by (4-20a). The coefficient e is then found to be

$$e = \frac{j\omega\mu}{D} \sum_{n_p} i_{n_p} v_{kn_p} = \frac{1}{D_1 \omega_k \varepsilon} \sum_{n_p} i_{n_p} v_{kn_p} \tag{3a}$$

$$D = -j\omega\omega_k \mu\varepsilon D_1 \tag{3b}$$

$$D_1 = j\Omega_\Delta + \frac{1}{Q_d} + \frac{1}{Q_w} \qquad \Omega_\Delta = \frac{\omega}{\omega_k} - \frac{\omega_k}{\omega} \tag{3c}$$

$$Q_d = \frac{\omega_k \varepsilon}{\sigma_d} \tag{4-21c}$$

$$\frac{1}{Q_w} = \frac{Z_s W}{\omega\mu} = \frac{1+j}{Q_c} \qquad Q_c = \frac{1}{W\delta_s} \tag{4-21d}$$

The general expression for the $\vec{E}$-field at any port is given by (4-2). Substituting the coefficient e [Eq. (3a)] and the potential function $\vec{U}_k$ [Eq. (4-3)] into (4-2), one obtains

$$\underline{\vec{E}} = \sum_k \vec{U}_k e \tag{4-2}$$

$$= \sum_k [\sum_m v_{km}\vec{e}_{tm}] \left[\frac{1}{D_1\omega_k\varepsilon} \sum_{n_p} \underline{i}_{n_p} v_{kn_p} \right] = \sum_m \vec{e}_{tm} \sum_{n_p} \underline{i}_{n_p} Z_{n_pm}$$

$$= \sum_m \vec{e}_{tm} \underline{V}_m \tag{4a}$$

$$\vec{U}_k = \sum_m v_{km}\vec{e}_{tm} \tag{4-3}$$

$$Z_{n_pm} = \frac{V_m}{\underline{i}_{n_p}} = \sum_k \frac{v_{km}v_{kn_p}}{D_1\omega_k\varepsilon} \tag{4b}$$

$$\underline{V}_m = \sum_{n_p} \underline{i}_{n_p} Z_{n_pm} \tag{4c}$$

Equation (4a) is the transverse electric field over the port surface S_{cp} and (4c) gives the potential on the same surface. Equation (4b) is a general expression for the impedance coefficient.

The above results indicate that a multiport cavity has similar relations as a network with multiple terminal-pairs. When only one of the ports is excited and the remaining ports are terminated by passive loads, these terminated ports will contribute to the Q and the resonance frequency of the cavity; and the problem is very much similar to that of a single-port cavity.

Consider the case when the cavity is excited by a single mode from a single port and the remaining ports are terminated by passive loads. Let the qth port be the exciting port with the qth mode, and each of the remaining ports with a propagating rth mode is terminated by an impedance Z_r. It is of interest to determine the input impedance of the qth port, which is defined by

$$Z_q = \frac{V_q}{\underline{i}_q} \tag{5}$$

The electric field at any port can be expressed in terms of the coefficient e.

$$\underline{\vec{E}} = \sum_k \vec{U}_k e = \sum_k e [\sum_m v_{km}\vec{e}_{tm}]$$

$$= \sum_m \vec{e}_{tm} \sum_k v_{km} e = \sum_m \vec{e}_{tm} \underline{V}_m \tag{6a}$$

$$\underline{V}_m = \sum_k v_{km} e \tag{6b}$$

and the input impedance at the mth port is

$$Z_m = \frac{\underline{V}_m}{\underline{i}_m} = \sum_k \frac{v_{km}}{\underline{i}_m} e \tag{6c}$$

In the neighborhood of the kth resonance frequency, the kth term in the summation of (6c) is large and varies rapidly with frequency and the remaining terms are small and slow varying; therefore, (6c) becomes

$$Z_m = \frac{v_{km}}{\underline{i}_m} e + Z_{km} = \text{input impedance of mth port} \tag{7a}$$

$$Z_{km} = \sum_{j \ne k} \frac{v_{jm}}{\underline{i}_m} e \tag{7b}$$

where small slow-varying terms are lumped together as Z_{km}. The output impedance of a passive mth port is $-Z_m$, i.e.,

$$Z_{m_{output}} = -Z_m = \frac{v_{km}}{\underline{i}_m} e + Z_{km}$$

or

$$\underline{i}_m = \frac{-v_{km} e}{Z_m + Z_{km}} \qquad \text{for} \qquad m \ne q \tag{8}$$

since the qth port is excited.

The coefficient e can be determined from (4-1).

$$\mu\varepsilon \frac{\partial^2 e}{\partial t^2} + c_k^2 e = -\mu \frac{\partial}{\partial t} \left[\int_V \vec{U}_k \cdot \underline{\vec{J}} \, dv \right.$$

$$\left. - \int_{S_{oc}} \vec{U}_k \cdot \hat{n} \times \underline{\vec{H}} \, da \right] - c_k^* \int_{S_{sc}} \hat{n} \times \underline{\vec{E}} \cdot \vec{W}_k \, da \tag{4-1}$$

$$= -\mu \frac{\partial}{\partial t} [I_1 - I_2] + I_3 \tag{9}$$

$$I_2 = \int_{S_{\infty}} \vec{U}_k \cdot \hat{n} \times \underline{\vec{H}} \, da = \sum_m \underline{i}_m v_{km} \tag{4-5}$$

$$= \sum_{m \neq q} v_{km} \frac{-v_{km}e}{Z_m + Z_{km}} + \underline{i}_q v_{kq} \tag{10a}$$

$$I_3 = -c_k^* \int_{S_{sc}} \hat{n} \times \underline{\vec{E}} \cdot \vec{W}_k \, da = -j\omega_k \varepsilon Z_s We \tag{10b}$$

by virtue of (4-18c). The substitution of (10) into (9) produces

$$\mu\varepsilon \frac{\partial^2 e}{\partial t^2} + c_k^2 e = -\mu \frac{\partial}{\partial t} \left[\int_V \vec{U}_k \cdot \underline{\vec{J}} \, dv \right.$$

$$\left. - \sum_{m \neq q} \frac{v_{km}^2 e}{Z_m + Z_{km}} - \underline{i}_q v_{kq} \right] + j\omega_k \varepsilon Z_s We \tag{11}$$

Let the solution of (11) be of the same form as (4-20a). Then one has

$$[-\omega^2 \mu\varepsilon + \omega_k^2 \mu\varepsilon] \, e$$

$$= -j\omega\mu \int_V \vec{U}_k \cdot \underline{\vec{J}} \, dv + j\omega\mu \sum_{m \neq q} \frac{v_{km}^2 e}{Z_m + Z_{km}} + j\omega\mu \underline{i}_q v_{kq} - j\omega_k \varepsilon Z_s We$$

or

$$\left[j\Omega_\Delta + \frac{Z_s W}{\omega\mu} + \frac{1}{\omega_k \varepsilon} \sum_{m \neq q} \frac{v_{km}^2}{Z_m + Z_{km}} \right] e = \frac{-1}{\omega_k \varepsilon} \int_V \vec{U}_k \cdot \underline{\vec{J}} \, dv + \frac{1}{\omega_k \varepsilon} \underline{i}_q v_{kq} \tag{12a}$$

$$\Omega_\Delta = \frac{\omega}{\omega_k} - \frac{\omega_k}{\omega} \tag{12b}$$

Let the current density be represented in the general form

$$\underline{\vec{J}} = \sigma_d \underline{\vec{E}} + \underline{\vec{J}}_e \tag{13}$$

where $\underline{\vec{J}}_e$ represents the equivalent current density other than the ohmic current density such as the moving electrons in an electronic beam. Then (12a) becomes

$$\left[j\Omega_\Delta' + \frac{1}{Q_w} + \sum_{m \neq q} \frac{1}{Q_{ekm}} \frac{Z_{om}}{Z_m + Z_{km}} + \frac{1}{Q_d} + \frac{1}{\omega_k \varepsilon} \frac{J}{e} \right] e = \frac{1}{\omega_k \varepsilon} \underline{i}_q v_{kq}$$

or

$$e = \cfrac{\cfrac{i_q v_{kq}}{\omega_k \varepsilon}}{j\Omega_\Delta' + \cfrac{1}{Q_k} + \displaystyle\sum_{m \neq q} \cfrac{1}{Q_{ekm}} \cfrac{Z_{om}}{Z_m + Z_{km}} + \cfrac{1}{\omega_k \varepsilon} \cfrac{J}{e}} \equiv \frac{1}{D_e} \frac{i_q v_{kq}}{\omega_k \varepsilon} \tag{14a}$$

$$\Omega_\Delta' = \frac{\omega}{\omega_k'} - \frac{\omega_k'}{\omega} \tag{14b}$$

$$\omega_k' = \omega_k + \Delta\omega_k \tag{4-15c}$$

$$\frac{1}{Q_k} = \frac{1}{Q_w} + \frac{1}{Q_d} \tag{14c}$$

$$\frac{1}{Q_{ekm}} = \frac{v_{km}^2}{\omega_k \varepsilon Z_{om}} \tag{4-14b}$$

$$J = \int_V \vec{J}_e \cdot \vec{U}_k \, dv \tag{14d}$$

It is to be noted that the modified resonance frequency ω_k' is used in Ω_Δ' since this factor varies rapidly with frequency, and $\omega_k' \approx \omega$ is used in other terms.

The input impedance (7) now becomes

$$Z_q = \frac{v_{kq}}{i_q} e + Z_{kq} = \frac{v_{kq}}{i_q} \frac{1}{D_e} \frac{i_q v_{kq}}{\omega_k \varepsilon} + Z_{kq}$$

$$= \cfrac{\cfrac{Z_{oq}}{Q_{ekq}}}{j\Omega_\Delta' + \cfrac{1}{Q_k} + \displaystyle\sum_{m \neq q} \cfrac{1}{Q_{ekm}} \cfrac{Z_{om}}{Z_m + Z_{km}} + \cfrac{1}{\omega_k \varepsilon} \cfrac{J}{e}} + Z_{kq} \tag{15}$$

Equation (15) is the general expression for the input impedance of the qth port. The effects of losses in the dielectric and in the nonperfect conducting wall are included in $1/Q_k$. The losses through various ports are shown in the summation term. Losses due to other current density $\vec{J}_e$ are represented by the last term in the denominator. It is to be noted that each mode in each of the output ports produces a correction to the frequency and a contribution to the total Q.

Equation (15) provides correct results near the resonance frequency of the kth mode. The results obtained from this equation will be in doubt when there are two resonant modes which are close together such that the two resonant peaks overlap. In such cases more elaborate techniques are required.

6. Electronic Discharge Within a Cavity

In some applications, electronic discharges or other current sources not governed by Ohm's law exist within the cavity. These types of current densities are denoted by $\vec{J_e}$. The ratio of coefficients, J/e, can be evaluated in the following manner.

Let the qth port be the only port which possesses such a current density $\vec{J_e}$, and all other ports are not excited. The qth port may be treated as a passive termination with an output impedance equal to the negative of (5-15). Then (5-14a), with $i_q = 0$, becomes

$$\left[j\Omega_\Delta' + \frac{1}{Q_k} + \sum_m \frac{1}{Q_{ekm}} \frac{Z_{om}}{Z_m + Z_{km}} + \frac{1}{\omega_k \varepsilon} \frac{J}{e} \right] e = 0 \tag{1}$$

The summation term includes the qth port since it is treated as a passive port without excitation, $i_q = 0$. The ratio of coefficients is therefore given by

$$-\frac{1}{\omega_k \varepsilon} \frac{J}{e} = j\Omega_\Delta' + \frac{1}{Q_k} + \sum_m \frac{1}{Q_{ekm}} \frac{Z_{om}}{Z_m + Z_{km}} \tag{2}$$

In some oscillators, such as a klystron, where the current density is confined to a region small compared to the wavelength, the left-hand side of (2) can be expressed in terms of lumped parameter coefficients. In a klystron, the electric field and the flow of electrons exist in the region between grids, which behaves practically like a lumped capacitor. Let the voltage between the grids be V and the total current flowing through the grid be I. Then

$$\vec{E} = \hat{E} \frac{V}{d} \qquad \hat{E} \equiv \frac{\vec{E}}{|\vec{E}|} \tag{3a}$$

$$\vec{J} = -\hat{E} \frac{I}{A} \tag{3b}$$

where d is the separation between grids and A is the cross-sectional area of the grids. The current density is directed opposite to the electric field within the source region.

The ratio of coefficients now becomes

$$-\frac{1}{\omega_k \varepsilon} \frac{J}{e} = -\frac{1}{\omega_k \varepsilon} \frac{\int_V \vec{J_e} \cdot \vec{U_k}\, dv}{\int_V \vec{E} \cdot \vec{U_k}\, dv} = -\frac{1}{\omega_k \varepsilon} \frac{\frac{-I}{A} \int_V \hat{E} \cdot \vec{U_k}\, dv}{\frac{V}{d} \int_V \hat{E} \cdot \vec{U_k}\, dv}$$

$$= \frac{1}{\omega_k \frac{\varepsilon A}{d}} \frac{I}{V} = \frac{1}{\omega_k C} \frac{I}{V} \tag{4a}$$

$$C = \frac{\varepsilon A}{d} \tag{4b}$$

Let the admittance be represented by

$$\frac{I}{\underline{V}} = g + jb \tag{5a}$$

for the case when the current and the potential gradient are co-directional, and

$$\frac{I}{\underline{V}} = -(g + jb) \tag{5b}$$

when the current and the potential gradient are contra-directional. Hence, (4a) becomes

$$-\frac{1}{\omega_k \varepsilon} \frac{J}{e} = -\frac{1}{\omega_k C} (g + jb) \tag{6}$$

The substitution of (6) into (2) gives

$$-\frac{g + jb}{\omega_k C} = j\Omega_\Delta' + \frac{1}{Q_k} + \sum_m \frac{1}{Q_{ekm}} \frac{Z_{om}}{Z_m + Z_{km}} = j\Omega_\Delta' + \frac{1}{Q_{kL}} \tag{7a}$$

$$\frac{1}{Q_{kL}} = \frac{1}{Q_k} + \sum_m \frac{1}{Q_{ekm}} \frac{Z_{om}}{Z_m + Z_{km}} \tag{7b}$$

where Q_{kL} is the loaded Q of the kth mode.

7. Problems

1. The electromagnetic fields in a rectangular cavity, Fig. 7-1, are

$$\vec{E}(\vec{r}, t) = \hat{y} E_o \sin \frac{\pi x}{a} \sin \frac{\pi z}{c} \cos \omega t$$

$$\vec{H}(\vec{r}, t) = \frac{E_o}{\omega \mu_o} \left[\hat{x} \frac{\pi}{c} \sin \frac{\pi x}{a} \cos \frac{\pi z}{c} - \hat{z} \frac{\pi}{a} \cos \frac{\pi x}{a} \sin \frac{\pi z}{c} \right] \sin \omega t$$

Show that the total electric and magnetic energy in the cavity is

$$w_e = \frac{abc}{8} \varepsilon E_o^2 \cos^2 \omega t$$

$$w_m = \frac{abc}{8} \left[\frac{1}{c^2} + \frac{1}{a^2} \right] \mu E_o^2 \sin^2 \omega t$$

2. Determine the losses of the cavity due to the imperfect conducting wall which has a surface resistance R_s.

3. An a x b air-filled rectangular waveguide of length L_1 is terminated by a short circuit. The other end of the guide is connected to a similar guide through a small hole of radius r_o on a perfectly conducting sheet.

The normalized reactance of the hole is given by

$$\hat{X} = \frac{8 \beta r_o^3}{3ab}$$

where β is the propagation constant for the dominant TE_{10} mode. Determine the resonance frequency of the structure.

4. Determine an approximate equivalent circuit for the structure described in Problem 3 at frequencies near the lowest resonance frequency.

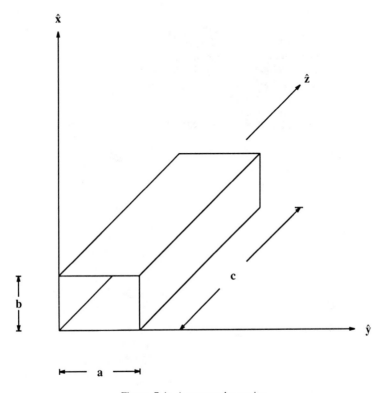

Figure 7-1: A rectangular cavity

1. Single-Port Cavity - Traveling Wave Approach

Microwave resonators may be investigated by either of the following approaches.

(α) Lumped equivalent circuit.

(β) Wave transmission through waveguide terminated by discontinuities.

Both techniques produce identical results for high-Q circuits. In the case of low Q, the lumped equivalent circuit techniques require complex higher-order terms to yield useful information. This is caused by the fact that when the Q is low, the adjacent resonance frequencies create interferences, making the one-pole approximation inadequate.

In the analysis of resonators, the plane of detuned-short is usually chosen as the reference plane for convenience. The plane of detuned-short is the plane where the reflection coefficient is equal to -1 at frequencies far away from resonance or where the reflection coefficient has its maximum negative value in cases where other poles interfere.

(a) General Analysis

A resonant cavity is composed of a coupling device and a section of waveguide. The coupling mechanism of a single-port resonator will now be investigated. A single-port resonator excited by a lossless device, such as an aperture or probe, is shown in Fig. 1-1. Let the input plane be 1-1' and the output plane be 2-2'. The scattering matrix of the coupling device is

$$[S] = \begin{bmatrix} s_{11} & s_{12} \\ s_{21} & s_{22} \end{bmatrix} \tag{1}$$

for the chosen reference planes. Let these reference planes be the detuned-short position; then

$$s_{11} = s_{22} = - |s_{11}| \tag{2a}$$

$$s_{12} = s_{21} \tag{2b}$$

for symmetrical and reciprocal devices. If the device is lossless, the scattering matrix also satisfies the unitary condition, (III.13-3).

$$s_{12} = s_{21} = \pm j \sqrt{1 - |s_{11}|^2} \equiv \pm jm \tag{3a}$$

$$m \equiv \sqrt{1 - |s_{11}|^2} = \text{a real number} \tag{3b}$$

$$s_{11} = s_{22} = - \sqrt{1 - m^2} \tag{3c}$$

where m is known as the coefficient of coupling.

The scattering matrix becomes

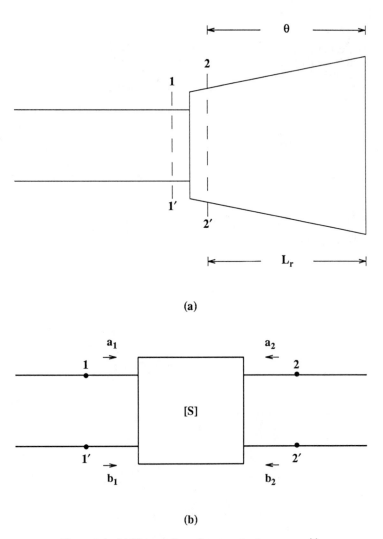

(a)

(b)

Figure 1-1: (a) The coupling of a resonator to a waveguide.
(b) An equivalent circuit for (a)

$$[S] = \begin{bmatrix} -\sqrt{1 - m^2} & jm \\ jm & -\sqrt{1 - m^2} \end{bmatrix} \tag{4}$$

Let the electrical length between the output plane 2-2' and the short-circuit terminating plane of the resonator be θ. The attenuation for waves traveling from plane 2-2' to the conducting end wall and back to plane 2-2' is denoted by α in nepers. The relation between the incident wave a_2 and the reflected wave b_2 at the output plane of the coupling device is

$$a_2 = -b_2 e^{\phi} \tag{5a}$$

$$\phi = -(\alpha + j2\theta) \tag{5b}$$

When the resonator is excited by an incident wave a_1 at plane 1-1', the reflected wave [b] is given by

$$\begin{bmatrix} b_1 \\ b_2 \end{bmatrix} = \begin{bmatrix} -\sqrt{1 - m^2} & jm \\ jm & -\sqrt{1 - m^2} \end{bmatrix} \begin{bmatrix} a_1 \\ -b_2 e^{\phi} \end{bmatrix}$$

$$= \begin{bmatrix} -a_1 \sqrt{1 - m^2} - jmb_2 e^{\phi} \\ jma_1 + b_2 \sqrt{1 - m^2}\, e^{\phi} \end{bmatrix} \tag{6a}$$

or

$$b_1 = -a_1 \left[\sqrt{1 - m^2} - \frac{m^2 e^{\phi}}{1 - \sqrt{1 - m^2}\, e^{\phi}} \right] \tag{6b}$$

$$b_2 = \frac{jma_1}{1 - \sqrt{1 - m^2}\, e^{\phi}} \tag{6c}$$

(b) Resonance

The reflected wave b_2 can be maximized by adjusting the electrical length θ. The maximum value of $|b_2|$ will be obtained if the denominator of (6c) is minimum, i.e., when

$$2\theta = 2n\pi \quad \text{or} \quad \theta = n\pi \quad n = 0, 1, 2, \ldots \tag{7}$$

For the electrical length θ specified by (7), $|b_1|$ assumes a minimum value. Equation (7) is the condition for resonance. Then

$$\theta = \frac{2\pi}{\lambda_{go}} L_1 = n\pi \quad \text{or} \quad L_1 = n \frac{\lambda_{go}}{2} \quad n = 0, 1, 2, \ldots \tag{8a}$$

$$b_1 = -a_1 \left[\sqrt{1 - m^2} - \frac{m^2 e^{-\alpha}}{1 - \sqrt{1 - m^2}\, e^{-\alpha}} \right]$$

$$= -a_1 \frac{\sqrt{1 - m^2} - e^{-\alpha}}{1 - \sqrt{1 - m^2}\, e^{-\alpha}} = b_{1_{min}} \tag{8b}$$

$$b_2 = a_1 \frac{jm}{1 - \sqrt{1 - m^2}\, e^{-\alpha}} = b_{2_{max}} \tag{8c}$$

At resonance, the axial length L_1 of the resonator is an integral multiple of a half-wavelength. For an exciting device which has very small dimensions in the axial direction, such as an aperture, the reference plane 2-2' is very close to the plane of the exciting structure. Hence, the actual axial length of the resonator is approximately equal to an integral number of half-wavelengths.

(c) Antiresonance

The reflected wave b_2 will assume a minimum value when

$$2\theta = (2n + 1)\pi \quad \text{or} \quad \theta = (2n + 1)\frac{\pi}{2} \quad n = 0, 1, 2, \ldots \tag{9a}$$

$$L_1 = (2n + 1)\frac{\lambda_{go}}{4} \quad n = 0, 1, 2, \ldots \tag{9b}$$

The corresponding values of b's are

$$b_1 = -a_1 \frac{\sqrt{1 - m^2} + e^{-\alpha}}{1 + \sqrt{1 - m^2}\, e^{-\alpha}} = b_{1_{max}} \tag{10a}$$

$$b_2 = \frac{jma_1}{1 + \sqrt{1 - m^2}\, e^{-\alpha}} = b_{2_{min}} \tag{10b}$$

The anti-resonance condition is specified by (9). At anti-resonance, the physical length L_1 is an odd multiple of a quarter-wavelength. The reflected parameter b_1 has its maximum value and the corresponding b_2 has its minimum value.

(d) Coupling Conditions

The reflected wave b_2 can be further maximized by adjusting the coefficient of coupling m. The magnitude of b_2 is obtained from (8c).

$$|b_2| = \sqrt{b_2 b_2^*} = \frac{m\,|a_1|}{D} \tag{11a}$$

$$D \equiv 1 - \sqrt{1 - m^2}\, e^{-\alpha} \tag{11b}$$

The value of $|b_2|$ will be maximized with respect to m by setting the derivative $d|b_2|/dm$ equal to zero.

$$\frac{d|b_2|}{dm} = |a_1| \left[\frac{1}{D} - \frac{m}{D^2} \frac{dD}{dm} \right] = \frac{|a_1|}{D^2} \left[D - \frac{m^2 e^{-\alpha}}{\sqrt{1 - m^2}} \right] = 0$$

$$D - \frac{m^2 e^{-\alpha}}{\sqrt{1 - m^2}} = 0$$

or

$$\sqrt{1 - m^2} - e^{-\alpha} = 0 \qquad \text{or} \qquad m = \pm \sqrt{1 - e^{-2\alpha}} \equiv m_c \qquad (12)$$

When the coefficient of coupling assumes the value m_c as specified by (12), it is called critical coupling. At critical coupling, the reflected waves, b_1 and b_2, become [Eq.(8)]

$$b_1 = 0 \qquad \text{at resonance} \qquad \text{and} \qquad m = m_c \qquad (13a)$$

$$b_2 = \pm j \frac{a_1}{\sqrt{1 - e^{-2\alpha}}} \qquad \text{at resonance} \qquad \text{and} \qquad m = m_c \qquad (13b)$$

At resonance with critical coupling, all incident power is transmitted to the resonator. If the attenuation α is made very small, b_2 can be made one or two orders greater than the magnitude of a_1. This is usually possible at microwave frequencies where components are made of extremely good conducting materials.

(d.1) Under-coupling is defined by $m < m_c$, or

$$|m| < \sqrt{1 - e^{-2\alpha}} \qquad \text{or} \qquad e^{-\alpha} < \sqrt{1 - |m|^2} \qquad (14)$$

At under-coupling, b_1 is given by

$$b_1 = -a_1 \frac{\sqrt{1 - m^2} - e^{-\alpha}}{1 - e^{-2\alpha}} < 0 \qquad \text{at resonance} \qquad (15)$$

Hence, the input plane 1-1′ is the location of minimum VSWR since

$$v_{total} = v_{inc} + v_{ref} = (1 - \Gamma) v_{inc}$$

(d.2) Over-coupling is specified by $m > m_c$; hence,

$$|m| > \sqrt{1 - e^{-2\alpha}} \qquad \text{or} \qquad e^{-\alpha} > \sqrt{1 - |m|^2} \qquad (16)$$

and the corresponding b_1 is

$$b_1 = -a_1 \frac{\sqrt{1 - m^2} - e^{-\alpha}}{1 - e^{-2\alpha}} > 0 \qquad \text{at resonance} \qquad (17)$$

The input plane 1-1' is now located at maximum VSWR.

It is to be noted that away from resonance, or at anti-resonance, the input plane 1–1' (the plane of detuned-short) is always a position of minimum VSWR and is independent of the condition of coupling.

(e) Concluding Remarks

The above normalized wave analysis would be satisfactory for both practical and theoretical applications if the junction scattering coefficients were directly measurable. Unfortunately, resonators are so well constructed to reduce their losses that they cannot be easily disassembled. Even if the coupling structure could be separated, the actual experimental determination of the scattering coefficient could still be extremely difficult because of the odd cross-sectional shape. However, if the resonator is made of a section of standard waveguide, then this approach is experimentally adequate.

Because of the above difficulty, resonators are analyzed in terms of equivalent RLC-circuits as will be discussed in the following sections.

2. Equivalent Circuit Representation

The external properties of a cavity resonator can be experimentally determined at discrete frequencies and extrapolated around frequencies near resonance. This is achieved by the concept of an equivalent circuit based upon the materials discussed in Chapter XII. The input impedance of a resonant cavity is given by Eq. (XII.4-11), and the cavity can therefore be represented by an equivalent cirucit as shown in Fig. 2-1.

$$Z_{11} = \sum_{n=1}^{N} Z_n = \frac{1}{j\Omega_\Delta} \frac{v_{k1}^2}{\omega_k \varepsilon} + Z_\Delta \qquad \text{for} \quad \omega \approx \omega_k \qquad \text{(XII.4-11c)}$$

$$Z_\Delta = \sum_{j \neq k} \frac{1}{j\Omega_j} \frac{v_{j1}^2}{\omega_j \varepsilon} \qquad \Omega_j = \frac{\omega}{\omega_j} - \frac{\omega_j}{\omega} \qquad \text{(XII.4-11d)}$$

The impedance of the coupling device is denoted by Z_c in Fig. 2-1.

When the resonance frequencies are sufficiently far apart, the equivalent circuit of the cavity at a particular resonance frequency can be approximated as shown in Fig. 2-2. The contributions from nonresonant modes are lumped together as Z_Δ.

The analysis and interpretation of some experimental results can be aided by choosing special reference planes. At frequencies far off resonance, both the cavity and its coupling device are highly reactive (for the lossless case). Such reactive termination causes a complete reflection of the incident wave and produces voltage nodes in the input guide. The position of these nodes is called the detuned-short position. At the plane of detuned-short, the cavity can be represented by the simple parallel resonant circuit in Fig. 2-3a.

The voltage standing wave will be maximum at a distance of $\lambda_{go}/4$ from the detuned-short plane, and this position is known as the detuned-open plane, Fig. 2-3b.

The parallel RLC-circuit is used when the detuned-short position is used as the reference plane. This circuit is effectively shorted by the capacitor at frequencies far above resonance and by the inductor at frequencies far below resonance.

The series RLC-circuit should be used when the detuned-open position is chosen as the reference plane. This circuit is effectively an open circuit at frequencies far above resonance (due to the inductor) and far below resonance (due to the capacitor).

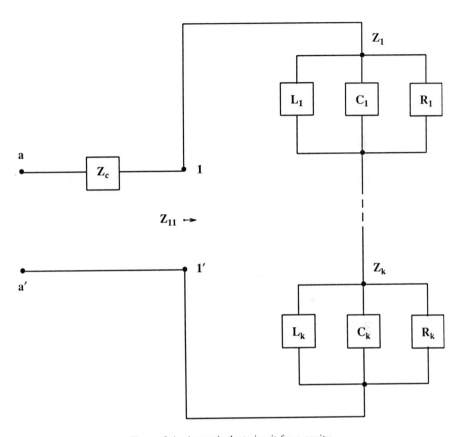

Figure 2-1: An equivalent circuit for a cavity.

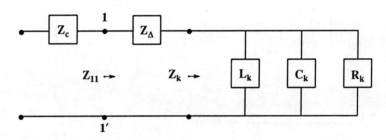

Figure 2-2: An approximate circuit for a cavity.

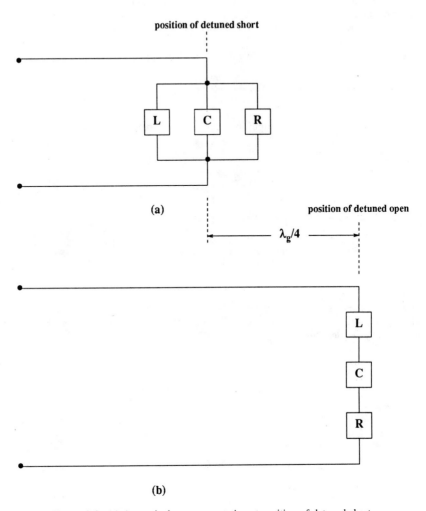

position of detuned short

(a)

position of detuned open

$\lambda_g/4$

(b)

Figure 2-3: (a) An equivalent representation at position of detuned-short.
(b) An equivalent representation at position of detuned-open.

3. Single-Port Resonator - Equivalent Circuit

(a) A single-port resonator is represented by an equivalent parallel resonant circuit, Fig. 3-1. The detuned-short plane 1-1' is chosen as the reference plane. This equivalent circuit is valid only near the resonance frequency for the particular cavity mode and for the particular waveguide mode.

At resonance, the reflection coefficient at plane 1-1' is [Eq. (1-8b)]

$$s_{11} = \frac{b_1}{a_1} = -\frac{\sqrt{1 - m^2} - e^{-\alpha}}{1 - \sqrt{1 - m^2} \, e^{-\alpha}} \tag{1}$$

and the corresponding normalized input impedance (purely resistive at resonance) is

$$\frac{R_o}{Z_o} = \frac{1 + s_{11}}{1 - s_{11}}$$

$$= \frac{1 - \sqrt{1 - m^2} \, e^{-\alpha} - [\sqrt{1 - m^2} - e^{-\alpha}]}{1 - \sqrt{1 - m^2} \, e^{-\alpha} + [\sqrt{1 - m^2} - e^{-\alpha}]} \tag{2\alpha}$$

$$= \frac{1 - \sqrt{1 - m^2}}{1 + \sqrt{1 - m^2}} \frac{1 + e^{-\alpha}}{1 - e^{-\alpha}} \tag{2}$$

The input impedance of a resonator is a function of the coefficient of coupling m and the attenuation constant α. For the convenience of analysis, these effects can be separated by introducing an ideal transformer, T (Fig. 3-2). The circuit elements are related to the original elements in Fig. 3-1 as follows.

$$C_o = \frac{C}{n^2} \tag{3a}$$

$$L_o = n^2 L \tag{3b}$$

$$R_o = n^2 R \tag{3c}$$

A comparison of (3c) with (2) reveals the following equivalences.

$$n^2 = \frac{1 - \sqrt{1 - m^2}}{1 + \sqrt{1 - m^2}} \tag{4a}$$

$$R = Z_o \frac{1 + e^{-\alpha}}{1 - e^{-\alpha}} \tag{4b}$$

Since the normalized impedances or admittances are measured experimentally, therefore a normalized circuit for Fig. 3-1 is useful and this is shown in Fig. 3-3. The normalized circuit elements are given by

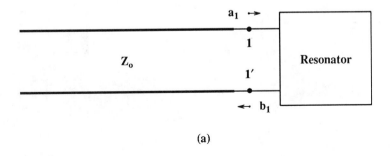

(a)

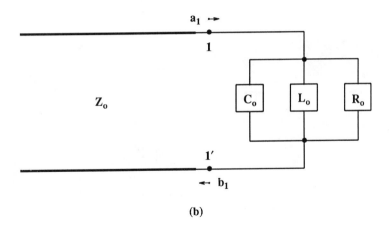

(b)

Figure 3-1: (a) A single-port cavity.
(b) An equivalent circuit for (a).

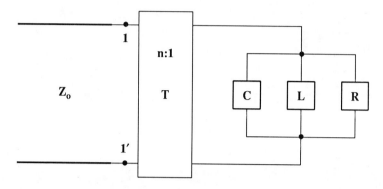

Figure 3-2: A resonator coupled by an ideal transformer.

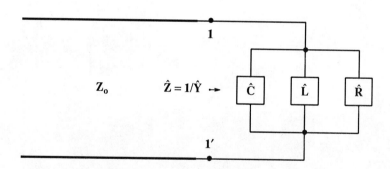

Figure 3-3: A normalized equivalent circuit.

$$\hat{L} = \frac{L_o}{Z_o} = n^2 \frac{L}{Z_o} \tag{5a}$$

$$\hat{C} = C_o Z_o = \frac{C Z_o}{n^2} \tag{5b}$$

$$\hat{R} = \frac{R_o}{Z_o} = n^2 \frac{R}{Z_o} \tag{5c}$$

(b) The normalized input admittance in Fig. 3-3 is

$$\hat{Y} = \hat{G} + j\hat{B} = \frac{1}{\hat{R}} + j \left[\omega\hat{C} - \frac{1}{\omega\hat{L}} \right] \tag{6a}$$

and the corresponding normalized impedance is

$$\hat{Z} = \frac{1}{\hat{Y}} = \frac{1}{\hat{G} + j\hat{B}} = \left[\frac{1}{\hat{R}} + j \left[\omega\hat{C} - \frac{1}{\omega\hat{L}} \right] \right]^{-1} \tag{6b}$$

The normalized susceptance is

$$\hat{B} = \omega\hat{C} - \frac{1}{\omega\hat{L}} = \frac{\hat{C}}{\omega} \left[\omega^2 - \frac{1}{\hat{L}\hat{C}} \right] = \frac{\hat{C}}{\omega} \frac{\omega_o}{\omega_o} \left[\omega^2 - \frac{1}{\hat{L}\hat{C}} \right]$$

$$= \sqrt{\frac{\hat{C}}{\hat{L}}} \frac{1}{\omega\omega_o} \left[\omega^2 - \omega_o^2 \right] \qquad \omega_o^2 = \frac{1}{\hat{L}\hat{C}} \tag{7}$$

Near resonance, $\omega \approx \omega_o$, one has

$$\hat{B} = \sqrt{\frac{\hat{C}}{\hat{L}}} \frac{1}{\omega\omega_o} (\omega + \omega_o)(\omega - \omega_o) \tag{8a}$$

$$\approx \sqrt{\frac{\hat{C}}{\hat{L}}} \frac{2\Delta\omega}{\omega_o} \tag{8b}$$

$$\Delta\omega = \omega - \omega_o \tag{8c}$$

The corresponding $\hat{Z}$ is

$$\hat{Z} \approx \left[\hat{G} + j \sqrt{\frac{\hat{C}}{\hat{L}}} \frac{2\Delta\omega}{\omega_o} \right]^{-1} \qquad \text{for} \quad \omega \approx \omega_o \tag{8d}$$

In the vicinity of resonance, the susceptance $\hat{B}$ is a linear function of frequency.

(c) The locus of the normalized impedance, $\hat{Z}$, is obtained by the inversion of the locus of the normalized admittance, $\hat{Y}$. First, the normalized admittance is plotted on the Smith admittance chart. This locus is a $\hat{G}$ = constant circle. The inversions are then carried out to obtain the locus of $\hat{Z}$ which is shown in Fig. 3-4. The locus of $\hat{Z}$ is an $\hat{R}$ = constant circle.

(d) The vector drawning from the center of the chart to any point on the $\hat{Z}$-locus represents the reflection coefficient, in magnitude and phase, of the corresponding point.

(d.1) Over-coupled case - This is the case when $\hat{R} > 1$. By imposing

$$m > m_c \qquad e^{-\alpha} > \sqrt{1 - m^2} \qquad (1\text{-}16)$$

in (2α) the locus of Γ encloses the origin. The phase of Γ begins at $+180°$ at $\omega = 0$ and decreases as ω increases. It is equal to $0°$ at resonance, $\omega = \omega_{o1}$. It continues to decrease as frequency increases and approaches $-180°$ as ω approaches infinity. The locus of the phase angle of Γ as a function of frequency is sketched in Fig. 3-5.

The magnitude of the voltage of the VSWR pattern is

$$|v| = | v_{inc} (1 + |\Gamma| e^{j\theta_r} e^{-j2\beta s}) |$$

where s is the distance measured from the load. Since θ_r decreases as ω increases, to maintain the same $|v|$, the position s must decrease. In other words, the position of the voltage minimum moves toward the load as frequency increases. This is known as the normal shift of the voltage minimum. By virtue of Fig. 3-5, the locus of minimum Γ as a function of the distance, s, moves continuously toward the load as frequency increases.

(d.2) Under-coupled case - In this case, $\hat{R} < 1$. Since

$$m < m_c \qquad e^{-\alpha} < \sqrt{1 - |m|^2} \qquad (1\text{-}14)$$

the locus of Γ excludes the origin. The phase of Γ begins at $+180°$ for $\omega = 0$ and decreases as frequency increases. It then increases and equals $+180°$ at $\omega = \omega_{o2}$. For $\omega > \omega_{o2}$, the phase continues to increase briefly before decreasing to $+180°$, (Fig. 3-6). The position of the voltage minimum shifts toward the load as θ_r decreases and then shifts toward the source when θ_r increases. This is called the reverse shift of the position of the minimum.

(d.3) Critical coupling - This is defined by $\hat{R} = 1$. The locus of Γ passes through the origin of the Smith chart. The phase of Γ is discontinuous at resonance, $\omega = \omega_{o3}$.

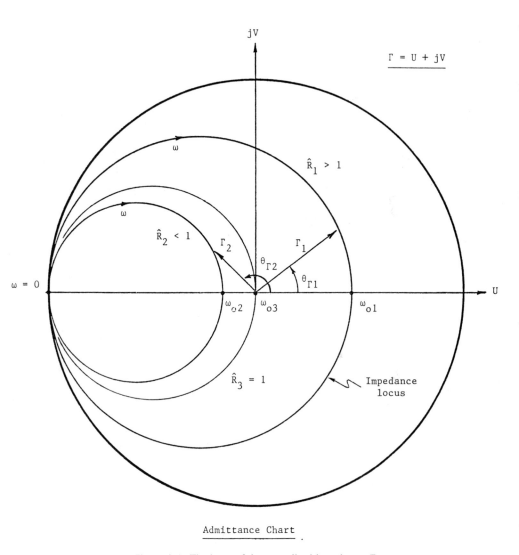

Admittance Chart

Figure 3-4: The locus of the normalized impedance, Z.

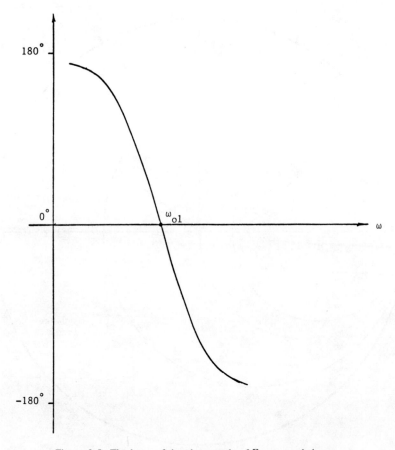

Figure 3-5: The locus of the phase angle of Γ-overcoupled case.

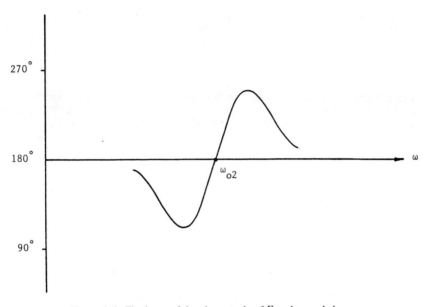

Figure 3-6: The locus of the phase angle of Γ-undercoupled case.

It changes from $+90°$ to $-90°$ when the limit is taken from the over-coupled case, $\hat{R} > 1$. It changes from $+90°$ to $+270°$ if the limit is taken from the under-coupled case, $\hat{R} < 1$ (Fig. 3-7).

(e) $\hat{R}$ can be determined by measuring the VSWR at resonance and identifying the condition of coupling by examining the shift of its minimum - normal or reverse. Once the condition of coupling is known, the normalized resistance $\hat{R}$ can be calculated from the VSWR.

$$\hat{R} = S_o \geq 1 \qquad\qquad \text{over–coupling} \qquad\qquad\qquad (9a)$$

$$\hat{R} = \frac{1}{S_o} \leq 1 \qquad\qquad \text{under–coupling} \qquad\qquad\qquad (9b)$$

$$S_o = \text{VSWR at} \quad \omega = \omega_o \qquad\qquad\qquad\qquad (9c)$$

With $\hat{R}$ known, one can then construct the locus of Γ. An additional impedance measurement at a frequency in the vicinity of resonance will determine $\sqrt{\hat{C}/\hat{L}}$ from (8), since $\omega_o^2 = 1/(\hat{L}\hat{C})$. Thus, the parameters $\hat{G}$, $\hat{L}$, and $\hat{C}$ are completely determined.

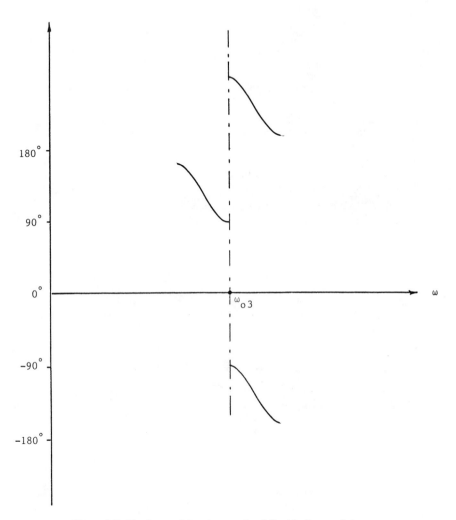

Figure 3-7: The locus of the phase angle of Γ-critically coupled case.

4. Unloaded Q

The general definition for the quality factor, Q, of a resonant circuit is

$$Q = \frac{<\text{energy stored}>}{<\text{energy dissipated per radian}>} \Bigg|_{\omega = \omega_o} \tag{1}$$

Since the definition of Q is based upon energies and frequency, it is therefore independent of the choice of the equivalent circuit.

(a) The unloaded Q is defined by modifying the general definition (1) as follows.

$$Q_u = \frac{<\text{energy stored in cavity}>}{<\text{energy dissipated within cavity per radian}>} \Bigg|_{\omega = \omega_o} \tag{2}$$

The unloaded Q depends on the losses in the cavity and its modes. It is independent of the coupling. When the cavity is represented by an equivalent parallel RLC-circuit, Fig. 4-1, the unloaded Q is found to be

$$\hat{Q}_u = \omega_o \hat{C} \hat{R} = \frac{1}{\hat{G}} \sqrt{\frac{\hat{C}}{\hat{L}}} \tag{3a}$$

$$\omega_o = \frac{1}{\sqrt{\hat{L}\hat{C}}} \qquad \hat{G} = \frac{1}{\hat{R}} \tag{3b}$$

The unloaded Q can be expressed in terms of the bandwidth of the cavity. The bandwidth is the range of frequencies between half-power frequencies where the power dissipation is one-half of that at resonance.

Let the following quantities be defined.

$$\omega^{\pm} = \text{upper and lower half-power frequencies} \tag{4a}$$

$$P^{\pm} = \text{power dissipated at } \omega^{\pm} = \frac{<P_o>}{2} \tag{4b}$$

$$<P_o> = \text{power dissipated at resonance} = \frac{|v_o|^2}{2\hat{R}} \tag{4c}$$

$$v_o = \text{voltage at resonance} \quad \text{and} \quad v^{\pm} = \text{voltage at } \omega^{\pm} \tag{4d}$$

$$\hat{Z} = \hat{R} = \text{normalized impedance at resonance} \tag{4e}$$

$$Z^{\pm} \equiv \text{normalized impedance at } \omega^{\pm} \tag{4f}$$

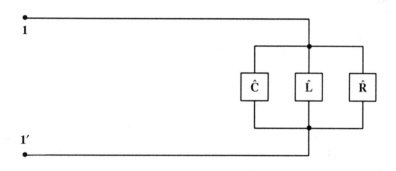

Figure 4-1: The parallel RLC equivalent circuit.

The dissipated power at half-power frequencies is also given by

$$< P^\pm > \; = \; \frac{|v^\pm|^2}{2\hat{R}} \tag{5}$$

The substitution of (4c) and (5) into (4b) yields

$$v^\pm \; = \; \frac{|v_0|}{\sqrt{2}} \tag{6}$$

With a constant-current source, one has

$$v^\pm \; = \; iZ^\pm \; = \; \frac{v_0}{\sqrt{2}} \; = \; \frac{1}{\sqrt{2}} \, i\hat{R}$$

or

$$Z^\pm \; = \; \frac{\hat{R}}{\sqrt{2}} \tag{7}$$

The input admittance of a resonator as shown in Fig. 4-1 is given by

$$\hat{Y} \; = \; \hat{G} + j\hat{B} \; = \; \frac{1}{\hat{R}} + j \left(\omega\hat{C} - \frac{1}{\omega\hat{L}} \right) \tag{8a}$$

$$\hat{Z} \; = \; \frac{1}{\hat{G} + j\hat{B}} \tag{8b}$$

At the half-power frequency, one has

$$Z^\pm \; = \; \frac{1}{\dfrac{1}{\hat{R}} + j\hat{B}^\pm} \; = \; \frac{\hat{R}}{1 + j\hat{R}\hat{B}^\pm} \qquad \hat{B}^\pm \; = \; \hat{B}(\omega = \omega^\pm) \tag{9}$$

For (9) to be equal to (7), one concludes that

$$\hat{R}\hat{B}^\pm \; = \; \pm 1 \tag{10}$$

Let the half-power frequencies be represented by

$$\omega^\pm \; = \; \omega_0 \pm \Delta\omega \tag{11a}$$

The substitution of (11a) into the expression for $\hat{B}$ in (8a) gives

$$\hat{B}^{\pm} = \hat{C} \left[(\omega_o \pm \Delta\omega) - \frac{\omega_o}{1 \pm \dfrac{\Delta\omega}{\omega_o}} \right]$$

$$\approx \hat{C} \left[\omega_o \pm \Delta\omega - \omega_o \left(1 \mp \frac{\Delta\omega}{\omega_o} \right) \right] = \pm \sqrt{\frac{\hat{C}}{\hat{L}}} \frac{2\Delta\omega}{\omega_o} \tag{11b}$$

The substitution of (11b) into (10) produces

$$\hat{R} = \pm \frac{1}{\hat{B}^{\pm}} = \pm \sqrt{\frac{\hat{L}}{\hat{C}}} \frac{\omega_o}{2\Delta\omega} \tag{12}$$

The unloaded Q can be expressed in terms of the bandwidth by eliminating $\hat{R}$ in (3a) by (12).

$$\hat{Q}_u = \omega_o \hat{C} \sqrt{\frac{\hat{L}}{\hat{C}}} \frac{\omega_o}{2\Delta\omega} = \frac{\omega_o}{BW} \tag{13a}$$

$$BW = 2\Delta\omega = \text{bandwidth} \tag{13b}$$

(b) The locus of $\hat{B} = \pm \hat{G}$ can be determined as follows. Consider the Smith chart as an impedance chart. The normalized impedance is related to the reflection coefficient in a loss-less guide by

$$\hat{Z} = \frac{1 + \Gamma e^{-j2\beta s}}{1 - \Gamma e^{-j2\beta s}} = \frac{1 + U + jV}{1 - U - jV} \tag{14a}$$

$$\Gamma e^{-j2\beta s} \equiv U + jV \tag{14b}$$

$$\hat{Z} = \hat{R} + j\hat{X} \tag{14c}$$

The corresponding normalized admittance is

$$\hat{Y} = \frac{1}{\hat{Z}} = \frac{1 - U - jV}{1 + U + jV} = \hat{G} + j\hat{B} \tag{15a}$$

$$\hat{G} = \frac{1}{D} (1 - U^2 - V^2) \tag{15b}$$

$$\hat{B} = \frac{-2V}{D} \tag{15c}$$

$$D = (1 + U)^2 + V^2 \tag{15d}$$

At a half-power point, (10) imposes $\hat{B} = \hat{G}$, and (15b) and (15c) give

$$1 - U^2 - V^2 = - 2V \qquad \text{or} \qquad U^2 + (V - 1)^2 = 2 \tag{16}$$

This is an equation for a circle with a radius equal to $\sqrt{2}$. The center of the circle is located at ($U = 0$, $V = 1$) in the impedance chart, Fig. 4-2.

For $\hat{B} = -\hat{G}$, one has

$$U^2 + (V + 1)^2 = 2 \tag{17}$$

This represents a circle whose center is located at ($U = 0$, $V = -1$).

The locus of half-power points of $\hat{Q}_u$ are the $\hat{B} = \pm \hat{G}$ circles in the admittance chart, Fig. 4-2.

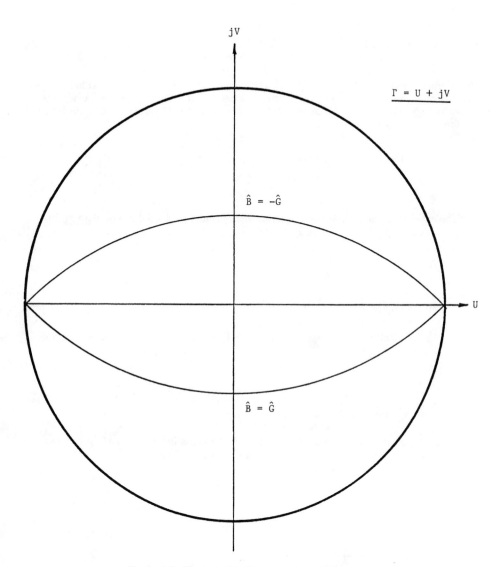

Figure 4-2: The loci of half-power points of Q.

5. External Q

The definition for the external Q, Q_e, is

$$Q_e = \left. \frac{< \text{energy stored in cavity} >}{< \text{energy dissipated in external circuit per radian} >} \right|_{\omega=\omega_0} \tag{1}$$

The external circuit is understood to be a passive circuit, i.e., with all sources killed but their internal impedances remain in circuit. The equivalent circuit for determining the external Q is shown in Fig. 5-1. In accordance with the Q for a parallel RLC-circuit, the external Q is therefore

$$\hat{Q}_e = \omega_0 \hat{C} \hat{R}_g = \omega_0 \hat{C} = \sqrt{\frac{\hat{C}}{\hat{L}}} \qquad \omega_0 = \frac{1}{\sqrt{\hat{L}\hat{C}}} \tag{2}$$

for matched source impedance, $\hat{R}_g = 1$.

The external Q can be expressed in terms of bandwidth by a similar procedure as for the loaded Q by using $\hat{R}_g$ instead of $\hat{R}$.

$$\hat{Q}_e = \frac{\omega_0}{2\Delta\omega_e} = \frac{\omega_0}{2BW_e} \tag{3a}$$

$$\omega_e^\pm = \omega_0 \pm \Delta\omega_e \tag{3b}$$

where $\omega_e^\pm$ are the half-power frequencies for Fig. 5-1, which satisfy the following condition.

$$\hat{B} = \pm 1 \qquad \text{at } \omega = \omega_e^\pm \tag{4a}$$

$$\hat{B} = \pm \sqrt{\frac{\hat{C}}{\hat{L}}} \frac{2\Delta\omega_e}{\omega_0} \qquad (4\text{–}11b) \tag{4b}$$

The locus of (4a) is shown in Fig. 5-2. The equivalent circuit for the coupling between the cavity and the exciting source is represented by Fig. 3-2. The normalized parameters are related to the intrinsic parameters as follows.

$$\hat{C} = \frac{Z_0 C_0}{n^2} \qquad \text{and} \qquad \hat{L} = \frac{n^2 L_0}{Z_0} \tag{5}$$

Hence (2) becomes

$$\hat{Q}_e = \frac{Z_0}{n^2} \sqrt{\frac{C_0}{L_0}} \tag{6}$$

Therefore, the external Q is a function which is dependent upon the coupling device.

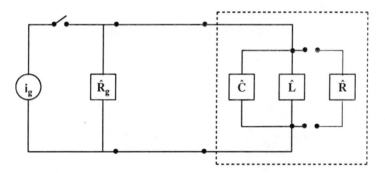

Figure 5-1: The equivalent circuit for determining the external Q.

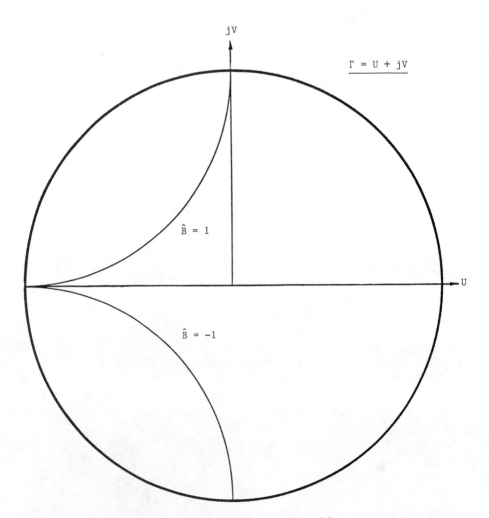

Figure 5-2: The loci of half-power points of Q_e.

6. Loaded Q

(a) The definition for the loaded Q, Q_L, is

$$Q_L = \frac{< \text{energy stored in cavity} >}{< \text{energy dissipated in both cavity \& external circuit} >} \Bigg|_{\omega=\omega_o} \tag{1}$$

The equivalent circuit for evaluating the loaded Q is as shown in Fig. 6-1. The loaded Q can be obtained from (4-3) by replacing $\hat{R}$ with the parallel combination of $\hat{R}$ and $\hat{R}_g$.

$$\hat{Q}_L = \omega_o \hat{C} \frac{\hat{R}}{\hat{R} + 1} = \sqrt{\frac{\hat{C}}{\hat{L}}} \frac{1}{1 + \hat{G}} \tag{2}$$

for a matched source, $\hat{R}_g = 1$.

The loaded Q may also be expressed in terms of the bandwidth.

$$\hat{Q}_L = \frac{\omega_o}{2\Delta\omega_L} \tag{3a}$$

$$\omega_L^{\pm} = \omega_o + \Delta\omega_L \tag{3b}$$

where $\omega_L^{\pm}$ are the half-power frequencies for the loaded cavity and satisfy the following condition.

$$\hat{B}^{\pm} = \sqrt{\frac{\hat{C}}{\hat{L}}} \frac{2\Delta\omega_L}{\omega_o} = \pm(1 + \hat{G}) \qquad \text{at } \omega = \omega_L^{\pm} \tag{4}$$

since the input admittance (Fig. 6-1) is

$$\hat{Y} = (1 + \hat{G}) + j\left(\omega\hat{C} - \frac{1}{\omega\hat{L}}\right) = 1 + \hat{G} + j\hat{B} \tag{5}$$

(b) The locus of $\hat{B}^{\pm} = \pm(1 + \hat{G})$ can be determined from (4-15).

$$\hat{G} = \frac{1}{D}(1 - U^2 - V^2) \tag{4-15b}$$

$$\hat{B} = \frac{-2V}{D} \tag{4-15c}$$

$$D = (1 + U)^2 + V^2 \tag{4-15d}$$

The substitution of (4-15) into (4) produces

$$1 + \hat{G} = \frac{1}{D}[1 - U^2 - V^2 + (1 + U)^2 + V^2] = \hat{B} = \frac{-2V}{D} \tag{6}$$

or

$$\frac{V}{U + 1} = -1 = \tan(-45°) \tag{7}$$

This is a straight line inclined at an angle of $-45°$ with respect to the $+U$-axis. It joins the point ($U = 0$, $V = -1$) and the point ($U = -1$, $V = 0$), Fig. 6-2.

For $\hat{B} = -(1 + \hat{G})$, one has

$$\frac{V}{U + 1} = 1 = \tan 45° \tag{8}$$

This is a straight line inclined at $45°$ with respect to the real axis. It connects the points $(-1, 0)$ and $(0, 1)$.

The loci of half-power points of $\hat{Q}_L$ are the straight lines, $\hat{B} = \pm(1 + \hat{G})$, in the admittance chart, Fig. 6-2.

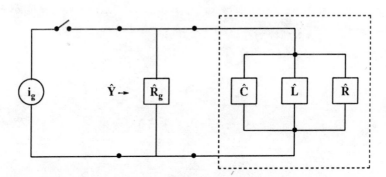

Figure 6-1: The equivalent circuit for determining the loaded Q.

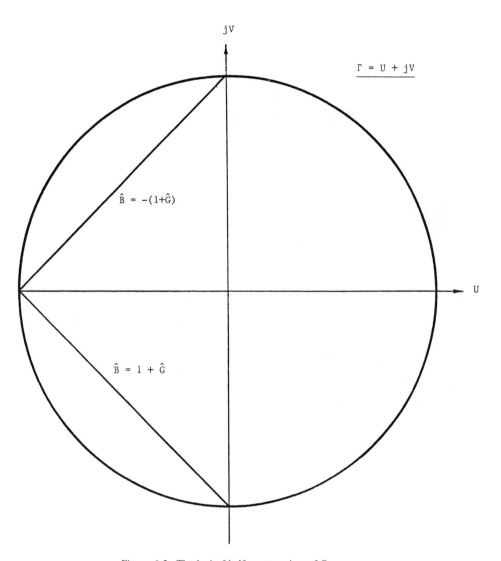

Figure 6-2: The loci of half-power points of Q_L.

7. Power Absorbed by a Cavity

Most microwave sources deliver constant power or may be made to do so. With constant input power P_i, the power absorbed by the cavity P_c is

$$P_c = P_i (1 - |\Gamma|^2) \tag{1a}$$

$$\Gamma = \frac{\hat{Z} - 1}{\hat{Z} + 1} \tag{1b}$$

The normalized impedance of a cavity is, from (4-8b) and (4-11),

$$\hat{Z} = \frac{1}{\hat{G} + j\hat{B}} = \frac{1}{\hat{G} + j\sqrt{\dfrac{\hat{C}}{\hat{L}}} \dfrac{2\Delta\omega}{\omega_o}} \qquad \text{at } \omega \approx \omega_o \tag{2}$$

The magnitude squared of the reflection coefficient is

$$|\Gamma|^2 = \text{magnitude of} \left[\frac{\dfrac{1}{\hat{G} + j\hat{B}} - 1}{\dfrac{1}{\hat{G} + j\hat{B}} + 1} \right]^2$$

$$= \frac{(1 - \hat{G})^2 + \hat{B}^2}{(1 + \hat{G})^2 + \hat{B}^2}$$

The power absorbed by the cavity is [Eq. (1a)]

$$P_c = P_i \frac{(1 + \hat{G})^2 - (1 - \hat{G})^2}{(1 + \hat{G})^2 + \hat{B}^2} = P_i \frac{4\hat{G}}{(1 + \hat{G})^2} \frac{1}{1 + \dfrac{\hat{B}^2}{(1 + \hat{G})^2}}$$

$$= P_i \frac{4\hat{Q}_L^2}{\hat{Q}_e \hat{Q}_u} \frac{1}{1 + \hat{Q}_L^2 \left[\dfrac{2\Delta\omega}{\omega_o} \right]^2} \qquad \text{for } P_i = \text{constant} \tag{3a}$$

$$\hat{G} = \frac{\hat{Q}_e}{\hat{Q}_u} \qquad \text{and} \qquad 1 + \hat{G} = \frac{\hat{Q}_e}{\hat{Q}_L} \tag{3b}$$

$$\hat{Q}_L = \sqrt{\frac{\hat{C}}{\hat{L}}} \frac{1}{1 + \hat{G}} \tag{6-2}$$

$$\hat{Q}_e = \sqrt{\frac{\hat{C}}{\hat{L}}} \tag{5-2}$$

$$\hat{Q}_u = \frac{1}{\hat{G}} \sqrt{\frac{\hat{C}}{\hat{L}}} \tag{4-3}$$

The power absorbed by the cavity can be expressed in terms of Q's and the frequency, with the lumped parameters of the cavity eliminated.

At resonance, $\Delta\omega = 0$, the power absorbed by the cavity is

$$P_c = P_i \frac{4\hat{Q}_L^2}{\hat{Q}_e \hat{Q}_u} \qquad \text{at } \omega = \omega_0 \tag{4a}$$

At half-power frequencies,

$$P_c^{\pm} \equiv P_c(\omega^{\pm}) = P_c = \frac{P_c \text{ at resonance}}{2} \qquad \text{for } P_i = \text{constant} \tag{4b}$$

For Eq. (3a) to produce (4b), one must have

$$\left[\hat{Q}_L \frac{2 \Delta\omega}{\omega_0} \right]^2 = 1 \qquad \text{or} \qquad \hat{Q}_L = \frac{\omega_0}{2\Delta\omega} = \frac{\omega_0}{BW} \tag{5}$$

The loaded Q is also defined by (6-3a)

$$\hat{Q}_L = \frac{\omega_0}{2\Delta\omega_L} = \frac{\omega_0}{BW_L} \tag{6-3a}$$

Hence, the bandwidth in (5) and that in (6-3a) must be identical, i.e.,

$$\Delta\omega = \Delta\omega_L$$

In other words, $\hat{Q}_L$ and its associated bandwidth provide a direct measure of the sharpness of power absorption from a constant-power source. It is to be noted that $2\Delta\omega_L$ as defined in (6-3a) includes both the power absorbed by the cavity and that in the internal impedance of the current source.

8. Experimental Determination of Q's

The measurment procedures for determining Q's experimentally are as follows:

(a) The detuned-short position is located by a standing-wave detector with the cavity tuned far off resonance.

(b) The cavity is then adjusted to resonance.

(c) The voltage standing-wave pattern is explored to determine whether the cavity is over-coupled or under-coupled - Section 3(e).

(d) The VSWR at resonance, S_o, is measured and the coefficient of coupling is calculated - Eq. (3-2).

(e) The locus of $\Gamma(\omega)$ is constructed - Section 3(d).

(f) The loci of

$$\hat{G} = \hat{X} ; \qquad \hat{B} = \pm (1 + \hat{G}) ; \qquad \text{and} \qquad \hat{B} = \pm 1$$

are constructed - Section (4).

(g) The half-power values of Q_L, Q_u, and Q_e are located on the plot - Section 6.

All half-power points derived in previous sections are summarized in Fig. 8-1.

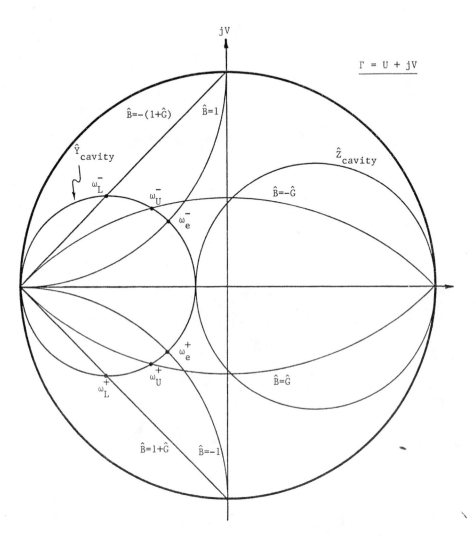

Figure 8-1: Some essential properties of a cavity-admittance representation.

9. Frequency Scale

The distribution of the measured points along the $\hat{R}$ = constant (or $\hat{G}$ = constant) circle is not a linear function of frequency. It is difficult to determine the exact half-power frequencies between data points. However, an auxiliary linear frequency scale can be established as follows.

Construct line MN perpendicular to the horizontal axis of the admittance chart, Fig. 9-1, (in this figure, line MN is chosen to be the jV-axis).

Let the experimental points be 1, 2, . . ., n, and the corresponding frequencies are f_1, f_2, . . ., f_n, respectively. Connect the R = 0 point to each data point and extend the lines to the vertical line MN. The intersections with line MN will be shown to vary linearly with frequency, and thereby providing an auxiliary frequency scale.

The reflection coefficient of the impedance data point is given by

$$\Gamma = \frac{\hat{Z} - 1}{\hat{Z} + 1} \tag{1a}$$

$$\hat{Z} = \frac{1}{\hat{G} + j\hat{B}} \tag{1b}$$

$$\hat{G} = \frac{1}{\hat{R}} \tag{1c}$$

$$\hat{B} = \omega\hat{C} - \frac{1}{\omega\hat{L}} \tag{1d}$$

$$\approx \sqrt{\frac{\hat{C}}{\hat{L}}} \frac{2\Delta\omega}{\omega_0} \qquad \text{at } \omega \approx \omega_0 \tag{1e}$$

The quantity $(1 + \Gamma)$ is then obtained from (1).

$$1 + \Gamma = \frac{2\hat{Z}}{\hat{Z} + 1} = \frac{2}{1 + \hat{G} + j\hat{B}}$$

$$= \frac{2(1 + \hat{G} - j\hat{B})}{(1 + \hat{G})^2 + \hat{B}^2} = |1 + \Gamma| e^{j\Phi} \tag{2a}$$

$$\Phi = \tan^{-1}\frac{-\hat{B}}{1 + \hat{G}} = \tan^{-1}\frac{-\sqrt{\dfrac{\hat{C}}{\hat{L}}}\dfrac{2\Delta\omega}{\omega_0}}{1 + \hat{G}} \tag{2b}$$

The interceptions along the MN line are proportional to

$$\tan\Phi = -2\sqrt{\frac{\hat{C}}{\hat{L}}}\frac{1}{1 + \hat{G}}\frac{\Delta\omega}{\omega_0} \qquad \omega \approx \omega_0 \tag{3}$$

and hence are linearly proportional to the frequency.

$$\Gamma = U + jV$$

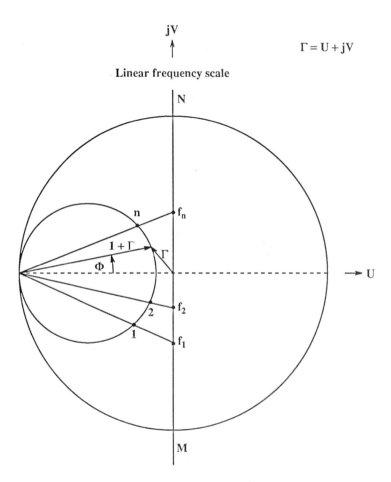

Figure 9-1: Frequency scale.

9.1. Example: Single-Port Resonator

A single-port resonator is made of a waveguide terminated by a short circuit at its output port. The input port is connected to a coupling junction, such as an aperture or a thin post. The distance between the input port and the short circuit is L_1 (Fig.9.1-1).

(a) If the coupling device is represented by a shunt normalized reactance, $j\hat{X}_d$, determine the coupling coefficient of the device.

(b) Determine the total length L_1 of the resonator.

(c) Determine the loaded and external Q's.

Solution :

(a) The scattering matrix of a coupling junction with respect to the detuned-short position is [Eq. (1-4)]

$$[\, S_m \,] = \begin{bmatrix} -\sqrt{1-m^2} & jm \\ jm & -\sqrt{1-m^2} \end{bmatrix} \tag{1}$$

where m is the coefficient of coupling.

The scattering coefficient of a shunt impedance, Fig. 9.1-2, is [Eq. (III.9.1-3)]

$$[\, S_Z \,] = \frac{1}{1+2\hat{Z}} \begin{bmatrix} -1 & 2\hat{Z} \\ 2\hat{Z} & -1 \end{bmatrix} \tag{2}$$

For a lumped element, the matrix $[S_Z]$ is valid at the location of the impedance element, i.e., there is no physical distance between planes 1-1′ and 2-2′, in Fig. 9.1-2.

For a purely reactive element, $\hat{Z} = j\hat{X}_d$, (2) becomes

$$[\, S_d' \,] = \frac{1}{1+j2\hat{X}_d} \begin{bmatrix} -1 & j2\hat{X}_d \\ j2\hat{X}_d & -1 \end{bmatrix} \tag{3}$$

The problem is to convert $[S_d']$ into the form of $[S_m]$. This requires the specification of the following conditions.

$$s_{11} = s_{22} = -|s_{11}| \qquad \text{and} \qquad s_{12} = s_{21} \tag{4}$$

The matrix $[S_d']$ already satisfies the second condition in (4). To conform to the first condition in (4), it is necessary to make a change of reference planes. Let the matrix with respect to the new reference planes be

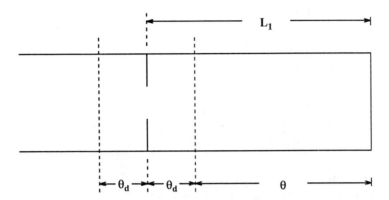

Figure 9.1-1: Arrangement of a single-port resonator.

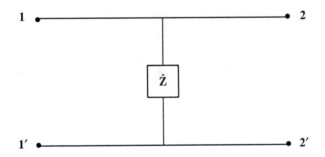

Figure 9.1-2: An impedance connected in shunt.

$$[S_d] = \frac{e^{j\theta_d}}{1 + j2\hat{X}_d} \begin{bmatrix} -1 & j2\hat{X}_d \\ j2\hat{X}_d & -1 \end{bmatrix} \tag{5}$$

where θ_d is the shift of reference plane in electrical length. The forward reflection coefficient is

$$s_{11} = -\frac{e^{j\theta_d}}{1 + j2\hat{X}_d} = \frac{-1}{|1 + j2\hat{X}_d|} e^{j\phi} \tag{6a}$$

$$\phi \equiv \theta_d - \tan^{-1} 2\hat{X}_d \tag{6b}$$

In order that $s_{11} = -|s_{11}|$, one must have

$$\phi = 2n\pi \quad \text{or} \quad \theta_d = 2n\pi + \tan^{-1} 2\hat{X}_d \quad n = 0, 1, 2, \ldots \tag{7a}$$

The minimum shift results when $n = 0$.

$$\theta_d = \tan^{-1} 2\hat{X}_d \tag{7b}$$

In accordance with this reference plane, (5) becomes

$$[S_d] = \frac{1}{\sqrt{1 + 4\hat{X}_d^2}} \begin{bmatrix} -1 & 2j\hat{X}_d \\ 2j\hat{X}_d & -1 \end{bmatrix} \tag{8}$$

The coupling coefficient can be obtained by the identification of corresponding terms between (1) and (8).

$$-\sqrt{1 - m^2} = \frac{-1}{\sqrt{1 + 4\hat{X}_d^2}} \quad \text{or} \quad m = \frac{2\hat{X}_d}{\sqrt{1 + 4\hat{X}_d^2}} = \frac{2}{\sqrt{4 + \hat{B}_d^2}} \tag{9a}$$

$$\hat{B}_d = \frac{1}{\hat{X}_d} \tag{9b}$$

The coupling coefficient is positive if the coupling device is inductive, and is negative when the device is capacitive.

(b) The condition of resonance, (1-7), requires that

$$\theta = n\pi \quad n = 0, 1, 2, \ldots \tag{1-7}$$

Therefore,

$$\beta L_1 = \theta + \theta_d \qquad \text{or} \qquad L_1 = \frac{\lambda_{go}}{2\pi} \left(n\pi + \tan^{-1} \frac{2}{\hat{B}_d} \right) \qquad n = 0, 1, 2, \ldots \qquad (10)$$

where λ_{go} is the guide wavelength at resonance.

(c) The expression for the total power flow, $< P_t >$, can be obtained as the product of the velocity of energy transport, v_g, and the total energy stored in the field per unit volume, $<w>$.

$$< P_t > = < w > v_g \qquad \text{or} \qquad < w > = \frac{< P_t >}{v_g} \qquad (11\#)$$

The total power flow $<P_t>$ is the sum of the power flows $<P_{a2}>$ and $<P_{b2}>$ of the a_2 and the b_2 wave, respectively.

$$< P_t > = < P_{a2} > + < P_{b2} >$$

If the cavity is constructed of a very good conductor, then

$$a_2 \approx b_2 \qquad \text{and} \qquad < P_{a2} > \approx < P_{b2} > \equiv < P >$$

or

$$< P_t > \approx < 2P >$$

The total stored energy per unit cross-sectional area in the cavity is

$$< W > = L_1 < w > = \frac{< 2P >}{v_g} L_1 \qquad (11)$$

If the wave is attenuated by a factor of α in the distance $2L_1$, the total length of travel, then the dissipated power per unit cross-sectional area per radian is

$$< w_{dis} > = \frac{2\alpha < P >}{\omega_0} \qquad (12)$$

The unloaded Q is

$$Q_u = \frac{< W >}{< w_{dis} >} = \frac{\omega_0 L_1}{\alpha v_g} = \frac{\pi \lambda_g}{\alpha \lambda_0^2} \qquad (13a)$$

$$v_g = \frac{\lambda_o}{\lambda_g} c \qquad \text{and} \qquad \frac{\omega_o}{v_g} = \frac{2\pi \dfrac{f_o}{c}}{\dfrac{\lambda_o}{\lambda_g}} = 2\pi \frac{\lambda_g}{\lambda_o^2} \tag{13b}$$

$$\bar{\alpha} = \frac{\alpha}{2L_1} = \text{average attenuation} \tag{13c}$$

The power dissipated per radian in the external circuit is

$$< P_e > = \frac{1}{2\omega_o} |b_1|^2 = \frac{1}{2\omega_o} |ma_2|^2 \approx \frac{m^2}{\omega_o} < P > \tag{14}$$

The external Q is

$$Q_e = \frac{< W >}{< P_e >} = \frac{2L_1 \omega_o}{m^2 v_g} = \frac{2L_1}{m^2} 2\pi \frac{\lambda_g}{\lambda_o^2}$$

$$= \pi L_1 \frac{\lambda_g}{\lambda_o^2} (4 + \hat{B}_d^2) \tag{15}$$

where m is given by (9).

It is assumed that the coupling device is thin and lossless and possesses no stored energy. A single dominant mode exists in the waveguide and other resonances are widely separated. The waveguide is also taken to be small.

10. Two-Port Cavity

(a) A two-port cavity has an input port connected to the source and an output port connected to the load.

The resonance behavior at the input looks like that of the one-port cavity. Most of the incident wave will be reflected by the discontinuity at the input port. A small portion of the incident energy will be trapped within the cavity and re-radiated back to the input guide. Power is absorbed by the cavity in the form of losses and transmission to the output port.

A two-port cavity can be represented by the equivalent circuit, Fig. 10-1. The intrinsic parameters of the cavity are C_o, L_o, and R_o. The coupling devices are represented by ideal transformers. The referemce planes 1–1′ and 2–2′ are planes of detuned-short where impedances are zero at frequencies far away from resonance.

Figure 10-1 can be simplified by the elimination of transformers, as is in Fig. 10-2. The normalized parameters are related to the intrinsic quantities as follows:

$$\hat{R} = n_1^2 \frac{R_o}{Z_{o1}} \tag{1a}$$

$$\hat{L} = n_1^2 \frac{L_o}{Z_{o1}} \tag{1b}$$

$$\hat{C} = \frac{1}{n_1^2} Z_{o1} C_o \tag{1c}$$

$$\hat{R}_L = \frac{n_1^2}{n_2^2} \frac{R_L}{Z_{o1}} \tag{1d}$$

The resonance frequency ω_o is defined as the frequency when the input impedance is real at the plane of detuned-short with the other port match-terminated.

$$\omega_o = \frac{1}{\sqrt{L_o C_o}} = \frac{1}{\sqrt{\hat{L}\hat{C}}} \tag{2}$$

The input impedance or admittance of a two-port cavity is identical to that of a single-port cavity provided the cavity resistance $\hat{R}$ in (3-6) is replaced by $\hat{R}'$, defined by

$$\hat{R}' \equiv \frac{\hat{R}\hat{R}_L}{\hat{R} + \hat{R}_L} \tag{3}$$

$\hat{R}'$ is the parallel combination of $\hat{R}$ and $\hat{R}_L$ in Fig. 10-2b. If the cavity is symmetrical,

$$n_1 = n_2 \qquad \text{and} \qquad Z_{o1} = Z_{o2}$$

then $\hat{R}_L = 1$ for a matched load so that

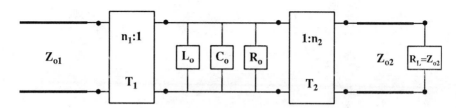

Figure 10-1: An equivalent circuit for a two-port cavity.

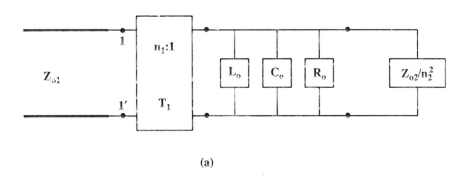

(a)

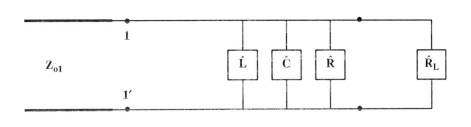

(b)

Figure 10-2: Simplified circuits for a two-port cavity.

$$\hat{R}' = \frac{\hat{R}}{\hat{R} + 1} < 1 \tag{3α}$$

and this is the under-coupled case.

The unloaded Q is defined by

$$Q_u = \omega_o \hat{C}\hat{R} \tag{4}$$

There are two external Q's, one for each port.

$$Q_{ei} = \omega_o \hat{C}\hat{R}_g = \text{external Q for the input port} = \omega\hat{C} \quad \text{for } \hat{R}_g = 1 \tag{5a}$$

$$Q_{eo} = \omega_o \hat{C} \hat{R}_L = \text{external Q for the output port} \tag{5b}$$

$$\frac{Q_{ei}}{Q_{eo}} = \frac{\omega_o \dfrac{Z_{o1}C_o}{n_1^2}}{\omega_o C_o \dfrac{R_L}{n_2^2}} = \frac{n_2^2}{n_1^2} \frac{Z_{o1}}{R_L} \tag{5c}$$

Equation (1) is used to obtain (5c). This ratio is a measure of the degree of coupling.

The loaded Q will involve the parallel combination of three resistors.

$$Q_L = \omega_o \hat{C}\hat{R}'' \tag{6a}$$

$$\hat{R}'' = \left[\frac{1}{1} + \frac{1}{\hat{R}} + \frac{1}{\hat{R}_L} \right]^{-1} \qquad \hat{R}_g = 1 \tag{6b}$$

Equation (6) can be rearranged as

$$Q_L = \frac{\omega_o \hat{C}}{\dfrac{1}{\hat{R}} + \dfrac{1}{1} + \dfrac{1}{\hat{R}_L}}$$

or

$$\frac{1}{Q_L} = \frac{1}{\omega_o \hat{C}} \left[\frac{1}{\hat{R}} + \frac{1}{1} + \frac{1}{\hat{R}_L} \right] = \frac{1}{Q_u} + \frac{1}{Q_{ei}} + \frac{1}{Q_{eo}} \tag{7}$$

The following relations involving Q's and R's are useful.

$$\hat{R} = \frac{Q_u}{Q_{ei}} \tag{8a}$$

$$\hat{R}_L = \frac{Q_{eo}}{Q_{ei}} \tag{8b}$$

for a matched source, $\hat{R}_g = 1$, and from (3) one has

$$\hat{R}' = \frac{\dfrac{Q_u}{Q_{ei}} \dfrac{Q_{eo}}{Q_{ei}}}{\dfrac{Q_u}{Q_{ei}} + \dfrac{Q_{eo}}{Q_{ei}}} = \frac{Q_u Q_{eo}}{Q_{ei}(Q_u + Q_{eo})} \tag{8c}$$

(b) The dissipated power P_{stru} in both the cavity and the load is obtained from (7.3) with an appropriate change of parameters.

$$P_{stru} = P_i \frac{4\hat{G}}{(1+\hat{G})^2} \frac{1}{1 + \dfrac{\hat{B}^2}{(1+\hat{G})^2}}$$

$$= P_i \frac{4\hat{G}}{(1+\hat{G})^2} \frac{1}{1 + \dfrac{1}{(1+\hat{G})^2} \dfrac{\hat{C}}{\hat{L}} \dfrac{(2\Delta\omega)^2}{\omega_0^2}}$$

$$= P_i \frac{4\hat{R}'}{(1+\hat{R}')^2 + (\omega_0\hat{C}\hat{R}')^2 \dfrac{(2\Delta\omega)^2}{\omega_0^2}} \tag{9a}$$

$$\hat{Z} = \frac{1}{\hat{G} + j\hat{B}} \qquad \hat{G} = \frac{1}{\hat{R}'} \qquad \hat{B} = \sqrt{\frac{\hat{C}}{\hat{L}}} \frac{2\Delta\omega}{\omega_0} \tag{9b}$$

(c) The power transmitted to the load, P_L, is

$$P_L = P_{stru} \frac{\hat{R}'}{\hat{R}_L} = P_{stru} \frac{\hat{R}}{\hat{R} + \hat{R}_L}$$

$$= \frac{\hat{R}}{\hat{R} + \hat{R}_L} \frac{4\hat{R}'}{(1+\hat{R}')^2 + (\omega_0\hat{C}\hat{R}')^2 \dfrac{(2\Delta\omega)^2}{\omega_0^2}} P_i \tag{10}$$

All normalized resistances can be eliminated from (10) by the use of (8).

$$\frac{\hat{R}}{\hat{R} + \hat{R}_L} = \frac{Q_u}{Q_u + Q_{eo}} \tag{11a}$$

$$1 + \hat{R}' = 1 + \frac{Q_u Q_{eo}}{Q_{ei}(Q_u + Q_{eo})} = \frac{Q_{ei}Q_{eo} + Q_u(Q_{ei} + Q_{eo})}{Q_{ei}(Q_u + Q_{eo})} \tag{11b}$$

$$\omega_0 \hat{C} \hat{R}' = \omega_0 \hat{C} \left[\frac{1}{\hat{R}} + \frac{1}{\hat{R}_L}\right]^{-1} = \left[\frac{1}{Q_u} + \frac{1}{Q_{eo}}\right]^{-1} = \frac{Q_u Q_{eo}}{Q_u + Q_{eo}} \tag{11c}$$

The substitution of (11) into (10) yields

$$P_L = \frac{4P_i \dfrac{Q_u}{Q_u + Q_{eo}} \dfrac{Q_u Q_{eo}}{Q_{ei}(Q_u + Q_{eo})}}{\left[\dfrac{Q_{ei}Q_{eo} + Q_u(Q_{ei} + Q_{eo})}{Q_{ei}(Q_u + Q_{eo})}\right]^2 + \left[\dfrac{Q_u Q_{eo}}{Q_u + Q_{eo}}\right]^2 \left[\dfrac{2\Delta\omega}{\omega_0}\right]^2}$$

$$= \frac{4P_i}{D_a D_b} \frac{Q_u^2 Q_{eo}}{Q_{ei}} = 4P_i \frac{Q_L^2}{Q_{ei}Q_{eo}} \frac{1}{1 + Q_L^2 \left[\dfrac{2\Delta\omega}{\omega_0}\right]^2} \tag{12α}$$

$$D_a = \left[\frac{Q_{ei}Q_{eo} + Q_u(Q_{ei} + Q_{eo})}{Q_{ei}}\right]^2 = \left[\frac{Q_u Q_{eo}}{Q_L}\right]^2 \tag{12β}$$

$$D_b = 1 + \left[\frac{Q_u Q_{ei}Q_{eo}}{Q_{ei}Q_{eo} + Q_u(Q_{ei} + Q_{eo})}\right]^2 \left[\frac{2\Delta\omega}{\omega_0}\right]^2 \tag{12γ}$$

$$= 1 + Q_L^2 \left[\frac{2\Delta\omega}{\omega_0}\right]^2$$

$$\frac{1}{Q_L} = \frac{Q_{ei}Q_{eo} + Q_u(Q_{ei} + Q_{eo})}{Q_u Q_{ei}Q_{eo}} \tag{7}$$

The loaded Q is also given by

$$Q_L = \frac{\omega_0}{2\Delta\omega_L} \qquad \text{or} \qquad \frac{Q_L}{\omega_0} = \frac{1}{2\Delta\omega_L}$$

Then the power absorbed by the load becomes

$$P_L = P_i \frac{4Q_L^2}{Q_{ei}Q_{eo}} \frac{1}{1 + \Omega_\Delta^2} \tag{12a}$$

$$\Omega_\Delta = \frac{2\hat{Q}_L}{\omega_o} \Delta\omega = \frac{\Delta\omega}{\Delta\omega_L} \tag{12b}$$

The transmission characteristics can be represented by a Bode plot, Fig. 10-3.

$$P_{db} \equiv 10 \log_{10} \frac{P_L}{P_i} = 10 \log \frac{4Q_L^2}{Q_{ei}Q_{eo}} - 10 \log \frac{1}{1 + \Omega_\Delta^2} \tag{13}$$

(d) For a symmetrical two-port cavity, $Q_{ei} = Q_{eo} = Q_e$, and with the unloaded Q much larger than either of the external Q's, the equivalent circuit for this special case is shown in Fig. 10-4. For high-Q_u, the circuit is approximated by lossless cavity, $\hat{R} = 0$. The loaded Q, (7), becomes

$$Q_L = \left[\frac{1}{Q_u} + \frac{2}{Q_e} \right]^{-1} = \frac{Q_u Q_e}{2 Q_u + Q_e} \approx \frac{Q_e}{2} \tag{14}$$

Then (12a) becomes

$$P_L = P_i \frac{4 \dfrac{Q_e^2}{4}}{Q_e^2} \frac{1}{1 + \Omega_\Delta^2} = \frac{P_i}{1 + \Omega_\Delta^2} \tag{15}$$

The input impedance of the resonator with a matched load, $\hat{R}_L = 1$, is

$$\hat{Z} = \frac{1}{1 + j\hat{B}} = \frac{1}{1 + j2\Omega_\Delta} \tag{16a}$$

$$\hat{B} = \sqrt{\frac{\hat{C}}{\hat{L}}} \frac{2\Delta\omega}{\omega_o} = \omega_o\hat{C} \frac{2\Delta\omega}{\omega_o} = Q_e \frac{2\Delta\omega}{\omega_o} \tag{16b}$$

$$= 2Q_L \frac{2\Delta\omega}{\omega_o} = 2 \frac{\omega_o}{2\Delta\omega_L} \frac{2\Delta\omega}{\omega_o} = 2\Omega_\Delta \tag{16b}$$

since $Q_e = 2Q_L$.

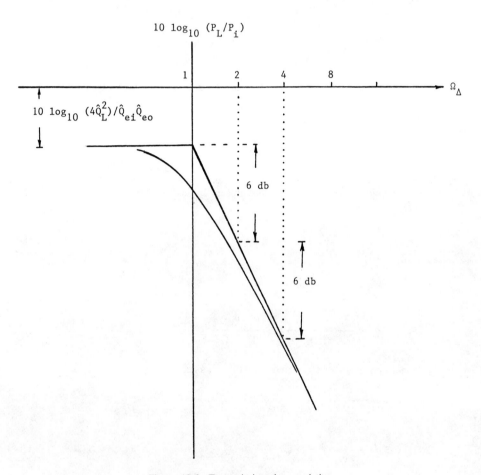

Figure 10-3: Transmission characteristics

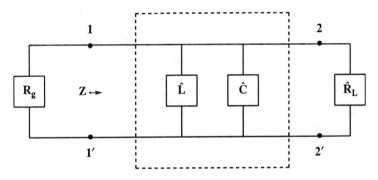

Figure 10-4: An equivalent circuit for a symmetrical, high Q two-port cavity.

The reflection coefficient at the input port is

$$\Gamma = \frac{\hat{Z} - 1}{\hat{Z} + 1} = \frac{-j2\Omega_\Delta}{2 + j2\Omega_\Delta} = \frac{-j\Omega_\Delta}{1 + j\Omega_\Delta} = s_{11} \tag{17}$$

The phase of the reflection coefficient is discontinuous at $\Omega_\Delta = 0$, which implies the coupling is critical.

The transmission coefficient is given by

$$s_{12} = \sqrt{1 - |\Gamma|^2} \tag{18}$$

s_{12} can be expressed in terms of Ω_Δ.

$$s_{12}s_{12}^* = 1 - \Gamma\Gamma^* = 1 - \frac{-j\Omega_\Delta}{1 + j\Omega_\Delta}\frac{j\Omega_\Delta}{1 - j\Omega_\Delta} = \frac{1}{(1 + j\Omega_\Delta)(1 - j\Omega_\Delta)}$$

or

$$s_{12} = \frac{\pm 1}{1 + j\Omega_\Delta} \tag{19}$$

The scattering matrix for the entire two-port symmetrical lossless cavity is therefore

$$[S] = \frac{1}{1 + j\Omega_\Delta}\begin{bmatrix} -j\Omega_\Delta & \pm 1 \\ \pm 1 & -j\Omega_\Delta \end{bmatrix} \tag{20}$$

(e) The concepts of this section can be extended to cavities with any number of ports by adding additional output ports (Fig. 10-5).

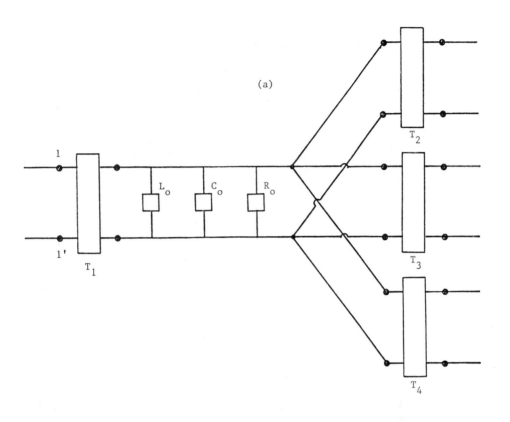

(a)

(b)

Figure 10-5: An equivalent circuit for a multi-port cavity.

11. Transmission Cavity

A symmetrical transmission cavity is constructed of a section of waveguide with an input and an output coupling device. These coupling devices are identical in a symmetrical cavity, Fig. 11-1. They are assumed to be lossless and are represented by the normalized susceptance $\hat{B}_d$.

At point c, just to the left of the junction q–q′, the normalized input admittance is

$$\hat{Y}_c(s = s_q^+) = 1 + j\hat{B}_d \tag{1}$$

for a match-terminated output, $Z_L = Z_o$. This admittance is transformed at point b, just to the right of the junction p–p′, to

$$\hat{Y}_b(s = s_p^-) = \frac{\hat{Y}_c + j \tan \beta L_1}{1 + j\hat{Y}_c \tan \beta L_1}$$

$$= \frac{(1 + j\hat{B}_d) + j \tan \beta L_1}{1 + j(1 + j\hat{B}_d) \tan \beta L_1}$$

$$= \frac{1 + j(\hat{B}_d + \tan \beta L_1)}{(1 - \hat{B}_d \tan \beta L_1) + j \tan \beta L_1} \tag{2}$$

The cavity is assumed lossless in the above evaluation.

The input admittance at point a, just to the left of the junction p–p′, is

$$\hat{Y}_a(s = s_p^+) = j\hat{B}_d + \hat{Y}_b = \frac{1}{D} [j\hat{B}_d D + 1 + j(\hat{B}_d + \tan \beta L_1)]$$

$$= \frac{1}{D} [(1 - \hat{B}_d \tan \beta L_1) + j\{ \tan \beta L_1 + j\hat{B}_d(2 - \hat{B}_d \tan \beta L_1)\}]$$

$$= 1 + j\frac{\hat{B}_d}{D}(2 - \hat{B}_d \tan \beta L_1) \tag{3a}$$

$$D = (1 - \hat{B}_d \tan \beta L_1) + j \tan \beta L_1 \tag{3b}$$

The cavity will be matched at resonance, i.e., $\hat{Y}_a = 1 + j0$, when

$$(2 - \hat{B}_d \tan \beta L_1) = 0 \quad \text{or} \quad \tan \beta L_1 = \frac{2}{\hat{B}_d} \quad \text{at } \omega = \omega_o \tag{4}$$

For a capacitive junction, $\hat{B}_d > 0$, the length L_1 is [Eq. (9.1-10)]

$$L_1 = \frac{\lambda_{go}}{2\pi} \left[m\pi + \tan^{-1} \frac{2}{\hat{B}_d} \right] \quad m = 1, 2, \ldots \tag{5a}$$

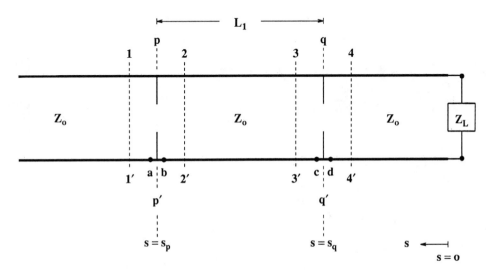

Figure 11-1: An equivalent circuit for a symmetrical two-port cavity with coupling devices.

For an inductive junction, $\hat{B}_d < 0$, one has

$$L_1 = \frac{\lambda_{go}}{2\pi} \left[m\pi - \tan^{-1} \frac{2}{|\hat{B}_d|} \right] \qquad m = 1, 2, \ldots \tag{5b}$$

The quantity $\tan^{-1} \frac{2}{|\hat{B}_d|}$ will be considered to be in the first quadrant in these expressions, and λ_{go} is the guided wavelength at resonance.

The loaded Q is defined by

$$\hat{Q}_L = \frac{\omega_o}{\Delta\omega_L} \tag{6}$$

where $\Delta\omega_L$ is the half-power bandwidth.

The electrical length, θ_o, of the cavity at resonance is

$$\theta_o = \beta_o L_1 = \frac{2\pi}{\lambda_{go}} L_1 \tag{7}$$

where β_o is the propagation coefficient at resonance. The bandwidth in electrical length is

$$\Delta\theta_L = \theta_2 - \theta_1 = (\beta_2 - \beta_1)L_1 = \Delta\beta_L L_1 \tag{8}$$

The propagation coefficient in a waveguide is

$$\beta^2 = \omega^2\mu\varepsilon - k_c^2 \tag{9}$$

then

$$2\beta \, d\beta = 2\mu\varepsilon\omega \, d\omega \equiv \frac{2\omega}{c^2} \, d\omega \qquad \text{or} \qquad d\omega = c^2 \frac{\beta}{\omega} \, d\beta$$

Near resonance, $\omega \approx \omega_o$, one has

$$d\omega \approx c^2 \frac{\beta_o}{\omega_o} \, d\beta = c \frac{\lambda_o}{\lambda_{go}} \, d\beta \tag{10a}$$

$$\beta_o = \beta(\omega = \omega_o) = \frac{2\pi}{\lambda_{go}} \tag{10b}$$

$$\omega_o = 2\pi f_o = 2\pi \frac{c}{\lambda_o} \tag{10c}$$

Therefore,

$$\Delta\omega_L \approx d\omega_L \approx c \frac{\lambda_o}{\lambda_{go}} \Delta\beta_L \tag{11}$$

The loaded Q is then obtained by substituting (10) and (11) into (6).

$$\hat{Q}_L = \frac{2\pi \dfrac{c}{\lambda_o}}{c \dfrac{\lambda_o}{\lambda_{go}} \Delta\beta_L} = 2\pi \frac{\lambda_{go}}{\lambda_o^2} \frac{1}{\Delta\beta_L}$$

$$= \left[\frac{\lambda_{go}}{\lambda_o}\right]^2 \frac{2\pi}{\lambda_{go}} \frac{1}{\Delta\beta_L} = \left[\frac{\lambda_{go}}{\lambda_o}\right]^2 \frac{\beta_o}{\Delta\beta_L}$$

$$= \left[\frac{\lambda_{go}}{\lambda_o}\right]^2 \frac{\theta_o}{\Delta\theta_L} \qquad (12a)$$

$$\theta_o = \beta_o L_1 = m\pi + \tan^{-1}\frac{2}{\hat{B}_d} \qquad (12b)$$

where θ_o is obtained from (5).

The loaded Q can be expressed in terms of the susceptance of the junction, $\hat{B}_d$, once the electrical length $\Delta\theta_L$ is related to $\hat{B}_d$. The value of $\Delta\theta_L$ will be investigated next.

The reflection coefficient of the cavity is given by

$$\Gamma = \frac{1 - \hat{Y}_a}{1 + \hat{Y}_a} = \frac{-j\hat{B}_d (2 - \hat{B}_d \tan\theta)}{2(1 - \hat{B}_d \tan\theta) + j[2\tan\theta + \hat{B}_d (2 - \hat{B}_d \tan\theta)]} \qquad (13a)$$

$$\theta = \beta L_1 \qquad (13b)$$

The corresponding $|\Gamma|^2$ is

$$|\Gamma|^2 = \frac{\hat{B}_d^2 (2 - \hat{B}_d \tan\theta)^2}{4(1 - \hat{B}_d \tan\theta)^2 + [2\tan\theta + \hat{B}_d (2 - \hat{B}_d \tan\theta)]^2}$$

$$= \frac{\hat{B}_d^2 (2 - \hat{B}_d \tan\theta)^2}{\hat{B}_d^2 (2 - \hat{B}_d \tan\theta)^2 + 4(1 + \tan^2\theta)}$$

$$= \frac{1}{1 + \dfrac{4(1 + \tan^2\theta)}{\hat{B}_d^2 (2 - \hat{B}_d \tan\theta)^2}} \qquad (14)$$

At the half-power point, $|\Gamma|^2 = 0.5$, which implies that

$$\frac{4 (1 + \tan^2 \theta)}{\hat{B}_d^2 (2 - \hat{B}_d \tan \theta)^2} = 1$$

or

$$(\hat{B}_d^4 - 4) \tan^2 \theta - 4\hat{B}_d^3 \tan \theta + 4 (\hat{B}_d^2 - 1) = 0 \tag{15a}$$

That is,

$$\tan \theta = \frac{1}{\hat{B}_d^4 - 4} [2\hat{B}_d^3 \pm 2 \sqrt{ \hat{B}_d^4 + 4\hat{B}_d^2 - 4 }] \tag{15b}$$

Note that $\tan \theta$ will be real if the terms under the square root are equal to or greater than zero. The limiting value is

$$\hat{B}_d^4 + 4 \hat{B}_d^2 - 4 = 0 \quad \text{or} \quad \hat{B}_d = \sqrt{2 (\sqrt{2} - 1)} = 0.91$$

That is , for $|\hat{B}_d| < 0.91$, $\tan \theta$ becomes imaginary and $|\Gamma|^2$ can never be 0.5.

The half-power electrical length is given by (15b)

$$\theta_1 = \tan^{-1} \frac{2 \hat{B}_d^3 - 2 \sqrt{\hat{B}_d^4 + 4\hat{B}_d^2 - 4}}{\hat{B}_d^4 - 4} \tag{16a}$$

$$\theta_2 = \tan^{-1} \frac{2 \hat{B}_d^3 + 2 \sqrt{\hat{B}_d^4 + 4\hat{B}_d^2 - 4}}{\hat{B}_d^4 - 4} \tag{16b}$$

The half-power bandwidth of the electrical length is therefore

$$\Delta\theta_L = \theta_2 - \theta_1 \tag{17}$$

$$\tan \Delta\theta_L = \frac{\tan \theta_2 - \tan \theta_1}{1 + \tan \theta_2 \tan \theta_1}$$

or

$$\Delta\theta_L = \tan^{-1} \frac{2b}{1 + a^2 - b^2} \tag{18a}$$

$$\theta_1 = \tan^{-1} (a - b) \tag{18b}$$

$$\theta_2 = \tan^{-1} (a + b) \tag{18c}$$

$$a = \frac{2\hat{B}_d^3}{\hat{B}_d^4 - 4} \tag{18d}$$

$$b = \frac{2\sqrt{\hat{B}_d^4 + 4\hat{B}_d^2 - 4}}{\hat{B}_d^4 - 4} \tag{18e}$$

This complicated expression can be simplified by the following approximation.

$$\frac{d(\tan\theta)}{d\theta} = \sec^2\theta = \lim_{\delta\theta \to 0} \frac{\delta(\tan\theta)}{\delta\theta} \qquad \text{or} \qquad \delta\theta \approx \frac{\delta(\tan\theta)}{\sec^2\theta} \tag{19a}$$

where $\delta\theta$ is a finite differential increment of θ. From (16), $\sec^2\theta$ is given by

$$\sec^2\theta = 1 + \tan^2\theta = 1 + \left[\frac{2\hat{B}_d^3 \pm 2\sqrt{\hat{B}_d^4 + 4\hat{B}_d^2 - 4}}{\hat{B}_d^4 - 4} \right]^2$$

Assume that $\hat{B}_d^4 \gg 4$; then

$$\sec^2\theta \approx 1 + \frac{4}{\hat{B}_d^8} \left[\hat{B}_d^3 \pm \hat{B}_d\sqrt{\hat{B}_d^2 + 4} \right]^2$$

$$= \frac{1}{\hat{B}_d^8} \left[\hat{B}_d^8 + 4\hat{B}_d^6 \pm 8\hat{B}_d^4\sqrt{\hat{B}_d^2 + 4} + \hat{B}_d^2(\hat{B}_d^2 + 4) \right]$$

$$= \frac{\hat{B}_d^2 + 4}{\hat{B}_d^2} \left[1 \pm \frac{8}{\hat{B}_d^2\sqrt{\hat{B}_d^2 + 4}} + \frac{1}{\hat{B}_d^4} \right] \approx \frac{\hat{B}_d^2 + 4}{\hat{B}_d^2} \tag{19b}$$

The finite increment in $\tan\theta$ can be approximated from (16)

$$\delta(\tan\theta) \approx \tan\theta_2 - \tan\theta_1 = \frac{4\sqrt{\hat{B}_d^4 + 4\hat{B}_d^2 - 4}}{\hat{B}_d^4 - 4} \approx \frac{4\sqrt{\hat{B}_d^2 + 4}}{\hat{B}_d^3} \tag{19c}$$

The substitution of (19b) and (19c) into (19a) gives

$$\delta\theta = \frac{4}{\hat{B}_d\sqrt{\hat{B}_d^2 + 4}} \tag{19d}$$

$$\Delta\theta_L \approx \delta\theta = \frac{4}{\hat{B}_d\sqrt{\hat{B}_d^2 + 4}} \qquad \text{for} \quad \hat{B}_d^4 \gg 4 \tag{19}$$

The loaded Q is then obtained by substituting (19) into (12).

$$\hat{Q}_L \approx \left[\frac{\lambda_{go}}{\lambda_o} \right]^2 \frac{1}{4} \hat{B}_d \sqrt{\hat{B}_d^2 + 4} \ (m\pi + \tan^{-1} \frac{2}{\hat{B}_d}) \tag{20}$$

In spite of the very approximate nature of the result, Eq. (20) provides reasonable values. For small values of $\hat{B}_d$, $|\Gamma|^2$ may not equal 0.5, the angle $\tan^{-1} \frac{2}{\hat{B}_d}$ approaches $\pm \frac{\pi}{2}$, and the loaded Q becomes

$$\hat{Q}_L \approx \frac{\hat{B}_d}{2} \left[\frac{\lambda_{go}}{\lambda_o} \right]^2 (m\pi \pm \frac{\pi}{2}) \qquad \text{for small } \hat{B}_d, \ \hat{B}_d \ll 4 \tag{21}$$

For $\hat{B}_d^2 \gg 4$, (20) becomes

$$\hat{Q}_L \approx \frac{\hat{B}_d^2}{4} \left[\frac{\lambda_{go}}{\lambda_o} \right]^2 (m\pi + \tan^{-1} \frac{2}{\hat{B}_d}) \qquad \text{for large } \hat{B}_d, \ \hat{B}_d \gg 4 \tag{22}$$

$|\Gamma|^2$ will be maximum when the denominator of (14) is minimum.

$$\frac{d}{d\theta} \left[1 + \frac{4 (1 + \tan^2 \theta)}{\hat{B}_d^2 (2 - \hat{B}_d \tan \theta)^2} \right] = 0 = \frac{d}{d\theta} \left[\frac{1 + \tan^2 \theta}{(2 - \hat{B}_d \tan \theta)^2} \right]$$

$$\frac{2 \tan \theta \sec^2 \theta}{(2 - \hat{B}_d \tan \theta)^2} + \frac{(1 + \tan^2 \theta) 2\hat{B}_d \sec^2 \theta}{(2 - \hat{B}_d \tan \theta)^3} = 0$$

$$(2 - \hat{B}_d \tan \theta) [(2 - \hat{B}_d \tan \theta) \tan \theta + \hat{B}_d + \hat{B}_d \tan^2 \theta] = 0$$

$$(2 - \hat{B}_d \tan \theta) (\hat{B}_d + 2 \tan \theta) = 0 \tag{23a}$$

or

$$\tan \theta = \frac{2}{\hat{B}_d} \qquad \text{or} \qquad \theta = \tan^{-1} \frac{2}{\hat{B}_d} \qquad \text{resonance,} \qquad |\Gamma| = |\Gamma_{min}| \tag{23b}$$

i.e., this corresponds to the condition of resonance and the reflection is minimum. The other solution is

$$\tan \theta = -\frac{\hat{B}_d}{2} \qquad \text{or} \qquad \theta = \tan^{-1} \frac{-\hat{B}_d}{2} \qquad \text{anti–resonance,} \qquad |\Gamma| = |\Gamma_{max}| \tag{23c}$$

This is the condition for anti-resonance and results in maximum reflection.

At anti-resonance, (23c) can be expressed as

$$\frac{2}{\hat{B}_d} = \frac{-1}{\tan \theta} = \tan (\pm m\pi + \theta \pm \frac{\pi}{2})$$

Then for positive $\hat{B}_d$,

$$L_1 = \frac{\lambda_g}{2\pi} (m\pi + \tan^{-1} \frac{2}{|\hat{B}_d|} \pm \frac{\pi}{2}) \tag{24a}$$

and for negative $\hat{B}_d$,

$$L_1 = \frac{\lambda_g}{2\pi} (m\pi - \tan^{-1} \frac{2}{|\hat{B}_d|} \pm \frac{\pi}{2}) \tag{24b}$$

Hence, the position of maximum $|\Gamma|$ is located at $\lambda_{go}/4$ or $\theta = \pi/2$ from the position of resonance.

The maximum of $|\Gamma|$ is given by substituting (23c) into (13a).

$$\Gamma_{max} = \frac{-j\hat{B}_d \left[2 - \hat{B}_d \left[\frac{-\hat{B}_d}{2} \right] \right]}{2 \left[1 - \hat{B}_d \left[\frac{-\hat{B}_d}{2} \right] \right] + j \left\{ 2 \left[\frac{-\hat{B}_d}{2} \right] + \hat{B}_d \left[2 - \hat{B}_d \left[\frac{-\hat{B}_d}{2} \right] \right] \right\}}$$

$$= \frac{-j\hat{B}_d \left[2 + \frac{\hat{B}_d^2}{2} \right]}{2 \left[1 + \frac{\hat{B}_d^2}{2} \right] + j\hat{B}_d \left[1 + \frac{\hat{B}_d^2}{2} \right]}$$

$$= \frac{\frac{-j\hat{B}_d}{2} (4 + \hat{B}_d^2)}{\left[1 + \frac{\hat{B}_d^2}{2} \right] (2 + j\hat{B}_d)}$$

$$= \frac{-j\hat{B}_d (4 + \hat{B}_d^2) (2 - j\hat{B}_d)}{2 \left[1 + \frac{\hat{B}_d^2}{2} \right] (4 + \hat{B}_d^2)} = \frac{-\hat{B}_d (\hat{B}_d + j2)}{(2 + \hat{B}_d^2)} = \Gamma_m' e^{j\phi_r} \tag{25a}$$

$$\Gamma'_m = \frac{\hat{B}_d \sqrt{\hat{B}_d^2 + 4}}{2 + \hat{B}_d^2} \tag{25b}$$

$$\phi_\Gamma = \pi + \tan^{-1} \frac{2}{\hat{B}_d} \tag{25c}$$

For $\hat{B}_d > 0$, $\hat{B}_d = |\hat{B}_d|$, then (25a) becomes

$$\Gamma_{max} = |\Gamma_m| \, e^{j\phi'_\Gamma} \tag{26a}$$

$$|\Gamma_m| = \frac{|\hat{B}_d| \sqrt{|\hat{B}_d|^2 + 4}}{2 + |\hat{B}_d|^2} \tag{26b}$$

$$\phi'_\Gamma = \pi + \tan^{-1} | \frac{2}{|\hat{B}_d|} \tag{26c}$$

For $\hat{B}_d < 0$, $\hat{B}_d = -|\hat{B}_d|$ and (25a) becomes

$$\Gamma_{max} = -\Gamma_m e^{j\phi} = \Gamma_m e^{j\phi''_\Gamma} \tag{27a}$$

$$\phi''_\Gamma = \pi + (\pi + \tan^{-1} \frac{2}{-|\hat{B}_d|}) = \tan^{-1} \frac{2}{-|\hat{B}_d|} = \pi - \tan^{-1} \frac{2}{|\hat{B}_d|} \tag{27b}$$

since $\tan^{-1} 2/(-|\hat{B}_d|)$ is in the fourth quardrant.

Equations (26) and (27) give the reflection coefficient at the input junction plane p. At an electrical length $\pm(1/2) \tan^{-1} \frac{|\hat{B}_d|}{2}$ from plane p, the reflection coefficient is negative and real. This is also the position of detuned-short as determined earlier.

12. Problems

1. An air-filled rectangular cavity resonator is shown in Fig. 12-1. Determine the dominant modes and the corresponding resonance frequencies for each of the cases:

(a) $a > b > c$
(b) $a > c > b$
(c) $a = b = c$

2. Design an air-filled cubic cavity ($a = b = c$) to be resonant at 10^{10} Hz for its dominant modes. If the cavity is made of copper, determine the Q of the cavity.

3. A two-port cavity is terminated by a reactive load. The load is made up of a short piece of guide, Z_0' , of length L_1 which is terminated by a short circuit.

Show that the resonance frequency of the cavity is shifted to

$$\omega_r = \omega_0 + d\omega$$

and

$$\frac{2\, d\omega}{\omega_0} = \frac{1}{Q_{E0}} \cot \frac{2\pi L_1}{\lambda_g'}$$

where λ_g' is the guide wavelength of the load and Q_{E0} is the external Q of the output port.

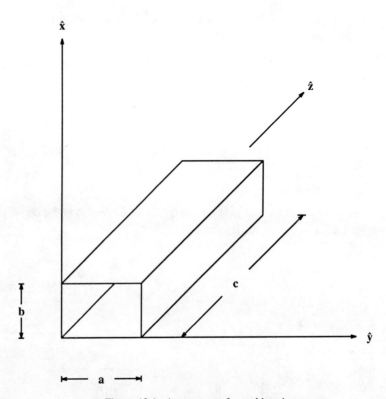

Figure 12-1: Arrangement for problem 1.

CHAPTER XIV   Filters

1. Introduction

A filter is a network that possesses nonuniform frequency response in the frequency range of interest. In an ideal filter, signals are transmitted without attenuation for all frequencies within the pass-band and there is zero transmission in the stop-band, as shown in Fig. 1-1, where $\hat{\omega} = \omega/\omega_c$ is the normalized frequency and ω_c is the cutoff frequency. Clearly, the ideal filter characteristic is not realizable, and the object in filter design is to obtain an approximation acceptable for a particular application.

Filter design for lumped-element networks has been well developed, and a number of synthesis procedures are available. Since distributed parameter elements are involved in microwave filters, the design procedure for these filter is much more complicated. However, in the special case of narrow-band filters, many microwave elements have frequency characteristics similar to those of lumped elements. Such microwave filters can be synthesized from a low-frequency filter by replacing lumped elements with appropriate microwave elements.

The two most commonly employed low-frequency filter synthesis techniques are the image-parameter method and the insertion-loss method. The image-parameter method furnishes the desired pass-band and stop-band properties; however, the exact frequency response for each band is not specified. In the insertion-loss method, a physically realizable frequency characteristic is postulated and an appropriate network is synthesized. The latter method is generally preferable since it offers more precise frequency response and involves less cut-and-try process.

Filter designs are usually carried out for lossless elements, and the inclusion of lossy elements, which will greatly increase the complexity of the synthesis, is rarely done. The losses in microwave elements can be made small enough such that the filter design based upon lossless elements does provide satisfactory performance.

Filter theory at lower frequencies is well established and is available in a wide range of textbooks on network theory. Some of these are listed in the bibliography [A]. At microwave frequencies both Altman [C-1] and Collin [C-2] provide some analysis at an advanced level and similar techniques will be used in some of the analysis in this chapter.

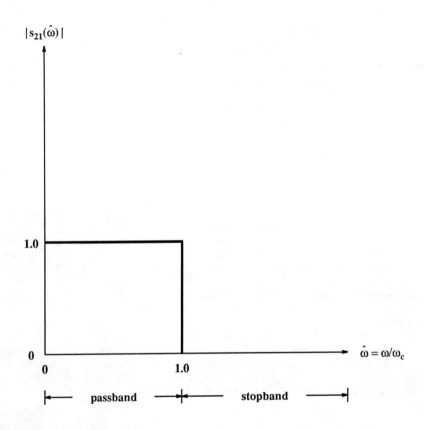

(a)

Figure 1-1: Ideal low-pass responses.
(a) Transmission response.
(b) Loss response.

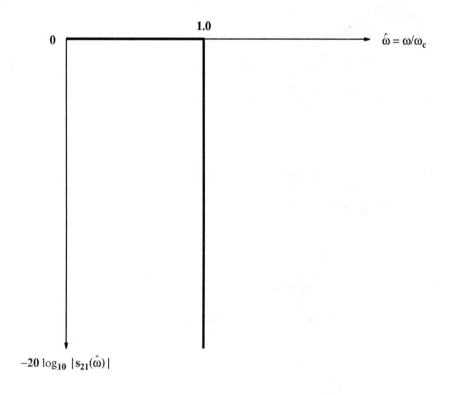

(b)

Figure 1-1: Ideal low-pass responses.
(a) Transmission response.
(b) Loss response.

2. Insertion Loss

The insertion loss is defined as the ratio of the available power to the actual power delivered to the load. The network under consideration is shown in Fig. 2-1. In Fig. 2-1a, the load Z_L is connected directly to the source, and in Fig.2-1b, a two-port network is inserted between the source and the load. The insertion power ratio is the ratio of the power $< P_{Lo} >$ absorbed by the load without the two-port network to the power $< P_L >$ absorbed by the load with the two-port network inserted.

$$K_{ip} \equiv \frac{< P_{Lo} >}{< P_L >} = \text{insertion power ratio} \tag{1}$$

Without the two-port network (Fig. 2-1a) the voltage and current at the load are

$$v_{Lo} = v_g \frac{Z_L}{Z_L + Z_g} \qquad \text{and} \qquad i_{Lo} = \frac{v_g}{Z_L + Z_g} \tag{2}$$

and the power dissipated in the load is

$$< P_{Lo} > = \frac{1}{2} \text{Re} [v_{Lo} i_{Lo}^*] = \frac{1}{2} \text{Re} \left[v_g \frac{Z_L}{Z_L + Z_g} \frac{v_g^*}{Z_L^* + Z_g^*} \right]$$

$$= \frac{1}{2} \frac{|v_g|^2}{|Z_L + Z_g|^2} \text{Re} [Z_L] \tag{3}$$

With the two-port network in place, let the voltage across the load be v_L and the current in the load is

$$i_L = \frac{v_L}{Z_L} \tag{4}$$

The corresponding power absorbed by the load is

$$< P_L > = \frac{1}{2} \text{Re} [v_L i_L^*] = \frac{|v_L|^2}{2} \text{Re} \left[\frac{1}{Z_L} \right] = \frac{|v_L|^2}{2 |Z_L|^2} \text{Re} [Z_L] \tag{5}$$

The insertion power ratio K_{ip} is given by

$$K_{ip} = \frac{< P_{Lo} >}{< P_L >} = \left[\frac{|v_g|}{|v_L|} \frac{|Z_L|}{|Z_L + Z_g|} \right]^2 \tag{6}$$

In terms of scattering elements, one also has

$$< P_L > = |s_{21}|^2 < P_A > = |s_{21}|^2 < P_{Lo} > \tag{7}$$

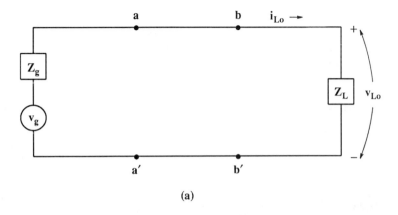

(a)

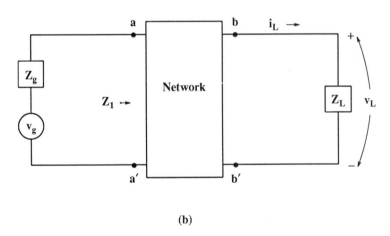

(b)

Figure 2-1: (a) Load connected directly to the source.
(b) Load connected through a two-port network.

where P_A is the available power from the source. The forward transmission coefficient for a two-port network is given by

$$s_{21} = \frac{2v_L}{v_g} \sqrt{\frac{R_{o1}}{R_{o2}}} \frac{Z_1 + Z_g}{Z_1 + R_{o1}} \qquad Z_L = R_{o2} \qquad \text{(III.3.3-5a)}$$

$$= \frac{2v_L}{v_g} \sqrt{\frac{R_{o1}}{R_{o2}}} \qquad Z_g = R_{o1} \quad \text{and} \quad Z_L = R_{o2} \qquad \text{(III.3.3-5b)}$$

For the special case when the load and source impedances are equal to the respective normalizing resistances

$$Z_g = R_{o1} \qquad \text{and} \qquad Z_L = R_{o2} \qquad (8)$$

then

$$K_{ip} = \left[\frac{|v_g|}{|v_L|} \frac{R_{o2}}{R_{o1} + R_{o2}} \right]^2 \qquad (9)$$

and

$$|s_{21}|^2 = \frac{4v_L^2}{v_g^2} \frac{R_{o1}}{R_{o2}} \qquad \text{or} \qquad \frac{v_L^2}{v_g^2} = |s_{21}|^2 \frac{R_{o2}}{4R_{o1}} \qquad (10)$$

Combining (10) with (9) yields

$$K_{ip} = \frac{1}{|s_{21}|^2} \frac{4R_{o1}}{R_{o2}} \frac{R_{o2}^2}{(R_{o1} + R_{o2})^2} = \frac{4R_{o1}R_{o2}}{(R_{o1} + R_{o2})^2} \frac{1}{|s_{21}|^2} \qquad (11a)$$

$$= \frac{1}{|s_{21}|^2} \qquad \text{for} \qquad R_{o1} = R_{o2} \qquad (11b)$$

The insertion power ratio is linearly proportional to the inverse of the magnitude squared of the forward transmission coefficient.

The insertion power ratio may be expressed as an exponential factor.

$$K_{ip} = \frac{<P_{Lo}>}{<P_L>} = e^{2\alpha} \qquad (12)$$

A bell is defined as the logarithm of a power ratio

$$\text{number of bells} \equiv \log_{10} \frac{P_1}{P_2} \qquad (13a)$$

A decibel is defined as

$$\text{number of db} \equiv 10 \log_{10} \frac{P_1}{P_2} \tag{13b}$$

Hence, the factor α in (12) can be measured in db by the definition

$$\alpha_{db} = 10 \log_{10} \frac{< P_{Lo} >}{< P_L >} \tag{14}$$

α_{db} is the insertion power loss of the two-port network in units of db.

3. Darlington's Filter Synthesis

Darlington proposed in 1939 (J. Math. Phys., vol. 18, p. 257-353, 1939) a filter synthesis procedure based upon the proportionality between the forward transmission coefficient and the insertion loss of a two-port network.

$$K_{ip} = \frac{<P_{Lo}>}{<P_L>} = \frac{4R_{o1}R_{o2}}{(R_{o1} + R_{o2})^2} \frac{1}{|s_{21}|^2} \tag{2-11}$$

$$= e^{2\alpha} \tag{2-12}$$

$$\alpha_{db} = 10 \log_{10} K_{ip} \tag{2-14}$$

The two-port network for the present analysis is the filter network.

This analysis will be confined to a low-pass filter which is terminated in equal source and load impedances, Fig. 3-1.

$$Z_g = R_o = Z_L \qquad R_{o1} = R_{o2} = R_o \tag{1}$$

Subject to (1), (2-11a) and (2-14) become

$$K_{ip} = \frac{1}{|s_{21}|^2} \tag{2a}$$

$$\alpha_{db} = 10 \log_{10} \frac{1}{|s_{21}|^2} = -20 \log_{10} |s_{21}| \tag{2b}$$

The specification of the amplitude-squared transmission coefficient, Fig. 3-2, is equivalent to the specification of the insertion loss, Fig. 3-3, in the design of filters.

The scattering matrix of a lossless network satisfies the unitary condition.

$$|s_{11}|^2 + |s_{21}|^2 = 1 \tag{III.13-3a}$$

For a symmetrical network, $s_{12} = s_{21}$, and

$$|s_{11}(j\omega)|^2 = 1 - |s_{21}(j\omega)|^2 \tag{3}$$

The input reflection coefficient is given by

$$s_{11} = \left[\frac{b_1}{a_1}\right]_{a_2=0} = \left[\frac{v_1 - i_1 R_o}{v_1 + i_1 R_o}\right]_{a_2=0} = \left[\frac{Z_1 - R_o}{Z_1 + R_o}\right]_{a_2=0} \tag{4a}$$

$$Z_1 = \frac{v_1}{i_1} \tag{4b}$$

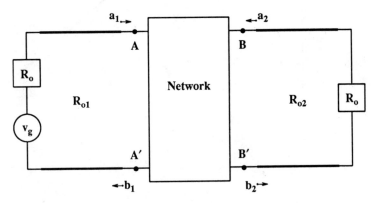

Figure 3-1: A two-port network terminated by identical source and load impedances.

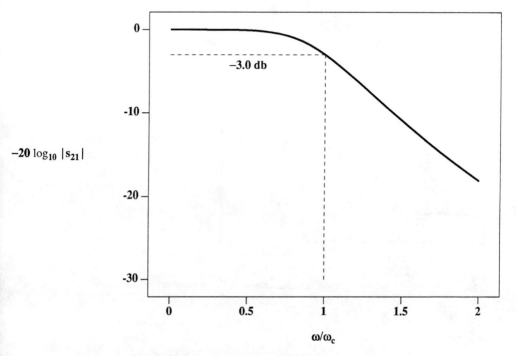

Figure 3-2: Response of $|s_{21}|^2$

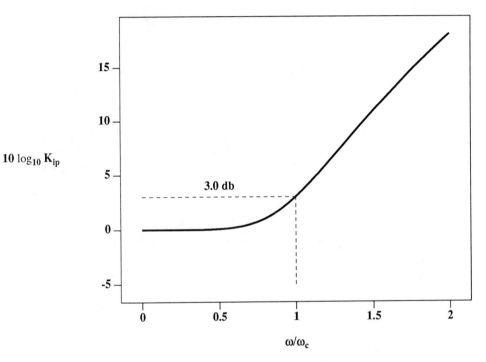

Figure 3-3: Response of insertion loss.

or

$$Z_1 = R_o \frac{1 + s_{11}}{1 - s_{11}} = \text{driving-point impedance of filter network} \tag{4c}$$

Equation (4c) implies that the specification of s_{11} will also specify the input impedance of the filter network. The filter synthesis problem is now reduced to the synthesis of the driving-point impedance of the filter.

The procedure of synthesis by Darlington's techniques can be outlined as follows:

(a) A frequency variation for the insertion loss or equivalently for the forward transmission coefficient $|s_{21}(j\omega)|^2$ is postulated.

$$|s_{21}(j\omega)|^2 \equiv |F(j\omega)|^2 \tag{5}$$

Then (3) becomes

$$|s_{11}(j\omega)|^2 = s_{11}(j\omega)s_{11}(-j\omega) = 1 - F(j\omega)F(-j\omega) \tag{6}$$

(b) The input reflection coefficient s_{11} is obtained by the method of analytic continuation - this technique will be discussed in the following section.

(c) The driving-point impedance of the filter network is obtained from (4c), using s_{11} as obtained in part (b).

(d) The driving-point impedance is synthesized by any one of the available realization techniques. Some of these techniques were reviewed in Sections X.7 and X.8.

There is an unlimited number of ways to prescribe the insertion loss function. Two widely used approximations are the Butterworth maximally flat response and the Chebycheff equal-ripple response. These will be investigated separately.

4. Analytic Continuation

The function $s_{11}(j\omega)$ can be extracted from $|s_{11}(j\omega)|^2$ by techniques of analytic continuation. The goal is to determine a rational function in $p \equiv j\omega$ which has the desired magnitude function. The square of the magnitude function is

$$| s_{11}(j\omega) |^2 = s_{11}(j\omega)s_{11}^*(j\omega) = s_{11}(j\omega)s_{11}(-j\omega) \tag{1}$$

Let $\omega = p/j$; then

$$| s_{11}(j\omega) |^2 = | s_{11}(p) |^2 = \Big[s_{11}(p)s_{11}(-p) \Big]_{\omega=\frac{p}{j}} \tag{2}$$

It should be noted that the notation $|s_{11}(j\omega)|^2$ is in terms of the imaginary variable $j\omega$, while the variable which appears in $|s_{11}(j\omega)|^2$ is the real variable ω^2. The desired generalization is made by substituting p/j for ω and (1) is still satisfied.

The separation of $[s_{11}(p)s_{11}(-p)]$ into two constituents is accomplished by a division of poles and zeros. Let $s_{11}(p)$ be represented by a general network function - a ratio of two polynomials in factored form.

$$s_{11}(p) = K \frac{(p - z_1) (p - z_2) \cdots (p - z_n)}{(p - p_1) (p - p_2) \cdots (p - p_m)} \tag{3}$$

Clearly, $s_{11}(-p)$ will have poles and zeros which are negative of those of $s_{11}(p)$; this can be shown by replacing p by $-p$ in (3). If $s_{11}(p)$ has poles and zeros as shown in Fig. 4-1a, then the poles and zeros of $s_{11}(-p)$ are the corresponding mirror images of those of $s_{11}(p)$, Fig. 4-1b. The function $[s_{11}(p)s_{11}(-p)]$ has the combined poles and zeros as shown in Fig. 4-1c.

Back to the problem at hand, the function $s_{11}(p)$ can be established by choosing those poles of $[s_{11}(p)s_{11}(-p)]$ which are located in the left-half plane so that $s_{11}(p)$ belongs to a stable network function. Such a restriction is not required in making the choice of zeros unless $s_{11}(p)$ is a driving-point function.

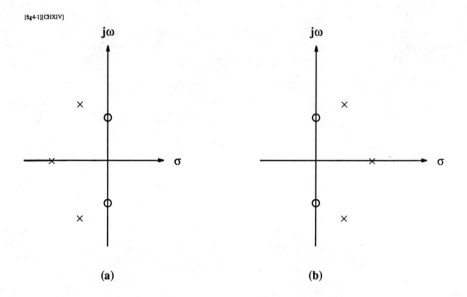

(a)

(b)

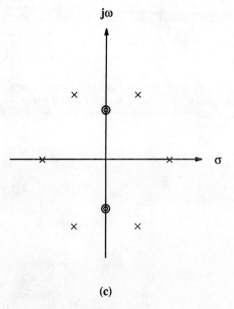

(c)

Figure 4-1: The poles and zeros of $|s_{11}|^2$

5. Butterworth Low-Pass Response

(a) The insertion loss function for a Butterworth maximally flat response may be chosen to be [see Section VII.4(b)]

$$K_{ip} = 1 + \Delta^2 \left[\frac{\omega}{\omega_c} \right]^{2n} = \frac{1}{|s_{21}|^2} = 1 + \Delta^2(\hat{\omega})^{2n} \tag{1a}$$

$$\hat{\omega} = \frac{\omega}{\omega_c} \tag{1b}$$

where Δ^2 is a constant known as pass-band tolerance and ω_c is the cutoff frequency. Δ is conventionally chosen to be unity. In the pass-band, $0 \le \omega \le \omega_c$, the maximum K_{ip} is equal to $1 + \Delta^2$. K_{ip} increases indefinitely for $\omega > \omega_c$.

The forward transfer scattering coefficient is [Eq. (3-2a)]

$$|s_{21}(j\omega)|^2 = |s_{21}|^2 = \frac{1}{1 + \Delta^2\hat{\omega}^{2n}} \tag{2}$$

(b) The gain in db may be expressed as

$$s_{21_{db}} = -10 \log_{10} |s_{21}(\omega)|^2 = -10 \log_{10} [1 + \Delta^2\hat{\omega}^{2n}] \tag{3a}$$

For small values of $\hat{\omega}$, one has

$$s_{21_{db}} \approx -10\log_{10} 1.0 = 0 \tag{3b}$$

For $\Delta^2\hat{\omega}^{2n} \gg 1$,

$$s_{21_{db}} \approx -10 \log_{10} [\Delta\hat{\omega}^n]^2 = -20 \log_{10} \Delta - 20n \log_{10} \hat{\omega} \tag{3c}$$

At large values of $\Delta^2(\hat{\omega})^{2n}$, the $s_{21_{db}}$ curve has a negative slope of 20n decibels per decade and approaches the ideal characteristic of a low-pass response, Fig. 1-1, as n approaches infinity.

(c) Let $p = j\omega$ or $\omega = p/j$; then (2) becomes

$$|s_{21}(j\omega)|^2 = |s_{21}(p)|^2 = \frac{1}{1 + \Delta^2(-\hat{p}^2)^n} \tag{4a}$$

$$\hat{p} = \frac{p}{\omega_c} \tag{4b}$$

and the input reflection coefficient is [Eq. (3-3)]

$$| s_{11}(p) |^2 = 1 - \frac{1}{1 + \Delta^2 (-\hat{p}^2)^n} = \frac{\Delta^2 (-\hat{p}^2)^n}{1 + \Delta^2 (-\hat{p}^2)^n} \tag{5}$$

The roots of the denominator of (4) are

$$\Delta^2 (-\hat{p}^2)^n = -1 = e^{j(2k-1)\pi} \qquad k = 1, 2, \ldots, 2n$$

The values of k are chosen such that there are 2n distinct roots.

$$\left[e^{\pm j\pi \hat{p}^2} \right]^n = \frac{1}{\Delta^2} e^{j(2k-1)\pi}$$

$$\hat{p}^{2n} = \frac{1}{\Delta^2} e^{j[(2k-1)\pi \mp n\pi]} = \frac{1}{\Delta^2} e^{j(2k+n-1)\pi}$$

The positive sign in front of $n\pi$ is chosen to avoid negative angles.

$$\hat{p}_k = \Delta^{-\frac{1}{n}} e^{j\theta_k} \equiv p_\Delta e^{j\theta_k} \tag{6a}$$

$$p_\Delta = \Delta^{-\frac{1}{n}} \tag{6b}$$

$$\theta_k = \frac{2k + n - 1}{2n} \pi \qquad k = 1, 2, \ldots, 2n. \tag{6c}$$

Equation (5) can be expressed as

$$s_{11}(p) \, s_{11}(-p) = \frac{\Delta^2 \hat{p}^n (-\hat{p})^n}{(\hat{p} - p_\Delta e^{j\theta_1})(\hat{p} - p_\Delta e^{j\theta_2}) \cdots (\hat{p} - p_\Delta e^{j\theta_{2n}})} \tag{7a}$$

The scattering functions $s_{11}(\pm p)$ are chosen as follows.

$$s_{11}(p) = \frac{\pm \Delta \hat{p}^n}{\prod_q (\hat{p} - p_\Delta e^{j\theta_q})} \tag{7b}$$

$$s_{11}(-p) = \frac{\mp \Delta \hat{p}^n}{\prod_r (\hat{p} - p_\Delta e^{j\theta_r})} \tag{7c}$$

where θ_q are angles located in the left half-plane and θ_r are angles located in the right half-plane. It is to be noted that the roots of $|s_{11}(p)|^2$ are located on a circle with a radius of unity and the angle between adjacent roots is π/n.

(d) The driving-point impedance is then given by [Eq. (3-4c)]

$$Z_{in} = R_o \frac{1 + s_{11}(p)}{1 - s_{11}(p)} \tag{8}$$

where $s_{11}(p)$ is defined by (7b). The driving-point impedance can be realized by one of the synthesis techniques such as Cauer expansion. This will be demonstrated by the following example.

5.1. Example: Design of a Maximally Flat Low-Pass Filter

Design a low-pass filter with maximally flat response. The load and source resistances are both equal to one ohm. The insertion loss function is specified by $n = 3$, $\omega_c = 1$, and $\Delta^2 = 1$ [Eq. (5-1)], i.e.,

$$K_{ip} = 1 + \hat{\omega}^6 = \frac{1}{|s_{21}(j\hat{\omega})|^2}$$

or the forward transmission coefficient is [Eq. (5-2)]

$$|s_{21}(\hat{p})|^2 = \frac{1}{1 + (-j\hat{p})^6} = \frac{1}{1 - \hat{p}^6}$$

Solution:

The input reflection coefficient is [Eq. (5-5)]

$$|s_{11}(\hat{p})|^2 = 1 - \frac{1}{|s_{21}(\hat{p})|^2} = \frac{-\hat{p}^6}{1 - \hat{p}^6}$$

The roots are [Eq. (5-6)]

$$\hat{p} = e^{j\frac{2k+n-1}{6}\pi} = e^{j\frac{2k+2}{6}\pi} = e^{j\frac{k+1}{3}\pi} = e^{j\theta_k} \qquad k = 1, 2, \ldots, 6$$

or

$$\theta_k = \frac{2\pi}{3}, \pi, \frac{4\pi}{3}, \frac{5\pi}{3}, 2\pi, \frac{7\pi}{3}$$

$$\theta_q = \frac{2\pi}{3}, \pi, \frac{4\pi}{3}$$

$$\theta_r = \frac{5\pi}{3}, 2\pi, \frac{7\pi}{3}$$

The desired input reflection coefficient is therefore

$$s_{11}(p) = \frac{-\hat{p}^3}{\left[\hat{p} - e^{\frac{j2\pi}{3}}\right](\hat{p} - e^{j\pi})\left[\hat{p} - e^{\frac{j4\pi}{3}}\right]} = \frac{-\hat{p}^3}{\hat{p}^3 + 2\hat{p}^2 + 2\hat{p} + 1}$$

The driving-point impedance is

$$Z_{in}(\hat{p}) = \frac{1 + s_{11}(\hat{p})}{1 - s_{11}(\hat{p})} = \frac{2\hat{p}^2 + 2\hat{p} - 1}{2\hat{p}^3 + 2\hat{p}^2 + 2\hat{p} + 1}$$

$$= \cfrac{1}{\hat{p} + \cfrac{1}{2\hat{p} + \cfrac{1}{\hat{p} + \cfrac{1}{1}}}}$$

The appropriate circuit is shown in Fig. 5.1-1.

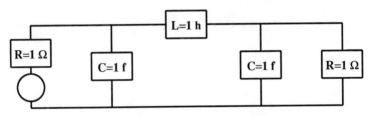

Figure 5.1-1: The realized circuit for example 5.1

6. Chebycheff Low-Pass Approximation

(a) This method makes use of Chebycheff polynomials (see Section VII.5-1) to approximate the ideal characteristic of a low-pass filter. The insertion power ratio is represented by [Eq. (2-11b)]

$$K_{ip} = \frac{1}{|\ s_{21}(j\hat{\omega})\ |^2} = \frac{1 + \Delta^2 C_n^2(\hat{\omega})}{g_n} \qquad \hat{\omega} \equiv \frac{\omega}{\omega_c} \tag{1}$$

where g_n is a constant factor and $g_n \leq 1$; Δ^2 is the constant of tolerance, and $C_n(\hat{\omega})$ is the Chebycheff polynomial of order n. These polynomials are defined as follows [Eq. (VII.5.1-1)].

$$C_n(x) = \cos(\ n \cos^{-1} x\ ) \qquad \text{for} \quad -1 \leq x \leq 1 \tag{2a}$$

$$C_n(x) = \cosh(\ n \cosh^{-1} x\ ) \qquad \text{for} \quad x \geq 1 \tag{2b}$$

For n = 0 and n = 1, (2a) yields

$$C_0(x) = 1 \qquad \text{and} \qquad C_1(x) = x \tag{3}$$

Higher-order Chebycheff polynomials can be obtained from the recursion formula, (VII.5.1-4f).

$$C_{n+1}(x) = 2x C_n(x) - C_{n-1}(x) \tag{4}$$

(b) The equal-ripple property in the pass-band can be visualized by rewriting (1) as

$$|\ s_{21}(j\hat{\omega})\ | = \frac{\sqrt{g_n}}{\sqrt{1 + \Delta^2 C_n^2(\hat{\omega})}} \tag{5a}$$

The maximum value of $|s_{21}|$ is when $C_n(x) = 0$, i.e.,

$$|\ s_{21}(j\hat{\omega})\ |_{max} = \sqrt{g_n} \tag{5b}$$

and $|s_{21}|$ is minimum when $C_n = 1$,

$$|\ s_{21}(j\hat{\omega})\ |_{min} = \frac{\sqrt{g_n}}{\sqrt{1 + \Delta^2}} \tag{5c}$$

since the magnitude of Chebycheff polynomials is bounded by ± 1 within the range of $-1 \leq x \leq 1$.

The ratio of the maximum to the minimum of the forward transmission coefficients is therefore

$$\frac{|\ s_{21}(j\hat{\omega})\ |_{max}}{|\ s_{21}(j\hat{\omega})\ |_{min}} = \sqrt{1 + \Delta^2} \tag{6}$$

(c) In the stop-band, $\Delta^2 C_n^2(\hat\omega) \gg 1$, the insertion loss function can be approximated as

$$| s_{21}(j\hat\omega) |^2 \approx \frac{g_n}{\Delta^2 C_n^2(\hat\omega)} \approx \frac{g_n}{\Delta^2 [\, 2^{n-1}\hat\omega^n \,]^2} \tag{7a}$$

$$C_n(x) \approx 2^{n-1}x^n \qquad \text{for large values of x} \tag{7b}$$

Equation (7b) is estimated from (VII.5.1-4).

The forward transmission coefficient in db in the stop-band is

$$s_{21_{db}} = 10 \log_{10} | s_{21}(j\hat\omega) |^2 = 10 \log g_n - 20 \log_{10} (2^{n-1}\Delta) - 20n \log_{10} \hat\omega \tag{8}$$

The db curve of the Chebycheff approximation has a slope of $-20n$ decibels per decade at large values of the normalized frequency.

(d) The input reflection coefficient is

$$| s_{11}(j\hat\omega) |^2 = 1 - | s_{21}(j\hat\omega) |^2 = \frac{1 - g_n + \Delta^2 C_n^2(\hat\omega)}{1 + \Delta^2 C_n^2(\hat\omega)}$$

$$= (1 - g_n) \frac{1 + \Delta_N^2 C_n^2(\hat\omega)}{1 + \Delta^2 C_n^2(\hat\omega)} \tag{9a}$$

$$\Delta_N = \frac{\Delta}{\sqrt{1 - g_n}} \tag{9b}$$

In terms of the variable $p = \dfrac{\omega}{j}$ or $\hat{p} = \dfrac{p}{\omega_c}$, one has

$$| s_{11}(\hat{p}) |^2 = (1 - g_n) \frac{1 + \Delta_N^2 C_n^2(-j\hat{p})}{1 + \Delta^2 C_n^2(-j\hat{p})} \equiv \frac{| N(\hat{p}) |^2}{| D(\hat{p}) |^2} \tag{9}$$

The poles of $| s_{11}(\hat{p}) |^2$ are the roots of the denominator of (9), $| D(\hat{p}) |^2 = 0$.

$$C_n^2(-j\hat{p}) = \frac{-1}{\Delta^2} \qquad \text{or} \qquad C_n(-j\hat{p}) = \frac{\pm j}{\Delta} \tag{10}$$

Let

$$\cos^{-1}(-j\hat{p}) \equiv u + jv \tag{11}$$

Then the substitution of (11) into (2a) yields

$$C_n(-j\hat{p}) = \cos[n(u+jv)] = \cos nu \cos (jnv) - \sin nu \sin (jnv)$$

$$= \cos nu \cosh nv - j \sin nu \sinh nv \tag{12}$$

Substituting (12) into (10) and separating the real and imaginary parts, one obtains

$$\cos nu \cosh nv = 0 \tag{13a}$$

$$\sin nu \sinh nv = \pm \frac{1}{\Delta} \tag{13b}$$

Since cosh nv can never be zero, (13a) imposes that

$$\cos nu = 0 \qquad \text{or} \qquad u = \pm \frac{2k-1}{n}\frac{\pi}{2} \qquad k = 1, 2, \ldots, 2n \tag{14a}$$

Consequently, one has

$$\sin nu = \pm 1 \tag{14b}$$

and (13b) yields

$$\sinh nv = \frac{1}{\Delta} \qquad \text{or} \qquad v = \frac{1}{n}\sinh^{-1}\frac{1}{\Delta} \tag{14c}$$

The parameter p can be obtained from (11)

$$p = j\cos(u+jv) = j\cos u \cosh v + \sin u \sinh v \tag{15}$$

Imposing the condition (14), one has

$$p_k = j\cos\left[\frac{2k-1}{2n}\pi\right]\cosh\left[\frac{1}{n}\sinh^{-1}\frac{1}{\Delta}\right]$$

$$+ \sin\left[\frac{2k-1}{2n}\pi\right]\sinh\left[\frac{1}{n}\sinh^{-1}\frac{1}{\Delta}\right] \tag{16}$$

The denominator of (9) is therefore given by

$$D(\hat{p}) = \sum_{k=1}^{2n}(\hat{p}-p_k) \tag{17}$$

where $\hat{p}_k$ are the roots given by (16).

The zeros of $|s_{11}(\hat{p})|^2$ are the zeros of $|N(\hat{p})|^2$ in (9). There are no known reasons to restrict the locations of the zeros of $s_{11}(\hat{p})$ except that the zeros of a complex-conjugate pair should be assigned together. The zeros of $s_{11}(\hat{p})$ can therefore be chosen arbitrarily. When the zeros are restricted to the LHS, the function is known as the minimum phase function.

It is convenient to express the reflection coefficient, (9), in the following form.

$$s_{11}(\hat{p})s_{11}(-\hat{p}) = (1 - g_n) \frac{\Delta_N^2 \, 2^{2n-2} \, q_N(\hat{p})q_N(-\hat{p})}{\Delta^2 2^{2n-2} \, q(\hat{p})q(-\hat{p})} = \frac{q_N(\hat{p})q_N(-\hat{p})}{q(\hat{p})q(-\hat{p})} \tag{18a}$$

where

$$|D(\hat{p})|^2 = 1 + \Delta^2 C_n^2(-j\hat{p}) = 0 \equiv \Delta^2 2^{2n-2} q(\hat{p})q(-\hat{p}) \tag{18b}$$

$$|N(\hat{p})|^2 = 1 + \Delta_N^2 C_n^2(-j\hat{p}) \equiv \Delta_N^2 2^{2n-2} q_N(\hat{p})q_N(-\hat{p}) \tag{18c}$$

$$\Delta_N = \frac{\Delta}{\sqrt{1 - g_n}} \tag{18d}$$

Then $s_{11}(\hat{p})$ can be extracted from (18a) as

$$s_{11}(\hat{p}) = \pm \frac{q_N(\hat{p})}{q(\hat{p})} \tag{19a}$$

$$q(\hat{p}) = \prod_{k=1}^{n} (\hat{p} - p_k) = \hat{p}^n + b_{n-1}\hat{p}^{n-1} + \cdots + b_1\hat{p} + b_o \tag{19b}$$

$$q_N(\hat{p}) = \prod_{k=1}^{n} (\hat{p} - p_k') = \hat{p}^n + d_{n-1}\hat{p}^{n-1} + \cdots + d_1\hat{p} + d_o \tag{19c}$$

where p_k and p_k' are LHS roots of (18b) and (18c), respectively. Hence, both $q(\hat{p})$ and $q_N(\hat{p})$ are Hurwitz polynomials. The $\pm$ sign is determined by the relative values of the source resistance and the load resistance. This will be demonstrated shortly.

The corresponding input impedance is [Eq. (3-4c)]

$$Z_{11}(\hat{p}) = Z_o \frac{1 + s_{11}(\hat{p})}{1 - s_{11}(\hat{p})} = Z_o \frac{q(\hat{p}) \pm q_N(\hat{p})}{q(\hat{p}) \mp q_N(\hat{p})} \tag{20}$$

Let the lossless two-port network be operating between a source with internal resistance R_g and a resistive load R_L (Fig. 6-1). Then the input impedance across 1–1' is

$$Z_{11}(\hat{p}) = R_g \frac{q(\hat{p}) + q_N(\hat{p})}{q(\hat{p}) - q_N(\hat{p})} \tag{21}$$

The quantities R_g, R_L, and g_n are related and cannot be chosen arbitrarily. A relation can be obtained at DC frequency, by setting $\hat{p} = 0$.

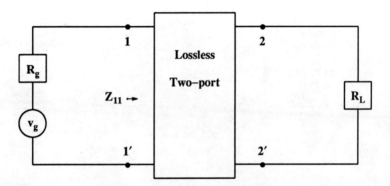

Figure 6-1: A terminated two-port network.

$$Z_{11}(\hat{p} = 0) = R_L = R_g \frac{q(0) \pm q_N(0)}{q(0) \mp q_N(0)}$$

or

$$\frac{R_L}{R_g} = \left[\frac{b_o + d_o}{b_o - d_o} \right]^{\pm 1} \tag{22}$$

and

$$b_o = 2^{1-n} \sinh nu = \frac{2^{1-n}}{\Delta} \qquad n \text{ odd} \tag{23a}$$

$$= 2^{1-n} \cosh nu = 2^{1-n} \sqrt{1 + \sinh^2 nu}$$

$$= 2^{1-n} \frac{\sqrt{1 + \Delta^2}}{\Delta} \qquad n \text{ even} \tag{23b}$$

$$d_o = 2^{1-n} \sinh nu_N = \frac{2^{1-n}}{\Delta_N} \qquad n \text{ odd} \tag{23c}$$

$$= 2^{1-n} \cosh nu_N = 2^{1-n} \frac{\sqrt{1 + \Delta^2} - g_n}{\Delta} \qquad n \text{ even} \tag{23d}$$

The derivation of the general expression, (23), is beyond the scope of this text. However, it will be demonstrated that these relations are valid for a given order n, by direct expansion of the following product (see Section 6.4).

$$b_o = (-1)^n \prod_{k=1}^{n} p_k$$

where p_k's are the roots of the polynomial, and n is the order of the polynomial. A similar relation applies to d_o.

The solution of

$$| N(\hat{p}) |^2 = 0$$

is obtained in a similar manner as for

$$| D(\hat{p}) |^2 = 0$$

by assuming

$$C_n(-\hat{p}) = \frac{\pm j}{\Delta_N} \qquad \text{and} \qquad \cos^{-1}(-j\hat{p}) \equiv u_N + jv_N$$

The substitution of (23) into (22) yields

$$\frac{R_L}{R_g} = \left[\frac{1 + \sqrt{1 - g_n}}{1 - \sqrt{1 - g_n}} \right]^{\pm 1} \qquad \text{n odd} \tag{24a}$$

$$= \left[\frac{\sqrt{1 + \Delta^2} + \sqrt{1 + \Delta^2 - g_n}}{\sqrt{1 + \Delta^2} - \sqrt{1 + \Delta^2 - g_n}} \right]^{\mp 1} \qquad \text{n even} \tag{24b}$$

The $\pm$ sign is determined according to whether $R_L \geq R_g$ or $R_L \leq R_g$. For a specified DC gain g_n, the R_L/R_g ratio is determined by (24) with appropriate n. On the other hand, when R_g and R_L are specified, then the DC gain can be determined from (24) and the pass-band tolerance may be chosen arbitrarily.

For example, for the case of maximum DC gain, $g_n = 1$, and n is odd, (24a) specifies that $R_g = R_L$. However, if $R_L/R_g = 3$ is specified, then the gain is $g_n = 3/4$.

6.1. Example: Determination of $s_{21}(\hat{p})$ for Chebycheff Low-Pass Approximation

Determine the appropriate forward transmission coefficient for a Chebycheff low-pass approximation with the following specification.

$$\Delta^2 = 0.5 \qquad n = 4$$

Solution:

The forward transmission coefficient is

$$| s_{21}(\hat{p}) |^2 = \frac{1}{1 + 0.5C_4^2(-j\hat{p})} \tag{1}$$

and the corresponding input reflection coefficient is

$$| s_{11}(\hat{p}) |^2 = 1 - | s_{21}(\hat{p}) |^2 = \frac{0.5C_4^2(-j\hat{p})}{1 + 0.5C_4^2(-j\hat{p})} \tag{2}$$

The parameter u_k is calculated from (6-14a).

$$u_k = \frac{2k-1}{4}\frac{\pi}{2} \qquad k = 1, 2, \ldots, 8$$

$$= \frac{\pi}{8}, \frac{3\pi}{8}, \frac{5\pi}{8}, \frac{7\pi}{8}, \frac{9\pi}{8}, \frac{11\pi}{8}, \frac{13\pi}{8}, \frac{15\pi}{8}$$

The parameter v is evaluated from (6.14c).

$$v = \frac{1}{4} \sinh^{-1} \frac{1}{\sqrt{0.5}} = 0.286$$

$$\sinh v = 0.290 \qquad \cosh v = 1.041$$

The root p_j is determined form (6-16).

$$p_1 = \sin \frac{\pi}{8} \sinh v + j \cos \frac{\pi}{8} \cosh v$$

$$= 0.383 \times 0.290 + j\, 0.924 \times 1.041 = 0.968 \; e^{j83.42°}$$

$$p_2 = \sin \frac{3\pi}{8} \times 0.290 + j \cos \frac{3\pi}{8} \times 1.041 = 0.481 \; e^{j56.11°}$$

$$p_3 = \sin \frac{5\pi}{8} \times 0.290 + j \cos \frac{5\pi}{8} \times 1.041 = 0.481 \; e^{-j56.11°}$$

$$p_4 = \sin \frac{7\pi}{8} \times 0.290 + j \cos \frac{7\pi}{8} \times 1.041 = 0.968 \, e^{-j83.42°}$$

$$p_5 = \sin \frac{-\pi}{8} \times 0.290 + j \cos \frac{-\pi}{8} \times 1.041 = -0.968 \, e^{-j83.42°} = -p_4$$

$$p_6 = \sin \frac{-3\pi}{8} \times 0.290 + j \cos \frac{-3\pi}{8} \times 1.041 = -0.481 \, e^{-j56.11°} = -p_3$$

$$p_7 = \sin \frac{-5\pi}{8} \times 0.290 + j \cos \frac{-3\pi}{8} \times 1.041 = -0.481 \, e^{j56.11°} = -p_2$$

$$p_8 = \sin \frac{-7\pi}{8} \times 0.290 + j \cos \frac{-7\pi}{8} \times 1.041 = -0.968 \, e^{j83.42°} = -p_1$$

The poles of $| \, s_{21}(\hat{p}) \, |^2$ are sketched in Fig. 6.1-1, and the appropriate poles for $s_{21}(\hat{p})$ should be located on the left half-plane; these are p_5 through p_8. Thus

$$s_{21}(\hat{p}) = \frac{1}{(\hat{p} - p_5)(\hat{p} - p_6)(\hat{p} - p_7)(\hat{p} - p_8)}$$

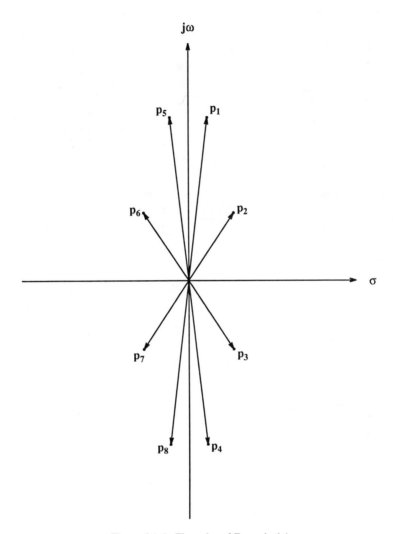

Figure 6.1-1: The poles of Example 6.1

6.2. Example: Determination of n and Tolerance Δ

Determine the order n and the tolerance Δ for a Chebycheff response which has a maximum ripple of 1 db within the pass-band and a minimum attenuation of 60 db at frequencies above five times that of cutoff.

Solution :

In the pass-band, the ripple is given by (6-6).

$$\text{ripple} = \frac{\mid s_{21}(j\hat{\omega}) \mid^2_{max}}{\mid s_{21}(j\hat{\omega}) \mid^2_{min}} = 1 + \Delta^2 \tag{1}$$

and

$$[\text{ripple}]_{db} = 10 \log_{10} [\text{ripple}] = 10 \log_{10} (1 + \Delta^2) = 1 \text{ db}$$

$$1 + \Delta^2 = 1.2589 \qquad \text{or} \qquad \Delta = 0.5088$$

In the stop-band, the forward transmission coefficient can be approximated by [Eq. (6-8)]

$$s_{21db} = 10 \log_{10} g_n - 20 \log_{10} \Delta + 20 \log_{10} 2 - 20 n \log_{10} 2\hat{\omega} \geq -60 \text{ db}$$

$$n \log_{10} (2 \times 5) = 3 + 0.5 \log_{10} g_n - \log_{10} 0.5088 + 0.301$$

or

$$n = 3 + 0.5 \log_{10} g_n + 0.2935 + 0.301 = 3.595 + 0.5 \log_{10} g_n$$

Thus, choosing n = 4 will be adequate since $g_n \leq 1$.

6.3. Example: Design of Chebycheff Filter

Design a low-pass Chebycheff filter which has a maximum pass-band ripple of 1 db. In the stop-band, the minimum attenuation should be 60 db at frequencies of $5\omega_c$ and above. The cutoff frequency is $\omega_c = 10^4$ rad/s. The source has an internal resistance of 50 ohms and the load resistance is 100 ohms.

Solution :

For 1-db ripple and 60-db attenuation at $\omega \geq 5\omega_c$, it was shown that (see Section 6.2)

$$n = 4 \qquad \text{and} \qquad \Delta = 0.5088$$

The constant factor g_n can be determined as follows. The constant g_n is related to the source and load resistances by (6-24). For $n = 4$, one has

$$\frac{R_L}{R_g} = \frac{100}{50} = 2 = \frac{1 + \sqrt{1 - G}}{1 - \sqrt{1 - G}} \tag{1a}$$

$$G \equiv \frac{g_n}{1 + \Delta^2} = \text{minimum gain} \tag{1b}$$

The plus sign is chosen to satisfy the given condition that the ratio on the right-hand side should be greater than unity. G is solved from (1a) and then substituted back into (1b) to evaluate g_n.

$$G = \frac{8}{9} = \frac{g_n}{1 + \Delta^2} \qquad \text{or} \qquad g_n = 1.12 > 1 \tag{2}$$

This is physically impossible to realize for a passive network, and the maximum value of unity will be used for g_n. Then (2) becomes

$$\frac{8}{9} = \frac{1}{1 + \Delta^2} \qquad \text{or} \qquad \Delta = 0.3535$$

The ripple for this tolerance is then

$$[\text{ripple}]_{db} = 10 \log_{10}(1 + \Delta^2) = 0.512 \text{ db}$$

The above choice of $g_n = 1$ does not violate the original specification on the ripple variation.

The current design parameters are:

$$n = 4, \qquad \Delta = 0.3535, \qquad \text{and} \qquad g_4 = 1$$

The forward transmission coefficient is [Eq. (6-5a)]

$$|s_{21}(j\hat{\omega})|^2 = \frac{1}{1 + \Delta^2 C_4^2(\hat{\omega})} \tag{3}$$

Let $\hat{p} = -j\hat{\omega}$; then

$$s_{21}(\hat{p})s_{21}(-\hat{p}) = \frac{1}{\mid D(\hat{p}) \mid^2} \tag{4a}$$

$$\mid D(\hat{p}) \mid^2 = 1 + \Delta^2 C_4^2(-j\hat{p}) \tag{4b}$$

The roots of $\mid D(\hat{p}) \mid^2 = 0$ are given by (6-16).

$$p_k = \sin u_k \sinh v + j \cos u_k \cosh v \tag{5a}$$

$$u_k = \pm \frac{2k-1}{4} \frac{\pi}{2} \qquad k = 1, 2, 3, \ldots, 8 \tag{5b}$$

$$v = \frac{1}{4} \sinh^{-1} \frac{1}{0.3535} = 0.4407 \tag{5c}$$

$$\sinh v = 0.4551 \qquad \text{and} \qquad \cosh v = 1.0987 \tag{5d}$$

Note that the u_k's are identical to those obtained in Section 6.1.

$$u_1 = \frac{\pi}{8}, \quad p_1 = 0.1742 + j1.015 = 1.0299 \, e^{j80.26°} \tag{6a}$$

$$u_2 = \frac{3\pi}{8}, \quad p_2 = 0.4205 + j0.4205 = 0.5946 \, e^{j45°} \tag{6b}$$

$$u_3 = \frac{5\pi}{8}, \quad p_3 = 0.4205 - j0.4205 = p_2^* \tag{6c}$$

$$u_4 = \frac{7\pi}{8}, \quad p_4 = 0.1742 - j1.015 = p_1^* \tag{6d}$$

$$u_5 = -\frac{\pi}{8}, \quad p_5 = -p_1^* \tag{6e}$$

$$u_6 = -\frac{3\pi}{8}, \quad p_6 = -p_2^* \tag{6f}$$

$$u_7 = -\frac{5\pi}{8}, \quad p_7 = -p_2 \tag{6g}$$

$$u_8 = -\frac{7\pi}{8}, \quad p_8 = -p_1 \tag{6h}$$

The associate Kurwitz polynomial for the Chebycheff response is defined by (6-18).

$$| D(\hat{p}) |^2 = \Delta^2 2^{2n-2} q(\hat{p})q(-\hat{p})$$

where $q(\hat{p})$ is composed of the roots on the LHS as given by (6).

$$q(\hat{p}) = (\hat{p} - p_5)(\hat{p} - p_5^*)(\hat{p} - p_6)(\hat{p} - p_6^*)$$

$$= (\hat{p} + 1.0299\ e^{-j80.26°})(\hat{p} + 1.0299\ e^{j80.26°})(\hat{p} + 0.5946\ e^{-j45°})$$

$$\times\ (\hat{p} + 0.5946\ e^{j45°})$$

$$= \hat{p}^4 + 1.189\hat{p}^3 + 1.707\hat{p}^2 + 1.015\hat{p} + 0.375 \tag{7}$$

The input reflection coefficient is given by (6-19a).

$$s_{11}(\hat{p}) = \pm\ \frac{q_N(\hat{p})}{q(\hat{p})} \tag{8}$$

where

$$q_N(\hat{p})q_N(-\hat{p}) \equiv C_4^2(-j\hat{p}) = [\ 8\ (-j\hat{p})^4 - 8\ (-j\hat{p})^2 + 1\ ]^2$$

$$= [\ 8\hat{p}^4 + 8\hat{p}^2 + 1\ ]^2$$

since $g_n = g_4 = 1$, or

$$q_N(\hat{p}) = \hat{p}^4 + \hat{p}^2 + 0.125 \tag{9}$$

where the coefficient of $\hat{p}^4$ has been normalized to unity.

The input impedance is given by [Eq. (6-20)]

$$Z_{11}(\hat{p}) = R_g\ \frac{q(\hat{p}) + q_N(\hat{p})}{q(\hat{p}) - q_N(\hat{p})} = R_g\ \frac{Z_N}{Z_D} \tag{10a}$$

$$Z_N \equiv q(\hat{p}) + q_N(\hat{p}) = 2\hat{p}^4 + 1.189\hat{p}^3 + 2.707\hat{p}^2 + 1.015\hat{p} + 0.500 \tag{10b}$$

$$Z_D \equiv q(\hat{p}) - q_N(\hat{p}) = 1.189\hat{p}^3 + 0.707\hat{p}^2 + 1.015\hat{p} + 0.250 \tag{10c}$$

The substitution of (10b) and (10c) into (10a) yields

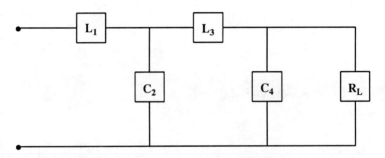

Figure 6.3-1: The realized circuit for Example 6.3-1.

$$\frac{Z_{11}}{R_g} = 1.682\hat{p} + \cfrac{1}{1.189\hat{p} + \cfrac{1}{2.381\hat{p} + \cfrac{1}{0.84\hat{p} + \cfrac{1}{2}}}} \tag{11}$$

The circuit elements are:

$$z_1 = \omega L_1 = 1.682 \frac{\omega}{\omega_c} R_g = 1.682 \frac{\omega}{10^4} 50 = \omega \times 8 \times 10^3 \tag{12a}$$

or

$$L_1 = 8\text{mh} \tag{12b}$$

$$y_2 = \omega C_2 = 1.189 \frac{\omega}{10^4} \frac{1}{50} = \omega (2.378 \times 10^{-6}) \quad \text{or} \quad C_2 = 2.378 \ \mu\text{f} \tag{12c}$$

$$z_3 = \omega \frac{2.381 \times 50}{10^4} = \omega (11.91 \times 10^{-3}) \quad \text{or} \quad L_3 = 11.91 \ \text{mh} \tag{12d}$$

$$y_4 = \frac{0.84 \times \omega}{50 \times 10^4} = \omega (1.68 \times 10^{-6}) \quad \text{or} \quad C_4 = 1.68 \ \mu\text{f} \tag{12e}$$

$$z_5 = 2 \times 50 = 100 \ \text{ohms} = R_L \tag{12f}$$

The filter circuit is as shown in Fig. 6.3-1.

6.4. Verification of Eq. (6-23)

Show the validity of Eq. (6-23) for $n = 2$ and $n = 3$.

Soltuion:

The Hurwitz polynomial for a Chebycheff response is given by

$$q(\hat{p}) = \prod_{k=1}^{n} (\hat{p} - p_k) = \hat{p}^n + b_{n-1}\hat{p}^{n-1} + \cdots + b_1\hat{p} + b_o \qquad (6\text{-}19\text{b})$$

By direct expansion of the product, the coefficient b_o is given by

$$b_o = (-1)^n \prod_{k=1}^{n} p_k' \qquad (1)$$

where p_k are roots in the left half-space.

$$p_k = \sinh v \sin u_k + j \cosh v \cos u_k \qquad (6\text{-}16)$$

$$v = \frac{1}{n} \sinh^{-1} \frac{1}{\Delta} \qquad (6\text{-}14\text{c})$$

$$u_k = \frac{2k - 1}{2n} \pi \qquad k = 1, 2, \ldots, 2n$$

(a) For $n = 2$,

$$u_k = \frac{\pi}{4}, \ \frac{3\pi}{4}, \ \frac{5\pi}{4}, \ \text{and} \ \frac{7\pi}{4} \qquad (2)$$

Roots on the left half-space correspond to

$$u_1 = \frac{5\pi}{4} \qquad \text{and} \qquad u_2 = \frac{7\pi}{4} \qquad (3)$$

Roots on the LHS are

$$p_1 = \sinh v \sin \frac{5\pi}{4} + j \cosh v \cos \frac{5\pi}{4} \qquad (4\text{a})$$

$$p_2 = \sinh v \sin \frac{7\pi}{4} + j \cosh v \cos \frac{7\pi}{4} \qquad (4\text{b})$$

Then

$$b_o = p_1 p_2 = \sinh^2 v \sin \frac{5\pi}{4} \sin \frac{7\pi}{4} - \cosh^2 v \cos \frac{5\pi}{4} \cos \frac{7\pi}{4}$$

$$+ j \cosh v \sinh v \left[\sin \frac{5\pi}{4} \cos \frac{7\pi}{4} + \cos \frac{5\pi}{4} \sin \frac{7\pi}{4} \right]$$

$$= \cosh^2 v \left[\sin \frac{5\pi}{4} \sin \frac{7\pi}{4} - \cos \frac{5\pi}{4} \cos \frac{7\pi}{4} \right] - \sin \frac{5\pi}{4} \sin \frac{7\pi}{4}$$

$$= \cosh^2 v - \frac{1}{2} \left[\cos \left[\frac{5\pi}{4} - \frac{7\pi}{4} \right] - \cos \left[\frac{5\pi}{4} + \frac{7\pi}{4} \right] \right]$$

$$= \cosh^2 v - \frac{1}{2} = \frac{1}{2} (2 \cosh v - 1) = \frac{1}{2} \cosh 2v$$

$$= \frac{1}{2^{n-1}} \cosh nv \qquad n = 2 \qquad\qquad (5)$$

(b) For n = 3, the parameter u_k has the following values:

$$u_k = \frac{\pi}{6}, \frac{3\pi}{6}, \frac{5\pi}{6}, \frac{7\pi}{6}, \frac{9\pi}{6}, \text{ and } \frac{11\pi}{6} \qquad\qquad (6)$$

Values of u_k corresponding to roots on the left half-space are

$$u_1 = \frac{7\pi}{6}, \qquad u_2 = 270°, \qquad u_3 = \frac{11\pi}{6} \qquad\qquad (7)$$

Roots on the LHS are

$$p_1 = \sinh v \sin \frac{7\pi}{6} + j \cosh v \cos \frac{7\pi}{6} \qquad\qquad (8a)$$

$$p_2 = \sinh v \sin 270° = -\sinh v \qquad\qquad (8b)$$

$$p_3 = \sinh v \sin \frac{11\pi}{6} + j \cosh v \cos \frac{11\pi}{6} \qquad\qquad (8c)$$

The coefficient b_o is then given by

$$b_o = -p_1p_2p_3 = \sinh v \left\{ \sinh^2 v \sin \frac{7\pi}{6} \sin \frac{11\pi}{6} - \cosh^2 v \cos\frac{7\pi}{6} \cos \frac{11\pi}{6} \right\}$$

$$+ j \cosh v \sinh v \left\{ \sin \frac{7\pi}{6} \cos \frac{11\pi}{6} + \cos \frac{7\pi}{6} \sin \frac{11\pi}{6} \right\}$$

$$b_o = \sinh v \left\{ \cosh^2 v \left[- \cos \left[\frac{7\pi}{6} + \frac{11\pi}{6} \right] \right] \right.$$

$$\left. - \frac{1}{2} \left[\cos \left[\frac{7\pi}{6} - \frac{11\pi}{6} \right] - \cos \left[\frac{7\pi}{6} + \frac{11\pi}{6} \right] \right] \right\}$$

$$= \sinh v \left[\cosh^2 v - \frac{1}{4} \right] = \sinh v \left[\sinh^2 v + 1 - \frac{1}{4} \right]$$

$$= \frac{1}{4} [4 \sinh^3 v + 3 \sinh v] = \frac{1}{4} \sinh 3v$$

$$= \frac{1}{2^{n-1}} \sinh nv \qquad n = 3 \qquad\qquad (9)$$

7. Low-Pass to Periodic Band-Pass Transformation

Basic principles in frequency transformation were discussed in Chapter X. The reactance functions in the reactance transformation need not be restricted to reactance of lumped circuit elements. Reactance functions of a circuit with distributed parameters may also be used. If the input reactance of a transmission line terminated by a short circuit is used, then one has

$$j\omega' = jZ_o \tan \beta L_1 = jZ_o \tan \frac{\omega L_1}{c} \tag{1a}$$

or

$$\Omega' = \tan \Omega \tag{1b}$$

$$\Omega' = \frac{\omega'}{Z_o} \qquad\qquad \Omega = \frac{\omega}{\dfrac{c}{L_1}} = \beta L_1 \tag{1c}$$

where Z_o is the characteristic impedance, L_1 is the length of the line, and c is the velocity of light.

The reactance of the inductive element in the LP network is

$$X_L = j\omega'L' = jZ_oL' \tan \Omega = jZ_{oL} \tan \Omega \tag{2a}$$

$$Z_{oL} = Z_oL' \tag{2b}$$

The reactance of the inductive element in the LP network is transformed into the reactance of a short-circuited line with an electrical length Ω and a characteristic impedance Z_{oL} (Fig. 7-1).

The reactance of the capacitive element in the LP network is

$$X_C = \frac{1}{j\omega'C'} = \frac{1}{jZ_oC' \tan \Omega} = -jZ_{oC} \cot \Omega \tag{3a}$$

$$Z_{oC} = \frac{1}{Z_oC'} \tag{3b}$$

The reactance of the capacitive element in the LP network is transformed into the input impedance of an open-circuited line with an electrical length of Ω and characteristic impedance Z_{oC} (Fig. 7-1).

The low-pass network is transformed by (1) into a periodic band-pass network as shown in Fig. 7-2. The graphical transformation of the low-pass response into the periodic band-pass response is shown in Fig. 7-3.

Filters constructed of transmission lines perform satisfactorily at frequencies where the junction effects are negligible.

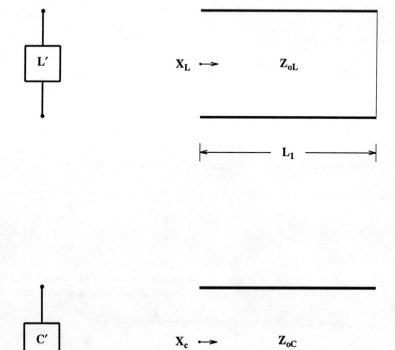

Figure 7-1: The circuit-element transformation: low-pass to periodic band-
pass.

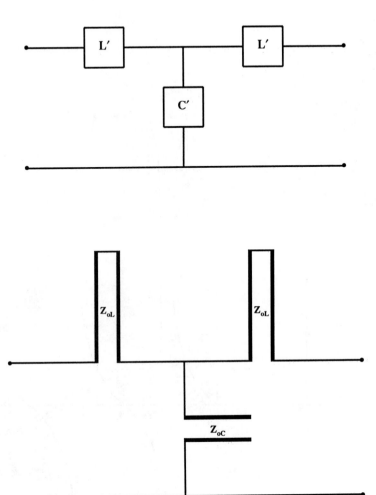

Figure 7-2: The network transformation: low-pass to periodic band-pass.

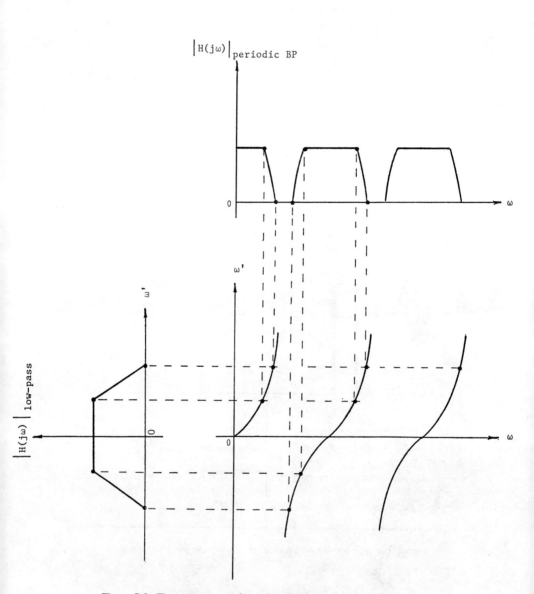

Figure 7-3: The response transformation: low-pass to periodic band-pass.

8. Resonators in Cascade - General

Two resonators are connected in cascade through a waveguide of length L_g as shown in Fig. 8-1a and its equivalent circuit is shown in Fig. 8-1b. The equivalent circuit to the right of plane 5-5' can be simplified by eliminating transformers T_3 and T_4 (Fig. 8-2). The new parameters and the intrinsic parameters are related as follows.

$$R_2' = n_3^2 R_{o2} \tag{1a}$$

$$L_2' = n_3^2 L_{o2} \tag{1b}$$

$$C_2' = \frac{1}{n_3^2} C_{o2} \tag{1c}$$

$$R_L' = \frac{n_3^2}{n_4^2} Z_{o3} \tag{1d}$$

The analysis of the general case with arbitrary length L_g of the coupling waveguide can be carried out by the impedance transformation relation (Fig. 8-3).

$$\hat{Z}_4 \equiv \frac{Z_{44'}}{Z_{o2}} = \frac{Z_{55'} + jZ_{o2} \tan \beta L_g}{Z_{o2} + jZ_{55'} \tan \beta L_g} = \frac{\hat{Z}_5 + j\tau}{1 + j\hat{Z}_5\tau} \tag{2a}$$

$$\tau \equiv \tan \beta L_g \tag{2b}$$

$$\hat{Z}_5 \equiv \frac{Z_{55'}}{Z_{o2}} \tag{2c}$$

$$Y_{55'} = \frac{1}{Z_{55'}} = \frac{1}{R_2'} + \frac{1}{R_L'} + j \left[\omega C_2' - \frac{1}{\omega L_2'} \right] \tag{2d}$$

where $Z_{55'}$ is the input impedance across terminals 5–5' and the coupling guide is assumed to be lossless.

One of the most interesting characteristics is the transmission behavior of the entire structure. The transmission coefficient can be evaluated from the reflection coefficient.

The input impedance across terminals 3-3', Fig. 8-1b, is

$$Z_{33'} = \frac{Z_{o2}}{n_2^2} \hat{Z}_4 = \frac{Z_{o2}}{n_2^2} \frac{\hat{Z}_5 + j\tau}{1 + j\hat{Z}_5\tau} \tag{3a}$$

and

$$Y_{33'} = \frac{1}{Z_{33'}} = \frac{n_2^2}{Z_{o2}} \frac{1 + j\hat{Z}_5\tau}{\hat{Z}_5 + j\tau} \tag{3b}$$

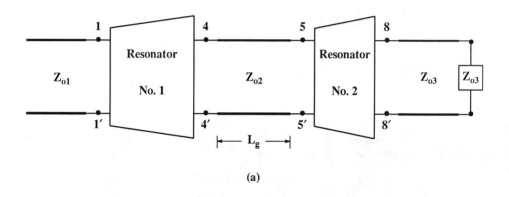

(a)

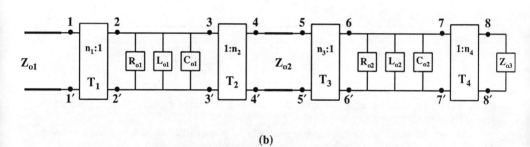

(b)

Figure 8-1: (a) The resonators in cascade.
(b) An equivalent circuit for (a).

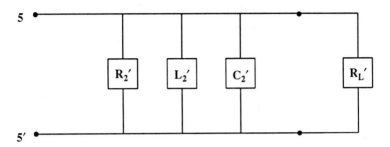

Figure 8-2: A simplified equivalent circuit for resonators in cascade.

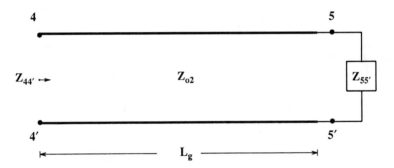

Figure 8-3: Waveguide terminated by an impedance.

The input admittances across terminals 2-2' and 1-1' are

$$Y_{22'} = \frac{1}{R_{o1}} + j \left[\omega C_{o1} - \frac{1}{\omega L_{o1}} \right] + Y_{33'}$$

$$= \frac{1}{R_{o1}} + j \left[\omega C_{o1} - \frac{1}{\omega L_{o1}} \right] + \frac{n_2^2}{Z_{o2}} \frac{1 + j\hat{Z}_5\tau}{\hat{Z}_5 + j\tau} \tag{4a}$$

$$Y_{11'} = n_1^{-2} Y_{22'} \tag{4b}$$

The reflection at the input port 1-1' is

$$\Gamma = \frac{1 - \hat{Y}_i}{1 + \hat{Y}_i} \tag{5a}$$

$$\hat{Y}_i = \frac{Y_{11'}}{Y_{o1}} \tag{5b}$$

The analysis of resonators coupled by a waveguide of arbitrary length is rather complicated. For practical purposes, two special cases will be investigated:

(a) $L_g = 0$, coincidental cascaded resonators.

(b) $L_g = \dfrac{\lambda_g}{4}$, quarter-wavelength cascaded resonators.

It will be shown that coincidental cascaded resonators have properties very similar to those of a single cavity. The transmission characteristics of coincidental coupled resonators are sharper than those of a single resonator due to the increase in the loaded Q.

It will also be shown that the quarter-wavelength coupled resonators provide much steeper transmission characteristics.

9. Coincidental Cascaded Resonators

(a) When resonators are in coincidental cascade, the equivalent circuit in Fig. 8-1b becomes that shows in Fig. 9-1. The parameters R_2', etc., are given by (8-1). Figure 9-1 can be further simplified by removing transformers T_1 and T_2 (Fig. 9-2). The normalized parameters are as follows.

$$\hat{R}_1 = \frac{n_1^2}{Z_{o1}} R_{o1} \tag{1a}$$

$$\hat{L}_1 = \frac{n_1^2}{Z_{o1}} L_{o1} \tag{1b}$$

$$\hat{C}_1 = \frac{Z_{o1}}{n_1^2} C_{o1} \tag{1c}$$

$$\hat{R}_2 = \left[\frac{n_1 n_3}{n_2} \right]^2 \frac{R_{o2}}{Z_{o1}} \tag{1d}$$

$$\hat{L}_2 = \left[\frac{n_1 n_3}{n_2} \right]^2 \frac{L_{o2}}{Z_{o2}} \tag{1e}$$

$$\hat{C}_2 = \left[\frac{n_2}{n_1 n_3} \right]^2 Z_{o1} C_{o2} \tag{1f}$$

$$\hat{R}_L = \left[\frac{n_1 n_3}{n_2 n_4} \right]^2 \frac{Z_{o3}}{Z_{o1}} \qquad R_L = Z_{o3} \tag{1g}$$

This is equivalent to a single resonator characterized by the following parameters, as shown in Fig. 9-3.

$$\hat{R} = \frac{\hat{R}_1 \hat{R}_2}{\hat{R}_1 + \hat{R}_2} \tag{2a}$$

$$\hat{L} = \frac{\hat{L}_1 \hat{L}_2}{\hat{L}_1 + \hat{L}_2} \tag{2b}$$

$$\hat{C} = \hat{C}_1 + \hat{C}_2 \tag{2c}$$

The resonance frequency is given by

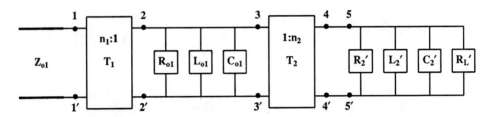

Figure 9-1: An equivalent circuit for coincidental cascade of resonators.

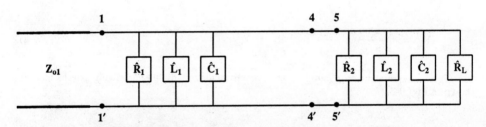

Figure 9-2: A simplified equivalent circuit for coincidentally cascaded resonators.

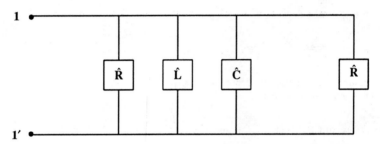

Figure 9-3: A single resonator representation for coincidentally cascaded resonators.

$$\omega_o = \frac{1}{\sqrt{\hat{L}\hat{C}}} = \frac{1}{\sqrt{\dfrac{q}{\omega_{o1}^2} + \dfrac{1-q}{\omega_{o2}^2}}} \tag{3a}$$

$$q = \frac{\hat{L}_2}{\hat{L}_1 + \hat{L}_2} \tag{3b}$$

$$\omega_{o1} = \frac{1}{\sqrt{\hat{L}_1 \hat{C}_1}} \tag{3c}$$

$$\omega_{o2} = \frac{1}{\sqrt{\hat{L}_2 \hat{C}_2}} \tag{3d}$$

When $\omega_{o1} = \omega_{o2}$, then

$$\omega_o = \frac{\omega_{o1}}{\sqrt{q + (1 - q)}} = \omega_{o1} \tag{4}$$

The resulting resonance frequency is independent of whether the resonators are identical or not. The combined unloaded Q is

$$\hat{Q}_u = \omega_o \hat{C}\hat{R} = \omega_o (\hat{C}_1 + \hat{C}_2) \frac{\hat{R}_1 \hat{R}_2}{\hat{R}_1 + \hat{R}_2} \tag{5a}$$

For the case when $\omega_{o1} = \omega_{o2} = \omega_o$, one has

$$\hat{Q}_u = q'\hat{Q}_{u1} + \hat{Q}_{u2}(1 - q') \tag{5b}$$

$$q' = \frac{\hat{R}_2}{\hat{R}_1 + \hat{R}_2} \qquad \hat{Q}_{uj} = \omega_o \hat{C}_j \hat{R}_j \quad j = 1, 2 \tag{5c}$$

The external Q with respect to the input port, $\hat{Q}_{ei}$, is

$$\hat{Q}_{ei} = \omega_o \hat{C}\hat{R}_{ext-in} = \omega_o (C_1 + C_2) \hat{R}_{ext-in} \equiv \hat{Q}_{ei1} + \hat{Q}_{ei2} \tag{6a}$$

where

$$\hat{Q}_{ei1} \equiv \omega_o C_1 \hat{R}_{ext-in} \qquad \hat{Q}_{ei2} \equiv \omega_o C_2 \hat{R}_{ext-in}$$

The external Q with respect to the output port, $\hat{Q}_{eo}$, is

$$\hat{Q}_{eo} = \omega_o \hat{C}\hat{R}_{ext-out} = \omega_o (C_1 + C_2) \hat{R}_{ext-out} \equiv \hat{Q}_{eo1} + \hat{Q}_{eo2} \tag{6b}$$

where $\hat{R}_{\text{ext-in}}$ and $\hat{R}_{\text{ext-out}}$ are the normalized external resistances of the input and the output port, respectively, of the resonator. For matched input and output ports, $\hat{R}_{\text{ex-in}} = 1$ and $\hat{R}_{\text{ex-out}} = 1$; then

$$\hat{Q}_{ei} = \hat{Q}_{eo} = \omega_o (C_1 + C_2) \tag{6c}$$

The total external Q, $\hat{Q}_e$, is

$$\hat{Q}_e = \hat{Q}_{ei} + \hat{Q}_{eo} = (\hat{Q}_{ei1} + \hat{Q}_{eo1}) + (\hat{Q}_{ei2} + \hat{Q}_{eo2}) \equiv \hat{Q}_{e1} + \hat{Q}_{e2}$$

$$= 2\omega_o (C_1 + C_2) \qquad \text{for matched ports} \tag{6d}$$

where $\hat{Q}_{e1}$ and $\hat{Q}_{e2}$ are the external Q of resonator 1 and resonator 2, respectively.

When two resonators are cascaded with coincidental reference planes, their stored energies are increased and hence they have a larger $\hat{Q}_e$ than that of a single resonator. The total unloaded Q, $\hat{Q}_u$, will lie between those of the individual resonators.

(b) For identical and symmetrical resonators in cascade, i.e.,

$$n_1 = n_2 = n_3 = n_4 = n \qquad \text{and} \qquad Z_{o1} = Z_{o3} = Z_o$$

$$R_{o1} = R_{o2} = R_o, \qquad L_{o1} = L_{o2} = L_o, \qquad C_{o1} = C_{o2} = C_o$$

the equivalent circuit becomes that in Fig. 9-4.

The total unloaded Q, $\hat{Q}_u'$, is

$$\hat{Q}_u' = \omega_o \hat{C} \hat{R} = \omega_o (2\hat{C}_1) \frac{\hat{R}_1}{2} = \omega_o \hat{C}_1 \hat{R}_1 = \hat{Q}_{u1} = \hat{Q}_u \tag{7}$$

where $\hat{Q}_u$ is the unloaded Q for a single resonator.

The total external Q, $\hat{Q}_e'$, for resonators with matched source and load is given by (6d),

$$\hat{Q}_e' = \hat{Q}_{ei}' + \hat{Q}_{eo}' = 2\hat{Q}_e \qquad \hat{Q}_{ei} = \hat{Q}_{eo} \equiv \hat{Q}_e \tag{8a}$$

$$\hat{Q}_{ek} = \hat{Q}_{eik} + \hat{Q}_{eok} \qquad k = 1, 2 \tag{8b}$$

where $\hat{Q}_e$ is the external Q for a single resonator.

The total loaded Q, $\hat{Q}_L'$, is

$$\frac{1}{\hat{Q}_L'} = \frac{1}{\hat{Q}_u'} + \frac{1}{\hat{Q}_{ei}'} + \frac{1}{\hat{Q}_{eo}'} = \frac{1}{\hat{Q}_u} + \frac{1}{\hat{Q}_e}$$

$$= \left[\frac{1}{\hat{Q}_u} + \frac{2}{\hat{Q}_e} \right] - \frac{1}{\hat{Q}_e} = \frac{1}{\hat{Q}_L} - \frac{1}{\hat{Q}_e}$$

or

$$\hat{Q}_L' = \frac{\hat{Q}_e \hat{Q}_L}{\hat{Q}_e - \hat{Q}_L} \tag{9}$$

where $\hat{Q}_u$, $\hat{Q}_L$, and $\hat{Q}_e$ are the respective normalized Q's for a single resonator.

The transmitted power is given by (XIII.10-12a).

$$P_L = P_i \frac{4\hat{Q}_L'^2}{\hat{Q}_e'^2} \frac{1}{1 + \left[\hat{Q}_L' \frac{2\Delta\omega}{\omega_o} \right]^2} \tag{10\alpha}$$

$$= P_i \frac{\hat{Q}_L^2}{(\hat{Q}_e - \hat{Q}_L)^2} \frac{1}{1 + \left[\frac{\hat{Q}_e \hat{Q}_L}{\hat{Q}_e - \hat{Q}_L} \frac{2\Delta\omega}{\omega_o} \right]^2} \tag{10a}$$

where (9) is used to obtain (10a), and P_i is the input power.

At resonance, $\Delta\omega = 0$, then

$$P_L = P_i \left[\frac{\hat{Q}_L}{\hat{Q}_e - \hat{Q}_L} \right]^2 \tag{10b}$$

and the transmission coefficient is

$$T = \frac{P_L}{P_i} = \left[\frac{\hat{Q}_L}{\hat{Q}_e - \hat{Q}_L} \right]^2 \tag{10c}$$

If the resonators are lossless, then

$$\frac{1}{\hat{Q}_L'} = \frac{1}{\hat{Q}_e'} = \frac{1}{2\hat{Q}_e} = \frac{1}{2\hat{Q}_L} \quad \text{or} \quad \hat{Q}_L' = 2\hat{Q}_L \tag{11\alpha}$$

With the use of (8) and (11α), (10α) becomes

$$P_L = P_i \frac{4(2\hat{Q}_L)^2}{4(2\hat{Q}_L)^2} \frac{1}{1 + (2\hat{Q}_L)^2 \left[\frac{2\Delta\omega}{\omega_o} \right]^2}$$

$$= P_i \frac{1}{1 + 4\hat{Q}_L^2 \left[\frac{2\Delta\omega}{\omega_o} \right]^2} \tag{11}$$

The transmission characteristic of the two coincidental cascaded resonators is very similar to, but sharper than, that of a single resonator. This is because the loaded Q of the cascaded resonators is almost twice as large as that of a single resonator.

(c) For identical resonators as specified in (b), the normalized input admittance at plane 1-1', Fig. 9-4, is

$$\hat{Y}_i = (1 + 2\hat{G}_1) + j2 \left[\omega\hat{C}_1 - \frac{1}{\omega\hat{L}_1} \right] = \hat{G}_i + j\hat{B}_i \tag{12a}$$

$$\hat{G}_i = 1 + 2\hat{G}_1 \tag{12b}$$

$$\hat{B}_i = 2 \left[\omega\hat{C}_1 - \frac{1}{\omega\hat{L}_1} \right] = 2\hat{B}_1 \tag{12c}$$

The reflection coefficient at the input port is

$$\Gamma = \frac{1 - \hat{Y}_i}{1 + \hat{Y}_i} = \frac{1 - \hat{G}_i - j\hat{B}_i}{1 + \hat{G}_i + j\hat{B}_i} \tag{13a}$$

and

$$| \Gamma |^2 = \frac{(1 - \hat{G}_i)^2 + \hat{B}_i^2}{(1 + \hat{G}_i)^2 + \hat{B}_i^2} \tag{13b}$$

The power absorbed by the entire structure is

$$P_{stru} = P_i (1 - | \Gamma |^2) = \frac{P_i}{D_i} [(1 + \hat{G}_i)^2 - (1 - \hat{G}_i)^2]$$

$$= \frac{4\hat{G}_i}{D_i} P_i \tag{14a}$$

$$D_i \equiv (1 + \hat{G}_i)^2 + \hat{B}_i^2 = 4 [(1 + \hat{G}_1)^2 + \hat{B}_1^2] \tag{14b}$$

The power transmitted to the load is

$$P_L = \frac{1}{1 + 2\hat{G}_1} P_{stru} = \frac{P_i}{(1 + \hat{G}_1)^2 + \hat{B}_1^2} \tag{15}$$

The transmission coefficient is therefore

$$T = \frac{P_L}{P_i} = \frac{1}{(1 + \hat{G}_1)^2 + \hat{B}_1^2} \tag{16a}$$

814

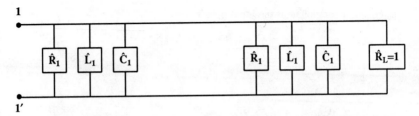

Figure 9-4: An equivalent circuit for coincidentally cascaded identical
resonators.

At resonance,

$$T = \frac{1}{(1 + \hat{G}_1)^2} \qquad \text{at resonance} \qquad (16b)$$

Also

$$\Gamma = \frac{1 - \hat{G}_i}{1 + \hat{G}_i} = \frac{-\hat{G}_1}{1 + \hat{G}_1} \qquad \text{at resonance} \qquad (17)$$

where (12b) is used to obtain (17).

(d) The properties of nearly coincidental coupled resonators may be analyzed by considering the case when the coupling guide is of incremental length, i.e., $L_g = \delta L$.

For the case of identical and symmetrical resonators as specified in (b), Fig. 8-1b becomes Fig. 9-5a; the latter can be rearranged as Fig. 9-5b and finally simplified to Fig. 9-5c. The circuit parameters in Fig. 9-5c are related to the corresponding intrinsic parameters as follows:

$$\overline{R}_1 = n^2 R_{o1} \qquad \overline{C}_1 = \frac{C_{o1}}{n^2} \qquad \overline{L}_1 = n^2 L_{o1} \qquad (18)$$

The input admittance across terminals 5-5′ is

$$Y_{55'} = \frac{1}{\overline{R}_1} + j\left[\omega\overline{C}_1 - \frac{1}{\omega\overline{L}_1}\right] + Y_L \equiv G_1 + jB_1 + Y_L \qquad (19a)$$

$$\hat{Y}_5 = \frac{Y_{55'}}{Y_o} = \hat{G}_1 + j\hat{B}_1 + \hat{Y}_L \qquad (19b)$$

$$B_1 = \omega\overline{C}_1 - \frac{1}{\omega\overline{L}_1} = \sqrt{\frac{\overline{C}_1}{\overline{L}_1}} \frac{2\Delta\omega}{\omega_o} \qquad (XIII.3\text{-}8b)$$

$$= 2Q_L \frac{2\Delta\omega}{\omega_o} \qquad (XIII.10\text{-}16c)$$

$$\Delta\omega \equiv \omega - \omega_o \qquad (19c)$$

The input impedance across terminals 4-4′ is

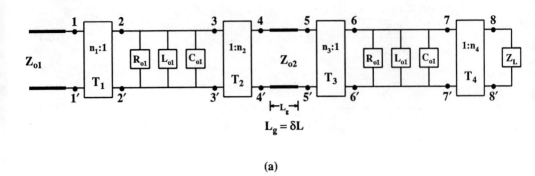

(a)

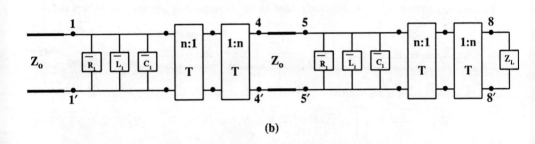

(b)

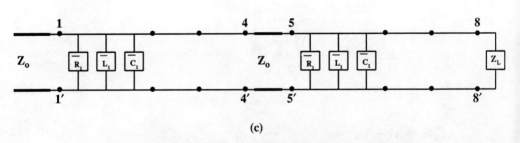

(c)

Figure 9-5: Arrangement of nearly coincidentally cascaded resonators and equivalent circuits for identical and symmetrical resonators.

$$Z_{44'} = Z_0 \frac{Z_{55'} + jZ_0\tau}{Z_0 + jZ_{55'}\tau} \tag{8-2}$$

$$\approx Z_0 \frac{Z_{55'} + jZ_0\Delta}{Z_0 + jZ_{55'}\Delta} \tag{20a}$$

$$\tau \equiv \tan \beta \, \delta L \approx \beta \, \delta L \equiv \Delta \tag{20b}$$

The normalized impedance across plane 4-4' is

$$\hat{Z}_4 = \frac{Z_{44'}}{Z_0} = \frac{1 + jY_{55'}Z_0\Delta}{Y_{55'}Z_0 + j\Delta} = \frac{1 + j\hat{Y}_5\Delta}{\hat{Y}_5 + j\Delta} \tag{21a}$$

and

$$\hat{Y}_4 = \frac{1}{\hat{Z}_4} = \frac{\hat{Y}_5 + j\Delta}{1 + j\hat{Y}_5\Delta} \tag{21b}$$

The total admittance across the input port 1-1' is

$$Y_{11'} = G_1 + jB_1 + \hat{Y}_4 Y_0$$

The normalized input admittance is

$$\hat{Y}_i \equiv \frac{Y_{11'}}{Y_0} = \hat{G}_1 + j\hat{B}_1 + \frac{\hat{Y}_5 + j\Delta}{1 + j\hat{Y}_5\Delta}$$

$$= \hat{G}_1 + j\hat{B}_1 + \frac{(\hat{G}_1 + j\hat{B}_1 + \hat{Y}_L) + j\Delta}{1 + j(\hat{G}_1 + j\hat{B}_1 + \hat{Y}_L)\Delta}$$

$$= \frac{1}{D} [(1 - \hat{B}_1\Delta)(\hat{G}_1 + j\hat{B}_1) + j(\hat{G}_1 + \hat{Y}_L)\hat{G}_1\Delta$$

$$- (\hat{G}_1 + \hat{Y}_L)\hat{B}_1\Delta + \hat{G}_1 + \hat{Y}_L + j(\hat{B}_1 + \Delta)] \tag{22a}$$

$$D \equiv (1 - \hat{B}_1\Delta) + j(\hat{G}_1 + \hat{Y}_L)\Delta \tag{22b}$$

For lossless resonators, $\hat{G}_1 = 0$, and (22) becomes

$$\hat{Y}_i = \frac{1}{D} [\hat{Y}_L (1 - \hat{B}_1 \Delta) + j(2\hat{B}_1 + \Delta - \hat{B}_1^2 \Delta)] \tag{23a}$$

$$D = (1 - \hat{B}_1 \Delta) + j\hat{Y}_L \Delta \tag{23b}$$

The reflection coefficient at the input port plane 1-1′ is given by

$$\Gamma = \frac{1 - \hat{Y}_i}{1 + \hat{Y}_i} \tag{8-5a}$$

The input port will be matched when $\Gamma = 0$ or $1 - \hat{Y}_i = 0$. One obtains from (23) that

$$\hat{Y}_i - 1 = \frac{1}{D} [(\hat{Y}_L - 1)(1 - \hat{B}_1 \Delta) + j(2\hat{B}_1 + \Delta - \hat{B}_1^2 \Delta - \hat{Y}_L \Delta)] \tag{24}$$

For a matched load, $\hat{Y}_L = 1$; then

$$\hat{Y}_i - 1 = \frac{1}{D} j\hat{B}_1 (2 - \hat{B}_1 \Delta) \tag{25a}$$

$$D = (1 - \hat{B}_1 \Delta) + j\Delta \tag{25b}$$

The conditions for no reflection at the input port are given by

$$\hat{B}_1 (2 - \hat{B}_1 \Delta) = 0 \qquad \text{for } \hat{Y}_L = 1$$

$$\text{(a)} \quad \hat{B}_1 = 0 = \frac{1}{Z_o} \left[\omega \overline{C}_1 - \frac{1}{\omega \overline{L}_1} \right] \quad \text{or} \quad \omega = \frac{1}{\sqrt{\overline{L}_1 \overline{C}_1}} = \omega_o \tag{26a}$$

$$\text{(b)} \quad \hat{B}_1 = \frac{2}{\Delta} = 2Q_L \frac{2\Delta\omega}{\omega_o} \tag{XIII.10-16c}$$

$$\Delta\omega = \omega - \omega_o = \frac{1}{2Q_L} \frac{\omega_o}{\Delta}$$

or

$$\omega = \omega_o \left[1 + \frac{1}{2Q_L} \frac{1}{\Delta} \right] = \omega_o \left[1 + \frac{1}{2Q_L} \frac{\lambda_g}{2\pi \, \delta L} \right] \equiv \omega_o' \tag{26b}$$

The structure will transmit maximum power at frequencies specified by (26). The frequency ω_o' varies inversely as δL and $\omega_o' \to +\infty$ as $+\delta L \to 0$. On the other hand, if δL is negative, $\omega_o' \to -\infty$ as $-\delta L \to 0$. A negative δL implies that the reference planes of the resonators are overlapped.

10. Quarter-Wavelength Coupled Resonators

(a) The equivalent circuit of quarter-wavelength coupled resonators is obtained by combining Figs. 8-1 and 8-2; the result is as shown in Fig. 10-1.

The input admittance across plane 5-5' is

$$Y_5 = \frac{1}{R_2'} + \frac{1}{R_L'} + j \left[\omega C_2' - \frac{1}{\omega L_2'} \right] = G_5 + jB_5 \tag{1a}$$

$$G_5 = \frac{1}{R_2'} + \frac{1}{R_L'} \tag{1b}$$

$$B_5 = \omega C_2' - \frac{1}{\omega L_2'} \tag{1c}$$

and

$$Z_5 = \frac{1}{Y_5} = \frac{1}{G_5 + jB_5} \tag{1d}$$

The primed parameters are given by (8-1).

Normalizing Z_5 with respect to Z_{o2}, one has

$$\hat{Z}_5 = \frac{Z_5}{Z_{o2}} = \frac{1}{Z_{o2} (G_5 + jB_5)} \tag{2}$$

The normalized input impedance at plane 4-4', Z_4, is the inverse of $\hat{Z}_5$ because of the quarter-wavelength transformer, (8-2a).

$$\hat{Z}_4 = \frac{1}{\hat{Z}_5} = Z_{o2} (G_5 + jB_5) \tag{3}$$

The denormalized input impedance at plane 4-4' is

$$Z_4 = Z_{o2}\hat{Z}_4 = Z_{o2}^2 (G_5 + jB_5)$$

$$= Z_{o2}^2 \left[\left[\frac{1}{R_2'} + \frac{1}{R_L'} \right] + j \left[\omega C_2' - \frac{1}{\omega L_2'} \right] \right] \tag{4}$$

The equivalent circuit is now as shown in Fig. 10-2a and the double-primed parameters are defined as follows.

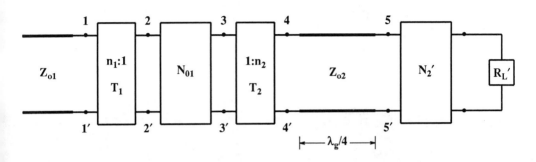

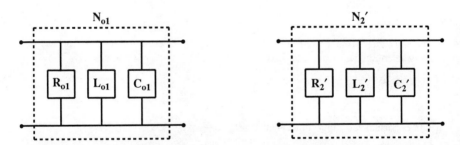

Figure 10-1: An equivalent circuit for quarter-wavelength coupled resonators.

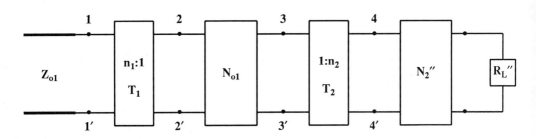

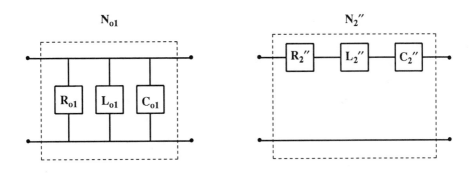

(a)

Figure 10-2: A simplified circuit for quarter-wavelength coupled resonators.

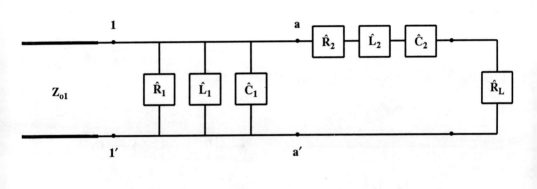

(b)

Figure 10-2: A simplified circuit for quarter-wavelength coupled resonators.

$$R_2'' = Z_{o2}^2 \frac{1}{R_2'} = \left[\frac{Z_{o2}}{n_3}\right]^2 \frac{1}{R_{o2}} \tag{5a}$$

$$L_2'' = Z_{o2}^2 C_2' = \left[\frac{Z_{o2}}{n_3}\right]^2 C_{o2} \tag{5b}$$

$$C_2'' = \frac{1}{Z_{o2}^2} L_2' = \left[\frac{n_3}{Z_{o2}}\right]^2 L_{o2} \tag{5c}$$

$$R_L'' = \frac{Z_{o2}^2}{R_L'} = \left[\frac{Z_{o2}n_4}{n_3}\right]^2 \frac{1}{Z_{o3}} \tag{5d}$$

The equivalent circuit can be further simplified to Fig. 10-2b by removing the transformers. The normalized parameters are:

$$\hat{R}_1 = \frac{n_1^2}{Z_{o1}} R_{o1} \tag{6a}$$

$$\hat{L}_1 = \frac{n_1^2}{Z_{o1}} L_{o1} \tag{6b}$$

$$\hat{C}_1 = \frac{Z_{o1}}{n_1^2} C_{o1} \tag{6c}$$

$$\hat{R}_2 = \frac{1}{Z_{o1}} \left[\frac{n_1}{n_2}\right]^2 R_2'' = \left[\frac{Z_{o2}n_1}{n_2 n_3}\right]^2 \frac{1}{Z_{o1}R_{o2}} \tag{6d}$$

$$\hat{L}_2 = \frac{1}{Z_{o1}} \left[\frac{n_1}{n_2}\right]^2 L_2'' = \left[\frac{Z_{o2}n_1}{n_2 n_3}\right]^2 \frac{C_{o2}}{Z_{o1}} \tag{6e}$$

$$\hat{C}_2 = Z_{o1} \left[\frac{n_2}{n_1}\right]^2 C_2'' = \left[\frac{n_2 n_3}{n_1 Z_{o2}}\right]^2 Z_{o1}L_{o2} \tag{6f}$$

$$\hat{R}_L = \frac{1}{Z_{o1}} \left[\frac{n_1}{n_2}\right]^2 R_L'' = \left[\frac{Z_{o2}n_1 n_4}{n_2 n_3}\right]^2 \frac{1}{Z_{o1}Z_{o3}} \tag{6g}$$

(b) When the resonators are symmetrical and identical, that is,

$$n_j = n \qquad \text{for} \quad j = 1, 2, 3, \text{ and } 4 \tag{7a}$$

$$Z_{oj} = Z_o \qquad \text{for} \quad j = 1, 2, \text{ and } 3 \tag{7b}$$

$$R_{o1} = R_{o2} = R_o \tag{7c}$$

$$L_{o1} = L_{o2} = L_o \tag{7d}$$

$$C_{o1} = C_{o2} = C_o \tag{7e}$$

then (6) becomes

$$\hat{R}_1 = \frac{n^2}{Z_o} R_o \tag{8a}$$

$$\hat{L}_1 = \frac{n^2}{Z_o} L_o \tag{8b}$$

$$\hat{C}_1 = \frac{Z_o}{n^2} C_o \tag{8c}$$

$$\hat{R}_2 = \frac{Z_o}{n^2} \frac{1}{R_o} = \frac{1}{\hat{R}_1} \tag{8d}$$

$$\hat{L}_2 = \frac{Z_o}{n^2} C_o = \hat{C}_1 \tag{8e}$$

$$\hat{C}_2 = \frac{n^2}{Z_o} L_o = \hat{L}_1 \tag{8f}$$

$$\hat{R}_L = 1 \tag{8g}$$

The normalized input impedance across plane a-a', Fig. 10-3, is

$$\hat{Z}_a = (\hat{R}_2 + \hat{R}_L) + j \left[\omega\hat{L}_2 - \frac{1}{\omega\hat{C}_2} \right]$$

$$= \left[\frac{1}{\hat{R}_1} + 1 \right] + j \left[\omega\hat{C}_1 - \frac{1}{\omega\hat{L}_1} \right] = \hat{R}_a + j\hat{X}_a \qquad \hat{R}_L = 1 \tag{9a}$$

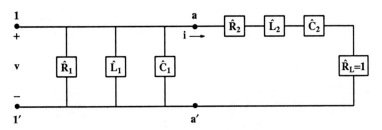

Figure 10-3: A simplified circuit for quarter-wavelength coupled resonators terminated with matched load.

$$\hat{R}_a = \frac{1}{\hat{R}_1} + 1 = \hat{G}_1 + 1 \tag{9b}$$

$$\hat{X}_a = \omega\hat{C}_1 - \frac{1}{\omega\hat{L}_1} = \hat{B}_1 \tag{9c}$$

$$= \sqrt{\frac{\hat{C}_1}{\hat{L}_1}} \frac{2\Delta\omega}{\omega_o} \qquad \text{at } \omega \approx \omega_o \tag{9d}$$

The normalized admittance at plane 1-1' is

$$\hat{Y}_1 = \hat{Y}_a + \left[\frac{1}{\hat{R}_1} + j \left[\omega\hat{C}_1 - \frac{1}{\omega\hat{L}_1} \right] \right] = \frac{1}{\hat{Z}_a} + (\hat{G}_1 + j\hat{B}_1)$$

$$= \frac{1}{(\hat{G}_1 + 1) + j\hat{B}_1} + \hat{G}_1 + j\hat{B}_1$$

$$= |\hat{Y}_a|^2 (\hat{G}_1 + 1 - j\hat{B}_1) + \hat{G}_1 + j\hat{B}_1 \tag{10a}$$

$$|\hat{Z}_a|^2 = (\hat{G}_1 + 1)^2 + \hat{B}_1^2 = \frac{1}{|\hat{Y}_a|^2} \tag{10b}$$

The reflection coefficient at the input is

$$\Gamma = \frac{1 - \hat{Y}_1}{1 + \hat{Y}_1} = \frac{1 - \hat{G}_1 - j\hat{B}_1 - |\hat{Y}_a|^2 (1 + \hat{G}_1 - j\hat{B}_1)}{1 + \hat{G}_1 + j\hat{B}_1 + |\hat{Y}_a|^2 (1 + \hat{G}_1 - j\hat{B}_1)}$$

$$= \frac{1 - \hat{G}_1 - |\hat{Y}_a|^2 (1 + \hat{G}_1) - j\hat{B}_1 (1 - |\hat{Y}_a|^2)}{1 + \hat{G}_1 + |\hat{Y}_a|^2 (1 + \hat{G}_1) + j\hat{B}_1 (1 - |\hat{Y}_a|^2)} \tag{11}$$

The power, P_{stru}, absorbed by the structure is

$$P_{stru} = P_i (1 - |\Gamma|^2) = \frac{4P_i}{D} [\hat{G}_1 + |\hat{Y}_a|^2 (1 + \hat{G}_1)] \tag{12a}$$

$$|\Gamma|^2 = \frac{[1 - \hat{G}_1 - |\hat{Y}_a|^2 (1 + \hat{G}_1)]^2 + \hat{B}_1^2 (1 - |\hat{Y}_a|^2)^2}{[1 + \hat{G}_1 + |\hat{Y}_a|^2 (1 + \hat{G}_1)]^2 + \hat{B}_1^2 (1 - |\hat{Y}_a|^2)^2} \tag{12b}$$

$$1 - |\Gamma|^2 = \frac{1}{D} \{ [(1 + \hat{G}_1) + |\hat{Y}_a|^2 (1 + \hat{G}_1)]^2$$

$$- [(1 - \hat{G}_1) - |\hat{Y}_a|^2 (1 + \hat{G}_1)]^2 \}$$

$$= \frac{4}{D} [\hat{G}_1 + |\hat{Y}_a|^2 (1 + \hat{G}_1)] \tag{12c}$$

$$D = [(1 + \hat{G}_1)(1 + |\hat{Y}_a|^2)]^2 + [\hat{B}_1 (1 - |\hat{Y}_a|^2)]^2 \tag{12d}$$

where P_i is the incident power. The power P_{stru} is the power absorbed by the entire structure, made up of both resonators and the load.

The power absorbed by the first cavity, P_1, is

$$P_1 = \frac{1}{2} \frac{|v|^2}{\hat{R}_1} = \frac{1}{2} |v|^2 \hat{G}_1 \tag{13}$$

The power, P_2, transmitted to the second cavity and the load is

$$P_2 = \frac{1}{2} |i|^2 \hat{R}_a = \frac{1}{2} \frac{|v|^2}{|\hat{Z}_a|^2} (1 + \hat{G}_1) \tag{14}$$

The fraction of P_{stru} that is transmitted through the second cavity is

$$P_t = \frac{P_2}{P_1 + P_2} P_{stru} \tag{15}$$

The power absorbed by the load is P_L,

$$P_L = \frac{|i|^2}{2} \hat{R}_L = \frac{1}{1 + \hat{G}_1} P_2 = \frac{1}{1 + \hat{G}_1} P_t \tag{16a}$$

The fraction of P_{stru} absorbed by the load is

$$P_L = \frac{1}{1 + \hat{G}_1} P_t = \frac{1}{1 + \hat{G}_1} \frac{P_2}{P_1 + P_2} P_{stru} \tag{16b}$$

The power transmission coefficient, T, is obtained by combining (12) through (16).

$$T = \frac{P_L}{P_i} = \frac{1}{1 + \hat{G}_1} \frac{P_2}{P_1 + P_2} \frac{P_{stru}}{P_i}$$

$$T = \frac{1}{1 + \hat{G}_1} \frac{|\hat{Y}_a|^2 (1 + \hat{G}_1)}{\hat{G}_1 + |\hat{Y}_a|^2 (1 + \hat{G}_1)} \frac{4}{D} [\hat{G}_1 + |\hat{Y}_a|^2 (1 + \hat{G}_1)]$$

$$= \frac{4 |\hat{Y}_a|^2}{[(1 + \hat{G}_1)(1 + |\hat{Y}_a|^2)]^2 + [\hat{B}_1 (1 - |\hat{Y}_a|^2)]^2} \tag{17a}$$

$$|\hat{Z}_a|^2 = |\hat{Y}_a|^{-2} = (1 + \hat{G}_1)^2 + \hat{B}_1^2 \tag{17b}$$

$$\hat{B}_1 = \sqrt{\frac{\hat{C}_1}{\hat{L}_1}} \frac{2\Delta\omega}{\omega_o} \qquad \text{for} \quad \omega \approx \omega_o \tag{17c}$$

At resonance, $\hat{B}_1 = 0$ and $|\hat{Z}_a|^2 = (1 + \hat{G}_1)^2$, the transmission coefficient becomes

$$T = \frac{\dfrac{4}{(1 + \hat{G}_1)^2}}{(1 + \hat{G}_1)^2 \left[1 + \dfrac{1}{(1 + \hat{G}_1)^2} \right]^2}$$

$$= \frac{4}{[1 + (1 + \hat{G}_1)^2]^2} \qquad \text{at resonance} \tag{18}$$

The power transmission coefficient can be expressed in terms of the Q's of the individual cavities. These Q's are

$$Q_u = \omega_o \hat{C}_1 \hat{R}_1 \tag{19a}$$

$$Q_e = \omega_o \hat{C}_1 \tag{19b}$$

$$\frac{1}{Q_L} = \frac{1}{Q_u} + \frac{1}{Q_{ei}} + \frac{1}{Q_{eo}} = \frac{1}{\omega_o \hat{C}_1 \hat{R}_1} + \frac{1}{\omega_o \hat{C}_1} + \frac{1}{\omega_o \hat{C}_1}$$

$$= \frac{1}{\omega_o \hat{C}_1} (\hat{G}_1 + 2)$$

$$Q_L = \frac{\omega_o \hat{C}_1}{\hat{G}_1 + 2} \tag{19c}$$

where $Q_{ei/o}$ are the external Q's for the input port, and the output port, respectively.

$$\frac{Q_e}{Q_L} = \hat{G}_1 + 2 \tag{19d}$$

Since

$$1 + (1 + \hat{G}_1)^2 = 1 + 1 + 2\hat{G}_1 + \hat{G}_1^2 = (\hat{G}_1 + 2)^2 - 2(\hat{G}_1 + 2) + 2$$

$$= \left[\frac{Q_e}{Q_L}\right]^2 - 2\frac{Q_e}{Q_L} + 2$$

hence,

$$T = \frac{1}{\left[\frac{1}{2}\right]^2 \left[\left[\frac{Q_e}{Q_L}\right]^2 - 2\frac{Q_e}{Q_L} + 2\right]^2}$$

$$= \frac{Q_L^2}{\left[\frac{Q_e^2}{2Q_L} + Q_L - Q_e\right]^2} \qquad \text{at} \quad \omega = \omega_o \tag{20}$$

The transmission coefficient at resonance for resonators arranged in coincidental cascade is

$$T' = \frac{Q_L^2}{(Q_e - Q_L)^2} \tag{9-10c}$$

A comparison between (20) and (9-10c) indicates that the quarter-wavelength coupled resonators provide a better transmission than coincidental coupled resonators (see problem 7).

At resonance, the reflection coefficient (11) becomes

$$\Gamma = \frac{(1 - \hat{G}_1) - (1 + \hat{G}_1)^{-1}}{(1 + \hat{G}_1) + (1 + \hat{G}_1)^{-1}} = \frac{-\hat{G}_1^2}{(1 + \hat{G}_1)^2 + 1}$$

$$= \frac{-\hat{G}_1}{\hat{G}_1 + 2 + \frac{2}{\hat{G}_1}} \qquad \text{at} \quad \omega = \omega_o \tag{21}$$

The corresponding reflection coefficient for coincidental cascaded resonators is

$$\Gamma' = \frac{-\hat{G}_1}{1 + \hat{G}_1} \qquad \text{at resonance} \tag{9-17}$$

It is quite obvious that the reflection coefficient for the quarter-wavelength coupled resonators can be lowered substantially with respect to that of the coincidental cascaded situation.

(c) The general expression for the power transmission coefficient, (17), is very complex, and it is difficult to visualize its behavior. For low-loss resonators, $\hat{G}_1 \ll 1$, (17) can be approximated as follows.

$$T \approx \frac{4 \mid \hat{Y}_a \mid^2}{(1 + \mid \hat{Y}_a \mid^2)^2 + [\hat{B}_1 (1 - \mid \hat{Y}_a \mid^2)]^2}$$

$$\approx \frac{\dfrac{4}{1 + \hat{B}_1^2}}{\left[1 + \dfrac{1}{1 + \hat{B}_1^2} \right]^2 + \hat{B}_1^2 \left[1 - \dfrac{1}{1 + \hat{B}_1^2} \right]^2}$$

$$= \frac{\dfrac{4}{1 + \hat{B}_1^2}}{\left[\dfrac{2 + \hat{B}_1^2}{1 + \hat{B}_1^2} \right]^2 + \hat{B}_1^2 \left[\dfrac{\hat{B}_1^2}{1 + \hat{B}_1^2} \right]^2}$$

$$= \frac{4 (1 + \hat{B}_1^2)}{4 (1 + \hat{B}_1^2) + \hat{B}_1^4 (1 + \hat{B}_1^2)}$$

$$= \frac{1}{1 + \dfrac{\hat{B}_1^4}{4}} \qquad \text{for } \hat{G}_1 \ll 1 \tag{22a}$$

$$\mid \hat{Y}_a \mid^2 \approx \frac{1}{1 + \hat{B}_1^2} \qquad \text{for } \hat{G}_1 \ll 1 \tag{22b}$$

$$\hat{B}_1 = \omega_o \hat{C}_1 \frac{2\Delta\omega}{\omega_o} \tag{22c}$$

The loaded Q is given by (19c), and for the case of low loss it reduces to

$$Q_L \approx \frac{1}{2} \omega_o \hat{C}_1 \qquad \text{for} \quad \hat{G}_1 \ll 1 \tag{23a}$$

or

$$\omega_o \hat{C}_1 = 2Q_L \qquad \text{for} \quad \hat{G}_1 \ll 1 \tag{23b}$$

and

$$\hat{B}_1 = 2Q_L \frac{2\Delta\omega}{\omega_o} \tag{23c}$$

The substitution of (23) into (22) yields

$$T \approx \frac{1}{1 + 4K_\Omega^4} \qquad \text{for} \quad \hat{G}_1 \ll 1 \tag{24a}$$

$$K_\Omega = Q_L \frac{2\Delta\omega}{\omega_o} \qquad \text{or} \qquad \hat{B}_1 = 2K_\Omega \tag{24b}$$

The corresponding transmission coefficient for the coincidental cascaded case is

$$T' \approx \frac{1}{1 + \hat{B}_1^2} = \frac{1}{1 + 4K_\Omega^2} \tag{9-16}$$

The transmission behavior of the quarter-wavelength coupled resonators is drastically different from that of the coincidental cascaded case. At large values of K_Ω, the transmission coefficient T decreases at the rate of 12 db per octave while T' has a decrement of 6 db per octave. For K_Ω less than unity, T is always greater than T'. The quarter-wavelength coupled system has a flatter transmission response in the pass-band and a steeper attenuation in the rejection band than the coincidental coupled resonators.

(d) The behavior of nearly quarter-wavelength coupled resonators can be investigated by a similar procedure as in the case of nearly coincidental coupled resonators. The equivalent circuit in Fig. 9-5c is valid for the present case with $L_g = \lambda_g/4 + \delta L$.

The input impedance across plane 4–4$'$ is [Eq. (8-2)]

$$Z_{44'} = Z_o \frac{\hat{Z}_5 - \dfrac{j}{\Delta}}{1 - j\dfrac{\hat{Z}_5}{\Delta}} = Z_o \frac{\Delta - j\hat{Y}_5}{\hat{Y}_5\Delta - j} \tag{25a}$$

$$\hat{Y}_4 = \frac{1}{\hat{Z}_4} = \frac{Z_o}{Z_{44'}} \tag{25b}$$

$$\hat{Y}_5 = \frac{Y_{55'}}{Y_o} = \hat{G}_1 + j\hat{B}_1 + \hat{Y}_L \tag{9-19b}$$

$$\tau = \tan \frac{2\pi}{\lambda_g} \left(\frac{\lambda_g}{4} + \delta L \right) \approx \frac{-1}{\beta \ \delta L} \equiv \frac{-1}{\Delta} \tag{25c}$$

The input admittance at the input port 1-1′ is

$$Y_{11'} = G_1 + jB_1 + Y_{44'} \tag{26a}$$

The normalized input admittance is

$$\hat{Y}_i = \frac{Y_{11'}}{Y_o} = \hat{G}_1 + j\hat{B}_1 + \frac{\hat{Y}_5 \Delta - j}{\Delta - j\hat{Y}_5} \tag{26b}$$

For lossless resonators, $\hat{G}_1 = 0$, and (26) becomes

$$\hat{Y}_i = j\hat{B}_1 + \frac{\hat{Y}_L \Delta + j(\hat{B}_1 \Delta - 1)}{\Delta + \hat{B}_1 - j\hat{Y}_L}$$

$$= \frac{1}{D} [\hat{Y}_L (\Delta + \hat{B}_1) + j(2\hat{B}_1\Delta + \hat{B}_1^2 - 1)] \tag{27a}$$

$$D = \Delta + \hat{B}_1 - j\hat{Y}_L \tag{27b}$$

For no reflection at the input port, one must have [Eq. (8-5a)]

$$\hat{Y}_i - 1 = 0 = \frac{1}{D} [(\hat{Y}_L - 1)(\hat{B}_1 + \Delta) + j(2\hat{B}_1\Delta + \hat{B}_1^2 - 1 + \hat{Y}_L)] \tag{28}$$

For matched load, $\hat{Y}_L = 1$; then (28) becomes

$$\hat{B}_1 (\hat{B}_1 + 2\Delta) = 0 \tag{29}$$

The conditions for maximum power transmission are:

(a) $\quad \hat{B}_1 = 0 \quad$ or $\quad \omega = \dfrac{1}{\sqrt{\hat{L}_1 \hat{C}_1}} = \omega_o \tag{30a}$

(b) $\quad \hat{B}_1 + 2\Delta = 0 \quad$ or $\quad -2\Delta = \hat{B}_1 = 2Q_L \dfrac{2\Delta\omega}{\omega_o} \tag{XIII.10-16c}$

$$\Delta\omega = \omega - \omega_o = \frac{-\omega_o}{2Q_L} \Delta$$

$$\omega = \omega_o \left[1 - \frac{\pi}{Q_L} \frac{\delta L}{\lambda_g} \right] \tag{30b'}$$

$$= \omega_o \left[1 - \frac{\pi}{Q_L} \left[\frac{L_g}{\lambda_g} - \frac{1}{4} \right] \right] \equiv \omega_o' \tag{30b}$$

$$L_g = \frac{\lambda_g}{4} + \delta L \qquad \text{or} \qquad \delta L = L_g - \frac{\lambda_g}{4} \tag{30c}$$

Equation (30b) implies that ω_o' is larger than the resonance frequency ω_o if the length L_g of the coupling waveguide is less than a quarter-wavelength and that it is small if $L_g > \lambda_g/4$.

10.1. Example: Quarter-Wavelength Coupled Nonsymmetrical Resonators

Two identical, nonsymmetrical lossless resonators are arranged as shown in Fig. 10.1-1. Determine the condition for maximum transmission of power to the load R_L.

Solution:

The equivalent circuit can be reduced to Fig. 10.1-2, and the parameters are given by (10-6) subject to the following conditions.

$$n_1 = n_4 \qquad \text{and} \qquad n_2 = n_3 \tag{1a}$$

$$L_{o1} = L_{o2} = L_o \qquad \text{and} \qquad C_{o1} = C_{o2} = C_o \tag{1b}$$

$$Z_{o1} = Z_{o2} = Z_{o3} = Z_o \tag{1c}$$

The parameters are

$$\hat{L}_1 = \frac{n_1^2}{Z_o} L_o \tag{2a}$$

$$\hat{C}_1 = \frac{Z_o}{n_1^2} C_o \tag{2b}$$

$$\hat{L}_2 = K_n Z_o \frac{C_o}{n_1^2} = K_n \hat{C}_1 \tag{2c}$$

$$\hat{C}_2 = \frac{1}{K_n} \frac{n_1^2}{Z_o} L_o = \frac{1}{K_n} \hat{L}_1 \tag{2d}$$

$$\hat{R}_L = \left[\frac{n_1}{n_2} \right]^4 \equiv K_n \tag{2e}$$

$$K_n \equiv \left[\frac{n_1}{n_2} \right]^4 \tag{2f}$$

The normalized input impedance across plane b-b' is

$$\hat{Z}_b = K_n + j \left[\omega\hat{L}_2 - \frac{1}{\omega\hat{C}_2} \right] = K_n + jK_n \left[\omega\hat{C}_1 - \frac{1}{\omega\hat{L}_1} \right]$$

$$= \hat{R}_b + j\hat{X}_b \tag{3a}$$

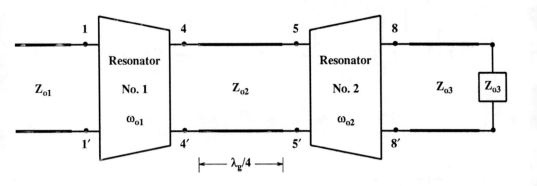

(a)

Figure 10.1-1: Arrangement of quarter-wavelength coupled identical but non-symmetrical lossless resonators and its equivalent circuit.

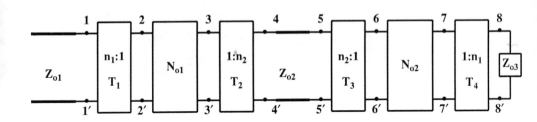

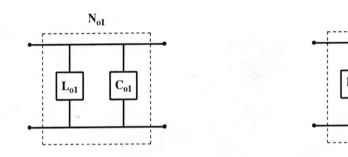

(b)

Figure 10.1-1: Arrangement of quarter-wavelength coupled identical but non-symmetrical lossless resonators and its equivalent circuit.

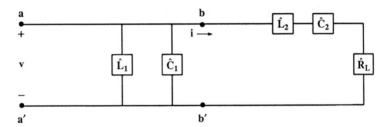

Figure 10.1-2: A simplified circuit for Example 10.1.

$$\hat{R}_b = K_n \tag{3b}$$

$$\hat{X}_b = K_n \hat{B}_1 = \Omega \qquad \text{at} \quad \omega \approx \omega_o \tag{3c}$$

$$\hat{B}_1 = \omega \hat{C}_1 \frac{2\Delta\omega}{\omega_o} \qquad \text{at} \quad \omega \approx \omega_o \qquad (10\text{--}22c) \tag{3d}$$

$$Q_e = \frac{\omega_o}{2\Delta\omega_e} = \omega_o \hat{C}_1 K_n \tag{3e}$$

$$\Omega = \frac{\Delta\omega}{\Delta\omega_e} \tag{3f}$$

where $\Delta\omega_e$ is the bandwidth of the external Q_e.

The input normalized admittance at plane a-a′ is

$$\hat{Y}_a = j \left[\omega \hat{C}_1 - \frac{1}{\omega \hat{L}_1} \right] + \hat{Y}_b = j\hat{B}_1 + \frac{1}{\hat{R}_b + j\hat{X}_b}$$

$$= j\hat{B}_1 + \frac{1}{K_n (1 + j\hat{B}_1)} = \frac{jK_n\hat{B}_1 (1 + j\hat{B}_1) + 1}{K_n (1 + j\hat{B}_1)}$$

$$= \frac{1 + jK_n\hat{B}_1 - K_n\hat{B}_1^2}{K_n (1 + j\hat{B}_1)} \tag{4}$$

Maximum transmission results when $\hat{Y}_a$ is matched to the input guide, i.e.,

$$\hat{Y}_a = 1 = \frac{1 + jK_n\hat{B}_1 - K_n\hat{B}_1^2}{K_n (1 + j\hat{B}_1)}$$

$$\hat{B}_1^2 = \frac{1 - K_n}{K_n} = \frac{1}{K_n} - 1$$

or

$$\hat{B}_1 = \pm\sqrt{\frac{1}{K_n} - 1} \qquad\qquad K_n = \left[\frac{n_1}{n_2} \right]^4 \tag{5}$$

Substituting (5) into (3c) yields

$$\Omega = \pm K_n \sqrt{\frac{1}{K_n} - 1} \qquad \text{for } \frac{1}{K_n} > 1 \tag{6}$$

This is the over-coupled case, $1/K_n > 1$. There will be total transmission at two distinct values of Ω.

$$\Omega_\pm = \pm K_n \sqrt{\frac{1}{K_n} - 1} \tag{7}$$

When $1/K_n = 1$,

$$\Omega_\pm = 0 \tag{8}$$

This is the critically coupled case treated in Section 10.

If $1/K_n < 1$, (5) or (6) does not have a real root, which implies that total transmission is impossible.

The frequency response of the power transmission can be obtained from [see Section 10]

$$T = \frac{P_L}{P_i} = 1 - |\Gamma|^2 \tag{9}$$

since, in the present case, both resonators are assumed to be lossless and hence, $P_L = P_{stru}$. The reflection coefficient is given by

$$\Gamma = \frac{1 - \hat{Y}_a}{1 + \hat{Y}_a}$$

$$= \frac{K_n(1 + \hat{B}_1^2) - 1}{K_n(1 - \hat{B}_1^2) + 1 + j1K_n\hat{B}_1} \tag{10a}$$

where $\hat{Y}_a$ is given by (4), and

$$1 - \hat{Y}_a = \frac{1}{D_1} [K_n(1 + \hat{B}_1^2) - 1] \tag{10b}$$

$$1 + \hat{Y}_a = \frac{1}{D_1} [K_n(1 - \hat{B}_1^2) + 1 + j2K_n\hat{B}_1] \tag{10c}$$

$$D_1 = K_n(1 + j\hat{B}_1) \tag{10d}$$

Therefore

$$|\Gamma|^2 = \frac{1}{D} [K_n(1 + \hat{B}_1^2) - 1]^2 \tag{11a}$$

$$D = [K_n(1 - \hat{B}_1^2) + 1]^2 + (2K_n\hat{B}_1)^2 \tag{11b}$$

Hence (9) becomes

$$T = \frac{4K_n}{D} \tag{12}$$

The frequency variation of T is sketched in Fig. 10.1-3 and $\hat{B}_1$ is given by

$$\hat{B}_1 = 2\omega_0\hat{C}_1 \left[\frac{\omega}{\omega_0} - 1 \right] \tag{13}$$

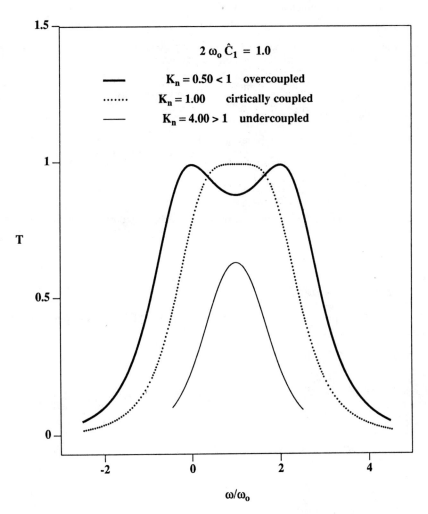

Figure 10.1-3: Responses of power transmission for Example 10.1.

10.2. Example: Scattering Matrix for Quarter-Wavelength Coupling

The coupling junction between quarter-wavelength coupled resonators is shown in Fig. 10.2-1. (a) Determine the scattering matrix between port 1-1′ and port 4-4′. (b) Determine the equivalent coupling-coefficient representation.

Solution :

(a) The entire structure in Fig. 10.2-1 can be represented by three modules in cascade, Fig. 10.2-2. The scattering matrix for the coupling device is

$$[\, S_m \,] \;=\; \begin{bmatrix} -\sqrt{1-m^2} & jm \\ jm & -\sqrt{1-m^2} \end{bmatrix} \qquad \text{(XIII.1--4)} \tag{1}$$

The scattering matrix for the $\dfrac{\lambda_g}{4}$-waveguide is

$$[\, S_\theta \,] \;=\; \begin{bmatrix} 0 & e^{-j\theta} \\ e^{-j\theta} & 0 \end{bmatrix} \;=\; \begin{bmatrix} 0 & -j \\ -j & 0 \end{bmatrix} \qquad \text{(IV.4.2--6)} \tag{2}$$

It is more convenient to use the transmission matrix in problems involving devices in cascade.

$$b_1 \;=\; s_{11}a_1 \,+\, s_{12}a_2 \tag{3a}$$

$$b_2 \;=\; s_{21}a_1 \,+\, s_{22}a_2 \tag{3b}$$

Equation (3) can be arranged to have the following form by solving for a_2 from (3a) and then substituting into (3b).

$$a_2 \;=\; t_{11}a_1 \,+\, t_{12}b_1 \tag{4a}$$

$$b_2 \;=\; t_{21}a_1 \,+\, t_{22}b_1 \tag{4b}$$

or

$$\begin{bmatrix} a_2 \\ b_2 \end{bmatrix} \;=\; [\, T \,] \begin{bmatrix} a_1 \\ b_1 \end{bmatrix} \tag{4'}$$

where

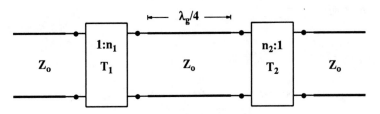

Figure 10.2-1: Arrangement for Example 10.2.

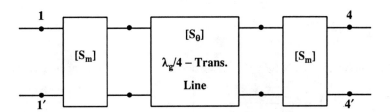

Figure 10.2-2: An equivalent circuit for Figure 10.2-1.

$$t_{11} = -\frac{s_{11}}{s_{12}} \tag{5a}$$

$$t_{12} = \frac{1}{s_{12}} \tag{5b}$$

$$t_{21} = s_{21} - \frac{s_{11}s_{22}}{s_{12}} \tag{5c}$$

$$t_{22} = \frac{s_{22}}{s_{12}} \tag{5d}$$

Conversely, (4) can be inverted to the form of (3) and

$$s_{11} = -\frac{t_{11}}{t_{12}} \tag{6a}$$

$$s_{12} = \frac{1}{t_{12}} \tag{6b}$$

$$s_{21} = t_{21} - \frac{t_{22}t_{11}}{t_{12}} \tag{6c}$$

$$s_{22} = \frac{t_{22}}{t_{12}} \tag{6d}$$

The transmission matrix for the coupling device is obtained by combining (1) and (5).

$$[T_m] = \begin{bmatrix} \dfrac{\sqrt{1-m^2}}{jm} & \dfrac{1}{jm} \\ \dfrac{-1}{jm} & \dfrac{-\sqrt{1-m^2}}{jm} \end{bmatrix} \tag{7}$$

The transmission matrix for the $\dfrac{\lambda_g}{4}$-waveguide is given by (2) and (5).

$$[T_\theta] = \begin{bmatrix} 0 & j \\ -j & 0 \end{bmatrix} \tag{8}$$

The transmission matrix for the entire structure is

$$[T] = [T_m] [T_\theta] [T_m] = \frac{-j}{m^2} \begin{bmatrix} -2\sqrt{1 - m^2} & -(2 - m^2) \\ 2 - m^2 & 2\sqrt{1 - m^2} \end{bmatrix} \tag{9}$$

The scattering matrix for the entire structure is obtained from (6) and (9).

$$[S] = \frac{1}{2 - m^2} \begin{bmatrix} 2\sqrt{1 - m^2} & -jm^2 \\ -jm^2 & -2\sqrt{1 - m^2} \end{bmatrix} \tag{10}$$

(b) If the entire structure is represented by an equivalent coupling device with a coupling coefficient M, then its scattering matrix is given by

$$[S_M] = \begin{bmatrix} -\sqrt{1 - M^2} & jM \\ jM & -\sqrt{1 - M^2} \end{bmatrix} \tag{11}$$

For (11) to be identical to (10), one must have

$$- \sqrt{1 - M^2} = \frac{2 \sqrt{1 - m^2}}{2 - m^2}$$

or

$$M^2 = 1 - \frac{4 (1 - m^2)}{(2 - m^2)^2} \qquad \text{i.e.,} \qquad M = \frac{\pm m^2}{2 - m^2} \tag{12}$$

The junctions coupled through a quarter-wavelength waveguide are equivalent to a coupling without the waveguide with the coefficient of coupling specified by (12).

It is to be noted that the above equivalence is strictly valid at a single frequency since the electrical length of the waveguide is frequency dependent.

The resultant overall length of each direct-coupled resonator will not be identical to that of the prototype resonator, since the reference planes for [S] of the entire structure will be different from the reference planes for the individual junctions.

11. Synthesis of Resonators in Cascade

(a) It was shown in Section 9 that the coincidental cascaded prototype resonators behave basically the same as a single resonator with an increase in Q_L.

The quarter-wavelength coupled resonators were treated in Section 10. The effect of the quarter-wavelength coupling can be represented by an equivalent direct coupling (Section 10.2).

The equivalent circuits for a quarter-wavelength coupling and its equivalent direct coupling are shown in Fig. 11-1.

The length of each resonator is given by [Eq. (XIII.11-5a)]

$$L_1 = \frac{\lambda_{go}}{2\pi} \left[m\pi + \frac{1}{2} \tan^{-1} \frac{2}{\hat{B}_d} + \frac{1}{2} \tan^{-1} \frac{2}{\hat{B}_d'} \right] \tag{1}$$

The length of the coupling waveguide is

$$L_L = \frac{\lambda_{go}}{2\pi} \left[\frac{\pi}{2} + \tan^{-1} \frac{2}{\hat{B}_d'} \right] \tag{2}$$

The coupling coefficient for a matched load is defined by [Eq. (XIII.10-5c)]

$$\frac{Q_{ei}}{Q_{eo}} = \frac{n_2^2}{n_1^2} \equiv n \tag{3}$$

where Q_{ei} and Q_{eo} are the external Q of the input port and the output port, respectively.

The equivalent direct-coupling coefficient of the quarter-wavelength coupling junction is

$$M = \frac{\pm m'^2}{2 - m'^2} \tag{10.2-12}$$

$$m' = \frac{2\hat{x}_d'}{\sqrt{1 + 4\hat{x}_d'^2}} \tag{XIII.9.1-9a}$$

where m' is the coupling coefficient of the junction characterized by the normalized reactance $\hat{x}_d'$, (XIII.9.1-9b).

The reactance $\hat{x}_d'$ can be obtained from (XIII.9.1-9a),

$$\hat{x}_d' = \frac{m'}{2\sqrt{1 - m'^2}} \tag{4a}$$

and the corresponding $\hat{x}_M$ for the equivalent direct-coupling coefficient is

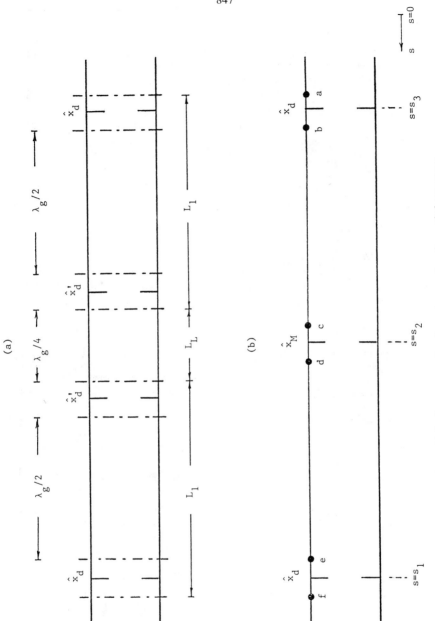

(a)

(b)

Figure 11-1: An arrangement of quarter-wavelength coupled resonators.

$$\hat{x}_M = \frac{M}{2\sqrt{1 - M^2}} \tag{4b}$$

$$= \frac{\pm m'^2}{4\sqrt{1 - m'^2}}$$

$$= \frac{\pm \hat{x}_d'^2}{\hat{x}_d' \sqrt{4 + \hat{B}_d'}} = \frac{\pm 1}{\hat{B}_d' \sqrt{4 + \hat{B}_d'^2}} \tag{4c}$$

$$\hat{B}_M = \frac{1}{\hat{x}_M} = \mp\hat{B}_d' \sqrt{4 + \hat{B}_d'^2} \tag{4d}$$

(b) The value of $\hat{B}_M$ can also be obtained from the Smith chart once $\hat{B}_d'$ and $\hat{B}_d$ are specified. For the critically coupled case, $\hat{B}_d' = \hat{B}_d$.

At $s = s_3^+$, Fig. 11-1b, just to the left of the output coupling device, the normalized admittance is

$$\hat{Y}_b = 1 + j\hat{B}_d \tag{5}$$

Point a, Fig. 11-1b, is located at the center of the Smith chart, Fig. 11-2. For inductive $j\hat{X}_d$, $j\hat{B}_d = -j/\hat{X}_d$, point b has the admittance specified by (5). At resonance, the input at point f is matched; hence, point f should coincide with point a on the chart, Fig. 11-2. The admittance at $s = s_1^-$, just to the right of the input coupling junction, is

$$\hat{Y}_e = \hat{Y}_a - j\hat{B}_d = 1 - j\hat{B}_d \tag{6}$$

Point e and point b are therefore symmetrically located, with respect to the $\hat{x} = 0$ axis, on the chart.

The difference between $\hat{Y}_d$ and $\hat{Y}_c$ is $\hat{B}_M$, and it is reasonable to expect that these points are also symmetrically positioned on the chart, since the structure is symmetrical.

The maximum value of $\hat{B}_M$ is the $\hat{B}$ = constant circle which is tangent to the $S = S_b$ circle; $S \equiv$ standing-wave ratio. This is point c, Fig. 11-2. The equivalent susceptance $\hat{B}_M$ is twice the susceptance at point c. This corresponds to critical coupling.

If $\hat{B}_M$ is chosen to be greater than $\hat{B}_c$, it will be impossible to reach the matched condition at point f. This corresponds to the under-coupled case.

If $\hat{B}_M$ is chosen to be smaller than $\hat{B}_c$, there would be two pairs of symmetrical locations, (c',d') and (c'',d''), corresponding to two different frequencies. This is the over-coupled case.

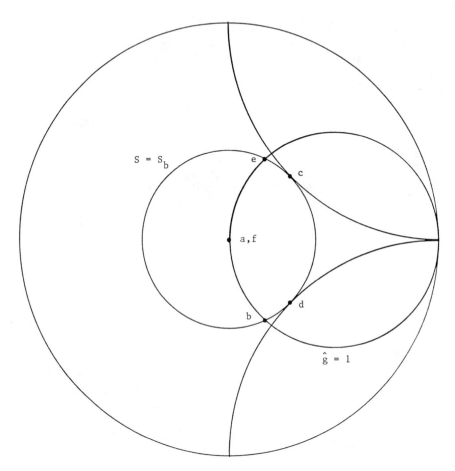

Figure 11-2: Immittance representation at various locations in Figure 11-1.

12. Immittance Inverters

At microwave frequencies, it is awkward to lump all branches of a filter at one location. To miminize mutual couplings, it is necessary to distribute circuit elements along the waveguide. The immittance inverter is one type of arrangement which will accomplish such a purpose.

A waveguide immittance inverter is an ideal quarter-wave transformer. A normalized load impedance (or admittance) connected at the output end is seen at the input as the inverse of the normalized load impedance (or load admittance), Fig. 12-1. These relations can be obtained by direct application of impedance transformation, (8-2).

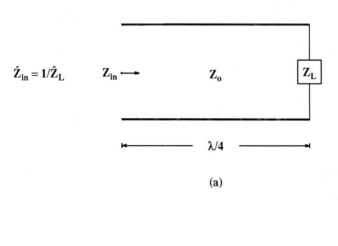

(a)

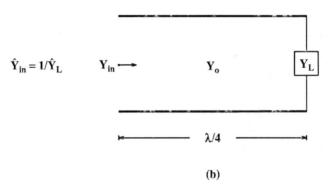

(b)

Figure 12-1: Waveguide immittance inverters.

13. Multiple Resonators in Cascade

The general case of m resonators in cascade with quarter-wavelength coupling is shown in Fig. 13-1a. The resonators are assumed to be lossless and symmetrical but they need not be identical.

When the coupling junctions are represented by ideal transformers, one has the equivalent circuit as shown in Fig. 13-1b. The input coupling transformer can be relocated with an appropriate change of parameters (Fig. 13-1c) and can then be eliminated. The simplified equivalent circuit is shown in Fig. 13-1d.

The quarter-wavelength transformers will next be removed.

When m = odd, the mth resonator has an even number of quarter-wavelength transformers located to its left. The mth resonator will thus be inverted an even number of times, and hence will return back to its original form, i.e., no transformation takes place for this resonator. Consequently, the mth resonator (m = odd) appears as it is at the input port.

For m = even, the mth resonator has an odd number of quarter-wavelength transformers on its left. Hence, the mth resonator (m = even) appears inverted at the input port.

Let each resonator be represented by an equivalent parallel LC-circuit, Fig. 13-2a, where the intrinsic parameters are C_{om} and L_{om}.

The same resonator is represented by Z_m' when the coupling transformers are removed; this is shown in Fig. 13-2b.

$$L_m = n_m^2 L_{om} \qquad \hat{L}_m = \frac{n_m^2}{Z_o} L_{om} \tag{1a}$$

$$C_m = \frac{C_{om}}{n_m^2} \qquad \hat{C}_m = \frac{Z_o}{n_m^2} C_{om} \tag{1b}$$

The inverted resonator is shown in Fig. 13-2c, and the parameters for the inverted circuit are

$$L''_m = C_m = \frac{C_{om}}{n_m^2} \qquad \hat{L}''_m = \hat{C}_m = \frac{Z_o}{n_m^2} C_{om} \tag{2a}$$

$$C''_m = L_m = n_m^2 L_{om} \qquad \hat{C}''_m = \hat{L}_m = \frac{n_m^2}{Z_o} L_{om} \tag{2b}$$

In other words, Fig. 13-1d can be represented by an equivalent circuit without quarter-wavelength transformers by replacing each two-port network with m = odd by the corresponding Z_m' (Fig. 13-2b) and the one with m = even by the corresponding Z_m'' (Fig. 13-2c). The resulting equivalent circuits are shown in Fig. 13-3.

The external Q is given by [Eq. (9-11α)]

$$Q_{em} = \omega_o \hat{C}_m = 2Q_{Lm} \qquad \hat{R}_{ext} = 1 \tag{3a}$$

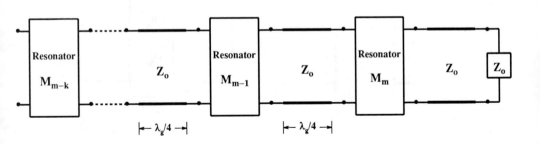

(a)

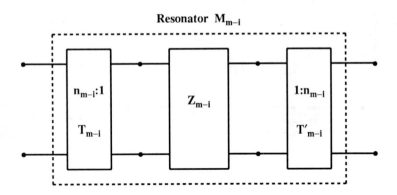

(b)

Figure 13-1: Quarter-wavelength coupled resonators and associated equivalent circuits.

Resonator M_{m-i}

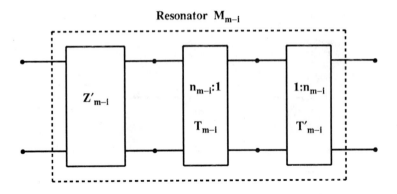

(c)

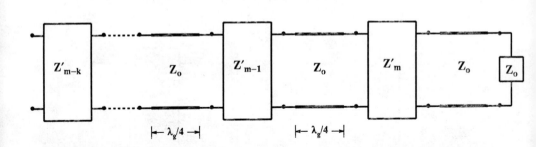

(d)

Figure 13-1: Quarter-wavelength coupled resonators and associated equivalent circuits.

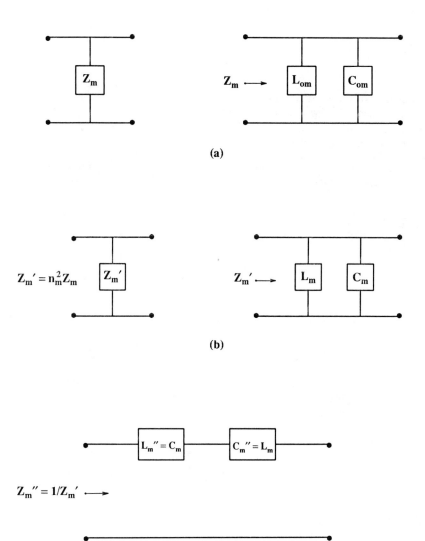

Figure 13-2: Immittance inversion.

or

$$\hat{C}_m = \frac{2Q_{Lm}}{\omega_o} \tag{3b}$$

The equivalent circuit, Fig. 13-3, is a general circuit for the prototype filter. The values of circuit elements will be specified by the prototype filter and (3) will dictate the required loaded Q_{Lm}.

The above analysis shows that the equivalent circuit of several resonators connected in cascade has the form of a prototype band-pass filter network. A practical problem in filter design at microwave frequencies is to transform a prototype filter network into an equivalent arrangement of resonators in cascade. This process will be illustrated in the following example.

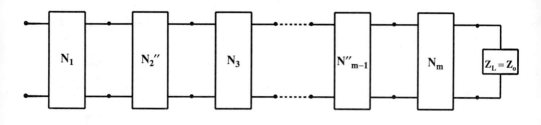

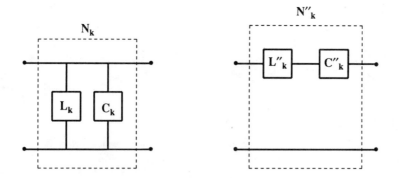

(a) m = odd

Figure 13-3: An equivalent circuit for resonators with invertors.

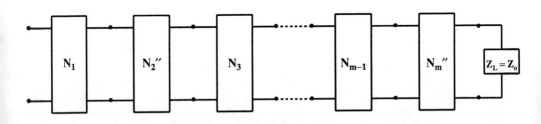

(b) m = even

Figure 13-3: An equivalent circuit for resonators with invertors.

13.1. Example: Resonators in Cascade

A prototype band-pass filter network is shown in Fig. 13.1-1. Determine the equivalent arrangement of resonators in cascade (Fig. 13.1-2). The quarter-wavelength transformers are chosen to have different characteristic impedances to provide the flexibility of adjusting impedance levels. The capacitances and inductances are chosen such that

$$\omega_0^2 = \frac{1}{\overline{L}_k\overline{C}_k} = \frac{1}{\overline{L}_j\overline{C}_j} \tag{1}$$

where parameters with bars are elements for the prototype circuit, Fig.13.1-1, and parameters without bars are elements for the circuit with inverters, Fig. 13.1-2.

Solution:

A typical module of Fig. 13.1-1 is shown in Fig. 13.1-3. The input terminals are terminals q-q', with q = 2k, k = 1, 2, ... The input admittance $\overline{Y}_q$ is

$$\overline{Y}_q = j\overline{B}_q + \frac{1}{\overline{Z}_{q-1}} = j\overline{B}_q + \cfrac{1}{j\overline{X}_{q-1} + \cfrac{1}{\overline{Y}_{q-2}}}$$

$$= j\Omega \sqrt{\frac{\overline{C}_q}{\overline{L}_q}} + \cfrac{1}{j\Omega \sqrt{\dfrac{\overline{L}_{q-1}}{\overline{C}_{q-1}}} + \cfrac{1}{\overline{Y}_{q-2}}}$$

$$= \sqrt{\frac{\overline{C}_q}{\overline{L}_q}} \left\{ j\Omega + \cfrac{\dfrac{\overline{C}_{q-1}}{\overline{C}_q}}{j\Omega + \sqrt{\dfrac{\overline{C}_{q-1}}{\overline{L}_{q-1}}}\dfrac{1}{\overline{Y}_{q-2}}} \right\}$$

$$\equiv \sqrt{\frac{\overline{C}_q}{\overline{L}_q}} \, \overline{Y}_q' \tag{2a}$$

where (1) is used to obtain the second to the last form.

$$j\overline{B}_q = j\left[\omega\overline{C}_q - \frac{1}{\omega\overline{L}_q}\right] = j\Omega\sqrt{\frac{\overline{C}_q}{\overline{L}_q}} \qquad \Omega = \frac{\omega}{\omega_0} - \frac{\omega_0}{\omega} \tag{2b}$$

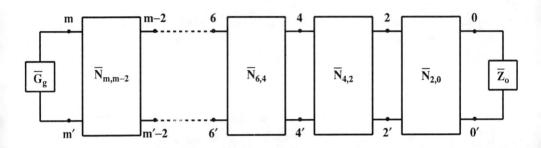

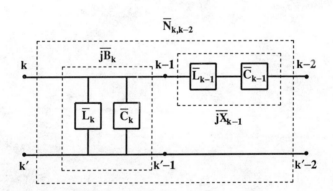

Figure 13.1-1: A prototype band-pass filter.

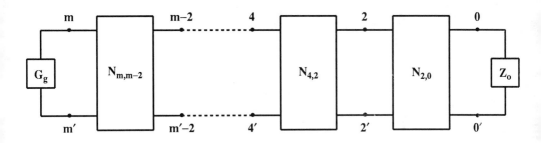

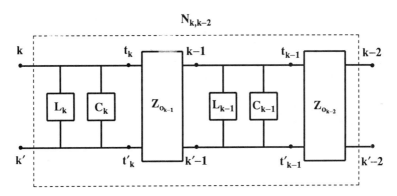

Figure 13.1-2: A band-pass filter with inverters.

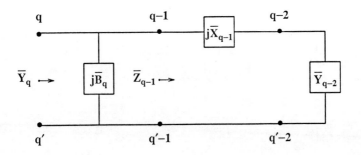

$$\overline{Y}_{q-2} = Y \, [\overline{C}_{q-2}/\overline{L}_{q-2}]^{1/2}$$

Figure 13.1-3: A typical module of Figure 13.1-1.

$$j\overline{X}_{q-1} = j\left[\omega\overline{L}_{q-1} - \frac{1}{\omega\overline{C}_{q-1}}\right] = j\Omega \sqrt{\frac{\overline{L}_{q-1}}{\overline{C}_{q-1}}} \tag{2c}$$

$$\overline{Y}'_q \equiv j\Omega + \frac{\dfrac{\overline{C}_{q-1}}{\overline{C}_q}}{j\Omega + \dfrac{\overline{L}_{q-2}}{\overline{L}_{q-1}}\overline{Z}'_{q-2}} \qquad \overline{L}_k = \frac{1}{\omega_o^2 \overline{C}_k} \qquad \text{for} \quad q \neq 2 \tag{2d}$$

$$\overline{Y}'_q \equiv j\Omega + \frac{\dfrac{\overline{C}_{q-1}}{\overline{C}_q}}{j\Omega + \sqrt{\dfrac{\overline{C}_{q-1}}{\overline{L}_{q-1}}}\,\overline{Z}_o} \qquad \text{for} \quad q = 2 \tag{2e}$$

Since these are identical modules, $\overline{Y}_{q-2}$ can be represented, for $q \neq 2$, by (2a) with q equal to $q - 2$.

$$\overline{Y}_{q-2} = \frac{1}{\overline{Z}_{q-2}} = \sqrt{\frac{\overline{C}_{q-2}}{\overline{L}_{q-2}}}\,\overline{Y}'_{q-2} \tag{2f}$$

$$\overline{Y}'_{q-2} = j\Omega + \frac{\dfrac{\overline{C}_{q-3}}{\overline{C}_{q-2}}}{j\Omega + \dfrac{\overline{L}_{q-4}}{\overline{L}_{q-3}}\overline{Z}'_{q-4}} \qquad q \neq 2 \tag{2g}$$

A typical module of Fig. 13.1-2 is shown in Fig. 13.1-4, and the corresponding input admittance Y_q, with $q = 2k$, $k = 1, 2, \ldots$, is

$$Y_q = jB_q + \frac{1}{Z_{t_{q-1}}} = jB_q + \frac{Z_{q-1}}{Z_{0_{q-1}}^2}$$

$$= j\sqrt{\frac{C_q}{L_q}}\,\Omega + \frac{\dfrac{1}{Z_{0_{q-1}}^2}}{j\sqrt{\dfrac{C_{q-1}}{L_{q-1}}}\,\Omega + \dfrac{1}{Z_{0_{q-2}}^2}\dfrac{1}{Y_{q-2}}} = \sqrt{\frac{C_q}{L_q}}\,Y'_q \tag{3a}$$

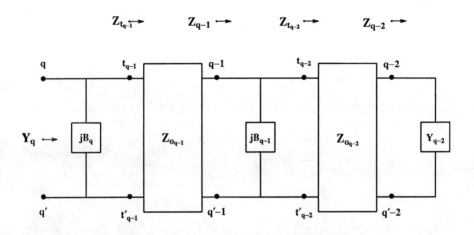

$$Y_{q-2} = Y'_{q-2}\,[C_{q-2}/L_{q-2}]^{1/2}$$

Figure 13.1-4: A typical module of Figure 13.1-2.

$$Y_q' \equiv j\Omega + \cfrac{\cfrac{1}{Z_{o_{q-1}}^2} \cfrac{L_q}{C_{q-1}}}{j\Omega + \cfrac{Z_{q-2}'}{Z_{o_{q-2}}^2} \cfrac{L_{q-1}}{C_{q-2}}} \qquad \text{for} \quad q \neq 2 \tag{3b}$$

$$Y_q' = j\Omega + \cfrac{\cfrac{1}{Z_{o_{q-1}}^2} \cfrac{L_q}{C_{q-1}}}{j\Omega + \cfrac{1}{Z_{o_{q-2}}^2} \sqrt{\cfrac{L_{q-1}}{C_{q-1}} Z_o}} \qquad \text{for} \quad q = 2 \tag{3c}$$

$$Y_{q-1} = jB_{q-1} + \frac{1}{Z_{t_{q-2}}} = jB_{q-1} + \frac{Z_{q-2}}{Z_{o_{q-2}}^2} = \sqrt{\frac{C_{q-1}}{L_{q-1}}} Y_{q-1}' \tag{3d}$$

$$Y'_{q-1} = j\Omega + \frac{1}{Z_{o_{q-2}}^2} \frac{L_{q-1}}{C_{q-2}} Z'_{q-2} \tag{3e}$$

$$jB_q = \sqrt{\frac{C_q}{L_q}} \; j\Omega \tag{3f}$$

$$jB_{q-1} = \sqrt{\frac{C_{q-1}}{L_{q-1}}} \; j\Omega \tag{3g}$$

The input admittances, $\overline{Y}_q'$ and Y_q', $q = 2k$ and $k = 1, 2, \ldots$, of the corresponding modules in Figs. 13.1-1 and 13.1-2 can be made identical if the following conditions are satisfied.

Condition a: $\qquad \dfrac{\overline{C}_{q-1}}{\overline{C}_q} = \dfrac{1}{Z_{o_{q-1}}^2} \dfrac{L_q}{C_{q-1}} \tag{A}$

$$Z_{o_{q-1}}^2 = \cfrac{L_q}{\cfrac{1}{\omega_0^2 L_{q-1}}} \cfrac{\cfrac{1}{\omega_0^2 \overline{L}_q}}{\overline{C}_{q-1}} = \frac{L_q L_{q-1}}{\overline{L}_q \overline{C}_{q-1}}$$

or

$$Z_{o_{q-1}} = \sqrt{\frac{L_q L_{q-1}}{\overline{L}_q \overline{C}_{q-1}}} \qquad q = 2k = \text{even}, \quad k = 1, 2, \ldots$$

Let $p = q - 1$; then

$$Z_{0_p} = \sqrt{\frac{L_{p+1}L_p}{\overline{L}_{p+1}\overline{C}_p}} \qquad p = q - 1 = 2k - 1 = \text{odd} \tag{4a}$$

Condition b: $\qquad \dfrac{\overline{L}_{q-2}}{\overline{L}_{q-1}} \overline{Z}'_{q-2} = \dfrac{Z'_{q-2}}{Z^2_{0_{q-2}}} \dfrac{L_{q-1}}{C_{q-2}} \qquad q \neq 2 \tag{B}$

$$Z^2_{0_{q-2}} = \frac{L_{q-1}\overline{L}_{q-1}}{C_{q-2}\overline{L}_{q-2}} = \sqrt{\frac{L_{q-1}L_{q-2}}{\overline{C}_{q-1}\overline{L}_{q-2}}}$$

since $\overline{Z}'_{q-2} = Z'_{q-2}$. Let $p = q - 1$; then

$$Z_{0_{p-1}} = \sqrt{\frac{L_p L_{p-1}}{\overline{C}_p \overline{L}_{p-1}}} \qquad p = 2k - 1 = \text{odd}$$

Let $p' = p - 1$, then

$$Z_{0_{p'}} = \sqrt{\frac{L_{p'+1}L_{p'}}{\overline{C}_{p'+1}\overline{L}_{p'}}} \qquad p' = 2k = \text{even} \qquad q \neq 2 \tag{4b}$$

Condition c: For $q = 2$,

$$\frac{1}{Z^2_{0_{q-2}}} \sqrt{\frac{L_{q-1}}{C_{q-1}}} Z_0 = \sqrt{\frac{\overline{C}_{q-1}}{\overline{L}_{q-1}}} \overline{Z}_0$$

or

$$Z^2_{0_{q-2}} = \sqrt{\frac{L_{q-1}\overline{L}_{q-1}}{C_{q-1}\overline{C}_{q-1}}} = \frac{L_{q-1}}{\overline{C}_{q-1}}$$

since $Z_0 = \overline{Z}_0$. Thus

$$Z_{0_{q-2}}|_{q=2} = Z_{0_0} = \frac{L_1}{C_1} \tag{4c}$$

The input of the prototype filter network (Fig. 13.1-1) has a shunt branch across the input terminals when $m = 2k$, $k = 1, 2, \ldots$, Fig. 13.1-5. The corresponding equivalent network for this case is shown in Fig. 13.1-6. The input admittances are given by (2a) and (3a) for $m = q$.

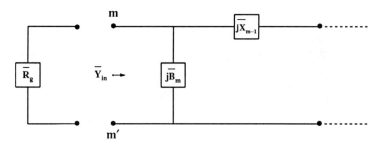

Figure 13.1-5: The input terminals of Figure 13.1-1 with a shunt element.

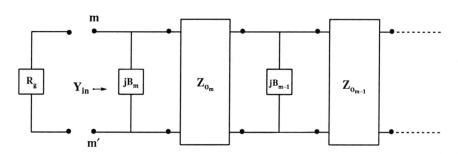

Figure 13.1-6: The input terminals of Figure 13.1-2 corresponding to Figure 13.1-5.

$$\overline{Y}_{in} = \overline{Y}_m = \sqrt{\frac{\overline{C}_m}{\overline{L}_m}} \, \overline{Y}'_m \tag{5a}$$

$$Y_{in} = Y_m = \sqrt{\frac{C_m}{L_m}} \, Y'_m \tag{5b}$$

where $Y'_m = \overline{Y}'_m$ is given by (2d) or (3b).

When the filter network in Fig. 13.1-5 is match-terminated by a source resistance $\overline{R}_g$, i.e.,

$$\frac{1}{\overline{R}_g} = \sqrt{\frac{\overline{C}_m}{\overline{L}_m}} \, \overline{Y}'_m \quad \text{or} \quad \overline{Y}'_m = \sqrt{\frac{\overline{L}_m}{\overline{C}_m}} \, \frac{1}{\overline{R}_g}$$

then

$$\frac{1}{R_g} = Y_{in} = \sqrt{\frac{C_m}{L_m}} \, Y'_m =: \sqrt{\frac{C_m}{L_m}} \, \overline{Y}'_m = \sqrt{\frac{C_m}{L_m} \frac{\overline{L}_m}{\overline{C}_m}} \, \frac{1}{\overline{R}_g}$$

Hence, the equivalent network in Fig. 13.1-6 will be match-terminated by R_g, which is given by

$$R_g = \overline{R}_g \sqrt{\frac{L_m \overline{C}_m}{\overline{L}_m C_m}} = \frac{L_m}{\overline{L}_m} \overline{R}_g \tag{6}$$

If the input of the prototype filter network (Fig. 13.1-1) is a series branch as shown in Fig. 13.1-7, the corresponding equivalent circuit is Fig. 13.1-8. In this case, the equivalent network requires additional quarter-wavelength transformers such that the network to the right of terminals m–m', in Figs. 13.1-7 and 13.1-8 will be equivalent. The corresponding input impedances are as follows.

$$\overline{Z}_{in} = \overline{Z}_{m+1} = j\overline{X}_{m+1} + \overline{Z}_m = j\sqrt{\frac{\overline{L}_{m+1}}{\overline{C}_{m+1}}} \, \Omega + \frac{1}{\sqrt{\frac{\overline{C}_m}{\overline{L}_m}} \, \overline{Y}'_m}$$

$$= \sqrt{\frac{\overline{L}_{m+1}}{\overline{C}_{m+1}}} \left[j\Omega + \frac{\overline{L}_m}{\overline{L}_{m+1}} \overline{Z}'_m \right] = \sqrt{\frac{\overline{L}_{m+1}}{\overline{C}_{m+1}}} \, \overline{Z}'_{m+1} \tag{7a}$$

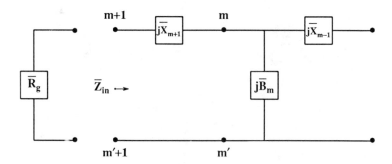

Figure 13.1-7: The input terminals of Figure 13.1-1 with a series element.

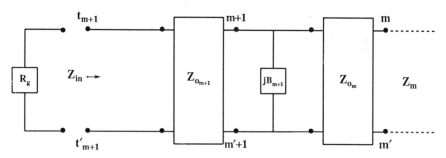

Figure 13.1-8: The input terminals of Figure 13.1-2 corresponding to Figure 13.1-7.

$$Z_{in} = Z_{t_{m+1}} = \frac{Z_{0_{m+1}}^2}{Z_{m+1}} = \frac{Z_{0_{m+1}}^2}{\dfrac{1}{jB_{m+1} + \dfrac{Z_m}{Z_{0_m}^2}}}$$

$$= Z_{0_{m+1}}^2 \left[\sqrt{\frac{C_{m+1}}{L_{m+1}}} \, j\Omega + \frac{1}{Z_{0_m}^2} \sqrt{\frac{L_m}{C_m}} \, Z'_m \right]$$

$$= Z_{0_{m+1}}^2 \sqrt{\frac{C_{m+1}}{L_{m+1}}} \, Z'_{m+1} \tag{7b}$$

$$Z'_{m+1} = \overline{Z}'_{m+1} = j\Omega + \frac{\overline{L}_m}{\overline{L}_{m+1}} \quad \text{and} \quad Z'_m = \overline{Z}'_m \tag{7c}$$

$$Z_{0_m}^2 = \frac{L_{m+1} L_m}{\overline{C}_{m+1} \overline{L}_m} \tag{4b}$$

If the original filter network, Fig. 13.1-7, is match-terminated by $\overline{R}_g$ at the source end, i.e.,

$$\overline{Z}_{in} = \sqrt{\frac{\overline{L}_{m+1}}{\overline{C}_{m+1}}} \, \overline{Z}'_{m+1} = \overline{R}_g$$

or

$$\overline{Z}'_{m+1} = \sqrt{\frac{\overline{C}_{m+1}}{\overline{L}_{m+1}}} \, \overline{R}_g \qquad m = 2k, \ k = 1, 2, \ldots \tag{8a}$$

then the appropriate source resistance R_g is

$$R_g = Z_{in} = Z_{0_{m+1}}^2 \sqrt{\frac{C_{m+1}}{L_{m+1}}} \sqrt{\frac{\overline{C}_{m+1}}{\overline{L}_{m+1}}} \, \overline{R}_g = Z_{0_{m+1}}^2 \frac{C_{m+1}}{L_{m+1}} \, \overline{R}_g \tag{8b}$$

$$Z_{0_{m+1}} = \sqrt{\frac{R_g}{\overline{R}_g} \frac{\overline{L}_{m+1}}{C_{m+1}}} \tag{8c}$$

The filter network in Fig. 13.1-1 can also be represented by an equivalent network containing only series resonance circuits by a similar procedure.

In this analysis, the quarter-wavelength transformers are assumed to be frequency independent, which is impossible. However, for a very narrow bandwidth, the theory provides acceptable results.

14. Problems

1. Determine the minimum order of a maximally flat filter with the following specifications:

(a) The 3-db cutoff frequency is $\omega_c = 1000$ r/s.

(b) The maximum attenuation is 0.1 db in the frequency range from $\omega = 0$ to $\omega = 250$ r/s.

(c) The minimum attenuation should be 60 db for frequencies above $\omega = 2000$ r/s.

2. Determine the order of a Chebycheff filter with the following specifications:

(a) Pass-band ranges from $\omega = 0$ to $\omega = 250$ r/s.

(b) The maximum attenuation in the pass-band is limited to 0.1 db.

(c) The minimum attenuation for frequencies above $\omega = 2000$ r/s should be 60 db.

3. Determine the normalized transfer function of a fourth-order Chebycheff response with 0.5-db ripple in the pass-band.

4. The transfer function of a given network is given by

$$|H(j\omega)|^2 = H_o \frac{\Delta^2 C_n^2(\Omega)}{1 + \Delta^2 C_n^2(\Omega)} \qquad \Omega \equiv \frac{\omega}{\omega_c}$$

where H_o is a constant and C_n is a Chebycheff polynomial of order n. Sketch the response of this function.

5. Determine whether any of the circuits in Fig. 14-1 can be used as an immittance inverter.

6. Design a three-section narrow-band Chebycheff band-pass filter. The center frequency of the pass-band is 1.01×10^9 Hz and the bandwidth is approximately $\pm 1\%$ of the center frequency. The ripple should be limited to no more than 0.1 db. The shunt elements, Fig. 14-2, are to be supplied by a short-circuited stub.

7. Show that a quarter-wavelength coupled resonator with high Q provides a steeper power transmission than coincidentally cascaded resonators. Hint: show that the power transmission coefficient for the quarter-wave coupled resonators can be expressed as

$$T = \frac{1}{D_1^2} \frac{Q_L^2}{(Q_L - Q_e)^2}$$

where

$$D_1 \equiv 1 + \frac{1}{2} \frac{M_Q^2}{1 - M_Q}$$

873

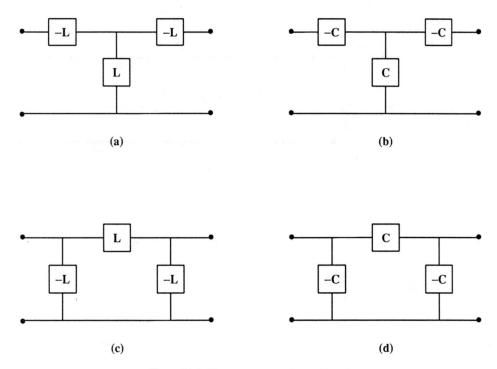

Figure 14-1: The arrangements for problem 5.

and

$$M_Q \equiv \frac{Q_e}{Q_L} = Q_e \left[\frac{1}{Q_u} + \frac{1}{Q_e} \right] \approx 1$$

for large value of Q_u.

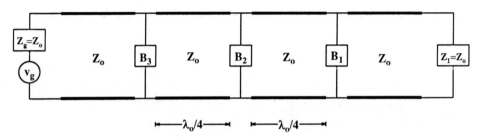

Figure 14-2: The arrangements for problem 6.

Bibliography

A. Network Theory

1. H. Baher, Synthesis of Electrical Networks, John Wiley and Sons, Inc., 1984.

2. W. Chen, Passive and Active Filters, John Wiley and Sons, Inc., 1986.

3. E.S. Kuh and D.O. Pederson, Principles of Circuit Synthesis, McGraw-Hill Book Co., Inc., 1959.

4. F.F. Kuo, Network Analysis and Synthesis, John Wiley and Sons, Inc., 1962.

5. M.E. Van Valkenburg, Modern Network Synthesis, John Wiley and Sons, Inc., 1960.

6. L. Weinberg, Network Analysis and Synthesis, Robert E. Krieger Publishing Co., 1975.

B. Transmission Line Theory

1. C.W. Davidson, Transmission Lines for Communications, The MacMillian Press Ltd., 1982.

2. R.W.P. King, Transmission-Line Theory, McGraw-Hill Book Co., Inc., 1955.

3. P.C. Magnusson, Transmission Lines and Wave Propagation, Second Edition, Allyn and Bacon, Inc., 1970.

4. R.K. Moore, Traveling-Wave Engineering, McGraw-Hill Book Co., Inc., 1960.

5. N.N. Rao, Elements of Engineering Electromagnetics, Second Edition, Prentice-Hall Inc., 1987.

C. Waveguide Theory

1. J.L. Altman, Microwave Circuits, D. Van Nostrand Co., Inc., 1964.

2. R.E. Collin, Foundations for Microwave Engineering, McGraw-Hill Book Co., Inc., 1966.

3. M. Javid and P.M. Brown, Field Analysis and Electromagnetics, McGraw-Hill Book Co., Inc., 1963.

4. K. Kurokawa, Theory of Microwave Circuits, Academic Press, Inc., 1969.

5. C.G. Montgomery, R.H. Dicke and E.M. Purcell, Principles of Microwave Circuits, McGraw-Hill Book Co., Inc., 1948.

6. D.T. Paris and F.K. Hurd, Basic Electromagnetic Theory, McGraw-Hill Book Co., Inc., 1969.

7. S. Ramo, J.R. Whinnery, and T. Van Duzer, Fields and Waves in Communication Electronics, Second Edition, John Wiley and Sons, Inc., 1984.

8. N.N. Rao, Elements of Engineering Electromagnetics, Second Edition, Prentice-Hall Inc., 1987.

9. J.C. Slater, Microwave Electronics, D. Van Nostrand Co., Inc., 1950.

D. Ferrites

1. P.J.B. Clarricoats, Microwave Ferrites, John Wiley and Sons, Inc., 1961.

2. J. Helszajn, Passive and Active Microwave Circuits, John Wiley and Sons, Inc., 1978.

3. J. Helszajn, Principles of Microwave Ferrite Engineering, Wiley-Interscience, 1969.

4. R.F. Soohoo, Theory and Application of Ferrites, Prentice-Hall Inc., 1960.

5. R.H. Waldron, Ferrites, D. Van Nostrand Company, Ltd., 1961.

Index